Networked Filtering and Fusion
in Wireless Sensor Networks

Networked Filtering and Fusion in Wireless Sensor Networks

Magdi S. Mahmoud
Yuanqing Xia

CRC Press
Taylor & Francis Group
Boca Raton   London   New York

CRC Press is an imprint of the
Taylor & Francis Group, an **informa** business

CRC Press
Taylor & Francis Group
6000 Broken Sound Parkway NW, Suite 300
Boca Raton, FL 33487-2742

First issued in paperback 2019

© 2015 by Taylor & Francis Group, LLC
CRC Press is an imprint of Taylor & Francis Group, an Informa business

No claim to original U.S. Government works

ISBN-13: 978-1-4822-5096-1 (hbk)
ISBN-13: 978-1-138-37493-5 (pbk)

Library of Congress Cataloging-in-Publication Data

Mahmoud, Magdi S.
  Networked filtering and fusion in wireless sensor networks / authors, Magdi S. Mahmoud, Yuanqing Xia.
      pages cm
  Includes bibliographical references and index.
  ISBN 978-1-4822-5096-1 (hardback)
  1. Multisensor data fusion. 2. Information filtering systems. I. Xia, Yuanqing. II. Title.

  TK7872.D48M34 2014
  004.6--dc23   2014027990

Visit the Taylor & Francis Web site at
http://www.taylorandfrancis.com

and the CRC Press Web site at
http://www.crcpress.com

To My Loving Wife **Salwa**,
To the "M" Family: Medhat, Monda, Mohamed, Menna and
Malak, Mostafa, Mohamed
and Ahmed Gouda

MsM

To My Honest and Diligent Wife **Wang Fangyu**,
To My lovely Daughter Xia Jingshu

YX

Contents

Notations and Symbols

$$
\begin{aligned}
\Re &= \text{set of real numbers} \\
\Re^n &= \text{set of all n-dimensional real vectors} \\
\Re^{n \times m} &= \text{set of } n \times m\text{-dimensional real matrices} \\
x^t \text{ or } A^t &= \text{transpose of vector x or matrix A} \\
A^{-1} &= \text{inverse of matrix A} \\
I &= \text{identity matrix of arbitrary order} \\
e_j &= \text{jth column of matrix I} \\
\lambda(A) &= \text{eigenvalue of matrix A} \\
\varrho(A) &= \text{spectral radius of matrix A} \\
\lambda_j(A) &= \text{jth eigenvalue of matrix A} \\
\lambda_m(A) &= \text{minimum eigenvalue of matrix A} \\
&\quad \text{where } \lambda(A) \text{ are real} \\
\lambda_M(A) &= \text{maximum eigenvalue of matrix A} \\
&\quad \text{where } \lambda(A) \text{ are real} \\
A^\dagger &= \text{Moore–Penrose-inverse of matrix A} \\
P > 0 &= \text{matrix P is real symmetric} \\
&\quad \text{and positive-definite} \\
P \geq 0 &= \text{matrix P is real symmetric} \\
&\quad \text{and positive semi-definite} \\
P < 0 &= \text{matrix P is real symmetric} \\
&\quad \text{and negative-definite} \\
P \leq 0 &= \text{matrix P is real symmetric} \\
&\quad \text{and negative semi-definite} \\
A(i,j), A_{ij} &= \text{ij-th element of matrix A} \\
\mathbf{det}(A) &= \text{determinant of matrix A} \\
\mathbf{trace}(A) &= \text{trace of matrix A}
\end{aligned}
$$

$$
\begin{aligned}
\mathbf{rank}(A) &= \text{rank of matrix } A \\
|a| &= \text{absolute value of scalar } a \\
||x|| &= \text{Euclidean norm of vector } x \\
||A|| &= \text{induced Euclidean norm of matrix } A \\
||x||_p &= \ell_p \text{ norm of vector } x \\
||A||_p &= \text{induced } \ell_p \text{ norm of matrix } A
\end{aligned}
$$

List of Acronyms

ACK	positive acknowledgment
CF	consensus filters
CIM	centralized integration method
CTS	clear to send
DBF	diffusion-based filtering
DC	distributed control
DCF	distributed coordination function
DF	distributed filtering
$DIFS$	distributed interframe space
DIP	distributed information processing
$DMPC$	distributed model predictive control
DPF	distributed particle filtering
EKF	extended Kalman filter
EMA	expected maximization algorithm
IUB	industrial utility boiler
LKF	Lyapunov–Krasovskii functional
LMI	linear matrix inequality
$MSDF$	multi-sensor data fusion
μKF	micro-Kalman filters
PE	performance evaluation
PKF	partitioned Kalman filter
RTS	request to send
$SIFS$	short interframe space
SNS	sensor networked systems
STF	self-tuning filtering
WSN	wireless sensor networks

Acknowledgments

We are indebted to the people who made the writing of this book possible. Professor Mahmoud thanks KFUPM management for the continuous encouragement and facilitating all sources of help. Particular appreciation goes to the deanship of scientific research (DSR) for providing a superb competitive environment of research activities through internal funding grants. It is a great pleasure to acknowledge the financial funding afforded by DSR through project no. IN131039 and for providing overall support of research activities at KFUPM. Professor Mahmoud owes a measure of gratitude to the National Natural Science Foundation of China (61225015) for fully sponsoring the technical visit to BIT, China during January–March, 2014.

During the past five years, we had the privilege of teaching various senior and graduate courses. The course notes, updated and organized, were instrumental in generating different chapters of this book, and valuable comments and/or suggestions by graduate students were greatly helpful, particularly those who attended the courses SE 537, SE 652 and SE 658 offered at the Systems Engineering Department over the period 2007–2011.

Most of all, however, we would like to express our deepest gratitude to all the members of our families for their enthusiastic support, without which this volume would not have been finished.

We would appreciate any comments, questions, criticisms, or corrections that readers may kindly provide to Professor Mahmoud at msmahmoud@kfupm.edu.sa or magdisadekmahmoud@gmail.com.

Magdi S. Mahmoud
Dhahran, Saudi Arabia

Yuanqing Xia
Beijing, China

Authors

MagdiSadek Mahmoud earned a BSc (Honors) in communication engineering, a MSc in electronic engineering and PhD in systems engineering, all from Cairo University in 1968, 1972 and 1974, respectively. He has been a professor of engineering since 1984. He is now a distinguished university professor at King Fahd University of Petroleum and Minerals (KFUPM), Saudi Arabia. He has been on the faculty at different universities world-wide including Egypt (CU, AUC), Kuwait (KU), UAE (UAEU), UK (UMIST), USA (Pitt, Case Western), Singapore (Nanyang Technological) and Australia (Adelaide). He has lectured in Venezuela (Caracas), Germany (Hanover), UK (Kent), USA (University of Texas at San Antonio), Canada (Montreal, Alberta) and China (BIT, Yanshan). He is the principal author of 34 books, inclusive book chapters and the author/co-author of more than 510 peer-reviewed papers. Dr. Mahmoud is the recipient of two national, one regional, and four university prizes for outstanding research in engineering and applied mathematics. He is a fellow of the IEE, a senior member of the IEEE, the CEI (UK), and a registered consultant engineer of information engineering and systems (Egypt). He is currently actively engaged in teaching and research in the development of modern methodologies of distributed control and filtering, networked control systems, triggering mechanisms in dynamical systems, fault-tolerant systems and information technology.

Yuanqing Xia graduated from the Department of Mathematics, Chuzhou University, Chuzhou, China, in 1991. He earned his MS degree in fundamental mathematics from Anhui University, China, in 1998 and his PhD degree in control theory and control engineering from Beijing University of Aeronautics and Astronautics, Beijing, China, in 2001. From 1991–1995, he was a teacher with Tongcheng Middle-School, Anhui, China. From January 2002 to November 2003, he was a postdoctoral research associate at the Institute of Systems Science, Academy of Mathematics and System Sciences, Chinese Academy of Sciences, Beijing, China, where he worked on navigation, guidance and control. From November 2003 to February 2004, he joined the National University of Singapore as a research fellow, where he worked on variable structure control. From February 2004 to February 2006, he was with the University of Glamorgan, Pontypridd, UK, as a research fellow, where he worked on networked control systems. From February 2007 to June 2008, he was a guest professor with Innsbruck Medical University, Innsbruck, Austria, where he worked on biomedical signal processing. Since July 2004, he has been with the Department of Automatic Control, Beijing Institute of Technology, Beijing, first as an associate professor, and then, since 2008, as a professor. His current research interests are in the fields of networked control systems, robust control, sliding mode control, active disturbance rejection control and biomedical signal processing.

Preface

In recent years, wireless sensor networks (WSN) have produced a large amount of data that need to be processed, delivered, and assessed according to the application objectives. The way these data are manipulated by the sensor nodes is a fundamental issue. Information fusion arises as a response to process data gathered by sensor nodes and benefits from their processing capability. By exploiting the synergy among the available data, information fusion techniques can reduce the amount of data traffic, filter noisy measurements, and make predictions and inferences about a monitored entity. The book introduces the subject of multi-sensor fusion as the method of choice for implementing distributed systems.

This book is about the current state-of-the-art of information fusion, presenting the known methods, algorithms, architectures, and models of information fusion, and discussing their applicability in the context of wireless sensor networks. Particular considerations are given to covering wide topics that were treated in the literature and presenting results of typical case studies. The key feature is to provide a teaching-oriented volume with research-supported elements and comprehensive references.

The book applies recently developed convex optimization theory and high efficient algorithms in estimation fusion, which opens a very attractive research subject on distributed estimation and fusion for sensor networks. Supplying powerful and advanced mathematical treatment of the fundamental problems, it will help to greatly broaden prospective applications of such developments in practice.

The ultimate vision of this work is that information-based control designers will be able to model parts of dynamic systems (much as control engineers model electrical and mechanical systems), and use those models to develop distributed fusion control algorithms based on a theory of feedback control. It is intended to present a cohesive overview of the key results of theory and applications of information fusion–related problems in networked systems in a unified framework.

Throughout this book, the following terminologies, conventions and notations have been adopted. All of them are quite standard in the scientific media and only vary in form or character. Matrices, if their dimensions are not explicitly stated, are assumed to be compatible for algebraic operations. In symmetric block matrices or complex matrix expressions, we use the symbol • to represent a term that is induced by symmetry.

Many modern large-scale systems are automatically managed through networks of computers that are tied to sensors and actuators, leading to networked control systems. The inter-relationships among communication, computation, and control in such systems are clearly a subject of great interest. Therefore, networked fusion and filtering are attracting increasing attention in view of their wide industrial implications. The idea for writing the book arose and developed through the consecutive visits of the first author to the second author at the BIT, China. In writing this volume, we took the approach of referring within the text to papers and/or books which we believe taught us some concepts, ideas, and methods. We further complemented this by adding remarks and notes within and at the end of each chapter to shed some light on other related results.

Chapter 1

Introduction

Information fusion arises as a response to process data gathered by sensor nodes and benefits from their processing capability. These nodes are the main entities of wireless sensor networks that produce a large amount of data, which in turn needs to be processed, delivered, and assessed according to the application objectives. By exploiting the synergy among the available data, information fusion techniques can reduce the amount of data traffic, filter noisy measurements, and make predictions and inferences about a monitored entity. The objective of this chapter is to provide a concise survey of the current state-of-the-art of information fusion by presenting the known methods, algorithms, architectures, and models of information fusion, and discussing their applicability in the context of wireless sensor networks.

1.1 Overview

A wireless sensor network (WSN) is a special type of ad hoc network composed of a large number of nodes equipped with different sensor devices. This network is supported by technological advances in low power wireless communications along with silicon integration of various functionalities such as sensing, communication, and processing. WSNs are emerging as an important computer class based on a new computing platform and networking structure. In turn, this will enable novel applications that are related to different areas such as environmental monitoring, industrial and manufacturing automation, health-care, and the military. Commonly, wireless sensor networks have strong constraints regarding power resources and computational capacity.

A WSN may be designed with different objectives. It may be designed to gather and process data from the environment in order to have a better understanding of the behavior of the monitored entity. It may also be designed to monitor an environment for the occurrence of a set of possible events, so that the proper action may be taken whenever necessary. A fundamental issue in WSNs is the way the collected data is processed. In this context, information fusion arises as a discipline that is concerned with how data gathered by sensors can be processed to increase the relevance of such a mass of data. In a nutshell, information fusion can be defined as the combination of multiple sources to obtain improved information (cheaper, greater quality, or greater relevance).

Information fusion is commonly used in detection and classification tasks in different application domains, such as robotics and military applications. Lately, these mechanisms have been used in new applications such as intrusion detection [1] and Denial of Service (DoS) detection [2]. Within the WSN domain, simple aggregation techniques (e.g., maximum, minimum, and average) have been used to reduce the overall data traffic to save energy [4, 6, 10].

Additionally, information fusion techniques have been applied to WSNs to improve location estimates of sensor nodes [13], detect routing failures [12], and collect link statistics for routing protocols [15] .

1.2   Fundamental Terms

Several different terms (e.g., data fusion, sensor fusion, and information fusion) have been used to describe aspects of the fusion subject (including theories, processes, systems, frameworks, tools, and methods). Consequently, there is a terminology confusion.

The terminology related to systems, architectures, applications, methods, and theories about the fusion of data from multiple sources is not unified. Different terms have been adopted, usually associated with specific aspects that characterize the fusion. For example, *sensor/multi–sensor fusion* is commonly used to specify that sensors provide the data being fused. Despite the philosophical issues about the difference between data and information, the terms *data fusion* and *information fusion* are usually accepted as overall terms.

Many definitions of data fusion have been provided through the years, most of them derived from military and remote sensing fields. In 1991, the data fusion work group of the Joint Directors of Laboratories (JDL) organized an effort to define a lexicon [14] with some terms of reference for data fusion. They define data fusion as a "multilevel, multifaceted process dealing with the automatic detection, association, correlation, estimation, and combination of data and information from multiple sources." We note that [5] generalizes this definition, stating that data can be provided by a single source or by multiple sources. Both definitions are general and can be applied in different fields, including remote sensing. Although they suggest the combination of data without specifying its importance or its objective, the JDL data fusion model provided by the U.S. Department of Defense [14] deals with quality improvement.

In [3], data fusion is defined as "the combination of data from multiple sensors, and related information provided by associated databases, to achieve improved accuracy and more specific inferences than could be achieved by the use of a single sensor alone." Here, data fusion is performed with an objective: accuracy improvement. However, this definition is restricted to data provided by sensors; it does not foresee the use of data from a single source.

Claiming that all previous definitions are focused on methods, means, and sensors, [31] changes the focus to the framework used to fuse data. Wald states that data fusion is a formal framework in which are expressed means and tools for the alliance of data originating from different sources. It aims at obtaining information of greater quality; the exact definition of greater quality will depend upon the application. In addition, Wald considers data taken from the same source at different instants as distinct sources. The word "quality" is a loose term intentionally adopted to denote that the fused data is somehow more appropriate to the application than the original data. In particular for WSNs, data can be fused with at least two objectives: accuracy improvement and energy-saving. Although the definition and terminology in [31] are well accepted by the *Geoscience and Remote Sensing Society*, and officially adopted by the *Data Fusion Server*, the term "Multi–sensor Fusion" has been used with the same meaning by other authors, such as [3].

Multi–sensor Integration is another term used in robotics/computer vision [7] and industrial automation [9]. According to [8], multi-sensor integration is the synergistic use of information provided by multiple sensory devices to assist in the accomplishment of a task by a system; and multi-sensor fusion deals with the combination of different sources of sensory information into one representational format during any stage in the integration process. Multi–sensor integration is a broader term than multi-sensor fusion. It makes explicit how the fused data is used by the whole system to interact with the environment.

1.3 Some Limitations

Information fusion should be considered a critical step in designing a wireless sensor network. The reason is that information fusion can be used to extend the network lifetime and is commonly used to fulfill application objectives, such as target tracking, event detection, and decision making. Hence, blundering information fusion may result in a waste of resources, and misleading assessments. Therefore, we must be aware of the possible limitations of information fusion to avoid blundering situations. Because of the resource rationalization needs of WSNs, data processing is commonly implemented as in-network algorithms [4, 11]. Hence, whenever possible, information fusion should be performed in a distributed (in-network) fashion to extend the network lifetime. Nonetheless, we must be aware of the limitations of distributed implementations of information fusion.

Considering the communication load, it was argued in the past that a centralized fusion system may outperform a distributed one since it has a global knowledge and access to all measurements. On the other hand, the distributed fusion is incremental and localized since it fuses measurements provided by a set of neighbor nodes and the result might be further fused by intermediate nodes until a sink node is reached. Such a drawback might often be present in WSNs wherein, due to resource limitations, distributed and localized algorithms are preferable to centralized ones. In addition, the lossy nature of wireless communication challenges information fusion because losses mean that input data may not be completely available.

Regarding information fusion, one might intuitively believe that in fusion processes, the more data the better, since the additional data should add knowledge, that is, to support decisions or filter embedded noise. However, it was shown that when the amount of additional incorrect data is greater than the amount of additional correct data, the overall performance of the fusion process can be reduced.

1.4   Information Fusion in Wireless Sensor Network

A Wireless Sensor Network (WSN) may be designed with different objectives. It may be designed to gather and process data from the environment in order to have a better understanding of the behavior of the monitored area. It may also be designed to watch an environment for the occurrence of a set of possible events, so the proper action may be taken whenever needed. A fundamental issue in WSN is the way to process the collected data. In this situation, information fusion arises as a discipline that is concerned with how data collected by sensors can be processed to increase the significance of such a mass of data [1]. Thus, data fusion can be defined as the combination of multiple sources to obtain improved data, that is, cheaper, greater quality, or of greater relevance.

Data fusion is commonly used in detection and classification tasks in different application domains, such as military applications and robotics [23]. Within the WSN domain, simple aggregation techniques i.e., maximum, minimum, and average have been used to reduce the overall data traffic to save energy [24], [25]. Additionally, data fusion techniques have been applied to WSNs to improve location estimates of sensor nodes, detect routing failures, and collect link statistics for routing protocols [26].

WSN is intended to be deployed in environments where sensors can be exposed to circumstances that might interfere with measurements provided [27]. Such circumstances include strong variations of pressure and temperature, radiation and electromagnetic noise. Thus, measurements may be imprecise in such scenarios. Even when environmental conditions are ideal, sensors may not give perfect measurements.

Basically, a sensor is a measurement device, and vagueness is usually associated with its observation. Such imprecision represents the imperfections of the technology and methods used to measure a physical incident. Failures are not an exception in WSN. As an example, consider a WSN that monitors a jungle to detect an event, such as fire or the presence of an animal. Sensor nodes can be destroyed by fire, animals, or even human beings; they might present manufacturing problems; and they might stop working due to a lack of energy. Each node that becomes inoperable might compromise the overall perception or the communication capability of the network. Here, perception ability is equivalent to the exposure concept. Both spatial and temporal coverage also pose limitations to WSN.

The sensing capability of a node is restricted to a limited area. For example, a thermometer in a room reports the temperature near the device but it might not represent fairly the overall temperature inside the room. Spatial coverage in WSN has been explored in different scenarios, such as node scheduling, target tracking, and sensor placement. Temporal coverage can be understood as the ability to fulfill the network purpose during its lifetime. For example, in a WSN for event detection, temporal coverage aims at assuring that no relevant event will be missed because there was no sensor perceiving the region at the specific time the event occurred. Thus, temporal coverage depends on the sensors sampling rate, nodes duty cycle, and communication delays. To overcome sensor failures, technological limitations, and spatial and temporal coverage problems, three properties must be ensured:

1. Complementarity
2. Cooperation
3. Redundancy

The area of interest can only be completely covered by the use of several sensor nodes, each cooperating with a partial view of the scene; and data fusion can be used to create the complete view from the pieces provided by each node. Redundancy makes the WSN less vulnerable to failure of a single node, and overlapping measurements can be fused to obtain more precise data.

Complementarity can be achieved by using sensors that observe different properties of the environment; data fusion can be used to combine complementary data so the resultant data allows inferences that might be not possible to be obtained from the individual measurements, e.g., angle and distance of an imminent threat can be fused to obtain its position. Due to redundancy and cooperation properties, WSN is often composed of a large number of sensor nodes, posing a new scalability challenge caused by possible collisions and transmissions of redundant data. Regarding the energy restrictions, communication should be reduced to increase the lifetime of the sensor nodes. Hence, data fusion is also important to reduce the overall communication load in the network by avoiding the transmission of redundant messages. In addition, any task in the network that handles signals or needs to make inferences can potentially use data fusion.

Data fusion should be considered a critical step in designing a wireless sensor network. The reason is that data fusion can be used to extend the network lifetime and is commonly used to fulfill the application objectives, such as event detection, target tracking, and decision making. Hence, careless data fusion may result in waste of resources and misleading assessments. Therefore, we must be aware of the possible limitations of data fusion to avoid blundering situations. Because of the resource rationalization needs of WSN, data processing is commonly implemented as in-network algorithms. Hence, data fusion should be performed in a distributed fashion to extend the network lifetime. Even so, we must be aware of the limitations of distributed implementations of data fusion.

Data fusion has established itself as an independent research area over the last decades, but a general formal theoretical framework to describe data fusion systems is still missing. One reason for this is the huge number of disparate research areas that utilize and illustrate some form of data fusion in their context of theory. For example, the concept of data or feature fusion, which forms together with classifier and decision fusion the three main divisions of fusion levels, initially occurred in multi–sensor processing. By now several other research fields have found its application useful. Besides the more classical data fusion approaches in statistics, control, robotics, computer vision, geosciences and remote sensing, artificial intelligence, and digital image/signal processing, the data retrieval community discovered some years ago its power in combining multiple data sources.

Several different terms have been used to illustrate the aspects regarding the fusion subject, that is, information fusion, sensor fusion, and data fusion. The expressions related to systems, applications, methods, architectures, and theories about the fusion of data from multiple sources are not unified yet.

Different terms have been adopted, usually associated with particular aspects that characterize the fusion, that is, sensor fusion is commonly used to specify that sensors provide the data being fused. Despite the theoretical issues about the difference between information and data, the terms information fusion and data fusion are usually accepted as overall terms.

Many definitions of data fusion have been provided along the years, most of them were found in military and remote sensing fields. The data fusion work group of the Joint Directors of Laboratories (JDL) organized an effort to define a dictionary with some terms of reference for data fusion [14]. They define data fusion as a multilevel process dealing with the automatic detection, estimation, association, correlation, and combination of data and data from several sources. The JDL data fusion model deals with quality improvement. Hall defines data fusion as a combination of data from multiple sensors to accomplish improved accuracy and more specific inferences that could be achieved by the use of a single sensor alone [30]. All the previous definitions are focused on means, methods, and sensors. Wald in [31] changes the attention of fuse data to the framework used. He defines data fusion as a formal framework in which is expressed means and tools for the alliance of data originating from different sources. He considers data taken from the same source at different instants as separate sources. For WSN, data can be fused with at least two objectives: accuracy improvement and energy saving.

Multi–sensor integration is another expression used in computer vision and industrial automation. Luo [32] defines multi-sensor integration as a synergistic use of data provided by multiple sensory devices to help in the accomplishment of a task by a system. However, multi-sensor fusion deals with the combination of different sources of sensory data into one representational format during any stage in the integration process. Multi-sensor integration is a broader term than multi-sensor fusion. It makes clear how the fused data is used by the whole system to interact with the environment. However, it might suggest that only sensory data is used in the fusion and integration processes.

The term data aggregation has become popular in the wireless sensor network community as a synonym for information fusion [33]. Data aggregation comprises the collection of raw data from pervasive data sources, the flexible, programmable composition of the raw data into less voluminous refined data, and the timely delivery of the refined data to data consumers. Aggregation is the ability to summarize data that is, the amount of data is reduced. However, for applications that require original and accurate measurements, such summarization may represent an accuracy loss [34].

Although many applications might be interested only in summarized data, we cannot always state whether or not the summarized data is more precise than the original data set. Because of that, the use of data aggregation as a general term should be avoided because it also refers to one example of data fusion, which is summarization. Figure 1.1 shows the relationship among the concepts of multi-sensor/sensor fusion, multi-sensor integration, data aggregation, information fusion, and data fusion. Here, we understand that both terms, information fusion and data fusion, can be used with the same meaning.

Multi-sensor/sensor fusion is the subset that operates with sensory sources. Data aggregation defines another subset of information fusion that means to reduce

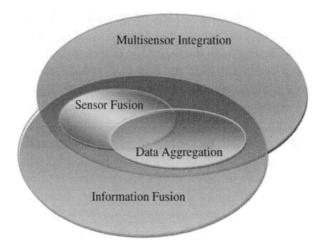

Figure 1.1: The relationship among the fusion terms.

the data volume, which can manipulate any type of information/data, including sensory data. Thus, multi-sensor integration is a slightly different term in the sense that it applies information fusion to make inferences using sensory devices and associated information to interact with the environment. Thus, multi-sensor/sensor fusion is fully contained in the intersection of multi–sensor integration and information/data fusion.

1.5 Classifying Information Fusion

Information fusion can be categorized based on several aspects. Relationships among the input data may be used to segregate information fusion into classes (e.g., cooperative, redundant, and complementary data). Also, the abstraction level of the manipulated data during the fusion process (measurement, signal, feature, decision) can be used to distinguish among fusion processes. Another common classification consists in making explicit the abstraction level of the input and output of a fusion process. These common classifications of information fusion are explored in this section [32].

1.5.1 Classification based on relationship among the sources

According to the relationship among the sources, information fusion can be classified as complementary, redundant, or cooperative [21]. Thus, according to the relationship among sources, information fusion can be:

Complementary: When information provided by the sources represents different portions of a broader scene, information fusion can be applied to obtain a piece of information that is more complete (broader). In Figure 1.2, sources S1 and S2 provide different pieces of information, a and b, respectively, that are fused to achieve a broader information, denoted by (a+b), composed of non-redundant pieces a and b that refer to different parts of the environment (e.g., temperature of west and east sides of the monitored area).

Fusion searches for completeness by compounding new information from different pieces. An example of complementary fusion consists in fusing data from sensor nodes (e.g., a sample from the sensor field) into a feature map that describes the whole sensor field [17], [18], [19], [20], hence broader information.

Redundant: If two or more independent sources provide the same piece of information, these pieces can be fused to increase the associated confidence. Sources S2 and S3 in Figure 1.2 provide the same information, b, which is fused to obtain more accurate information, (b). Redundant fusion might be used to increase the reliability, accuracy, and confidence of the information. In WSNs, redundant fusion can provide high quality information and prevent sensor nodes from transmitting redundant information data.

Cooperative: Two independent sources are cooperative when the information provided by them is fused into new information (usually more complex than the original data) that, from the application perspective, better represents the reality. Sources S4 and S5, in Figure 1.2, provide different information, c and c^*, that are fused into (c), which better describes the scene compared to c and c^* individually.

A traditional example of cooperative fusion is the computation of a target location based on angle and distance data. Cooperative fusion should be carefully applied since the resultant data is subject to the inaccuracies and imperfections of all participating sources [35], [36].

1.5.2 Classification based on levels of abstraction

The work of [8] uses essentially four levels of abstraction to classify information fusion: signal, pixel, feature, and symbol.

Signal level fusion deals with single or multidimensional signals from sensors. It can be used in real-time applications or as an intermediate step for further fusions.

Pixel level fusion operates on images and can be used to enhance image-processing tasks.

Feature level fusion deals with features or attributes extracted from signals or images, such as shape and speed.

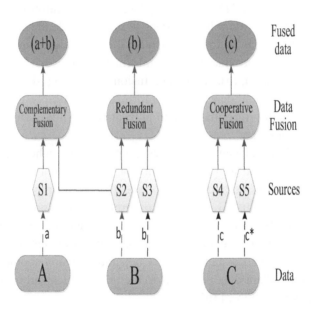

Figure 1.2: Types of data fusion based on the relationship among the sources.

In *Symbol level fusion*, information is a symbol that represents a decision, and it is also referred to as decision level. Typically, the feature and symbol fusions are used in object recognition tasks.

Such a classification presents some drawbacks and is not suitable for all information fusion applications:

1. Both signals and images are considered raw data usually provided by sensors, so they might be included in the same class.

2. Raw data may not be only from sensors, since information fusion systems might also fuse data provided by databases or human interaction.

3. It suggests that a fusion process cannot deal with all levels simultaneously.

According to the level of abstraction of the manipulated data, data fusion can be classified into four categories:

1. *Low-level fusion*: Raw data are provided as inputs and combined into new data that are more accurate than the individual inputs. In [29], an example of low-level fusion is given by applying a moving average filter to estimate ambient noise and determine whether or not the communication channel is clear.

2. *Medium-level fusion*: Features and attributes of an entity are fused to obtain a feature map that may be used for other tasks. It is also known as feature/attribute level fusion.

3. *High-level fusion*: It is known as symbol or decision level fusion. It takes decisions or symbolic representations as input and combines them to obtain a more confident and/or a global decision. An example of high-level fusion is the Bayesian approach for binary event detection proposed by Krishnamachari in [39] that detects and corrects measurement faults.

4. *Multilevel fusion*: Fusion process encompasses data of different abstraction levels and both input and output of fusion can be of any level. For example, a measurement is fused with a feature to provide a decision.

1.5.3 Classification based on input and output

Another classification was introduced in [28] that considers the abstraction level. Data fusion processes are categorized based on the level of abstraction of the input and output data [28]. Five categories are identified accordingly:

1. *Data In Data Out* (DAI-DAO): In this class, data fusion deals with raw data and the result is also raw data, possibly more accurate or reliable.

2. *Data In Feature Out* (DAI-FEO): Data fusion uses raw data from sources to extract features or attributes that describe an entity. Entity here means any object, situation, or world abstraction.

3. *Feature In Feature Out* (FEI-FEO): It works on a set of features to improve/ refine a feature, or extract new ones.

4. *Feature In Decision Out* (FEI-DEO): Data fusion takes a set of features of an entity generating a symbolic representation or a decision.

5. *Decision In Decision Out* (DEI-DEO): Decisions can be fused in order to obtain new decisions or give emphasis on previous ones.

In comparison to the classification presented before, this classification can be seen as an extension of the earlier one with a finer granularity where DAI-DAO corresponds to Low Level Fusion, FEI-FEO to Medium Level Fusion, DEI-DEO to High Level Fusion, DAI-FEO and FEI-DEO are included in Multilevel Fusion.

1.6   Outline of the Book

During the past several years with the explosive advances in digital technology, considerable research investigations have been directed to the interlinks among three main fields: communication networking, computation techniques and control design. One fundamental issue that arises from these investigations is the increasingly significant role of "information". This brings about the need to grasp several new notions like "networking", "wireless sensor networks", "networked control/filtering", "information fusion", to name a few. An integral part linking these functions is networked filtering and fusions, which arose from different views. This book is written about the networked filtering and fusion and the wide-spectrum of its applications in wireless sensor networks.

1.6.1   Methodology

Throughout the book, our methodology in each chapter/section is composed of five steps:

- **Modeling and/or Representation**
  In which we discuss the main ingredients of the model under consideration.

- **Definitions and/or Assumptions**
  Here we state the definitions and/or constraints on the model variables to pave the way for subsequent analysis.

- **Analysis and Examples**
  This signifies the core of the respective sections and subsections which contains some solved examples for illustration.

- **Results**
  These are provided most of the time in the form of theorems, lemmas and corollaries.

- **Remarks**
  Which are given to shed some light on the relevance of the developed results vis-a-vis published work.

In the sequel, theorems (lemmas, corollaries) are keyed to chapters and stated in *italic* font with **bold titles**, for example, **Theorem 5.2** means Theorem 2 in Chapter 5 and so on. For convenience, we have grouped the reference in one major bibliography cited toward the end of the book. Relevant notes and research issues are offered at the end of each chapter for the purpose of stimulating the reader. We hope that this way of articulating the information will attract the attention of a wide spectrum of readership. There is a separate section about suggested topics for further research.

In brief, the main features of the book are:

1. It provides an overall assessment of networked filtering and fusion for wireless sensor networks over the past several years,

2. It addresses several issues that occur at the interlinks of communication, computation and control, with focus on the role of information,

3. It presents key concepts with their proofs followed by efficient computational methods,

4. It gives suggested topics for further research,

5. It treats representative simulation examples.

1.6.2   Chapter organization

Recent advances in micro-electro-mechanical systems (MEMS) technology, wireless communications, and digital electronics have enabled the development of low-cost, low-power, multifunctional sensor nodes that are small in size and communicate untethered in short distances. These tiny sensor nodes, which consist of sensing, data processing, and communicating components, leverage the idea of sensor networks based on collaborative effort of a large number of nodes. On another avenue, real-time systems have been investigated for a long time in the control literature and have attracted increasingly more attention for more than three decades. The literature grew progressively and quite a number of fundamental concepts and powerful tools have been developed from various disciplines. Many fundamental problems are brought up that call for further exploration. Among the core issues is the development of novel control systems applications for wireless sensor networks. In particular, there still is not a unified framework that can cope with the core issues in a systematic way. This motivated us to write the current book on networked filtering and fusion for wireless sensor networks

The book is primarily intended for researchers and engineers in the systems, control and communication community. It can also serve as complementary reading for elective courses for networked filtering and fusion and/or wireless sensor networks at the post-graduate level. The material in the book is divided into twelve chapters:

Chapter 1 is an introductory chapter in which the different terms, issues classifications and limitations pertaining to information fusion and wireless sensor networks are described.

Chapter 2 is devoted exclusively to wireless sensor networks where some definitions, common characteristics, and required mechanisms are presented. It also covers types of sensor networks, and lists recent applications. It discusses the relevant topics like routing protocols, sensor selection schemes, and security requirements as well as the quality of service management.

Chapter 3 deals with methods for distributed sensor fusion with focus on consensus problems in networked systems and consensus filters. Important topics like assessment of distributed state estimation, multi-sensor management as well as distributed estimation for adaptive sensor selection are also examined in depth.

Chapter 4 provides an overview of distributed Kalman estimation methods and their wide applications. The purpose is to have a catalog of tabulation and referencing to various methods.

Chapter 5 establishes a detailed characterization of expected maximization based forward-backward Kalman filter and its application to information fusion fault diagnosis. It gives simulation results of lab-scale industrial processes.

Chapter 6 traces the development of wireless control and estimation methods as distinct applications of wireless sensor networks, introduces sources of wireless communication errors, and discusses the structure as well as the networked control design.

Chapter 7 presents recent results on multi-sensor data-fusion techniques using model-based, model-free and probabilistic schemes. Detailed simulations of typical processes are incorporated.

Chapter 8 outlines the multi-sensor data fusion approaches with particular emphasis on discrete-time unscented Kalman filtering and gives detailed simulation scenarios.

Chapter 9 focuses on Kalman filter fusion with and without complete prior information. Technical analyses of modified Kalman fusion filter are provided in terms of lower-bound, upper bound and convergence behavior.

Chapter 10 details algorithms of distributed estimation via information matrix and presents the fundamental attributes in cases of complete feedback and partial feedback. The content includes the weighted covariance algorithm, Kalman-like particle filter algorithm, and measurement fusion algorithm.

Chapter 11 contains a systematic treatment of distributed cooperative filtering, distributed consensus filtering and distributed fusion in sensor networks with improved performance.

Chapter 12 is an appendix containing mathematical analysis of relevant topics, fundamental lemmas, and basic algebraic inequalities that are used throughout the book.

Several illustrative numerical examples are provided within the individual chapters.

1.7 Notes

This chapter briefly discusses the concepts and modules of data fusion and provides a detailed survey of the existing data fusion techniques and methods. It is emphasized that data fusion is generally defined as the use of techniques that combine data from multiple sources and gather this information in order to achieve inferences, which will be more efficient and potentially more accurate than if they were achieved by means of a single source. The term "efficient", in this case, can mean more reliable delivery of accurate information, that is more complete, and more dependable. By and large, the data fusion can be implemented in both centralized and distributed systems. In a centralized system, all raw sensor data would be sent to one node, and the data fusion would all occur at the same location. In a distributed system, the different fusion modules would be implemented on distributed components.

1.8   Proposed Topics

1. It is known that "Role-based model" represents a change of focus on how data fusion systems can be modeled and designed [27]. Data fusion systems are specified based on the fusion roles and the relationships among them providing a more fine-grained model for the fusion system. The two members of this generation are the Object–Oriented Model and the Frankel–Bedworth architecture. Investigate the merits and demerits of both members items of their components, fusion tasks, and their use in WSN.

2. Provide a detailed answer on the following questions about information fusion:

   - What is information fusion?
   - Why should a designer use it?
   - What are the available techniques? and
   - How should a designer use such techniques?

3. Centralized versus distributed was a major quandary in information fusion. Give your view about the challenges to assure temporal and spatial correlation among the sources while the data is fused and disseminated at the same time.

Chapter 2

Wireless Sensor Networks

A wireless sensor network (WSN) has important applications such as remote environmental monitoring and target tracking. This has been enabled by the availability, particularly in recent years, of sensors that are smaller, cheaper, and intelligent. These sensors are equipped with wireless interfaces so that they can communicate with one another to form a network. The design of a WSN depends significantly on the application, and it must consider factors such as the environment, the application's design objectives, cost, hardware, and system constraints.

2.1   Some Definitions

A wireless sensor network (WSN) typically has little or no infrastructure. It consists of a number of sensor nodes (a few tens to thousands) working together to monitor a region to obtain data about the environment. A schematic view of a WSN is depicted in Figure 2.1. There are two types of WSNs: structured and unstructured. An unstructured WSN is one that contains a dense collection of sensor nodes. Sensor nodes may be deployed in an ad hoc manner in the field. *In ad hoc deployment, sensor nodes may be randomly placed in the field.* Once deployed, the network is left unattended to perform monitoring and reporting functions. In an unstructured WSN, network maintenance such as managing connectivity and detecting failures is difficult, since there are so many nodes. In a structured WSN, all or some of the sensor nodes are deployed in a pre-planned manner. *In pre-planned deployment, sensor nodes are pre-determined to be placed at fixed locations.* The advantage of a structured network is that fewer nodes can be deployed with lower network maintenance and management cost. Fewer nodes can be deployed since, the nodes are placed at specific locations to provide coverage while ad hoc deployment can have uncovered regions.

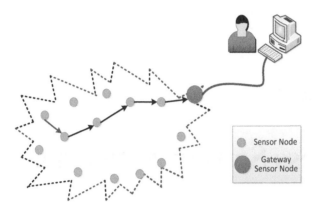

Figure 2.1: View of wireless sensor networks.

2.2   Common Characteristics

There is a wide range of application types of WSN, which possess certain common-
alities especially with respect to the characteristics and the required mechanisms
of such systems. Realizing these characteristics with new mechanisms is the major
challenge of the vision of wireless sensor networks. The following characteristics
are shared among most of the application examples discussed above:

- **Type of service** The service type rendered by a conventional communication
  network is evident; it moves bits from one place to another. For a WSN,
  moving bits is only a means to an end, but not the actual purpose. Rather,
  a WSN is expected to provide meaningful information and/or actions about
  a given task: "People want answers, not numbers" [59]. Additionally, concepts like scoping of interactions to specific geographic regions or to time
  intervals will become important. Hence, new paradigms of using such a network are required, along with new interfaces and new ways of thinking about
  the service of a network.

- **Fault tolerance** Since nodes may run out of energy or might be damaged,
  or since the wireless communication between two nodes can be permanently
  interrupted, it is important that the WSN as a whole is able to tolerate such
  faults. To tolerate node failure, redundant deployment is necessary, using
  more nodes than would be strictly necessary if all the nodes functioned correctly.

- **Quality of service** Closely related to the type of a network's service is the quality of that service. Traditional quality of service requirements usually coming from multimedia-type applications like bounded delay or minimum bandwidth, are irrelevant when applications are tolerant to latency or the bandwidth of the transmitted data is very small in the first place. In some cases, only occasional delivery of a packet can be more than enough; in other cases, very high reliability requirements exist. In yet other cases, delay is important when actuators are to be controlled in a real-time fashion by the sensor network. The packet delivery ratio is an insufficient metric; what is relevant is the amount and quality of information that can be extracted at given sinks about the observed objects or area. Therefore, adapted quality concepts like reliable detection of events or the approximation quality of say, a temperature map is important.

- **Lifetime** In many scenarios, nodes will have to rely on a limited supply of energy (using batteries). Replacing these energy sources in the field is usually not practicable, and simultaneously, a WSN must operate at least for a given mission time or as long as possible. Hence, the lifetime of a WSN becomes very important. Evidently, an energy-efficient way to operate the WSN is necessary. As an alternative or supplement to energy supplies, a limited power source (via power sources like solar cells, for example) might also be available on a sensor node. Typically, these sources are not powerful enough to ensure continuous operation but can provide some recharging of batteries. Under such conditions, the lifetime of the network should ideally be infinite.

  By and large, the lifetime of a network also has direct trade-offs against quality of service: investing more energy can increase quality but decrease the lifetime. Concepts to harmonize these trade-offs are required. The precise definition of lifetime depends on the application at hand. A simple option is to use the time until the first node fails (or runs out of energy) as the network lifetime. Other options include the time until the network is disconnected in two or more partitions, the time until 50% (or some other fixed ratio) of nodes have failed, or the time when for the first time a point in the observed region is no longer covered by at least a single sensor node (when using redundant deployment, it is possible and beneficial to have each point in space covered by several sensor nodes initially).

- **Scalability** Since a WSN might include a large number of nodes, the employed architectures and protocols must be able scale to these numbers.

- **Wide range of densities** In a WSN, the number of nodes per unit area, the

density of the network, can vary considerably. Different applications will
have very different node densities. Even within a given application, density
can vary over time and space because nodes fail or move; the density also
does not have to be homogeneous in the entire network (because of imperfect
deployment, for example) and the network should adapt to such variations.

- **Programmability** Not only will it be necessary for the nodes to process
  information, but also they will have to react flexibly to changes in their
  tasks. These nodes should be programmable, and their programming must
  be changeable during operation when new tasks become important. A fixed
  way of information processing is insufficient.

- **Maintainability** As both the environment of a WSN and the WSN itself
  change (depleted batteries, failing nodes, new tasks), the system has to adapt.
  It has to monitor its own health and status to change operational parameters
  or to choose different trade-offs (e.g., to provide lower quality when the energy resource become scarce). In this sense, the network has to maintain itself; it could also be able to interact with external maintenance mechanisms
  to ensure its extended operation at a required quality [60].

2.3   Required Mechanisms

To realize the foregoing requirements, innovative mechanisms for a communication network have to be found, as well as new architectures, and protocol concepts.
A particular challenge here is the need to find mechanisms that are sufficiently specific to the idiosyncrasies of a given application to support the specific quality of
service, lifetime, and maintainability requirements [246]. On the other hand, these
mechanisms also have to generalize to a wider range of applications lest a complete
from-scratch development and implementation of a WSN becomes necessary for
every individual application. This would likely render WSNs as a technological
concept economically infeasible. Some of the mechanisms that will form typical
parts of WSNs are:

- **Multihop wireless communication** While wireless communication will be
  a core technique, a direct communication between a sender and a receiver
  is faced with limitations. In particular, communication over long distances
  is only possible using prohibitively high transmission power. The use of intermediate nodes as relays can reduce the total required power. Hence, for
  many forms of WSNs, so-called multihop communication will be a necessary ingredient.

- **Energy-efficient operation** To support long lifetimes, energy-efficient operation is a key technique. Options to look into include energy-efficient data transport between two nodes (measured in J/bits) or, more importantly, the energy-efficient determination of a requested information. Also, non-homogeneous energy consumption, the forming of "hotspots", is an issue.

- **Auto-configuration** A WSN will have to configure most of its operational parameters autonomously, independent of external configuration because the sheer number of nodes and simplified deployment will require that capability in most applications. As an example, nodes should be able to determine their geographical positions only using other nodes of the network, or so-called "self-location". Also, the network should be able to tolerate failing nodes (because of a depleted battery, for example) or to integrate new nodes (because of incremental deployment after a failure, for example).

- **Collaboration and in-network processing** In some applications, a single sensor is not able to decide whether an event has happened, so several sensors have to collaborate to detect an event, and only the joint data of many sensors provides enough information. Information is processed in the network itself in various forms to achieve this collaboration, as opposed to having every node transmit all the data to an external network and processing it "at the edge" of the network.

  An example is to determine the highest or the average temperature within an area and to report that value to a sink. To solve such tasks efficiently, readings from individual sensors can be aggregated as they propagate through the network, reducing the amount of data to be transmitted and hence improving the energy efficiency. How to perform such aggregation is an open question.

- **Data centric** Traditional communication networks are typically centered around the transfer of data between two specific devices, each equipped with (at least) one network address; the operation of such networks is thus address-centric. In a WSN, where nodes are typically deployed redundantly to protect against node failures or to compensate for the low quality of a single node's actual sensing equipment, the identity of the particular node supplying data becomes irrelevant. What is important are the answers and values themselves, not which node has provided them. Hence, switching from an address-centric paradigm to a data-centric paradigm in designing architecture and communication protocols is promising.

  An example for such a data-centric interaction would be to request the average temperature in a given location area, as opposed to requiring temperature readings from individual nodes. Such a data-centric paradigm can also be used to set conditions for alerts or events ("raise an alarm if temperature exceeds a threshold"). In this sense, the data-centric approach is closely related to query concepts known from databases; it also combines well with collaboration, in-network processing, and aggregation.

- **Locality** By and large, the principle of locality will have to be embraced extensively to ensure, in particular, scalability. Nodes, which are very limited in resources like memory, should attempt to limit the state that they accumulate during protocol processing to only information about their direct neighbors. It is hoped that this will allow the network to scale to large numbers of nodes without having to rely on powerful processing at each single node. How to combine the locality principle with efficient protocol designs is still an open research topic, however.

- **Exploit trade-offs** Similar to the locality principle, WSNs will have to rely to a large degree on exploiting various inherent trade-offs between mutually contradictory goals, both during system/protocol design and at run-time. Examples for such trade-offs have been mentioned already: higher energy expenditure allows higher result accuracy, or a longer lifetime of the entire network trades off against the lifetime of individual nodes. Another important trade-off is node density: depending on application, deployment, and node failures at run-time, the density of the network can change considerably, the protocols will have to handle very different situations, possibly present at different places in a single network. Again, not all the research questions are solved here.

Harnessing these mechanisms such that they are easy to use, yet sufficiently general, is a major challenge for an application programmer. Departing from an address-centric view of the network requires new programming interfaces that go beyond the simple semantics of the conventional socket.

2.4   Related Ingredients

Wireless sensor networks (WSNs) have gained worldwide attention in recent years, particularly with the proliferation of micro-electro-mechanical systems (MEMS) technology which has facilitated the development of smart sensors. These sensors are small, with limited processing and computing resources, and they are inexpensive compared to traditional sensors. These sensor nodes can sense, measure, and gather information from the environment and, based on some local decision process, they can transmit the sensed data to the user.

Smart sensor nodes are low power devices equipped with one or more sensors, a processor, memory, a power supply, a radio, and an actuator. In this regard, *an actuator is an electro-mechanical device that can be used to control different components in a system. In a sensor node, actuators can actuate different sensing devices, adjust sensor parameters, move the sensor, or monitor power in the sensor node.* A variety of mechanical, thermal, biological, chemical, optical, and magnetic sensors may be attached to the sensor node to measure properties of the environment. Since the sensor nodes have limited memory and are typically deployed in difficult-to-access locations, a radio is implemented for wireless communication to transfer the data to a base station (e.g., a laptop, a personal hand held device, or an access point to a fixed infrastructure). Battery is the main power source in a sensor node. A secondary power supply that harvests power from the environment such as solar panels may be added to the node depending on the appropriateness of the environment where the sensor will be deployed. Depending on the application and the type of sensors used, actuators may be incorporated in the sensors.

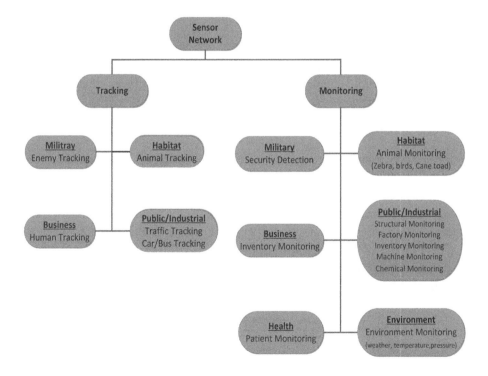

Figure 2.2: Overview of sensor applications.

2.4.1 Key issues

Current state-of-the-art sensor technology provides a solution to design and develop many types of wireless sensor applications. Available sensors in the market include generic (multi-purpose) nodes and gateway (bridge) nodes. A generic (multi-purpose) sensor node's task is to take measurements from the monitored environment. It may be equipped with a variety of devices which can measure various physical attributes such as light, temperature, humidity, barometric pressure, velocity, acceleration, acoustics, magnetic field, etc. Gateway (bridge) nodes gather data from generic sensors and relay them to the base station. Gateway nodes have higher processing capability, battery power, and transmission (radio) range. A combination of generic and gateway nodes is typically deployed to form a WSN.

To enable wireless sensor applications using sensor technologies, the range of tasks can be broadly classified into three groups as shown in Figure 2.3:

1. The first group is the system, where each sensor node is an individual subsystem. In order to support different application software on a sensor system, development of new platforms, operating systems, and storage schemes are needed.

2. The second group is communication protocols, which enable communication both between the application and sensors, and among sensor nodes.

3. The third group is services which are developed to enhance the application and to improve system performance and network efficiency.

From application requirements and network management perspectives, it is important that sensor nodes are capable of self-organizing themselves. That is, the sensor nodes can organize themselves into a network and subsequently are able to control and manage themselves efficiently. As sensor nodes are limited in power, processing capacity, and storage, new communication protocols and management services are needed to fulfill these requirements.

The communication protocol consists of five standard protocol layers for packet switching:

- application layer,

- transport layer,

- network layer,

- data-link layer, and

- physical layer.

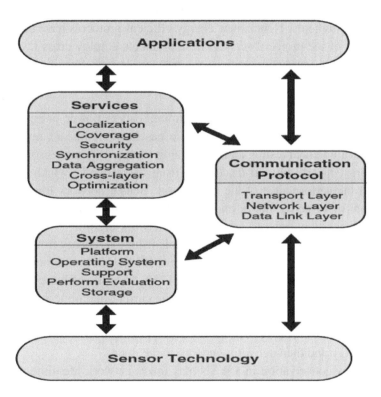

Figure 2.3: Classification of various issues in wireless sensor networks.

In what follows, we study how protocols at different layers address network dynamics and energy efficiency. Functions such as localization, coverage, storage, synchronization, security, and data aggregation and compression are explored as sensor network services.

Implementation of protocols at different layers in the protocol stack can significantly affect energy consumption, end-to-end delay, and system efficiency. It is important to optimize communication and minimize energy usage. Traditional networking protocols do not work well in a WSN since they are not designed to meet these requirements. Hence, new energy-efficient protocols have been proposed for all layers of the protocol stack. These protocols employ cross-layer optimization by supporting interactions across the protocol layers. Specifically, protocol state information at a particular layer is shared across all the layers to meet the specific requirements of the WSN.

Because sensor nodes operate on limited battery power, energy usage is a very important concern in WSNs; and there has been significant research focus that revolves around harvesting and minimizing energy. When a sensor node is depleted of energy, it will die and disconnect from the network, which can significantly impact the performance of the application. Sensor network life–time depends on the number of active nodes and connectivity of the network, so energy must be used efficiently in order to maximize the network lifetime.

Energy harvesting involves nodes replenishing its energy from an energy source. Potential energy sources include solar cells [172], [173], vibration [174], fuel cells, acoustic noise, and a mobile supplier [175]. In terms of harvesting energy from the environment [176], the solar cell is the current mature technique that harvests energy from light. There is also work in using a mobile energy supplier such as a robot to replenish energy. The robots would be responsible for charging themselves with energy and then delivering energy to the nodes.

Energy conservation in a WSN maximizes network life–time and is addressed through efficient reliable wireless communication, intelligent sensor placement to achieve adequate coverage, security and efficient storage management, and through data aggregation and data compression. The above approaches aim to satisfy the energy constraint and to provide quality of service (QoS) for the application *QoS defines parameters such as end-to-end delay which must be guaranteed to an application/user.*

For reliable communication, services such as congestion control, active buffer monitoring, acknowledgments, and packet-loss recovery are necessary to guarantee reliable packet delivery. Communication strength is dependent on the placement of sensor nodes. Sparse sensor placement may result in long-range transmission and higher energy usage while dense sensor placement may result in short-range transmission and less energy consumption. Coverage is interrelated with sensor placement. The total number of sensors in the network and their placement determine the degree of network coverage. Depending on the application, a higher degree of coverage may be required to increase the accuracy of the sensed data. In what follows, we review new protocols and algorithms developed in these areas.

2.4.2 Types of sensor networks

Current WSNs are deployed on land, under–ground, and under–water. Depending on the environment, a sensor network faces different challenges and constraints. There are five types of WSNs: terrestrial WSN, underground WSN, underwater WSN, multi-media WSN, and mobile WSN.

Terrestrial WSNs [165] typically consist of hundreds to thousands of inexpensive wireless sensor nodes deployed in a given area, either in an ad hoc or in a pre-planned manner. In ad hoc deployment, sensor nodes can be dropped from a plane and randomly placed into the target area. In pre-planned deployment, there are grid placement, optimal placement [177], 2-D and 3-D placement [178], [179] models.

In a terrestrial WSN, reliable communication in a dense environment is very important. Terrestrial sensor nodes must be able to effectively communicate data back to the base station. While battery power is limited and may not be rechargeable, terrestrial sensor nodes, however, can be equipped with a secondary power source such as solar cells. In any case, it is important for sensor nodes to conserve energy. For a terrestrial WSN, energy can be conserved with multi-hop optimal routing, a short transmission range, in-network data aggregation, eliminating data redundancy, minimizing delays, and using low duty-cycle operations.

Underground WSNs [180], [181] consist of a number of sensor nodes buried underground or in a cave or mine used to monitor underground conditions. Additional sink nodes are located above ground to relay information from the sensor nodes to the base station. An underground WSN is more expensive than a terrestrial WSN in terms of equipment, deployment, and maintenance. Underground sensor nodes are expensive because appropriate equipment parts must be selected to ensure reliable communication through soil, rocks, water, and other mineral contents. The underground environment makes wireless communication a challenge due to signal losses and high levels of attenuation.

Unlike terrestrial WSNs, the deployment of an underground WSN requires careful planning, and energy and cost considerations. Energy is an important concern in underground WSNs. Like terrestrial WSNs, underground sensor nodes are equipped with a limited battery power and once deployed into the ground, it is difficult to recharge or replace a sensor node's battery. As before, a key objective is to conserve energy in order to increase the lifetime of network which can be achieved by implementing an efficient communication protocol.

Underwater WSNs [182, 183] consist of a number of sensor nodes and vehicles deployed under–water. As compared to terrestrial WSNs, underwater sensor nodes are more expensive and fewer sensor nodes are deployed. Autonomous underwater vehicles are used for exploration or gathering data from sensor nodes. Compared to a dense deployment of sensor nodes in a terrestrial WSN, a sparse deployment of sensor nodes is placed underwater. Typical underwater wireless communications are established through transmission of acoustic waves. A challenge in underwater acoustic communication is the limited bandwidth, long propagation delay, and signal fading issue. Another challenge is sensor node failure due to environmental conditions. Underwater sensor nodes must be able to self-configure and adapt to harsh ocean environments. Underwater sensor nodes are equipped with a limited battery which cannot be replaced or recharged. The issue of energy conservation for underwater WSNs involves developing efficient underwater communication and networking techniques.

Multi-media WSNs [184] have been proposed to enable monitoring and tracking of events in the form of multimedia such as video, audio, and imaging. Multimedia WSNs consist of a number of low cost sensor nodes equipped with cameras and microphones. These sensor nodes interconnect with each other over a wireless connection for data retrieval, processing, correlation, and compression. Multimedia sensor nodes are deployed in a pre-planned manner in the environment to guarantee coverage. Challenges in multi-media WSN include high bandwidth demand, high energy consumption, quality of service (QoS) provisioning, data processing and compressing techniques, and cross-layer design.

Multi-media content such as a video stream requires high bandwidth in order for the content to be delivered. As a result, high data rate leads to high energy consumption. Transmission techniques that support high bandwidth and low energy consumption have to be developed. QoS provisioning is a challenging task in a multi-media WSN due to the variable delay and variable channel capacity. It is important that a certain level of QoS must be achieved for reliable content delivery. In-network processing, filtering, and compression can significantly improve network performance in terms of filtering and extracting redundant information and merging contents. Similarly, cross-layer interaction among the layers can improve the processing and the delivery process.

Mobile WSNs consist of a collection of sensor nodes that can move on their own and interact with the physical environment. Mobile nodes have the ability to sense, compute, and communicate like static nodes. A key difference is that mobile nodes have the ability to reposition and organize themselves in the network. A mobile WSN can start off with some initial deployment and nodes can then spread out to gather information. Information gathered by a mobile node can be communicated to another mobile node when they are within range of each other.

Another key difference is data distribution. In a static WSN, data can be distributed using fixed routing or flooding while dynamic routing is used in a mobile WSN. Challenges in mobile WSNs include deployment, localization, self-organization, navigation and control, coverage, energy, maintenance, and data processing.

Mobile WSN applications include but are not limited to environment monitoring, target tracking, search and rescue, and real-time monitoring of hazardous material. For environmental monitoring in disaster areas, manual deployment might not be possible. With mobile sensor nodes, they can move to areas of events after deployment to provide the required coverage. In military surveillance and tracking, mobile sensor nodes can collaborate and make decisions based on the target. Mobile sensor nodes can achieve a higher degree of coverage and connectivity compared to static sensor nodes. In the presence of obstacles in the field, mobile sensor nodes can plan ahead and move appropriately to obstructed regions to increase target exposure.

2.4.3 Main advantages

Recent advances in micro-electro-mechanical systems (MEMS) technology, wireless communications, and digital electronics have enabled the development of low-cost, low-power, multifunctional sensor nodes that are small in size and can communicate untethered for short distances. These tiny sensor nodes, which consist of sensing, data processing, and communicating components, leverage the idea of sensor networks based on collaborative effort of a large number of nodes. Sensor networks represent a significant improvement over traditional sensors, which are deployed in the following two ways [142]:

- Sensors can be positioned far from the actual phenomenon of *something known by sense perception*. In this approach, large sensors that use some complex techniques to distinguish the targets from environmental noise are required.

- Several sensors that perform only sensing can be deployed. The positions of the sensors and communications topology are carefully engineered. They transmit time series of the sensed phenomenon to the central nodes where computations are performed and data are fused.

Thus a sensor network is composed of a large number of sensor nodes, which are densely deployed either inside the phenomenon or very close to it. In general, the position of sensor nodes need not be engineered or pre-determined. This allows random deployment in inaccessible terrains or disaster relief operations. On the other hand, this also means that sensor network protocols and algorithms must possess self-organizing capabilities. Another unique feature of sensor networks is the cooperative effort of sensor nodes. Sensor nodes are fitted with an on-board processor. Instead of sending the raw data to the nodes responsible for the fusion, sensor nodes use their processing abilities to locally carry out simple computations and transmit only the required and partially processed data.

2.5   Sensor Networks Applications

Sensor networks may consist of many different types of sensors such as seismic, low sampling rate magnetic, thermal, visual, infrared, acoustic and radar, which are able to monitor a wide variety of ambient conditions that include the following [143]:

- temperature,

- humidity,

- vehicular movement,

- lightning condition,

- pressure,

- soil makeup,

- noise levels,

- the presence or absence of certain kinds of objects,

- mechanical stress levels on attached objects, and

- the current characteristics such as speed, direction, and size of an object.

Sensor nodes can be used for continuous sensing, event detection, event ID, location sensing, and local control of actuators. The concept of micro-sensing and wireless connection of these nodes promise many new application areas. We categorize the applications into military, environment, health, home, and other commercial areas. It is possible to expand this classification with more categories such as space exploration, chemical processing and disaster relief.

2.5.1 Military applications

Wireless sensor networks can be an integral part of military command, control, communications, computing, intelligence, surveillance, reconnaissance and targeting (C4ISRT) systems. The rapid deployment, self-organization and fault tolerance characteristics of sensor networks make them a very promising sensing technique for the military (C4ISRT). Since sensor networks are based on the dense deployment of disposable and low-cost sensor nodes, destruction of some nodes by hostile actions does not affect a military operation as much as the destruction of a traditional sensor, which makes the sensor networks concept a better approach for battlefields. Some of the military applications of sensor networks are monitoring friendly forces, equipment and ammunition; battlefield surveillance; reconnaissance of opposing forces and terrain; targeting; battle damage assessment; and nuclear, biological and chemical (NBC) attack detection and reconnaissance.

- **Monitoring friendly forces, equipment and ammunition:**

Leaders and commanders can constantly monitor the status of friendly troops, the condition and the availability of the equipment and the ammunition in a battlefield, by the use of sensor networks. Every troop, vehicle, equipment and critical ammunition can be attached with small sensors that report the status. These reports are gathered in sink nodes and sent to the troop leaders. The data can also be forwarded to the upper levels of the command hierarchy while being aggregated with the data from other units at each level.

- **Nuclear, biological and chemical attack detection and reconnaissance:**

In chemical and biological warfare, being close to ground zero is important for timely and accurate detection of the agents. Sensor networks deployed in the friendly region and used as a chemical or biological warning system can provide the friendly forces with critical reaction time, which drops casualties drastically. We can also use sensor networks for detailed reconnaissance after an NBC attack is detected. For instance, we can make a nuclear reconnaissance without exposing a recon team to nuclear radiation.

- **Battlefield surveillance:**

Critical terrains, approach routes, paths and straits can be rapidly covered with sensor networks and closely watched for the activities of the opposing forces. As the operations evolve and new operational plans are prepared, new sensor networks can be deployed anytime for battlefield surveillance.

- **Reconnaissance of opposing forces and terrain:**

Sensor networks can be deployed in critical terrains, and some valuable, detailed, and timely intelligence about the opposing forces and terrain can be gathered within minutes before the opposing forces can intercept them.

- **Targeting:**

Sensor networks can be incorporated in guidance systems of the intelligent ammunition. Battle damage assessment: Just before or after attacks, sensor networks can be deployed in the target area to gather the battle damage assessment data.

2.5.2   Environmental applications

Some environmental applications of sensor networks include tracking the movements of birds, small animals, and insects; monitoring environmental conditions that affect crops and livestock; monitoring irrigation; using macro-instruments for large-scale Earth monitoring and planetary exploration; chemical/ biological detection; precision agriculture; biological, Earth, and environmental monitoring in marine, soil, and atmospheric contexts; forest fire detection; meteorological or geophysical research; flood detection; bio-complexity mapping of the environment; and pollution study [144], [145], [146], [142], [148], [160].

- **Forest fire detection:**

Since sensor nodes may be strategically, randomly, and densely deployed in a forest, sensor nodes can relay the exact origin of the fire to the end users before the fire is spread and uncontrollable. Millions of sensor nodes can be deployed and integrated using radio frequencies/optical systems. Also, they may be equipped with effective power scavenging methods [149], such as solar cells, because the sensors may be left unattended for months and even years. The sensor nodes will collaborate with each other to perform distributed sensing and overcome obstacles, such as trees and rocks, that block wired sensors' line of sight.

- **Biocomplexity mapping of the environment:**

A biocomplexity mapping of the environment requires sophisticated approaches to integrate information across temporal and spatial scales [154]. The advances of technology in the remote sensing and automated data collection have enabled higher spatial, spectral, and temporal resolution at a geometrically declining cost per unit area [150]. Along with these advances, the sensor nodes also have the ability to connect with the Internet, which allows remote users to control, monitor, and observe the biocomplexity of the environment.

Although satellite and airborne sensors are useful in observing large biodiversity, e.g., spatial complexity of dominant plant species, they are not fine grain enough to observe small size biodiversity, which makes up most of the biodiversity in an ecosystem [147]. As a result, there is a need for ground level deployment of wireless sensor nodes to observe the biocomplexity [155]. One example of biocomplexity mapping of the environment is done at the *James Reserve in Southern California* [145]. Three monitoring grids with each having 25100 sensor nodes will be implemented for fixed view multimedia and environmental sensor data loggers.

• **Flood detection:**

An example of flood detection is the ALERT system [161] deployed in the US. Several types of sensors deployed in the ALERT system are rainfall, water level, and weather sensors. These sensors supply information to the centralized database system in a pre-defined way. Research projects, such as the "COUGAR" *Device Database Project at Cornell University* [152] and the Data Space project at Rutgers [156], are investigating distributed approaches in interacting with sensor nodes in the sensor field to provide snapshot, and long-running queries.

• **Precision Agriculture:**

Some of the benefit is the ability to monitor the pesticides level in the drinking water, the level of soil erosion, and the level of air pollution in real time.

2.5.3   Health applications

Some of the health applications for sensor networks are providing interfaces for the disabled; integrated patient monitoring; diagnostics; drug administration in hospitals; monitoring the movements and internal processes of insects or other small animals; telemonitoring of human physiological data; and tracking and monitoring doctors and patients inside a hospital [153],[160].

• **Telemonitoring of human physiological data:**

The physiological data collected by the sensor networks can be stored for a long period of time [164], and can be used for medical exploration [162]. The installed sensor networks can also monitor and detect elderly people's behavior, e.g., a fall [151]. These small sensor nodes allow the subject a greater freedom of movement and allow doctors to identify pre-defined symptoms earlier [157]. Also, they facilitate a higher quality of life for the subjects compared to the treatment centers [163]. A "Health Smart Home" is designed in the Faculty of Medicine in Grenoble–France to validate the feasibility of such systems [158].

• **Tracking and monitoring doctors and patients inside a hospital:**

Each patient has small and light weight sensor nodes attached to them. Each sensor node has its specific task. For example, one sensor node may be detecting the heart rate while another is detecting the blood pressure. Doctors may also carry a sensor node, which allows other doctors to locate them within the hospital.

• **Drug administration in hospitals:**

If sensor nodes can be attached to medications, the chance of getting and prescribing the wrong medication to patients can be minimized, because patients will have sensor nodes that identify their allergies and their required medications. Computerized systems as described in [159] have shown that sensor nodes can help minimize adverse drug events.

2.5.4   Application trends

WSNs have great potential for many applications in scenarios such as military target tracking and surveillance [166, 167], natural disaster relief [168], biomedical health monitoring [169, 170], and hazardous environment exploration and seismic sensing [171]. In military target tracking and surveillance, a WSN can assist in intrusion detection and identification. Specific examples include spatially-correlated and coordinated troop and tank movements. With natural disasters, sensor nodes can sense and analyze the environment to forecast disasters before they occur. In biomedical applications, surgical implants of sensors can help monitor a patient's health. For seismic sensing, ad hoc deployment of sensors along the volcanic area can detect the development of earthquakes and eruptions. Unlike traditional networks, a WSN has its own design and resource constraints. Resource constraints include a limited amount of energy, short communication range, low bandwidth, and limited processing and storage in each node. Design constraints are application dependent and are based on the monitored environment. The environment plays a key role in determining the size of the network, the deployment scheme, and the network topology. The size of the network varies with the monitored environment. For indoor environments, fewer nodes are required to form a network in a limited space, whereas outdoor environments may require more nodes to cover a larger area. An ad hoc deployment is preferred over pre-planned deployment when the environment is inaccessible by humans or when the network is composed of hundreds to thousands of nodes. Obstructions in the environment can also limit communication between nodes, which in turn affects the network connectivity (or topology). Research in WSNs aims to meet the above constraints by introducing new design concepts, creating or improving existing protocols, building new applications, and developing new algorithms.

2.5.5   Hardware constraints

A sensor node is made up of four basic components as shown in Figure 2.4:

- a sensing unit,

- a processing unit,

- a transceiver unit and

- a power unit.

They may also have application dependent additional components such as a location finding system, a power generator and a mobilizer. Sensing units are usually composed of two subunits: sensors and analog to digital converters (ADCs). The analog signals produced by the sensors based on the observed phenomenon are converted to digital signals by the ADC, and then fed into the processing unit. The processing unit, which is generally associated with a small storage unit, manages the procedures that make the sensor node collaborate with the other nodes to carry out the assigned sensing tasks. A transceiver unit connects the node to the network.

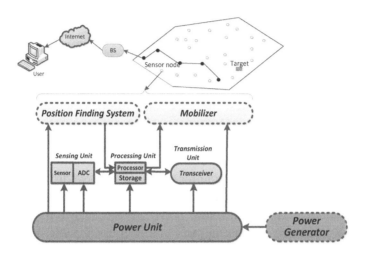

Figure 2.4: The components of a sensor node.

Most of the sensor network routing techniques and sensing tasks require the knowledge of location with high accuracy. It is common that a sensor node has a location finding system. A mobilizer may sometimes be needed to move sensor nodes when it is required to carry out the assigned tasks. All of these subunits may need to fit into a matchbox-sized module [142]. The required size may be smaller than even a cubic centimeter [186] which is light enough to remain suspended in the air. Apart from the size, there are also some other stringent constraints for sensor nodes. These nodes must [185]:

- consume extremely low power,

- operate in high volumetric densities,

- have low production cost and be dispensable,

- be autonomous and operate unattended,

- be adaptive to the environment.

2.6   Routing Protocols

Micro-sensors have been developed in recent years as a result of advances in micro-electro-mechanical systems and low power, and highly integrated digital electronics. Such sensors are generally equipped with communication and data processing capabilities [192]. The sensing circuitry measures ambient conditions related to the environment surrounding the sensor and transforms them into an electric signal. Processing such a signal reveals some properties about objects located and/or events happening in the vicinity of the sensor. The sensor sends such collected data, usually via radio transmitter, to a command center (sink) either directly or through a data concentration center (a gateway).

The decrease in the size and cost of sensors, resulting from such technological advances, has fueled interest in the possible use of large sets of disposable unattended sensors. Such interest has motivated intensive research in the past few years addressing the potential of collaboration among sensors in data gathering and processing, and the coordination and management of the sensing activity and data flow to the sink. A natural architecture for such collaborative distributed sensors is a network with wireless links that can be formed among the sensors in an ad hoc manner. Networking unattended sensor nodes are expected to have a significant impact on the efficiency of many military and civil applications such as combat field surveillance, security and disaster management. These systems process data gathered from multiple sensors to monitor events in an area of interest. For example, in a disaster management setup, a large number of sensors can be dropped by a helicopter. Networking these sensors can assist rescue operations by locating survivors, identifying risky areas and making the rescue crew more aware of the overall situation. Such applications of sensor networks not only can increase the efficiency of rescue operations but also ensure the safety of the rescue crew. On the military side, applications of sensor networks are numerous. For example, the use of a networked set of sensors can limit the need for personnel involvement in the usually dangerous reconnaissance missions. In addition, sensor networks can enable a more civic use of landmines by making them remotely controllable and target-specific in order to prevent harming civilians and animals.

Security applications of sensor networks include intrusion detection and criminal hunting. However, sensor nodes are constrained in energy supply and bandwidth. Such constraints, combined with a typical deployment of large number of sensor nodes, have posed many challenges to the design and management of sensor networks. These challenges necessitate energy awareness at all layers of networking protocol stacks. The issues related to physical and link layers are generally common for all kind of sensor applications, therefore the research on these areas has been focused on system-level power awareness such as dynamic voltage scaling, radio communication hardware, low duty cycle issues, system partitioning, and energy-aware MAC protocols [187]-[191].

At the network layer, the main aim is to find ways for energy-efficient route setup and reliable relaying of data from the sensor nodes to the sink so that the lifetime of the network is maximized. Routing in sensor networks is very challenging due to several characteristics that distinguish them from contemporary communication and wireless ad hoc networks:

1. It is not possible to build a global addressing scheme for the deployment of the sheer number of sensor nodes. Therefore, classical IP-based protocols cannot be applied to sensor networks.

2. Contrary to typical communication networks, almost all the applications of sensor networks require the flow of sensed data from multiple regions (sources) to a particular sink.

3. Generated data traffic has significant redundancy in it since multiple sensors may generate the same data within the vicinity of a phenomenon. Such redundancy needs to be exploited by the routing protocols to improve energy and bandwidth utilization.

4. Sensor nodes are tightly constrained in terms of transmission power, onboard energy, processing capacity and storage, and thus require careful resource management. Due to such differences, many new algorithms have been proposed for the problem of routing data in sensor networks.

These routing mechanisms have considered the characteristics of sensor nodes along with the application and architecture requirements. Almost all of the routing protocols can be classified as data-centric, hierarchical, or location-based, although there are few distinct ones based on network flow or quality of service (QoS) awareness.

Data-centric protocols are query-based and depend on the naming of desired data, which helps in eliminating many redundant transmissions. Hierarchical protocols aim at clustering the nodes so that cluster heads can do some aggregation and reduction of data in order to save energy. Location-based protocols utilize the position information to relay the data to the desired regions rather than to the whole network.

The last category includes routing approaches that are based on general network-flow modeling and protocols that strive for meeting some QoS requirements along with the routing function. In this chapter, we will explore the routing mechanisms for sensor networks developed in recent years. Each routing protocol is discussed under the proper category.

2.6.1 System architecture and design issues

Depending on the application, different architectures and design goals/constraints have been considered for sensor networks. Since the performance of a routing protocol is closely related to the architectural model, in this section we strive to capture architectural issues and highlight their implications.

- **Network dynamics:** There are three main components in a sensor network. These are the sensor nodes, sink, and monitored events. Aside from the very few setups that utilize mobile sensors [193], most of the network architectures assume that sensor nodes are stationary. On the other hand, supporting the mobility of sinks or cluster-heads (gateways) is sometimes deemed necessary [194]. Routing messages from or to moving nodes is more challenging since route stability becomes an important optimization factor, in addition to energy, bandwidth, etc.

- **Node deployment:** Another consideration is the topological deployment of nodes. This is application dependent and affects the performance of the routing protocol. The deployment is either deterministic or self-organizing. In deterministic situations, the sensors are manually placed and data is routed through pre-determined paths. However, in self-organizing systems, the sensor nodes are scattered randomly creating an infrastructure in an ad hoc manner [195],[196]-[198].

- **Energy considerations:** During the creation of an infrastructure, the process of setting up the routes is greatly influenced by energy considerations. Since the transmission power of a wireless radio is proportional to the distance squared or an even higher order in the presence of obstacles, multi-hop routing will consume less energy than direct communication. However, multi-hop routing introduces significant overhead for topology management and medium access control.

- **Data delivery models:** Depending on the application of the sensor network, the data delivery model to the sink can be continuous, event-driven, query-driven, and hybrid. In the continuous delivery model, each sensor sends data periodically. In event-driven and query-driven models, the transmission of data is triggered when an event occurs or a query is generated by the sink. Some networks apply a hybrid model using a combination of continuous, event-driven and query-driven data delivery. The routing protocol is highly influenced by the data delivery model, especially with regard to the minimization of energy consumption and route stability.

- **Node capabilities:** In a sensor network, different functionalities can be associated with the sensor nodes. In earlier works [199], [200], [201], all sensor nodes are assumed to be homogenous, having equal capacity in terms of computation, communication and power. However, depending on the application, a node can be dedicated to a particular special function such as relaying, sensing and aggregation since engaging the three functionalities at the same time on a node might quickly drain the energy of that node. Some of the hierarchical protocols proposed in the literature designate a cluster-head different from the normal sensors.

- **Data aggregation/fusion:** Since sensor nodes might generate significant redundant data, similar packets from multiple nodes can be aggregated so that the number of transmissions is reduced. Data aggregation is the combination of data from different sources by using functions such as suppression (eliminating duplicates), min, max, and average [205]. Some of these functions can be performed either partially or fully in each sensor node, by allowing sensor nodes to conduct in-network data reduction [200], [202], [206]. Recognizing that computation would be less energy consuming than communication [196], substantial energy savings can be obtained through data aggregation. This technique has been used to achieve energy efficiency and traffic optimization in a number of routing protocols [200], [202], [206]–[209].

In data-centric routing, the sink sends queries to certain regions and waits for data from the sensors located in the selected regions. Since data is being requested through queries, attribute-based naming is necessary to specify the properties of data. SPIN [207] is the first data-centric protocol, which considers data negotiation between nodes in order to eliminate redundant data and save energy. Later, Directed Diffusion [200] was developed and has become a breakthrough in data-centric routing. Then, many other protocols were proposed either based on Directed Diffusion [208]–[210] or following a similar concept [198], [206], [211], [212]. In this section, we will describe these protocols in detail and highlight the key ideas.

2.6.2 Flooding and gossiping

Flooding and gossiping [213] are two classical mechanisms to relay data in sensor networks without the need for any routing algorithms and topology maintenance. In flooding, each sensor receiving a data packet broadcasts it to all of its neighbors and this process continues until the packet arrives at the destination or the maximum number of hops for the packet is reached. On the other hand, gossiping is a slightly enhanced version of flooding where the receiving node sends the packet

to a randomly selected neighbor, which picks another random neighbor to forward
the packet to and so on.

Although flooding is very easy to implement, it has several drawbacks, see
Figures 2.5 and 2.6 redrawn from [192]. In Figure 2.5, node A starts by flooding
its data to all of its neighbors. D gets two same copies of data eventually, which is
not necessary. In Figure 2.6, two sensors cover an overlapping geographic region
and C gets same copy of data from these sensors.

Such drawbacks include the implosion caused by duplicated messages sent
to same node, the overlap when two nodes sensing the same region send similar
packets to the same neighbor, and resource blindness caused by consuming large
amounts of energy without consideration for the energy constraints [207]. Gossiping avoids the problem of implosion by just selecting a random node to send the
packet to rather than broadcasting. However, this causes delays in propagation of
data through the nodes.

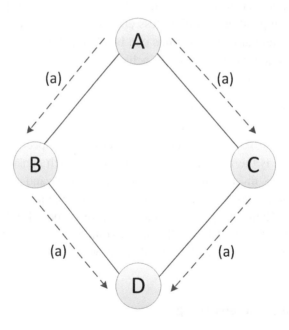

Figure 2.5: The implosion problem.

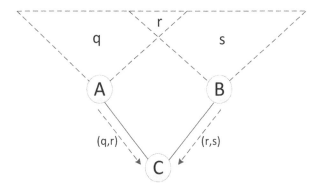

Figure 2.6: The overlap problem.

2.6.3 Sensor protocols for information via negotiation

Sensor protocols for information via negotiation (SPIN) [222] is among the early work to pursue a data-centric routing mechanism. The idea behind SPIN is to name the data using high-level descriptors or meta-data. Before transmission, meta-data are exchanged among sensors via a data advertisement mechanism, which is the key feature of SPIN. Each node upon receiving new data, advertises it to its neighbors and interested neighbors, i.e., those who do not have the data, retrieve the data by sending a request message. SPIN's meta-data negotiation solves the classic problems of flooding such as redundant information passing, overlapping of sensing areas and resource blindness, thus achieving a lot of energy efficiency.

There is no standard meta-data format and it is assumed to be application specific, that is, using an application level framing. There are three messages defined in SPIN to exchange data between nodes. These are: the ADV message to allow a sensor to advertise a particular meta-data, a REQ message to request the specific data, and the DATA message that carries the actual data. Figure 2.7, redrawn from [222], summarizes the steps of the SPIN protocol. One of the advantages of SPIN is that topological changes are localized since each node needs to know only its single-hop neighbors.

2.6.4 Directed diffusion

Directed diffusion [200], [201] is an important milestone in the data-centric routing research of sensor networks. The idea aims at diffusing data through sensor nodes by using a naming scheme for the data. The main reason behind using such a scheme is to get rid of unnecessary operations of network layer routing in order to save energy. Directed Diffusion suggests the use of attribute-value pairs for the data, and queries the sensors in an on demand basis by using those pairs. In order to create a query, an interest is defined using a list of attribute-value pairs such as the name of objects, interval, duration, geographical area, etc. The interest is broadcast by a sink through its neighbors. Each node receiving the interest can do caching for later use. The nodes also have the ability to do in-network data aggregation, which is modeled as a minimum Steiner tree problem.

Directed Diffusion differs from SPIN in terms of the on demand data querying mechanism it has. In Directed Diffusion the sink queries the sensor nodes if a specific data is available by flooding some tasks. In SPIN, sensors advertise the availability of data allowing interested nodes to query that data. Directed Diffusion has many advantages. Since it is data–centric, all communication is neighbor-to-neighbor with no need for a node addressing mechanism. Each node can do aggregation and caching, in addition to sensing. Caching is a big advantage in terms of energy efficiency and delay. In addition, direct diffusion is highly energy efficient since it is on demand and there is no need for maintaining global network topology.

2.6.5   Geographic and energy-aware routing

Geographic and energy-aware routing is abbreviated as (GEAR). It proposes to use a set of sub-optimal paths occasionally to increase the lifetime of the network. These paths are chosen by means of a probability function, which depends on the energy consumption of each path. Network survivability is the main metric that the approach is concerned with. The approach argues that using the minimum energy path all the time will deplete the energy of nodes on that path. Instead, one of the multiple paths is used with a certain probability so that the whole network lifetime increases. The protocol assumes that each node is addressable through class-based addressing which includes the location and types of the nodes.

2.6.6   Gradient-based routing

Gradient-based routing is abbreviated as (GBR). The idea is to keep the number of hops when the interest is diffused through the network. Hence, each node can discover the minimum number of hops to the sink, which is called the height of the node. The difference between a node's height and that of its neighbor is considered the gradient on that link. A packet is forwarded on a link with the largest gradient.

2.6.7 Constrained anisotropic diffusion routing

Constrained anisotropic diffusion routing (CADR) is a protocol which strives to be a general form of Directed Diffusion. Two techniques, namely information-driven sensor querying (IDSQ) and constrained anisotropic diffusion routing, are proposed. The idea is to query sensors and route data in a network in order to maximize the information gain, while minimizing the latency and bandwidth. This is achieved by activating only the sensors that are close to a particular event, and dynamically adjusting data routes. The major difference from Directed Diffusion is the consideration of information gain in addition to the communication cost. In CADR, each node evaluates an information/cost objective and routes data based on the local information/cost gradient and the end-user requirements. The information utility measure is modeled using standard estimation theory.

2.6.8 Active query forwarding

A fairly new data-centric mechanism for querying sensor networks is ACtive QUery forwarding In sensoR nEtworks (ACQUIRE). The approach views the sensor network as a distributed database and is well-suited for complex queries which consist of several sub queries. The querying mechanism works as follows: the query is forwarded by the sink and each node receiving the query tries to respond partially by using its pre-cached information and forwarding it to another sensor. If the pre-cached information is not up-to-date, the node gathers information from its neighbors within a look-ahead of the hops. Once the query is resolved completely, it is sent back through either the reverse or shortest path to the sink.

2.6.9 Low-energy adaptive clustering hierarchy

Low-energy adaptive clustering hierarchy (LEACH) is one of the most popular hierarchical routing algorithms for sensor networks. The idea is to form clusters of the sensor nodes based on the received signal strength and use local cluster heads as routers to the sink. This will save energy, since the transmissions will only be done by such cluster heads rather than all the sensor nodes. The optimal number of cluster heads is estimated to be 5% of the total number of nodes. All the data processing, such as data fusion and aggregation, is local to the cluster. Cluster heads change randomly over time in order to balance the energy dissipation of the nodes.

2.6.10   Power-efficient gathering

Power-efficient GAthering in Sensor Information Systems (PEGASIS) [202] is an improvement of the LEACH protocol. Rather than forming multiple clusters, PE-GASIS forms chains from sensor nodes so that each node transmits and receives from a neighbor and only one node is selected from that chain to transmit to the base station (sink). Gathered data moves from node to node, is aggregated, and eventually sent to the base station. The chain construction is performed in a greedy way.

2.6.11   Adaptive threshold sensitive energy efficient network

The Adaptive Threshold sensitive Energy Efficient sensor Network protocol [225] (APTEEN) aims at both capturing periodic data collections and reacting to time critical events. The architecture is same as in TEEN. When the base station forms the clusters, the cluster heads broadcast the attributes, the threshold values, and the transmission schedule to all nodes. Cluster heads also perform data aggregation in order to save energy. APTEEN supports three different query types: historical, to analyze past data values; one-time, to take a snapshot view of the network; and persistent to monitor an event for a period of time.

2.6.12   Minimum energy communication network

A minimum energy communication network (MECN) [224] sets up and maintains a minimum energy network for wireless networks by utilizing low power GPS. Although the protocol assumes a mobile network, it is most applicable to sensor networks, which are not mobile. A minimum power topology for stationary nodes including a master node is found. MECN assumes a master site as the information sink, which is always the case for sensor networks. MECN identifies a relay region for every node. The relay region consists of nodes in a surrounding area, where transmitting through those nodes is more energy efficient than direct transmission.

2.6.13   Geographic adaptive fidelity

Geographic adaptive fidelity (GAF) [223] is an energy-aware location-based routing algorithm designed primarily for mobile ad hoc networks, but may be applicable to sensor networks as well. GAF conserves energy by turning off unnecessary nodes in the network without affecting the level of routing fidelity. It forms a virtual grid for the covered area. Each node uses its GPS-indicated location to associate itself with a point in the virtual grid. Nodes associated with the same point on the grid are considered equivalent in terms of the cost of packet routing. Such equivalence is exploited in keeping some nodes located in a particular grid area in sleeping state in order to save energy. Thus, GAF can substantially increase the network's lifetime as the number of nodes increases.

2.7 Sensor Selection Schemes

Sensor networks consist of a large number of small sensor devices that have the capability to take various measurements of their environment. These measurements can include seismic, acoustic, magnetic, IR and video information. Each of these devices is equipped with a small processor and wireless communication antenna and is powered by a battery, making it very resource constrained. To be used, sensors are scattered around a sensing field to collect information about their surroundings. For example, sensors can be used in a battlefield to gather information about enemy troops, detect events such as explosions, and track and localize targets. Upon deployment in a field, they form an ad hoc network and communicate with each other and with data processing centers.

Sensor networks are usually intended to last for long periods of time, such as months or even years. However, due to the limited energy available on board, if a sensor remains active continuously, its energy will be depleted quickly, leading to its death. To prolong the network's lifetime, sensors alternate between being active and sleeping. There are several sensor selection algorithms used to achieve this while still achieving the goal of deployment. The decision as to which sensor should be activated takes into account a variety of factors depending on the algorithm such as residual energy, required coverage, or the type of information required. Sensors are selected to do one or multiple missions. These missions can be general and related to the function of the network, such as monitoring the whole field by ensuring complete coverage, or more specific and application-oriented, such as tracking a target's movement. At a given time, the system might be required to do multiple missions, such as monitoring an event and, at the same time, tracking a single or multiple moving objects.

2.7.1 Sensor selection problem

The sensor selection problem can be defined as follows:

Given a set of sensors $S = \{S_1, ..., S_n\}$, we need to determine the best subset S^ℓ of k sensors to satisfy the requirements of one or multiple missions.

The best subset is one which achieves the required accuracy of information with respect to a task while meeting the energy constraints of the sensors.

So, we have two conflicting goals:

1. to collect information of high accuracy, and

2. to lower the cost of operation.

This trade-off is usually modeled using the notions of utility and cost:

- *Utility:* accuracy of the gathered information and its usefulness to a mission,

- *Cost:* this consists mainly of the energy expended activating and operating the sensors, which is directly proportional to number of selected sensors, k. Another cost factor that can be considered here is the risk of detecting a sensor which may increase for active sensors, especially if wireless communication is used.

The goal of a sensor selection scheme is to select a subset S^ℓ of k sensors such that the total utility is maximized while the overall cost is less than a certain budget.

In what follows in this section, we categorize the schemes used for sensor selection based on the purpose of the selection as follows:

- *Coverage schemes:* include selection schemes that are used to ensure sensing coverage of the location or targets of interest.

- *Target tracking and localization schemes:* include schemes that select sensors for target tracking and localization purposes.

- *Single mission assignment schemes:* include schemes that select sensors for a single specific mission.

- *Multiple mission assignment schemes:* include schemes that select sensors so that multiple specific missions are collectively accomplished.

In the sequel, we provide a brief account of these schemes.

2.7.2   Coverage schemes

In this section, we discuss schemes in which sensors are selected in order to ensure complete coverage of the field. This means that every point in the field must be in the sensing range of at least one sensor. Both static sensors and mobile sensors, are included.

If the sensor nodes are densely deployed, such that there is redundancy in coverage, then only a subset of sensors needs to be active in order to achieve full coverage, while the rest can enter sleep mode. This conserves energy and hence prolongs the network lifetime. Selection schemes are used to decide which sensors are to be turned on and for how long. If, during the course of operation, a node fails resulting in a coverage hole, the selection protocols can be rerun to activate one or more of the redundant nodes to restore the coverage.

If sensors are mobile (e.g. placed on robots), more options arise. A sensor node can now move to a new location in order to fill a coverage hole. This makes deployment easier, because even if a random deployment results in incomplete coverage, the nodes can then relocate themselves to ensure coverage. Similarly, if a node fails, then the network can be dynamically reconfigured to make up for it. There are also new challenges posed by mobility for selection schemes because the decision of which node to move has an additional energy cost due to the movement.

2.7.3 Target tracking and localization schemes

Methods in which sensors are selected for the purposes of target tracking and target localization. These schemes can be further classified based on the approach used in the selection algorithm. There are three categories:

1. Entropy-based solutions, where the selection schemes aim to minimize the entropy of measurement,

2. Dynamic information-driven solutions, where the aim is to maximize the information gain based on the dynamic information gathered and

3. Mean squared error-based solutions, where the aim is to minimize the mean squared error of measurements.

2.7.4 Single mission assignment schemes

In a sensor network which must perform a specific mission repeatedly over time, sensors need to be selected such that the mission is accomplished in the most efficient manner. The objective of such selection schemes is to select the sensor nodes that are most useful for the mission. This notion of usefulness is quantified using a utility value for an application-layer task.

2.7.5 Multiple mission assignment schemes

In dealing with a multiple mission scenario, it must be noted that, by mission, we refer to a specific (application-level) task and not a network level function, such as coverage or data dissemination. These multiple missions may belong to the one big operations, or may belong to multiple operations that the sensor network is responsible form.

2.8   Quality of Service Management

WSNs are designed to gather information about the state of the physical world and transmit the sensed data to interested users. They are typically used in applications like habitat monitoring, military surveillance, agriculture and environmental sensing, and health monitoring. In most cases, they are unable to affect the physical environment. In many applications, however, only observing the state of the physical system is not sufficient; they are also expected to respond to the sensed events/data by performing corresponding actions upon the system. For instance, in a fire handling system, the actuators need to turn on the water sprinklers upon receipt of a report of fire. This need for actuation heralds the emergence of wireless sensor/actuator networks (WSANs), a substantial extension of sensor networks, that feature the coexistence of sensors and actuators. WSANs enable the application systems to sense, interact with, and change the physical world, to monitor and manipulate the temperature and lighting in a smart office, or the change speed and direction of a mobile robot. It is envisioned that WSANs will be one of the most critical technologies for building the network infrastructure of future cyber-physical systems [66]. They will revolutionize the way we interact with the physical world.

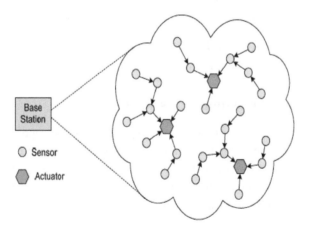

Figure 2.7: A wireless sensor and actuator network.

As shown in Figure 2.7, a WSAN is a networked system of geographically distributed sensor and actuator nodes that are interconnected via wireless links. Both sensor and actuator nodes are normally equipped with certain data processing and wireless communication capabilities, as well as a power supply. In most situations, sensor nodes are stationary whereas actuator nodes, e.g., mobile robots and unmanned aerial vehicles, are mobile. Sensors gather information about the state of the physical world and transmit the collected data to actuators through single-hop or multi-hop communications. Upon receiving the required information, the actuators make the decision about how to react to the information and perform corresponding actions to change the behavior of the physical environment. The base station is principally responsible for monitoring and managing the overall network through communications with sensors and actuators.

2.8.1 QoS requirements

From an end user's perspective, real-world WSAN applications have their specific requirements on the QoS of the underlying network infrastructure [62]. For instance, in a fire handling system, sensors need to report the occurrence of a fire to actuators in a timely and reliable fashion; then, the actuators, equipped with water sprinklers, will react by a certain deadline so that the situation will not become uncontrollable. It is intuitive that different applications may have different QoS requirements. For instance, for a safety-critical control system, a long delay in transmitting data from sensors to actuators and packet loss occurring during the course of transmission may not be allowed, while such delays may be acceptable for an air-conditioning system that maintains the temperature inside an office.

- *Throughput* is the effective number of data flow transported within a certain period of time, also specified as bandwidth in some situations. In general, the bigger the throughput of the network, the better the performance of the system is. Those nodes that generate high-speed data streams, such as a camera sensor node used to transmit images for target tracking, often require high throughput. In order to improve the resource efficiency, furthermore, the throughput of WSAN should often be maximized.

- *Delay* is the time elapsed from the departure of a data packet from the source node to the arrival at the destination node, including queuing delay, switching delay, propagation delay, etc. Delay sensitive applications usually require WSANs to deliver the data packets in real time. Notice that real time does not necessarily mean fast computation or communication [63],[64],[65]. A real-time system is unique in that it needs to execute at a speed that fulfills the timing requirements.

- *Jitter* is generally referred to as variations in delay, despite many other definitions. It is often caused by the difference in queuing delays experienced by consecutive packets.

- *Packet loss rate* is the percentage of data packets that are lost during the process of transmission. It can be used to represent the probability of packets being lost. A packet may be lost due to e.g., congestion, bit error, or bad connectivity. This parameter is closely related to the reliability of the network.

2.8.2   Challenges

WSANs cannot be simply regarded as WSNs due to the co-existence of sensors and actuators, as mentioned previously. In this section, some of the major features of WSANs that challenge QoS provisioning will be discussed [61],[67].

Resource constraints

As in WSNs, sensor nodes are usually low-cost, low-power, small devices that are equipped with only limited data processing capability, transmission rate, battery energy, and memory. For example, the MICAz mote from Crossbow is based on the Atmel ATmega128L 8-bit microcontroller that provides only up to 8 MHz clock frequency, 128-KB flash program memory and 4-KB EEPROM; the transmit data rate is limited to 250 Kbps. Due to the limitation on transmission power, the available bandwidth and the radio range of the wireless channel are often limited. In particular, energy conservation is critically important for extending the lifetime of the network, because it is often infeasible or undesirable to recharge or replace the batteries attached to sensor nodes once they are deployed. Actuator nodes typically have stronger computation and communication capabilities and more energy budget relative to sensors. Resource constraints apply to both sensors and actuators, notwithstanding.

Platform heterogeneity

Sensors and actuators do not share the same level of resource constraints, as mentioned above. Possibly designed using different technologies and with different goals, they are different from each other in many aspects such as computing/communication capabilities, functionality, and number. In a large-scale system of systems, the hardware and networking technologies used in the underlying WSANs may differ from one subsystem to another. This is true because of the lack of relevant standards dedicated to WSANs and hence commercially available products often have disparate features. This platform heterogeneity makes it very difficult to make full use of the resources available in the integrated system. Consequently, resource efficiency cannot be maximized in many situations. In addition, the platform heterogeneity also makes it challenging to achieve real-time and reliable communication between different nodes.

Dynamic network topology

Unlike WSNs where (sensor) nodes are typically stationary, the actuators in WSANs may be mobile. In fact, node mobility is an intrinsic nature of many applications such as, among others, intelligent transportation, assisted living, urban warfare, planetary exploration, and animal control. During run time, new sensor/actuator nodes may be added; the state of a node is possibly changed to or from sleeping mode by the employed power management mechanism; some nodes may even die due to exhausted battery energy. All of these factors may potentially cause the network topologies of WSANs to change dynamically.

Dealing with the inherent dynamics of WSANs requires QoS mechanisms to work in dynamic and even unpredictable environments. In this context, QoS adaptation becomes necessary; that is, WSANs must be adaptive and flexible at run time with respect to changes in available resources. For example, when an intermediate node dies, the network should still be able to guarantee real-time and reliable communication by exploiting appropriate protocols and algorithms.

Mixed traffic

Diverse applications may need to share the same WSAN, inducing both periodic and aperiodic data. This feature will become increasingly evident as the scale of WSANs grows. Some sensors may be used to create the measurements of certain physical variables in a periodic manner for the purpose of monitoring and/or control. Meanwhile, some others may be deployed to detect critical events. For instance, in a smart home, some sensors are used to sense the temperature and lighting, while some others are responsible for reporting events like the entering or leaving of a person. Furthermore, disparate sensors for different kinds of physical variables, e.g., temperature, humidity, location, and speed, generate traffic flows with different characteristics (e.g., message size and sampling rate). This feature of WSANs necessitates the support of service differentiation in QoS management.

2.9 Wireless Sensor Network Security

Due to the fact that wireless sensor networks are potentially low cost solutions to a variety of real-world challenges, they are quickly gaining popularity [137]. Their low cost provides a means to deploy large sensor arrays in a variety of conditions capable of performing both military and civilian tasks. But sensor networks also introduce severe resource constraints due to their lack of data storage and power. Both of these represent major obstacles to the implementation of traditional computer security techniques in a wireless sensor network. The unreliable communication channel and unattended operation make the security defenses even harder.

The main aspects of wireless sensor network security can be classified into four major categories: *the obstacles to sensor network security*, *the requirements of a secure wireless sensor network*, *attacks*, and *defensive measures*. A brief account on each will now be presented.

2.9.1   Obstacles of sensor security

A wireless sensor network is a special network which has many constraints compared to a traditional computer network. Due to these constraints it is difficult to directly employ the existing security approaches to the area of wireless sensor networks. Therefore, it is necessary to know and understand these constraints.

1. *Very limited resources:* All security approaches require a certain amount of resources for the implementation, including data memory, code space, and energy to power the sensor. However, currently these resources are very limited in a tiny wireless sensor.

2. *Unreliable communication:* Certainly, unreliable communication is another threat to sensor security. The security of the network relies heavily on a defined protocol, which in turn depends on communication.

3. *Unattended operation:* Depending on the function of the particular sensor network, the sensor nodes may be left unattended for long periods of time. There are three main caveats to unattended sensor nodes: exposure to physical attacks, managed remotely and no central management.

   Perhaps most important, the longer that a sensor is left unattended the more likely it is that an adversary has compromised the node.

2.9.2   Security requirements

The requirements of wireless sensor network security will encompass both those of typical networks and the unique requirements suited solely to wireless sensor networks. These can be summarized below:

- *Data confidentiality:* relates to the following [138]:

  1. A sensor network should not leak sensor readings to its neighbors.

  2. In many applications nodes communicate highly sensitive data such as key distribution, therefore it is extremely important to build a secure channel in a wireless sensor network.

  3. Public sensor information, such as sensor identities and public keys, should also be encrypted to some extent to protect against traffic analysis attacks.

- *Data integrity:*

  Data loss or damage can occur even without the presence of a malicious node, due to a harsh communication environment. Thus, data integrity ensures that any received data has not been altered in transit.

- *Data freshness:*

  Even if confidentiality and data integrity are assured, we also need to ensure the freshness of each message. Informally, data freshness suggests that the data is recent, and it ensures that no old messages have been replayed. This requirement is especially important when there are shared-key strategies employed in the design.

- *Availability:*

  Adjusting the traditional encryption algorithms to fit within the wireless sensor network is not free, and will introduce some extra costs. Some approaches choose to modify the code to reuse as much code as possible. Other approaches try to make use of additional communication to achieve the same goal. The requirement of security not only affects the operation of the network, but also is highly important in maintaining the availability of the whole network.

- *Self-organization:*

  A wireless sensor network is a typically an ad hoc network, which requires every sensor node be independent and flexible enough to be self-organizing and self-healing according to different situations. There is no fixed infrastructure available for the purpose of network management in a sensor network. This inherent feature brings a great challenge to wireless sensor network security as well. If self-organization is lacking in a sensor network, the damage resulting from an attack or even from the hazardous environment may be devastating.

- *Time synchronization:*

  Most sensor network applications rely on some form of time synchronization in order to conserve power. Furthermore, sensors may wish to compute the end-to-end delay of a packet as it travels between two pairwise sensors. A more collaborative sensor network may require group synchronization for tracking applications.

- *Secure localization:*

  Often, the utility of a sensor network will rely on its ability to accurately and automatically locate each sensor in the network. A sensor network designed to locate faults will need accurate location information in order to pinpoint the location of a fault. There are several techniques to handle this issue, see [139].

- *Authentication:*

  An adversary is not just limited to modifying the data packet. It can change the whole packet stream by injecting additional packets. So the receiver needs to ensure that the data used in any decision-making process originates from the correct source. On the other hand, when constructing the sensor network, authentication is necessary for many administrative tasks (e.g., network reprogramming or controlling sensor node duty cycle). From the above, we can see that message authentication is important for many applications in sensor networks. Informally, data authentication allows a receiver to verify that the data really is being sent by the claimed sender. In the case of two-party communication, data authentication can be achieved through a purely symmetric mechanism: the sender and the receiver share a secret key to compute the message authentication code (MAC) of all communicated data.

  In [138], a key-chain distribution system for their TESLA secure broadcast protocol is proposed. The basic idea of the TESLA system is to achieve asymmetric cryptography by delaying the disclosure of the symmetric keys. In this case, a sender will broadcast a message generated with a secret key. After a certain period of time, the sender will disclose the secret key. The receiver is responsible for buffering the packet until the secret key has been disclosed. After disclosure the receiver can authenticate the packet, provided that the packet was received before the key was disclosed.

  One limitation of μTESLA is that some initial information must be unicast to each sensor node before authentication of broadcast messages can begin. Liu and Ning [140], [141] propose an enhancement to the μTESLA system that uses broadcasting of the key chain commitments rather than μTESLA's unicasting technique. They present a series of schemes starting with a simple pre-determination of key chains, finally settling on a multi-level key chain technique. The multi-level key chain scheme uses pre-determination and broadcasting to achieve a scalable key distribution technique that is designed to be resistant to denial of service attacks, including jamming.

2.10   Notes

The topic of wireless sensor networks has important applications such as remote environmental monitoring and target tracking. We have surveyed some issues in three different categories. To prop further, WSN applications such as communication architectures, security, and management need to be looked at.

We have examined wireless sensor/actuator networks (WSAN), see Figure 2.7, which constitute a group of sensors and actuators that are geographically distributed and interconnected by wireless networks in the sense that: Sensors gather information about the state of physical world. Actuators react to this information by performing appropriate actions.

We have analyzed the problem of sensor selection in wireless sensor networks. We discussed four different classes of schemes, namely

1. Coverage schemes,

2. Target tracking and localization schemes,

3. Single mission assignment schemes and

4. Multiple mission assignment schemes.

We also looked at solutions to similar selection and matching problems in other fields and discussed their applicability to sensor networks. We believe that there are some important open research problems in this area and we have discussed some of them here.

2.11   Proposed Topics

1. Discuss the pros and cons of a "mixed" routing protocol that integrates sensor networks with wired networks (i.e. the Internet). Recall that a wide class of applications in security and environmental monitoring requires the data collected from the sensor nodes to be transmitted to a server so that further analysis can be done. On the other hand, the requests from the user should be made to the sink through the Internet. Since the routing requirements of each environment are different, further research is necessary for handling these kinds of situations.

2. The concept of service-oriented architecture (SOA) is by no means new and has been widely used in, for example, the web services domain. However, many of its elegant potentials have not ever been explored in WSANs, though

SOA will undoubtedly have a major impact in many branches of technology. Identifying and specifying services are crucial for exploiting SOA in WSANs. In this regard, attempt to answer the following: how many categories of services should be classified in the context of WSAN? What are the functionality, interface, and properties of each service? What are its quality levels relevant to performance requirements? In particular, how to deal with the difference between sensors and actuators when specifying services?

3. Many of the papers that propose these sensor selection schemes look at the problem from a theoretical stand point. But another issue that needs to be studied is the dynamics of these different schemes and how they would perform in realistic settings in which messages can be dropped and nodes can fail. Also, the different performance aspects of these schemes such as convergence times, communication overhead and how they affect sensor network lifetime need to be studied.

4. An interesting topic is to study whether and how the purpose of the selection (like monitoring, answering a query, disseminating data, fault tolerance, etc.) may affect the selection scheme. For instance, the selection schemes for choosing sensors for monitoring purposes may face different requirements and challenges when compared to choosing sensors for answering a query.

Chapter 3

Distributed Sensor Fusion

This chapter examines the problem of distributed state estimation for two kinds of important systems: the plant wide processes and the sensor networks. Then it introduces a distributed filtering scheme that allows the nodes of a sensor network to track the average of n sensor measurements using an average consensus–based distributed filter called a *consensus filter*. This consensus filter plays a crucial role in solving a data fusion problem that allows implementation of a scheme for distributed Kalman filtering in sensor networks. Next, we develop a systematic procedure leading to a distributed estimation in WSN using the extended information filter for target tracking and the adaptive sensor selection (ASS) algorithm. The algorithm incorporates a cost function based on the geometrical dilution of precision (GDOP) for sensor selection. Finally, multi-sensor management is formally described as a system or process that seeks to manage or coordinate the usage of a suite of sensors or measurement devices in a dynamic, uncertain environment, to improve the performance of data fusion, and ultimately that of perception.

3.1 Assessment of Distributed State Estimation

State estimation is an important topic in the study of dynamical systems. The problem of estimation can be structured into three categories:

1. Centralized Scheme,

2. Decentralized Scheme, or

3. Distributed Scheme.

Distributed Estimation is a compromise between completely centralized and decentralized versions of estimation. In the following, we provide an assessment of distributed estimation based on Kalman filtering techniques for large-scale or sensor networks. In simulation, a second order dynamical system is employed in a scenario of ten sensor nodes. The sensor nodes attempt to estimate the states of the dynamical system with embedded consensus filters. The results show that the distributed estimation algorithm effectively approximates the central Kalman filter.

3.1.1   Introduction

For the last few years, distributed estimation has emerged as an important research problem in the control community. Advancement in communication technology encouraged the control community to research and devise new methods of estimation and control, considering communication between sensing nodes for distributed and relatively large systems. Due to the size and communication issues and the application problem, the distributed estimation brings new challenges to propose a unified estimation and control strategy for the distributed system. In this section, we have gathered some recent work on distributed estimation based on the Kalman filter.

Power system grids are distributed systems with interconnected operation. These grids offer two challenging problems during the operation: (i) dimension of the problem, and (ii) lack of information and measurement exchange between the distributed areas of the grid [510]. Such systems require the distributed nature of state estimators instead of completely centralized or completely decentralized estimators.

A distributed state estimator to monitor such power mega grids is discussed in [510]. Mega grids consist of cooperative operation between several power system areas in order to manage power system transactions over distributed and larger areas. The problems of larger dimension and lack of information are both addressed by [510]. In this distributed setup, individual area state estimators don't have to share measurements with any neighbors, while coordination is achieved through a central coordinator.

Consensus for a better estimate is a critical issue in distributed estimation techniques. In [511], the authors investigate the consensus problem for a team of agents with inconsistent information. Kalman filtering approaches to the consensus problem are discussed where convergences are observed for strongly connected networks. A modification to the basic Kalman filtering algorithm is also discussed to get unbiased estimates in case of static and dynamic communication networks [511].

Research in distributed estimation has taken two routes from the main stream:

1. Distributed estimation for plant wide processes (commonly known as large scale systems), and

2. Distributed estimation for sensor networks.

Large scale systems in process control are the plant wide processes which require decentralized and distributed estimation and control algorithms for optimum performance, integrated safety and stable operation. Due to the complex hierarchy, nonlinear process model, and optimization constraints, the distributed and decentralized structure of estimation and control have been seen as potential solution in research community. Recall the distributed estimation and control methods proposed specifically for plant wide large scale processes in [512], [513] and [514].

The design of distributed estimation techniques depends upon the type of application. The application of tracking a signal (or monitoring for environmental variable like dust storm) using a large number of sensors has a small number for estimation variables (e.g., one or two). While in the case of a distributed estimation of a dynamic distributed system (e.g., a process control application) has a relatively equivalent number of sensors and estimation variables (i.e., system states). The first category also considers the control signal along with the state and measurement in the problem definition as follows [512]:

$$x(k) = \Phi x(k-1) + Bu(k-1) + w(k-1), \tag{3.1}$$

$$y(k) = Hx(k) + v(k), \tag{3.2}$$

while the second category of sensing networks considers only the estimation problem of state via the measurement equation, and assumes the system dynamic, as follows [515]:

$$x(k) = \Phi x(k-1) + Bw(k-1), \tag{3.3}$$

$$z(k) = Hx(k) + v(k), \tag{3.4}$$

where, $x(k)$ is the system state, Φ is the state transition matrix, B is the input matrix, $w(k)$ is the process noise, $v(k)$ is the measurement noise, $y(k)$ is the measurement and H is the sensing model.

In [527] the performance of a distributed Kalman filter is evaluated experimentally using an ultrasound based positioning application with seven sensor nodes. All the sensor nodes need an estimate of the full state of the observed system while there is no centralized computation. Communication between the sensor nodes takes place between neighbors, once per sampling interval. The new estimate is the weighted average of the individual estimates. Optimization is used with proper weights to guarantee small estimation error covariance in stationary setting. Robustness of the estimation is also evaluated against packet loss.

In [528], the problem of estimating the state of a scalar linear system from distributed noisy measurements is considered. Local estimate is calculated at each node using local measurements and received estimates from the neighbors.

Integrated radar stations offer a challenging distributed estimation scenario. In [529], a multi-static radar system with mobile receivers is explored from an estimation view point. In this system, a transmitter (at a known location), is emitting a radar signal that bounces off a target. This echo is received by a group of Unmanned Aerial Vehicles (UAVs). These UAVs estimate the time-delay and Doppler from the echo (received) signal. A diffusion strategy for distributed Kalman filtering is proposed in [530], where node data and estimates are diffused across the network.

Two modified distributed Kalman filtering algorithms are presented in [531]. These algorithms are trust based distributed Kalman filtering approaches to estimate oscillation modes in power systems having false measurements and sudden failures.

The performance of distributed Kalman filter is analyzed in [532] where it is reported that the algorithm presented in [515] performs closely to ideal filter under the strong connectivity conditions. The neighborhood communication approach reduces the communication requirement and this makes the algorithm to be easily deployed . In [533], the problem of distributed Kalman filter is resolved as a fusion problem of two consensus filters. The consensus filters are distributed algorithms that allow calculation of average-consensus of time-varying signals.

A minimum variance based distributed estimation for sensor networks to track a noisy signal is discussed in [534], where each node computes the estimate as a weighted sum of its own and its neighbors' measurements and estimates. These weights are adaptively updated to minimize the variance of the estimation error. It is noted that since [534] does not include any stability analysis of the proposed estimation method, stability analysis will be a good extension of [534]. Moreover, the conditions are assumed to be ideal, thus consideration of delay and packet loss between the estimation nodes, along with experimental validation, will also be a good extension as mentioned in [534].

Distributed estimation and tracking is one of the fundamental problems in wireless sensor networks (WSN). Fusion of estimates from multiple sensors and tracking problems have been extensively researched in signal processing, control theory, and robotics (see for example [516], [517], [518], [520], and [524]). The problem of estimation is further tackled in view of packet loss in the wireless network in [525], and [521]. In decentralized Kalman filtering [526], and [523], state estimation is carried out using a set of local Kalman filters that communicate with all other nodes. The information flow is all-to-all with communication complexity of $O(n^2)$, while in scalable or distributed Kalman filtering algorithms, each node only communicates with its neighbors on a network, thus showing a potential advantage.

3.1.2 Consensus-based distributed Kalman filter

In distributed estimation algorithms, there are several sensing nodes with good enough computational power, along with the sensing equipment. These sensor nodes are equipped with estimation algorithm, based on the application and communication policy with neighboring nodes. The sensor nodes can be used to estimate states of a dynamic process or to estimate a target (i.e., tracking a signal). Thus, the objective of a filter is to perform distributed state estimation (or tracking) for a process/target that evolves according to

$$x(k + 1) = A_k x(k) + B_k w(k); \quad x(0) \sim N(\bar{x}(0), P_0). \tag{3.5}$$

In the case of a state estimation of a closed loop dynamic system, the matrix A_k is the closed loop matrix.

To represent network structure and connectivity, graph theory is used to describe the sensor network, assuming a sensor network with an ad hoc topology $G = (V, E)$ and n nodes. The graph G is undirected i.e., balanced (degree of a node is equal to the number of incident edges on that node), $V = \{1, 2, \ldots, n\}$, and $E \subset V \times V$. The sensing model of the ith sensor is

$$z_i(k) = H_i(k)x(k) + v_i(k), \quad z_i \in R^p \tag{3.6}$$

where, w_k and v_k are zero-mean white Gaussian noise (WGN) and $x(0) \in R^m$ is the initial state. The statistics of the measurement noise is given by

$$E[w(k)w(l)^T] = Q(k)\delta_{kl}, \tag{3.7}$$

$$E[v_i(k)v_j(l)^T] = R_i(k)\delta_{kl}\delta_{ij}, \tag{3.8}$$

where $\delta_{kl} = 1$ if $k = l$, and $\delta_{kl} = 0$, otherwise.

Let $z(k) = col(z_1(k), \ldots, z_n(k)) \in R^{np}$ be the collective sensor data of the entire sensor network at time k. Given the information $Z_k = \{z(0), z(1), \ldots, z(k)\}$, the estimates of the state of the process can be expressed as

$$\hat{x}_k = E(x_k|Z_k), \bar{x}_k = E(x_k|Z_{k-1}), \tag{3.9}$$

$$P_k = \Sigma_{k|k-1}, M_k = \Sigma_{k|k} \tag{3.10}$$

where $\Sigma_{k|k-1}$ and $\Sigma_{k|k}$ are the estimation error covariance matrices and $\Sigma_{0|-1} = P_0$. Defining an output matrix $H = col(H_1, H_2, \ldots, H_n)$, one can define a central estimate $\hat{x}(k)$ associated with the data $z(k)$ given by

$$\hat{x}(k) = \bar{x}(k) + K_k(z(k) - H\bar{x}(k)). \tag{3.11}$$

Assuming that the measurement noise of the sensors is uncorrelated, the covariance matrix of the noise $v = col(v_1, \ldots, v_n)$ is $R = diag(R_1, \ldots, R_n)$ (the time indices are dropped). Let us define two network-wide aggregate quantities: the fused inverse-covariance matrices

$$S(k) = \frac{1}{n} \sum_{i=1}^{n} H_i'^T(k) R_i^{-1}(k) H_i(k) \tag{3.12}$$

and the fused sensor data

$$y(k) = \frac{1}{n} \sum_{i=1}^{n} H_i^T(k) R_i^{-1}(k) z_i(k). \tag{3.13}$$

Theorem 3.1.1 *(Micro-KF Iterations, [515]) Suppose every node of the network applies the following iterations*

$$
\begin{aligned}
M_i(k) &= (P_i(k)^{-1} + S(k))^{-1}, \\
\hat{x}(k) &= \bar{x}(k) + M_i(k)[y(k) - S(k)\bar{x}(k)], \\
P_i(k+1) &= A_k M_i(k) A_k^T + B_k Q_i(k) B_k^T, \\
\bar{x}(k+1) &= A_k \hat{x}(k).
\end{aligned}
\tag{3.14}
$$

where $Q_i(k) = nQ(k)$ and $P_i(0) = nP_0$. Then, the local and central state estimates for all nodes are the same, i.e. $\hat{x}_i(k) = \hat{x}(k)$ for all i.

Computing the averages $y(k)$ and $S(k)$ with the consensus of sensor nodes results in a distributed Kalman filter. The problem of dynamic consensus has been solved with different consensus filters (see [515], [522]). The main idea of consensus filters is based on following formulation:

Assuming $N_i = \{j : (i, j) \in E\}$ is the set of neighbors of node i on graph G. Let $L = D - A$ be the Laplacian matrix of G and $\lambda_2 = \lambda_2(L)$ denote its algebraic connectivity. The high-pass consensus filter is a linear system in the form

$$
\begin{cases}
\dot{q}_i = \beta \sum_{j \in N_i}(q_j - q_i) + \beta \sum_{j \in N_i}(u_j - u_i); \quad \beta > 0 \\
y_i = q_i + u_i
\end{cases}
\tag{3.15}
$$

where u_i is the input of node i, q_i is the state of the consensus filter, and y_i is its output. The gain $\beta > 0$ is relatively large ($\beta \sim O(1/\lambda_2)$) for randomly generated ad hoc topologies that are rather sparse. Note that when $\beta = 1$, the consensus filters of [522] become similar to the consensus filters described in [515]. The collective dynamics of the consensus filters is given by

$$
\begin{cases}
\dot{q} = -\beta \hat{L}q - \beta \hat{L}u \\
p = q + u
\end{cases}
\tag{3.16}
$$

where $\hat{L} = L \otimes I_m$ is the m-dimensional graph Laplacian and $\otimes$ stands for the Kronecker product, see the Apprndix. For a connected network, $p_i(t)$ asymptotically converges to $1/n \sum_i u_i(t)$ as $t \to \infty$. The equation of the dynamic consensus algorithm is given as $\dot{p} = -Lp + \dot{u}$ which reduces to (3.16) for $m = 1$ by defining $q = p - u$ and assuming the graph is weighted with weights in $\{0, \beta\}$.

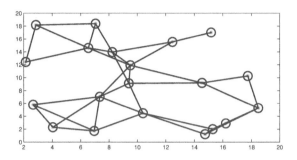

Figure 3.1: The network topology in case of 20 sensors.

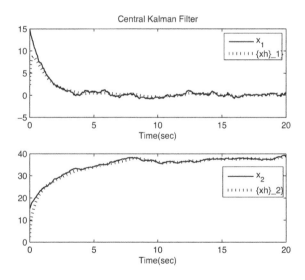

Figure 3.2: Central Kalman filter.

• Distributed Kalman Filtering Algorithm using consensus filtering. [515]–[522]

STATE Initialization: $q_i = 0, X_i = 0_{m \times m}, P_i = nP_0, \bar{x}_i = x(0)$
WHILE new data exists

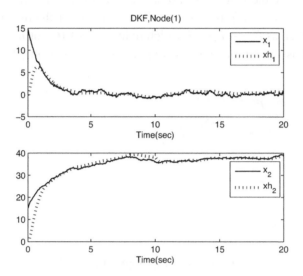

Figure 3.3: Distributed Kalman filter at node 1.

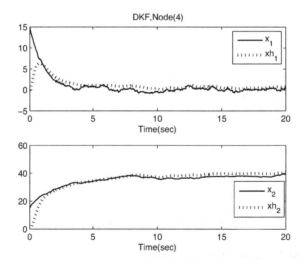

Figure 3.4: Distributed Kalman filter at node 4.

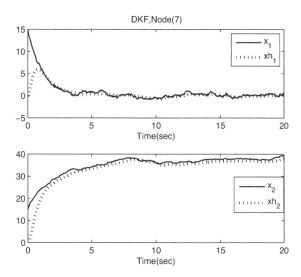

Figure 3.5: Distributed Kalman filter at node 7.

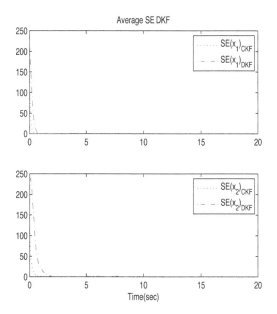

Figure 3.6: Average SE DKF.

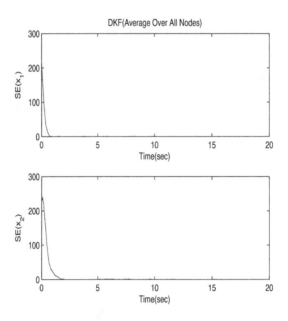

Figure 3.7: DKF (Average over all nodes).

STATE Update the state of the data CF:

$u_j = H_j^T R_j^{-1} z_j, \ \forall j \in N_i \cup \{i\}$

$q_i \leftarrow q_i + \epsilon\beta \sum_{j \in N_i}[(q_j - q_i) + (u_j - u_i)]$

$y_i = q_i + u_i$ **STATE** Update the state of the covariance CF:

$U_j = H_j^T R_j^{-1} H_j, \ \forall j \in N_i\{i\}$

$X_i \leftarrow X_i + \epsilon\beta \sum_{j \in N_i}[(X_j - X_i) + (U_j - U_i)]$

$S_i = X_i + U_i$ **STATE** Estimate the target state using Micro-KF:

$M_i = (P_i^{-1} + S_i)^{-1}$

$\hat{x}_i = \bar{x}_i + M_i(y_i - S_i \bar{x}_i)$

STATE Update the state of the Micro-KF:

$P_i \leftarrow AM_i A^T + nBQB^T$

$\bar{x}_i \leftarrow A\hat{x}_i$ **ENDWHILE**

In the above algorithm, each node uses the inputs $u_i(k) = H_i^T(k)R_i^{-1}(k)z_i(k)$ and $U_i(k) = H_i^T(k)R_i^{-1}(k)H_i(k)$ with zero initial states $q_i(0) = 0$ and $X_i(0) = 0$. The outputs $y_i(k)$ and $S_i(k)$ of the consensus filters asymptotically converge to $y(k)$ and $S(k)$.

3.1.3 Simulation example 1

The dynamic system is observed by 10 sensor nodes. These nodes run an individual distributed Kalman filter which is presented in [533]. The estimation results from nodes 1, 4 and 7 are given below from Figure 3.1 to 3.5.

Consider a target with dynamics

$$\dot{x} = \begin{bmatrix} -1 & 0 \\ 1 & 0 \end{bmatrix} x + I_2 w$$

The network has 10 sensors and the following data is used in simulation: $R_i = 100(i^{0.5} I_2)$, $Q = 25$, $P_0 = I_2$ and the step size is 0.02. The estimation at node 1, 4 and 7 when the sensor nodes have high connectivity is shown in Figures 3.8, 3.9, 3.10, 3.11 and 3.12.

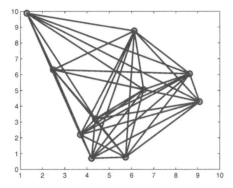

Figure 3.8: Network topology with high connectivity between sensor nodes (10 sensor nodes).

The estimation at nodes 1, 4 and 7 when the connectivity between the sensor nodes is low is illustrated in Figures 3.13, 3.14 and 3.12.

3.2 Distributed Sensor Fusion

Consensus algorithms for networked dynamic systems provide scalable algorithms for sensor fusion in sensor networks. This chapter introduces a distributed filter that allows the nodes of a sensor network to track the average of n sensor measurements using an average consensus-based distributed filter called a consensus filter. This consensus filter plays a crucial role in solving a data fusion problem that allows implementation of a scheme for distributed Kalman filtering in sensor

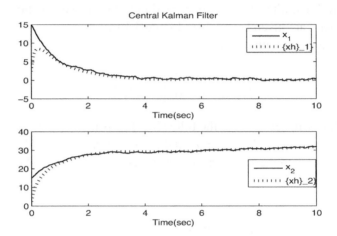

Figure 3.9: Central Kalman filter.

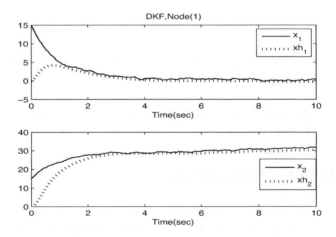

Figure 3.10: High connectivity between sensor nodes: DKF at node 1.

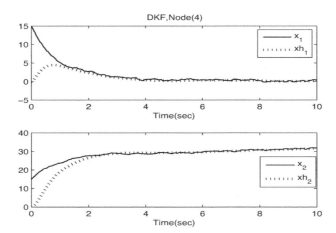

Figure 3.11: High connectivity between sensor nodes: DKF at node 4.

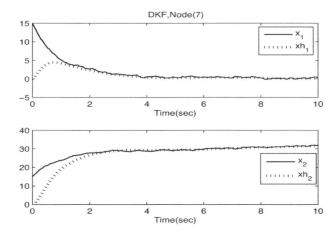

Figure 3.12: High connectivity between sensor nodes: DKF at node 7.

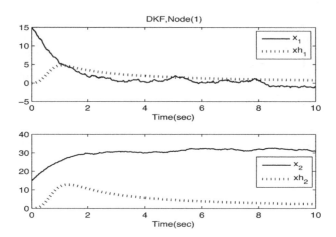

Figure 3.13: Low connectivity between sensor nodes: DKF at node 1.

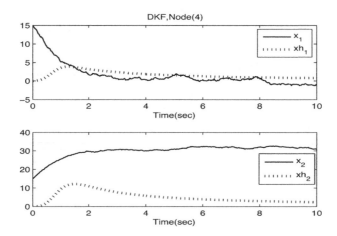

Figure 3.14: Low connectivity between sensor nodes: DKF at node 4.

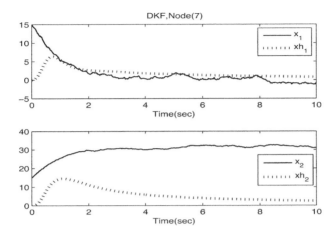

Figure 3.15: Low connectivity between sensor nodes: DKF at node 7.

networks. The analysis of the convergence, noise propagation reduction, and the ability to track fast signals are provided for the consensus filters. As a byproduct, a novel critical phenomenon is found that relates the size of a sensor network to its tracking and sensor fusion capabilities. This performance limitation as a tracking uncertainty principle will be conveniently characterized and in turn, this answers a fundamental question regarding how large a sensor network must be for effective sensor fusion. Moreover, regular networks emerge as efficient topologies for the distributed fusion of noisy information, though arbitrary overlay networks can be used.

3.2.1 Introduction

A fundamental problem in sensor networks is to solve detection and estimation problems using scalable algorithms. This requires the development of novel distributed algorithms for estimation and in particular, Kalman filtering that is currently unavailable. In [40] a scalable sensor fusion scheme was proposed that requires fusion of sensor measurements combined with local Kalman filtering. The key element of this approach is to develop a distributed algorithm that allows the nodes of a sensor network to track the average of all of their measurements. We refer to this problem as dynamic average-consensus. Consensus problems [41], [42] for networked dynamic systems have been extensively used by many researchers as part of the solution of more complex problems. In what follows, the average-consensus algorithm for n constant values in [41], [42] is generalized to the case of n measurements of noisy signals obtained from n sensors in the form of a distributed low-pass filter called the Consensus Filter. The role of this consensus filter is to perform the distributed fusion of sensor measurements that is necessary for implementation of a scalable Kalman filtering scheme proposed in [40]. We show that consensus filters can also be used independently for distributed sensor fusion.

3.2.2 Consensus problems in networked systems

In what follows, we introduce some elements of algebraic graph theory. Let $G = (V, E)$ be a graph with a nonnegative adjacency matrix $A = [a_{ij}]$ that specifies the interconnection topology of a network of dynamic systems, sensors, or agents. The set of nodes is denoted by $V = \{1, \ldots, n\}$. For complex networks, we refer to $|V|$ and $|E|$ as the scale and size of the network, respectively. Let $N_i = \{i \in V : a_{ij} \neq 0\}$ denote the set of neighbors of node i and $J_i = N_i \cup \{i\}$ denote the set of inclusive neighbors of node i. A consensus algorithm can be expressed in the form of a linear system

$$\dot{x}_i(t) = \sum_{j \in N_i} a_{ij}(x_j(t) - x_i(t)), x(0) = c \in \Re^n. \tag{3.17}$$

Given a connected network G, all the solutions of system (3.5) converge to an aligned state $x^* = (\mu, \ldots, \mu)^T$ with identical elements equal to $\mu = bar x(0) = \frac{1}{n} \sum_i c_i$. This explains why the term "average-consensus" is used to refer to the distributed algorithm in (3.17). In a more compact form, system (3.17) can be expressed as

$$\dot{x} = -Lx, \tag{3.18}$$

where L is the Laplacian matrix [411] of graph G and is defined as

$$L = \Delta - A \tag{3.19}$$

where $\Delta = diag(A1)$ is the degree matrix of G with diagonal elements $d_i = \sum_j a_{ij}$. Here, $1 = (1, \ldots, 1)^T \in R^n$ denotes the vector of ones that is always a right eigenvector of L corresponding to $\lambda_1 = 0$ (i.e. $L1 = 0$). The second smallest eigenvalue λ_2 of L determines the speed of convergence of the algorithm.

3.2.3 Consensus filters

Consider a sensor network of size n with an information flow. We keep in mind that the information flow in a sensor network might (or might not) be the same as the overlay network (i.e., communication network), G. Assume each sensor is measuring a signal $s(t)$ that is corrupted by noise v_i that is a zero-mean white Gaussian noise (WGN). Thus, the sensing model of the network is

$$u_i(t) = r(t) + v_i(t), i = 1, \ldots, n \tag{3.20}$$

or $u(t) = r(t)1 + v(t)$. Let R_i denote the covariance matrix of v_i for all i.

The objective here is to design the dynamics of a distributed low-pass filter with state $x = (x_1, \ldots, x_n)^T \in R^n$ that takes u as the input and $y = x$ as the output with the property that asymptotically all nodes of the network reach an ϵ-consensus regarding the value of signal $r(t)$ in all time t. By ϵ-consensus, we mean there is a ball of radius ϵ that contains the state of all nodes (i.e., approximate agreement). In most applications, $r(t)$ is a low-to-medium frequency signal and $v(t)$ is a high-frequency noise. Thus, the consensus filter must act as a low-pass filter.

We propose the following dynamic consensus algorithm

$$\dot{x}_i(t) = \sum_{j \in N_i} a_{ij}(x_j(t) - x_i(t)) + \sum_{j \in J_i} a_{ij}(u_j(t) - x_i(t)), \qquad (3.21)$$

as a candidate for a distributed low-pass consensus filter. The remainder of the chapter is devoted to establishing the properties of this distributed filter. Note that the algorithm in (3.21) only requires communication among neighboring nodes of the network, and thus is a distributed algorithm.

Remark 3.2.1 *In discrete-time, the dynamic consensus algorithm in (3.21) can be stated as follows:*

$$x_i^+ = x_i + \delta[\sum_{j \in N_i} a_{ij}(x_j - x_i) + \sum_{j \in J_i} a_{ij}(u_j - x_i)], \qquad (3.22)$$

where x_i is the current state of node i, x_i^+ is the next state, and δ is the step-size of iterations. We will conduct all of our analysis in continuous-time.

Proposition 3.2.1 *The distributed algorithm in (3.21) gives a consensus filter with the following collective dynamics*

$$\dot{x} = -(I_n + \Delta + L)x + (I_n + A)u \qquad (3.23)$$

that is an LTI system with specification $A = -(I + \Delta + L), B = I_n + A, C = I_n$ and a proper MIMO transfer function.

Proof 3.2.1 *First, let us rewrite the system in (3.21) as*

$$\begin{aligned}
\dot{x}_i &= \sum_{j \in N_i} a_{ij}(x_j - x_i) + \sum_{j \in J_i} a_{ij}(u_j - u_i + u_i - x_i), \\
&= \sum_{j \in N_i} a_{ij}(x_j - x_i) + \sum_{j \in N_i} a_{ij}(u_j - u_i) \\
&\quad + |J_i|(u_i - x_i).
\end{aligned}$$

Noting that $|J_i| = 1 + d_i$, *from the definition of graph Laplacian, we get*

$$\begin{aligned} \dot{x} &= -Lx - Lu + (I_n + \Delta)(u - x), \\ &= -(I_n + \Delta + L)x + (I_n + \Delta - L)u \end{aligned}$$

But $\Delta - L = A$ *and therefore* $\dot{x} = Ax + Bu, y = Cx$ *with matrices that are defined in the question.*

The transfer function of the consensus filter is given by

$$H(s) = [sI_n + (I_n + \Delta + L)]^{-1}(I_n + A) \qquad (3.24)$$

Applying the Gershgorin theorem to matrix $A = I_n + 2\Delta + A$ guarantees that all poles of $H(s)$ are strictly negative and fall within the interval $[-(1 + d_{\min}), -(1 + 3d_{\max})]$ with $d_{\max} = \max_i d_i$ and $d_{\min} = \min_i d_i$. i.e. $1 + d_{\min} \leq \lambda_i(A) \leq (1 + 3d_{\max})$ for all i. This immediately implies the following stability property of the consensus filter.

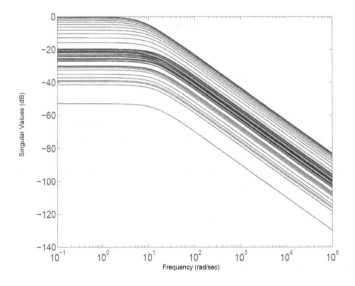

Figure 3.16: The singular value plots of the low-pass consensus filter for a regular network.

Corollary 3.2.1 *The consensus filter in (3.24) is a distributed stable low-pass filter.*

Proof 3.2.2 *Apparently, all the poles of $H(s)$ are strictly negative and thus the filter is stable. On the other hand, $H(s)$ is a proper MIMO transfer function satisfying $\lim_{s\to\infty} H(s) = 0$ which means it is a low-pass filter. Figure 3.16 shows the singular value plots of the low-pass consensus filter (or CF_{lp}) for a regular network with $n = 100$ nodes and degree $k = 6$.*

Remark 3.2.2 *The following dynamic consensus algorithm [107]*

$$\dot{x} = -Lx + \dot{u}(t)$$

gives a high-pass consensus filter (CF_{hp}) that is useful for distributed data fusion applications with low-noise data.

It remains to establish that all nodes asymptotically can reach an ϵ-consensus regarding $r(t)$.

Proposition 3.2.2 *Let $r(t)$ be a signal with a uniformly bounded rate $|\dot{r}(t)| \leq \nu$. Then, $x^*(t) = r(t)1$ is a globally asymptotically ϵ-stable equilibrium of the dynamics of the consensus filter given by*

$$\dot{x} = -Lx - Lu + (I_n + \Delta)(u - x) \tag{3.25}$$

with input $u = r(t)1$ and

$$\epsilon = \frac{\nu\sqrt{n}}{\lambda_{\min}(A)} \tag{3.26}$$

Proof 3.2.3 *Given the input $u = r(t)1$, the dynamics of the system in (3.25) reduces to*

$$\dot{x} = -Lx + (I_n + \Delta)(r(t)1 - x) \tag{3.27}$$

with an obvious equilibrium at $x = r(t)1$ that is an aligned state with elements that are identical to the signal $r(t)$. This is due to the fact that $L1 = 0$. Defining the error variable $\eta = x - r(t)1$ gives

$$\dot{\eta} = -A\eta + \dot{r}(t)1 \tag{3.28}$$

where $A = I_n + \Delta + L$ is a positive definite matrix with the property that

$$1 + d_{\min} \leq \lambda_{\min}(A) \leq \lambda_{\max}(A) \leq 1 + 3d_{\max}. \tag{3.29}$$

Let us define the Lyapunov function $\varphi(\eta) = \frac{1}{2}\eta^T\eta$ for the perturbed linear system in (3.28). We have

$$\begin{aligned} \dot{\varphi} &= -\eta^T A\eta + \dot{r}(t)(1^T\eta) \\ &\leq -\lambda_{\min}(A)\|\eta\|^2 + \nu\sqrt{n}\|\eta\|. \end{aligned}$$

This is because $|1^T \eta| \le \|1\| \|\eta\| = \sqrt{n} \|\eta\|$. As a result, one obtains

$$\dot{\varphi}(\eta) \le -\lambda_{\min}(A)(\|\eta\| - \frac{\nu\sqrt{n}}{2\lambda_{\min}(A)})^2 + \left(\frac{\nu\sqrt{n}}{2\lambda_{\min}(A)} \right)^2$$

Let B_ρ be a closed ball centered at $\eta = 0$ with radius

$$\rho = \frac{\nu\sqrt{n}}{\lambda_{\min}(A)} \tag{3.30}$$

and let $\Omega_c = \{\eta : \varphi(\eta) \le c\}$ be a level set of the Lyapunov function $\varphi(\eta)$ with $c = \frac{1}{2}\rho^2$. Then, $\Omega_c = B_\rho$ because

$$\|\eta\| \le \rho \Rightarrow \varphi(\eta) = \frac{1}{2}\eta^T \eta \le \frac{1}{2}\rho^2 = c.$$

As a result, any solution of (3.28) starting in $\Re^n$ Ω_c satisfies $\dot{\varphi} < 0$. Thus, it enters Ω_c in some finite time and remains in Ω_c thereafter (i.e. Ω_c is an invariant level-set). This guarantees global asymptotic ϵ-stability of $\eta = 0$ with a radius $\epsilon = \rho$. Of course, ϵ-stability of $\eta = 0$ implies ϵ-tracking of $r(t)$ by every node of the network (i.e., ϵ- consensus is asymptotically reached).

The following result describes the occurrence of a critical phenomenon in regular complex networks.

Proposition 3.2.3 *Consider a regular network G of degree k. Let $r(t)$ be a signal with a finite rate $|\dot{r}| \le \nu$. Then, the dynamics of the consensus filter in the form*

$$\dot{x} = -Lx - Lu + (I + \Delta)(u - x) \tag{3.31}$$

satisfies the following properties:

- *The mean $\mu(t) = barx(t)$ of the state of all nodes is the output of a scalar low-pass filter*

$$\dot{\mu} = (k+1)(\bar{u}ht) - \mu) \tag{3.32}$$

  with an input $\bar{u}(t) = r(t) + w(t)$ and a zero-mean noise $w(t) = \frac{1}{n}\sum_i v_i(t)$.

- *Assume the network node degree $k = \beta n^\gamma$ is exponentially scale-dependent. Then, there exists a critical exponent $\gamma_c = \frac{1}{2}$ such that for all $\gamma > \gamma_c$ (or networks with more than $O(n^{1.5})$ links), the radius of ϵ-tracking vanishes as the scale n becomes infinity large for any arbitrary $\nu, \beta(\epsilon$ is defined in Proposition 3.2.2).*

Proof 3.2.4 *Part i) follows from the fact that $\mu = \frac{1}{n}(1^T x)$ and $1^T L = 0$. Moreover, for regular networks with degree k, $I_n + \Delta = (k + 1)I_n$ and $\lambda_{\min}(A) = (k + 1)$. To show part ii), note that for a k-regular network, the expression for ϵ greatly simplifies as*

$$\epsilon_n = \frac{\nu\sqrt{n}}{1 + k} = \frac{\nu\sqrt{n}}{1 + \beta n^\gamma} \tag{3.33}$$

Thus, for all $\gamma > \gamma_c = \frac{1}{2}$, $\epsilon_n \to 0$ as $n \to 1$ regardless of the values of $\beta, \nu < \infty$. In other words, ϵ_n-tracking of $r(t)$ is achieved asymptotically by every node with a vanishing ϵ_n for large-scale regular networks of size (i.e. $nk/2$) greater than $O(n^{1.5})$.

Remark 3.2.3 *The white noise $w(t) = \frac{1}{n}\sum_i v_i(t)$ has a covariance matrix $\frac{1}{n}\bar{R}$ that is n times smaller that the average covariance $\bar{R} = \frac{1}{n}\sum_i R_i$ of all (uncorrelated) v_i's. For a large-scale network, $w(t)$ can possibly become multiple orders of magnitude weaker than all the v_i's.*

Proposition 3.2.4 *(scale-uncertainty principle) A regular complex network with density $\sigma = (2|E| + n)/n^3/2$ and tracking uncertainty $\varepsilon = \epsilon/\nu$ that runs the dynamic consensus algorithm in (3.21) satisfies the following uncertainty principle*

$$(network\ density) \times (tracking\ uncertainty) = 1, \tag{3.34}$$

or $\sigma \times \varepsilon = 1$.

Proof 3.2.5 *The proof follows from (3.33) and the identity $2|E| := \sum_i d_i = nk$.*

Defining the performance of tracking as $1/\varepsilon$, we get the following trade-off between tracking performance and network density:

$$(network\ density) \propto (tracking\ performance).$$

The most common application is to track a signal that has a single, or multiple, sinusoidal components.

3.2.4 Simulation example 2

We examine here the problem of tracking of sinusoidal signals. Consider the case of a signal $r(t) = a\sin(\omega t)$ with $a, \omega > 0$ that is being measured by every sensor in a sensor network. This signal could possibly represent the x-coordinate of the position of a moving object that goes in circles. The main question of interest is how large must the sensor network be? This is important for the purpose of tracking $r(t)$ within a tube of radius $\epsilon \leq \delta a$ (e.g. $\delta = 0.1$).

Notice that $\nu = a\omega$ and therefore the tracking uncertainty satisfies. To guarantee $\epsilon \leq \delta a$, we must have $\varepsilon = \epsilon/\nu \leq \delta/\omega$. Using the uncertainty principle, $\sigma \times \varepsilon = 1$ and thus $\omega \leq \delta \times \sigma$.

For a network with $n = 1000$ nodes and weighted degree $k = \beta n^\gamma$ with $\beta = 10, \gamma = 0.6 > \gamma_c$ (all weights of the graph are in $\{0, \beta\}$), we get $k = 631$ and $\omega \leq 2$ (rad/sec) for $\epsilon = 0.1a$ accuracy. This is a relatively conservative bound, and in practice the network is capable of tracking much faster signals with only 100 nodes. Finding a less conservative uncertainty principle is a real challenge.

One cannot arbitrarily increase β because based on the low-pass filter with state μ, this is equivalent to using a high-gain observer for $\bar{u}$ that amplifies noise.

3.2.5   Simulation example 3

In this section, we present simulation results for sensor networks with two type of topologies:

a) a regular network of degree $k = 6$ and

b) a random network obtained as a spatially induced graph from $n = 400$ points with coordinates $\{q_i\}_{i \in V}$ that are distributed uniformly at random in an $n \times n$ square region with a set of neighbors $N_i = \{q_j : \|q_i - q_i\| < \rho_0\}$ and a radio range of $\rho_0 = 2\sqrt{n}$. These networks are shown in Fig 3.17.

Networks (a) and (b), shown in Figure 3.17, have an average-degree of 6 and 7.1, respectively. Apparently, the random network is irregular.

We use the following three test signals

$$
\begin{aligned}
r_1(t) &= \sin(2t); \\
r_2(t) &= \sin(t) + \sin(2t+3) + \sin(5t+4), \\
r_3(t) &= \sin(10t).
\end{aligned}
$$

For r_1 and r_2, we set the covariance matrix to $R_i = 0.3$ for all nodes and for $r_3, R_i = 0.6$ for all i.

Figure 3.18 demonstrates sensor fusion using a low-pass consensus filter with a regular network topology for sensor measurements $r_1(t) + v_i(t)$ obtained from $n = 100$ nodes. The fused measurements Figure 3.18 (b) have a covariance that is almost 100 times smaller than the covariance of the sensor data.

Similarly, Figure 3.19 demonstrates sensor fusion using a distributed low-pass consensus filter for sensor data $r_2(t) + v_i(t)$ obtained from $n = 100$ nodes. Again, the network topology is regular. All nodes are apparently capable of tracking $r_3(t)$ within a radius of uncertainty that is determined by $|\dot{r}_3|$ and the noise covariance R_i.

Now, to demonstrate tracking capabilities of larger networks, we consider tracking $r_3(t)$ that is 5 times faster than $r_1(t)$ using a consensus filter in a network with random topology. The results of the sensor fusion are shown in Figure 3.20.

3.2.6 Some observations

We introduced consensus filters as a tool for distributed sensor fusion in sensor networks. The consensus filter is a dynamic version of the average-consensus algorithm that has been extensively used for sensor fusion as well as other applications that involve networked dynamic systems and collaborative decision making. It was mentioned that based on a new scalable Kalman filtering scheme, a crucial part of the solution is to estimate the average of n signals in a distributed way. It was shown that consensus filters effectively solve this dynamic average-consensus problem. This distributed filter acts as a low-pass filter induced by the information flow in the sensor network. In addition, ϵ- tracking properties of consensus filters for sensor fusion was analyzed in detail. The byproduct of this analysis was a novel type of critical phenomenon in complex networks that relates the size of the sensor network to its capability to track relatively fast signals. This limitation was characterized as a tracking uncertainty principle. Simulation results for large regular and random sensor network were presented.

3.3 Estimation for Adaptive Sensor Selection

Wireless sensor networks (WSNs) are usually deployed for monitoring systems using distributed detection and the estimation of sensors. Sensor selection in wireless sensor networks (WSN) for target tracking is considered in this section. A distributed estimation scenario is considered based on the extended information filter. A cost function using the geometrical dilution of precision (GDOP) measure is derived for active sensor selection. A consensus based estimation method is proposed in this chapter for heterogeneous WSNs with two types of sensors. The convergence properties of the proposed estimators are analyzed under time-varying inputs. Accordingly, a new adaptive sensor selection (ASS) algorithm is presented in which the number of active sensors is adaptively determined based on the absolute local innovations vector.

3.3.1 Introduction

Wireless sensor networks have attracted much research attention in recent years [68] and can be utilized in many different applications, including battlefield surveillance, biological detection, machine failure diagnosis, environmental monitoring,

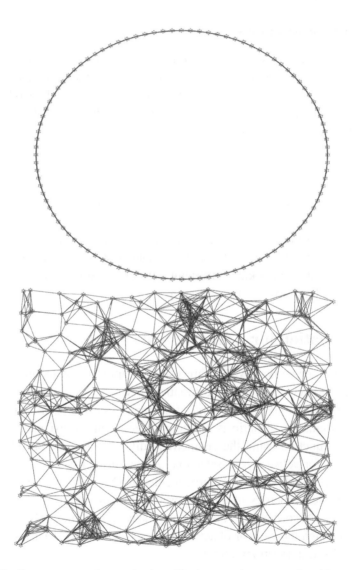

Figure 3.17: Sensor network topologies: Top) a regular network with n = 100 and degree k = 6 and Bottom) a random network with n = 400 and 2833 links.

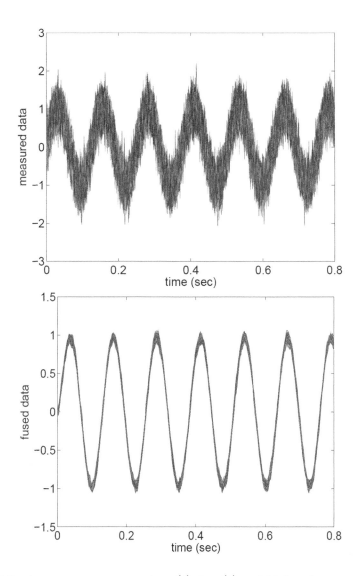

Figure 3.18: a) sensor measurements $r_1(t) + v_i(t)$ and b) fused sensor data via a low-pass consensus filter in a regular network.

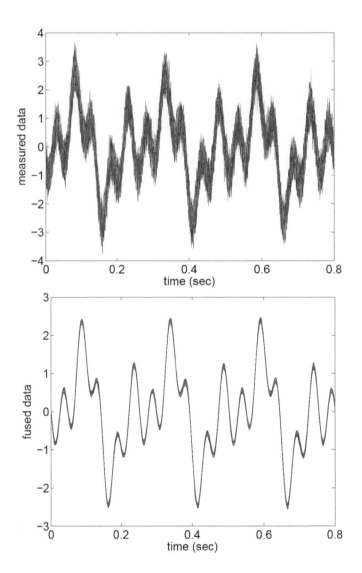

Figure 3.19: a) sensor measurements $r_2(t) + v_i(t)$ and b) fused sensor data via a low-pass consensus filter with a regular network topology.

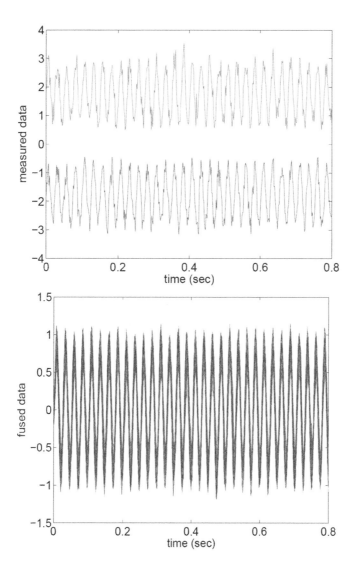

Figure 3.20: (a) upper and lower envelopes $(\max_i u_i(t), \min_i u_i(t))$ of sensor measurements $r_3(t) + v_i(t)$, and b) fused measurements (i.e. states x_i) after consensus filtering in a sensor network with randomly distributed nodes.

home security, and so on. Wireless sensor networks are usually composed of hundreds or thousands of sensors [69]. These sensors are small, with limited computing and processing resources. These sensor nodes can sense, measure, and gather information from the environment and, based on some local decisions, relevant information can be processed. Moreover, they can transmit the sensed data to the user. One of the recent applications of WSNs is the distributed estimation of unknown signals through local data cooperatively, [70],[71].

According to the capability of sensors, there are two types of network: homogeneous and heterogeneous WSNs [73]. A heterogeneous WSN is more complex, compared to, homogeneous WSN and it consists of a number of sensor nodes of different types deployed in a particular area, which are collectively working together to achieve a particular aim [75]. The aim may be to monitor any of the physical or environmental conditions. In a heterogeneous WSN some sensors have a larger sensing range and more power to achieve a longer transmission range. Typically, a large number of inexpensive nodes perform sensing, while a few expensive nodes (perhaps embedded PCs) provide data filtering, fusion, and transport. This partitioning of tasks ensures a cost-effective design as well as a more efficient implementation of the overall sensing application. However, realizing the full potential of heterogeneity requires careful network engineering, including careful placement of the heterogeneous resources and the design of resource-aware protocols. Effective exploitation of heterogeneity can impact the economic feasibility of applications such as industrial equipment monitoring [76]. The heterogeneous WSN increases the detection probability for a given intrusion detection distance, [73].

In [110], the distributed finite-horizon filtering problem was considered for a class of discrete time-varying systems with RVNs over lossy sensor networks involving quantization errors and successive packet dropouts. A distributed moving horizon estimation algorithm for constrained discrete-time linear systems is proposed in [78]. Using suitable assumptions we proved convergence of the estimates to a common value. Robust static output-feedback controllers are designed that achieve consensus in networks of heterogeneous agents modeled as nonlinear systems of relative degree two [79]. Both ideal communication networks and networks with communication constraints are considered, e.g., with limited communication range or heterogeneous communication delays. Distributed consensus algorithms are low-complexity iterative algorithms where neighboring nodes communicate with each other to reach an agreement regarding a function of the measurements [41].

A distributed channel estimation problem in a sensor network which employs a random sleep strategy to conserve energy is considered in [80]. A distributed algorithm for achieving globally optimal decisions, either estimation or detection, is analyzed in [81] through a self-synchronization mechanism among linearly coupled integrators initialized with local measurements. The distributed H_∞ filtering problem is addressed in [82, 113] for a class of polynomial nonlinear stochastic systems in sensor networks. In [91], the distributed estimation of an unknown vector signal is considered in a resource constrained sensor network with a fusion center. Due to power and bandwidth limitations, each sensor compresses its data in order to minimize the amount of information that needs to be communicated to the fusion center.

In [86], a distributed implementation of the Kalman filter is presented for sparse large-scale systems monitored by sensor networks. In this solution, the communication, computing, and storage is local and distributed across the sensor network; no single sensor processes, dimensional vectors or matrices, where usually a large number is the dimension of the state vector representing the random field. In [90], a distributed least mean-square (LMS)-type of adaptive algorithm is developed for WSN based tracking applications, where inter-sensor communications are constrained to single-hop neighboring sensors and are challenged by the effects of additive receiver noise. Starting from a well-posed convex optimization problem defining the desired estimator, [90] reformulated it into an equivalent constrained form whose structure lends itself naturally to decentralized implementation. Potential applications of distributed estimation were reported in [108], [109]. Distributed estimation problems were widely investigated for sensor networks with incomplete measurements.

State estimation via wireless sensors over fading channels is studied in [92]. Packet loss probabilities depend upon the time-varying channel gains, packet lengths, and transmission power levels of the sensors. Measurements are coded into packets by using either independent coding or distributed zero-error coding. At the gateway, a time-varying Kalman filter uses the received packets to provide the state estimates. The problem of selecting the best nodes for localizing (in the mean squared (MS) position error sense) a target in a distributed wireless sensor network is considered in [93]. Each node consists of an array of sensors that are able to estimate the direction of arrival (DOA) to a target. Different computationally efficient node selection approaches that use global network knowledge are introduced.

An energy-efficient adaptive sensor scheduling scheme for target tracking in WSNs was presented in [83] by jointly selecting the next tasking sensor and determining the sampling interval based on the predicted tracking accuracy and tracking energy cost. Energy-based source localization applications were considered in [84], where distributed sensors quantize acoustic signal energy readings and transmit quantized data to a fusion node, which then produces an estimate of the source location. A new information consensus filter was presented in [85] for distributed dynamic-state estimation. Estimation is handled by the traditional information filter, while communication of measurements is handled by a consensus filter. The method was unbiased and conservative, and the actual covariance of local estimates in this filter is close to the centralized filter.

The high-capability sensors have a long transmission range, while low capability sensors have a shorter transmission range. Thus, a high-capability sensor packets may reach the low-capability sensor, while the low capability sensor packets may not be able to reach the corresponding high-capability sensor [74]. Two types of sensors are considered there, that is, type-I sensors with high processing ability and type-II sensors with low ability. Each sensor carries out local estimation and adjusts its estimate through communication with its neighbors. In this work, we investigate the stability and convergence properties of the heterogeneous estimator.

In this section, we develop a systematic procedure leading to a distributed estimation in WSN using the extended information filter for target tracking and the adaptive sensor selection (ASS) algorithm. The algorithm incorporates a cost function based on the geometrical dilution of precision (GDOP) for sensor selection. The convergence properties of a distributed estimator for heterogeneous WSNs are addressed. Accordingly, the ASS algorithm is presented which includes the absolute error between the measured and predicted observations in the selection of the number of active sensors as well as the best topology for the least possible tracking error. The number of active sensors was adaptively determined based on the absolute local innovations vector. Simulation results revealed that the ASS with SS achieves the similar performance as the ASS with GNS.

3.3.2 Distributed estimation in dynamic systems

A heterogeneous WSN is composed of two types of sensors: type-I sensors are of high-quality and type-II sensors are low-end ones [41]. Our goal is to design a distributed algorithm taking heterogeneity into account and use it to track an unknown signal $\xi(t)$ based on the sensing signals

$$y_i(t) \;=\; b_i \xi(t) + w_i(t) \tag{3.35}$$

where $b_i \geq 0$ and w_i are independent white Gaussian noises with zero mean and variances σ_i^2. Note that wireless sensor networks (WSNs) are usually deployed for monitoring systems with distributed detection and estimation of sensors. In this case the sensing signal is $y_i(t)$ containing the desired signal $\zeta(t)$ which is unknown ($\zeta(t)$ is fading by b_i, thus b_i cannot be zero.) and some noise. Our goal here is to track this signal if we can sense it, whatever the behavior of this signal [105].

A non-linear motion model is considered for target tracking with a distributed estimation algorithm in the information space [94], [95]. The state transition and the observation vector are given by

$$x(t) = f(x(t-1), t) + G(t)v(t) \tag{3.36}$$
$$z(t) = h(x(t), t) + w(t) \tag{3.37}$$

where $x(t)$ is the state vector of time t, $f(.)$ and $h(.)$, are the non-linear state transition function and non-linear observation model, respectively. Here $v(t) \, N(0, Q(t))$ is an additive noise vector, and $w(t) \, N(0, R(t))$ is the zero mean white observation noise vector with the covariance matrix $R(t)$. The mean square error (MSE) estimate of the state $x(i)$ and the covariance matrix at time i, given the information up to and including time j, is given by

$$\hat{x}(i|j) = E\left[x(i)|z(1) \ldots z(j)\right] \tag{3.38}$$
$$P(i|j) = E\left[\left(x(i) - \hat{x}(i|j)\right)\left(x(i) - \hat{x}(i|j)\right)^T |z(1) \ldots z(j)\right] \tag{3.39}$$

It must be observed that the updating of $\hat{x}_s$ is performed much the same way as a Kalman filter, see Section II in [87],[88],[89]. The two key information-analytic variables are the information matrix and information state vector where the former is the inverse of the covariance matrix,

$$Y(t|t') = P^{-1}(t|t') \tag{3.40}$$

The information state vector is given by the product of the information matrix and the state estimate vector

$$\hat{y}(t|t') = P^{-1}(t|t')\hat{x}(t|t') = Y(t|t')\hat{x}(t|t') \tag{3.41}$$

The information filter can be extended to a linear estimation algorithm [96] for non-linear systems leading to the extended information filter (EIF) [95]. This filter estimates the information about non-linear state parameters given non-linear observations and non-linear system dynamics. When the EIF is distributed and the model distribution is applied, the distributed extended information filter (DEIF) is obtained as [94].

Prediction stage:

$$\hat{y}_s(t|t-1) = Y_s(t|t-1)f(t, \hat{x}_s(t-1|t-1)) \tag{3.42}$$
$$\hat{Y}_S(t|t-1) = \left[\nabla f(t)Y_s^{-1}(t-1|t-1)\nabla f^T(t)\right.$$
$$\left. + \; G(t)Q(t)G^T(t)\right]^{-1} \tag{3.43}$$

Estimation stage:

$$\hat{y}_s(t|t) \ = \ \hat{y}_s(t|t-1) + \sum_{j=1}^{S} i_j(t) \tag{3.44}$$

$$Y_S(t|t) \ = \ Y_S(t|t-1) + \sum_{j=1}^{S} I_j(t) \tag{3.45}$$

where the associated information matrix and state information contribution from the local observation are calculated as

$$I_j(t) \ = \ \nabla h^T(t) R_j^{-1} \nabla h(t) \tag{3.46}$$

$$i_j(t) \ = \ \nabla h^T(t) R_j^{-1} \big[v_j(t) + \nabla h(t)\hat{x}_j(t|t-1)\big] \tag{3.47}$$

and $v_j(t)$ is the local innovations vector as

$$v_j(t) \ = \ z_j(t) - h_j(t)\hat{x}_j(t|t-1) \tag{3.48}$$

Let $\mathcal{I}$ and the complement $\mathcal{I}^c$ be the sets of type-I and type-II sensors respectively. The total number of sensors is N. The estimator implemented by sensor i is as follows:

$$\dot{x}_i(t) \ = \ \alpha_i[y_i(t) - x_i(t)] + \beta_i x_i(t)$$
$$+ \ \frac{\sigma}{c_i} \sum_{j \in \mathcal{N}_i(t)} a_{ij}(t)[x_j(t) - x_i(t)], i \ \in \mathcal{I} \tag{3.49}$$

$$x_i(t) \ = \ \sum_{j \in \mathcal{N}_i(t)} \delta_{ij}(t)x_j(t) + \delta_i i(t)y_i(t), \ \ i \in \mathcal{I}^c \tag{3.50}$$

where $x_i \in \Re$ is the estimate of the unknown signal $\xi(t)$; $\alpha_i > 0$ is a tuning parameter, $\beta_i > 0$ is a scalar associated with the network topology, $\sigma > 0$ is the estimator gain, $c_i > 0$ quantifies the confidence that sensor i has in its current estimate; $a_{ij}(t)$ is defined as,

$$a_{ij}(t) \ = \ \sqrt{\frac{p_{T_j}(t)|h_{ij}(t)|^2}{d_{ij}^\gamma}}, \quad 1 \le i, \, j \le N, \tag{3.51}$$

to imply the signal power decay, where $p_{Tj}(t)$ is the transmit power of sensor j, $h_{ij}(t)$ is the fading coefficient, $2 \le \gamma \le 4$ is the path loss exponent and d_{ij} is the distance between sensor i and sensor j; $\delta_{ij}(t)$ are non-negative weights satisfying $\sum_{j \in \mathcal{N}_i(t) \cup \{i\}} \delta_{ij}(t) = 1$, $i \in \mathcal{I}$, $1 \leq j \leq N$. $\mathcal{N}_i(t)$ is the neighbors of sensor i at time t. The underlying interaction relations can be represented as a directed dynamic graph $\mathcal{G}(t)$, where vertices denote the sensors and edges refer to interactive links with the property that $a_{ij}(t) > 0$ if and only if there is information transmitted from sensor i to sensor j at time t.

For simplicity, the sensors are labeled from 1 to m. Set $\mathcal{I} = 1, 2, \ldots, m$. Let $\tilde{\mathcal{I}}(t)$ be the subset sensors of $\mathcal{I}$ that have direct communications with sensors in $\mathcal{I}^c$ and denote $\check{\mathcal{I}}(t) = \tilde{\mathcal{I}}(t) \bigcup \mathcal{I}^c$. Define the positive and negative neighbors of sensor i as $\mathcal{N}_i^+(t) = \mathcal{N}_i(t) \bigcap \mathcal{I}$ and $\mathcal{N}_i^-(t) = \mathcal{N}_i(t) \bigcap \mathcal{I}^c$, respectively.

The following lemma is important to derive our main results.

Lemma 3.3.1 *If* $0 < \delta_{kj}(t) \leq 1$, $m < k, j \leq N$, $j \neq k$, *then* $\mathcal{I} - \Sigma_2(t)$ *is invertible, where* $\Sigma_2(t) = [l_{ij}(t)] \in \Re^{(N-m) \times (N-m)}$,

$$
l_{ij}(t) = \begin{cases} 0, & j = i, \\ \delta_{(m+i)(m+j)}(t), & j \neq i. \end{cases}
$$

Proof: It suffices to show that $\rho(\Sigma_2(t)) < 1$, where $\rho(\cdot)$ is the spectral radius.

Let $N - m$ sensors in $\mathcal{I}^c$ and a virtual sensor v correspond to the states of a Markov chain and the one-step transition probability from state j to state i be $p_{ij}(t)$, where $p_{jj}(t) = 0$, $p_{ij}(t) = \delta_{ij}(t)$, if $j \neq i$ and $p_{vv}(t) = 1$, $p_{vj}(t) = 1 - \sum_{k \in \mathcal{N}_j^-} \delta_{kj}(t)$, if $j \neq v$. Then the transition probability matrix can be expressed as

$$
P(t) = \begin{bmatrix} \Sigma_2(t) & p_v(t) \\ 0 & 1 \end{bmatrix}, \quad p_v(t) = [p_{v(m+1)}(t), \ldots, p_{vN}(t)]^T
$$

We claim that v is a globally reachable state, namely, v can be reached from all the other states in finite steps with positive probabilities.

Let $\mathcal{V}_1(t)$ and $\mathcal{V}_2(t)$ be the sets of states from which v is reachable and not reachable at time t, respectively. Then for every state $v_1 \in \mathcal{V}1(t)$, it is not reachable from any state in $\mathcal{V}_2(t)$ with probability 1, since otherwise, v is reachable from some state $v' \in \mathcal{V}_2(t)$ with nonzero probability. It is a contradiction to the definition of $\mathcal{V}_2(t)$. Hence, for any two states $v_1 \in \mathcal{V}_1(t)$ and $v_2 \in \mathcal{V}_2(t)$, they are not reachable from each other with nonzero probability at time t recalling that $\delta_{ij} \neq 0$ if and only if $\delta_{ji} \neq 0$. Thus there exists a permutation matrix $\mathcal{P}$ such that

$$
\mathcal{P}^T P(t) \mathcal{P} = \begin{bmatrix} \Sigma_{V_1}(t) & 0 & p_{\tilde{v}}(t) \\ 0 & \Sigma_{V_2}(t) & 0 \\ 0 & 0 & 1 \end{bmatrix}
$$

where $p_{\tilde{v}}(t)$ is the stacked vector of $p_{vi}(t)$, $i \in \mathcal{V}_1(t)$.

On the other hand, since $\mathcal{G}(t)$ is connected, every two sensors $j \in \mathcal{V}_2(t)$ and $i \in \mathcal{I}$ are joined by a path at time t. Noting that as one state of the Markov chain, j is not reachable from any state in $\mathcal{V}_1(t)$, thus there must be some sensor $i' \in \mathcal{I}$ such that i' and j are adjacent in graph $\mathcal{G}(t)$, i.e. $\delta_{i'j} \neq 0$, which contradicts $\sum_{k \in \mathcal{N}_j^-} \delta_{kj}(t) = 1$. Therefore, it follows from [226] that $\rho(\Sigma_2(t)) < 1$.

Based on **Lemma** 3.3.1, we obtain from (3.50) that

$$\tilde{x}(t) = (I - \Sigma_2(t))^{-1}[\Sigma_1(t)x(t) + \Upsilon(t)\tilde{y}(t)] \tag{3.52}$$

where $x = [x_1, \ldots, x_m]^T$, $\tilde{x} = [x_{m+1}, \ldots, x_N]^T$, $\tilde{y} = [y_{m+1}, \ldots, y_N]^T$; $\Upsilon(t) = \mathrm{diag}\{\delta_{(m+1)(m+1)}(t), \ldots, \delta_{NN}(t)\}$ and $\Sigma_1(t) = [\tilde{l}_{ij}(t)] \in \Re^{(N-m) \times m}, \tilde{l}_{ij}(t) = \delta_{(m+i)j}(t)$.

Substituting (3.50) into (3.46) yields

$$\dot{x}(t) \quad = \quad (\Theta - \Lambda + F(t))x(t) + [\Lambda, B(t)]y(t), \tag{3.53}$$

where $y = [y_1, \ldots, y_N]^T$, $\Theta = \mathrm{diag}\{\beta_1, \ldots, \beta_m\}$,$\Lambda = \mathrm{diag}\{\alpha_1, \ldots, \alpha_m\}$, $F(t) = \sigma C^{-1}(A_1(t) + A_2(t)(I - \Sigma_2(t))^{-1}\Sigma_1(t))$, $B(t) = \sigma C^{-1}A_2(t)(I - \Sigma_2(t))^{-1}\Upsilon(t)$, $C = \mathrm{diag}\{c_1, \ldots, c_m\}$, $A_2(t) = [h_{ij}(t)] \in \Re^{m \times (N-m)}$, $A_1(t) = [\tilde{h}_{ij}(t)] \in \Re^{m \times m}, h_{ij} = a_{i(m+j)}(t)$ and

$$\tilde{h}_{ij}(t) = \begin{cases} -\sum_{k \in \mathcal{N}_i} a_{ik}(t), & j = i, \\ a_{ij}(t), & j \neq i. \end{cases}$$

The estimator (3.53) is a low-pass estimator, because if $a_{ij}(t)$ and $\delta_{ij}(t)$ are constants, then the transfer function from input y to output x is

$$T(s) \quad = \quad (sI - (\Theta - \Lambda + F))^{-1}[\Lambda, B],$$

which tends to zero as $s \to +\infty$. Thus the estimator is not appropriate for a distributed estimation with low-noise data. In most applications, x(t) is a low-to-medium frequency signal and y(t) is a high-frequency noise. Thus, the consensus estimator must act as a low-pass estimator [406].

3.3.3   Convergence properties

In this section, convergence results of the proposed estimator are presented to show that it can be used to track the ambiance signal $\xi(t)$ with an error bound which can be minimized to a desirable level through a minimization problem.

Lemma 3.3.2 *If $a_{ij}(t)$ and $\delta_{kj}(t)$, $1 \leq i \leq m$, $m < k \leq N$, $1 \leq j \leq N$, are piecewise continuous. Moreover, each nonzero $a_{ij}(t)$ satisfies $\underline{a} \leq a_{ij}(t) \leq \bar{a}$, where $\underline{a}, \bar{a}$ are positive constants and $0 < \delta_{kj}(t) \leq 1$, $j \neq k$, then $F(t)$ and $B(t)$ are bounded.*

Proof: Since $\underline{a} \leq a_{ij}(t) \leq \bar{a}$ all entries of $A_1(t)$ and $A_2(t)$ are bounded. Furthermore, from Lemma 3.3.1, we know that $(I - \Sigma_2(t))^{-1}$ exists which implies that $\det(\Sigma_2(t)) \neq 0$. Thus it follows that $\det(\Sigma_2(t))$ is bounded away from zero and all entries of the adjoint matrix of $\Sigma_2(t)$ are also bounded in terms of their continuous dependence on $\delta_{ij}(t)$. Therefore, the boundedness of $(I - \Sigma_2(t))^{-1}$ is guaranteed. Moreover $\Sigma_1(t)$ and $\Upsilon(t)$ are bounded.

Here, The error variables are defined as

$$e_i(t) = x_i(t) - \xi(t), \quad i = 1, 2, \ldots, m, \tag{3.54}$$

then the dynamics, with $\Pi(t) = \Theta - \Lambda + F(t)$, are

$$\dot{e}(t) = \Pi(t)e(t) + \xi(t)\Pi(t)\mathbf{1} - \dot{\xi}(t)\mathbf{1} + [\Lambda, B(t)]y(t), \tag{3.55}$$

Definition 3.3.1 *For the noise-free case, i. e. $w_i(t) = 0$, $i \in \mathcal{V}$, $e(t)$ is said to be uniformly exponentially convergent to a ball $\mathbf{B}(\epsilon) = \{e(t)\Re^m : \|e(t)\| \leq \epsilon\}$ with decay rate $\theta > 0$, if given $\vartheta > 0$, $\|e(t_0)\| \leq \vartheta$, then there exists $\varrho(\vartheta) > 0$ independent of t_0 such that*

$$\|e(t)\| \leq \epsilon + \varrho(\vartheta)\exp(-\theta(t - t_0)), \quad t \geq t_0.$$

Theorem 3.3.1 *Suppose that there exist constants $\nu \geq 0$, $\mu \geq 0$ such that $\|\xi(t)\| \leq \mu$, $\|\dot{\xi}(t)\|\nu$. If the conditions of Lemma 3.3.2 are satisfied, then the error $e(t)$ is exponentially convergent to the ball $\mathbf{B}(\epsilon_{\min}) = \{e : \|e\| \leq \epsilon_{min}\}$ with decay rate θ and radius*

$$\epsilon_{min} = \sqrt{\frac{1}{2\theta}(\rho\tau_1\mu + m\tau_2\nu)\lambda_{max}(Q)},$$

where ρ is the upper bound of $\|\Pi(t)\mathbf{1} + [\Lambda, B(t)]b\|^2$, Q is a positive definite matrix and τ_1, τ_2 are the solutions of the following minimization problem

$$\begin{array}{ll} \text{Minimize} & \iota_1\tau_1 + \iota_2\tau_2, \\ \text{subject to} & \begin{bmatrix} \Phi(t) & \sqrt{\mu}I & \sqrt{\nu}I \\ \sqrt{\mu}I & -\tau_1 I & 0 \\ \sqrt{\nu}I & 0 & -\tau_2 I \end{bmatrix} < 0 \end{array} \tag{3.56}$$

where $\Phi(t) = \Pi(t)Q + Q\Pi^T(t) + 2\theta Q$ and ι_1, ι_2 are positive weighting factors. And the convergence time t^* for $e(t)$ to reach and stay in $\mathbf{B}(\epsilon_{\min} + \epsilon)$ is given by

$$t^* \leq \frac{1}{\theta}\ln\frac{\omega}{\epsilon}\sqrt{\frac{\lambda_{\max}(Q)}{\lambda_{\min}(Q)}},$$

where ϵ, ω are defined in (3.57).

Remark 3.3.1 *Since it takes time for the sensors to communicate with their neighbors and compute the estimation, consensus is not expected for tracking rapidly-changed signals. Because of that we assume both the ambiance signal and its derivative are bounded.*

Proof: Consider the Lyapunov functional candidate $V(t) = e^T(t)Pe(t)$, where P is positive definite, then for any $\theta > 0$,

$$\begin{aligned}\dot{V}(t) &= e^T(t)(P\Pi(t) + \Pi^T(t)P + 2\theta P)e(t) - 2\theta e^T(t)Pe(t)\\ &\quad 2\xi(t)e^T(t)P(\Pi(t)\mathbf{1} + [\Lambda, B(t)]b) - 2\dot{\xi}e^T(t)P\mathbf{1}\\ &\leq e^T(t)\Omega(t)e(t) - 2\theta e^T(t)Pe(t) + (\rho\tau_1\mu + m\tau_2\nu),\end{aligned}$$

where $\Omega(t) = P\Pi(t) + \Pi^T(t)P + 2\theta P + (\tau_1^{-1}\mu + \tau_2^{-1}\nu)P^2$, ρ is the upper bound of $\|\Pi(t)\mathbf{1} + [\Lambda, B(t)]b\|^2$ and $b = [b_1, \ldots, b_N]^T$.

If (3.56) holds, let $P = Q^{-1}$ and pre-multiplying and post-multiplying by

$$diag\{P, I, I\}$$

to (3.56) yields $\Omega(t) < 0$. thus

$$\dot{V}(t) < -2\theta V(t) + (\rho\tau_1\mu + m\tau_2\nu),$$

Based on Gronwall's lemma, it is obtained that

$$V(t) \leq \frac{(\rho\tau_1\mu + m\tau_2\nu)}{2\theta} + V(t_0)\exp(-2\theta(t - t_0)), t \geq t_0.$$

As a consequence,

$$\begin{aligned}\|e(t)\| &\leq \sqrt{\frac{1}{2\theta}(\rho\tau_1\mu + m\tau_2\nu)\lambda_{\max}(Q)}\\ &\quad + \sqrt{V(t_0)\lambda_{\max}(Q)}\exp(-2\theta(t - t_0)), t \geq t_0.\end{aligned}$$

It follows from Definition 3.3.1 that $e(t)$ exponentially converges to the ball $\mathbf{B}(\epsilon_{\min})$ with decay rate θ.

Let

$$\omega \; > \; \max\{\|e(t_0)\|, \epsilon_{\min} + \epsilon\}, \quad \epsilon > 0. \tag{3.57}$$

From (3.57), it can easily be shown that when $t \geq t_0 + t^*$, $e(t)$ will enter and remain in the ball $\mathbf{B}(\epsilon_{\min} + \epsilon)$. This completes the proof.

It must be asserted that **Theorem** 3.3.1 is valid only in the noise free case. When noise is present, the distribution of error e(t) will change and it is worthwhile to study this point in our future work.

Corollary 3.3.1 *For type-I sensors, if $\beta_i + \phi > \alpha_i$, a necessary condition for (3.56) is as follows*

$$\sum_{j \in \mathcal{N}_i(t)} a_{ij}(t) - \sum_{j \in \mathcal{N}_i^-(t)} a_{ij}(t) \tilde{f}_{ji} > \frac{c_i}{\sigma}(\beta_i + \theta - \alpha_i),$$

where $\tilde{f}_{ji}$ are the entries of $(I - \Sigma_2(t))^{-1}\Sigma_1(t)$.

Corollary 3.3.1 characterizes the effect of network topology on the convergence of the estimator (3.49) and (3.50). It reveals that the number of positive neighbors of each type-I senor should be greater than a threshold related to its confidence, estimator gain, and decay rate. And this provides some useful suggestions on the deployment of sensors in practice. We will show this in a simulation study.

3.3.4 Sensor selection for target tracking

In WSN, it is desired to track a target [417, 418] with the least error under practical constraints resulted from the sensors used to process power measurements. However, due to the constraints of full connections, energy limitations and sensors' lifetime, a smaller set of sensors may be used at the cost of a slight increase in the position estimate error. In the GNS, M is constant in time and no criterion is used to determine its value [419]. We present a new algorithm to find the best possible set N_a with M active sensors in the network N_s with N nodes by minimizing the mean square (MS) localization error. To do so, at each time instant, the best active set is selected using the GDOP, and the size of an active set is determined using the absolute local innovations vector. This procedure determines the topology and the number of active sensors according to the tracking error.

3.3.5   Selection of best active set

To define a cost function in order to select an active set, the state transition and observation models are required. The target is considered as a point-object moving in a two dimensional plane. We consider the non-linear coordinated turn rate model [420], wherein the target moves with a nearly constant speed and an unknown turn rate. Also, $x(t) := \{x_1(t), x_2(t), \dot{x}_1(t), \dot{x}_2(t), \rho(t)\}$ is the target state vector representing the coordinates x_1, x_2, the velocities $\dot{x}_1, \dot{x}_2$, and the turn rate $\rho(t)$. Then, we have

$$
x(t+1) = \begin{bmatrix}
1 & 0 & \frac{\sin \rho(t)\Delta\tau}{\rho(t)} & -\frac{1-\cos \rho(t)\Delta\tau}{\rho(t)} & 0 \\
0 & 1 & -\frac{1-\cos \rho(t)\Delta\tau}{\rho(t)} & \frac{\sin \rho(t)\Delta\tau}{\rho(t)} & 0 \\
0 & 0 & \cos \rho(t)\Delta\tau & -\sin \rho(t)\Delta\tau & 0 \\
0 & 0 & \sin \rho(t)\Delta\tau & \cos \rho(t)\Delta\tau & 0 \\
0 & 0 & 0 & 0 & 1
\end{bmatrix} x(t)
$$

$$
+ \begin{bmatrix}
\frac{\Delta\tau^2}{2} & 0 & 0 \\
0 & \frac{\Delta\tau^2}{2} & 0 \\
\Delta\tau & 0 & 0 \\
0 & \Delta\tau & 0 \\
0 & 0 & 1
\end{bmatrix} v(t) \tag{3.58}
$$

where $v(t)\ N(0,Q)$ is the motion noise and Dt denotes the length of a time step. The process noise level $Q = diag[\sigma_1^2, \sigma_2^2, \sigma_\rho^2]$ is assumed to be known. The correct noise level depends on the expected turn rate range. Also, it controls the trade-off between tracking of a constant velocity and a maneuvering target.

The measurements are a function of the relative distance between the sensor (x_1^s, x_2^s) and the target. We assume that the sensors can measure the power of the target's signal which decays exponentially relative to the distance. The target-originated measurements follow the log-normal shadowing model defined as [421].

$$
\begin{aligned}
z_s(t) &= h_s(d_s(x(t)), t) + w_s(t) \\
&= K - 10\gamma \log((x_1 - x_1^s)^2 + (x_2 - x_2^s)^2)^{1/2} + w_s(t)
\end{aligned}
\tag{3.59}
$$

where $w_s(t)$ is a zero-mean i.i.d. Gaussian observation noise with variance R_s which accounts for the shadowing effects and other uncertainties. Also, the sensor noise is uncorrelated, K is the transmission power, and $\gamma \in [2,5]$ is the path loss exponent. These parameters depend on the radio environment, antenna characteristics, terrain, and so on.

Assuming M active sensors at time t, $z(t)$ is partitioned to M sub-vectors corresponding to the observations made by each individual sensor

$$z(t) \;=\; [z_1^T(t), \ldots, z_M^T(t)]^T \tag{3.60}$$

and the covariance matrix is updated as

$$P^{-1}(t|t) \;=\; P^{-1}(t|t-1) + \underbrace{\sum_{j=1}^{M} I_j(t)}_{I} \tag{3.61}$$

where the information matrix I is

$$I \;=\; \sum_{j=1}^{M} \nabla h_j^T(t) R_j^{-1} \nabla h_j(t) \tag{3.62}$$

The observation noise variance is equal for all sensors. If the position of the jth sensor relative to the predicted target location in polar coordinates, (r_j, ϕ_j) is defined as

$$\begin{cases} \phi_j = \arctan \frac{x_2 - x_2^j}{x_1 - x_1^j} \\ r_j = ((x_1 - x_1^s)^2 + (x_2 - x_2^s)^2)^{1/2} \end{cases} \tag{3.63}$$

then ∇h_j is given by

$$\nabla h_j \;=\; -10\gamma \log e \times \begin{bmatrix} \frac{\cos \phi_j}{r_j} & \frac{\sin \phi_j}{r_j} & 0 & 0 & 0 \end{bmatrix} \tag{3.64}$$

As a result, I is obtained as

$$I \;=\; \begin{bmatrix} J & 0 \\ 0 & 0 \end{bmatrix};$$

$$J \;=\; \frac{(10\gamma \log e)^2}{\sigma^2} \sum_{j=1}^{M} \frac{1}{r_j^2} \begin{bmatrix} \cos^2 \phi_j & \cos \phi_j \sin \phi_j \\ \cos \phi_j \sin \phi_j & \sin^2 \phi_j \end{bmatrix}$$

$$\tag{3.65}$$

where J is the measurement Fisher information matrix, which represents the inverse state covariance error if no prior information is used. Then, the cost function for the sensor selection is

$$\rho(N_a) \;=\; \frac{\mathrm{tr}\{J\}}{\det\{J\}} \tag{3.66}$$

In (3.66), it is assumed that the target position as well as the locations of all the sensors with respect to the target is already known. In practice, the locations of all the sensors are, a priori, known and the predicted target state obtained from the tracking algorithm is used to approximate the distances between the target and the sensors. By ignoring the prior information, it is desirable to locate the sensors as close as possible to the target with the best angular diversity Bar-Shalom:01. The simplest method is an exhaustive search that evaluates the $\rho(N_a)$ via (3.66) for all the possible active sets of size M given a set of possible sensors of size N. Then, the set N_a with the smallest $\rho(N_a)$ is chosen. The number of combinations is

$$N_c \;=\; \frac{N!}{M!(N-M)!} \tag{3.67}$$

This algorithm is only viable if N is small. Otherwise, the suboptimal approaches described below may be used.

3.3.6   Global node selection

The GNS algorithm [419] incorporates a greedy strategy called 'add one sensor node at a time' for bearing-only sensors. This approach first selects the optimal active subset of two sensors by an exhaustive search that minimizes (3.66). Then, one sensor is added at a time to minimize (3.66). This process is repeated until the active subset contains M sensors. The greedy approach is near optimal for bearing-only sensors such that non-optimality of active sensor selection is approximately 5%. In order to evaluate the performance of this search method for power measurement sensors, we consider a network of 16 sensors randomly deployed in a $50 \times 50 m^2$ field with 1000 different configurations.

3.3.7   Spatial split

The spatial split (SS) algorithm is a modification of the Closest approach [419] in which M active sensors are selected using the closeness of sensors' locations to the target location estimate. This algorithm does not exploit the angular diversity of sensors in computations. Here, the SS algorithm is modified by incorporating the angular diversity into account. To do so, a sensing region is defined by a circle centered at the target location estimate with a radius of the maximum sensing range. Then, the following stages are performed:

- Since the GDOP measure is too low for farther apart sensors, the sensing region is split into some sectors equal to M.

- Considering the increase of MS position error by the sensors apart from the target, the closest sensor of each sector to the target position estimate is selected.

3.3.8 Computational complexity

In the GNS, the complexity of the initial search which first selects two active sensors is $O(N^2)$ for N possible sensors. Then, the added sensor at each time step considers the complexity of $O(N^2 - m)$ in which m is the number of selected sensors until the present time. Then, the total number of search points is given by

$$\frac{N(N-1)}{2} + \sum_{m=2}^{M-1} N - m \tag{3.68}$$

In the SS, the complexity of the selection of one sensor in each sector is $O(N/M)$ and its overall complexity is approximately $O(N)$.

3.3.9 Number of active sensors

The ASS algorithm is introduced in which in each instant the number of active sensors is selected using $v_j(t)$ in (3.48). In this way, considering the initial value M_0 for the size of the active set and determining the best active set with this size, the target tracking is begun. Then, $|v_j(t)|$ is compared to the maximum allowed absolute error, v_{max}. When $|v_j(t)|$ is larger than v_{max}, in the next instant the number of active sensors is increased by one. Otherwise, it is decreased and the target tracking continues.

3.3.10 Simulation results

It is assumed that the sensors can communicate with each other, share some cumulative statistics with no communication loss, are synchronized, know their positions, and share common information such as, a prior, density and motion model. The network consists of 32 sensors randomly scattered over a field of $100 \times 100 m^2$.

Also, $K = 9$ dBm, $\gamma = 3$, $\Delta \tau = 1s$, $R_s = 0.4$, and $Q = 0.002^2 \text{diag}[1, 1, 1]$. The prior estimate is $\hat{x}(0|0) = x(0) + x_{bias}$, where x_{bias} presents a strong bias on the position drawn from a uniform distribution on a square of length $30m$ centered at [0,0].

The energy consumption provides the energy required for a sensor to reach another sensor in a single hop [68]. Figure 3.21 compares different approaches for 1000 Monte Carlo trials against the average number of active sensors for both trajectories. In the ASS, to find the best active set of size M the GNS and SS algorithms are considered. In all cases, the DEIF is applied for target tracking. From the results, the average number of active sensors increases and the RMS error decreases for all the approaches. As seen, in the ASS, the minimum RMS error is achieved in average using four active sensors, while the other algorithms must activate more than 10 sensors for the same error.

In a similar way, Figure 3.22 compares the localization performance against the energy consumption. It is shown that there is a trade-off between the RMS error and the energy usage. Also, for a minimum RMS error, the energy consumption of ASS is less than that of the others. Moreover, the SS shows the worst performance; however, the ASS with SS compared to the ASS with GNS performs very similar.

3.4 Multi-Sensor Management

Multi-sensor systems are becoming increasingly important in a variety of military and civilian applications. Since a single sensor generally can only perceive limited partial information about the environment, multiple similar and/or dissimilar sensors are required to provide sufficient local pictures with different focus and from different viewpoints in an integrated manner. Further, information from heterogeneous sensors can be combined using data fusion algorithms to obtain synergistic observation effects. Thus, the benefits of multi-sensor systems are to broaden machine perception and enhance awareness of the state of the world compared to what could be acquired by a single sensor system.

With the advancements in sensor technology, sensors are becoming more agile. Also, more of them are needed in a scenario in response to the increasingly intricate nature of the environment to be sensed. The increased sophistication of sensor assets along with the large amounts of data to be processed has pushed the information acquisition problem far beyond what can be handled by a human operator. This motivates the emerging interest in research into automatic and semiautomatic management of sensor resources for improving overall perception performance beyond the basic fusion of data.

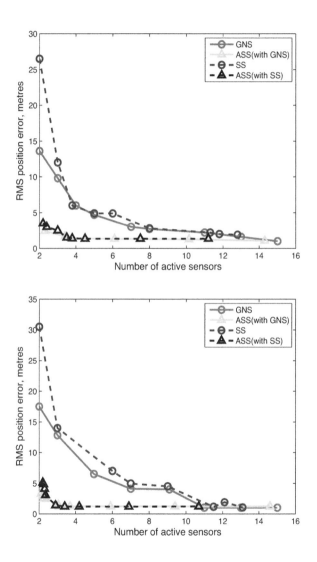

Figure 3.21: RMS position error against the average number of active sensors for a straight trajectory (top) and a maneuvering trajectory (bottom).

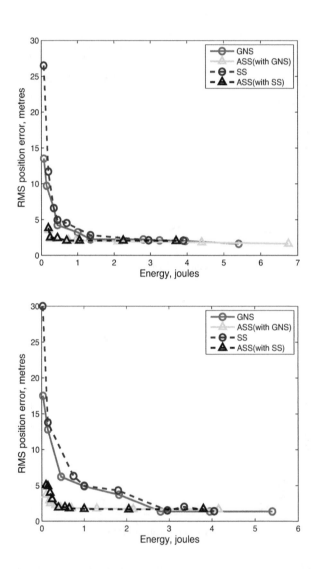

Figure 3.22: RMS position error against the energy consumption for a straight trajectory (top) and a maneuvering trajectory (bottom).

3.4.1 Primary purpose

Multi-sensor management is formally described as a system or process that seeks to manage or coordinate the usage of a suite of sensors or measurement devices in a dynamic, uncertain environment, to improve the performance of data fusion and ultimately that of perception. It is also beneficial to avoid overwhelming storage and computational requirements in a sensor and data rich environment by controlling the data gathering process that only the truly necessary data are collected and stored [131]. The why and what issues of both single-sensor and multi-sensor management were thoroughly discussed in [114], [115], [127], [128] and [129]. To reiterate, the basic objective of sensor management is to select the right sensors to do the right service on the right object at the right time. The sensor manager is responsible for answering questions like:

- Which observation tasks are to be performed and what are their priorities?

- How many sensors are required to meet an information request?

- When are extra sensors to be deployed and in which locations?

- Which sensor sets are to be applied to which tasks?

- What is the action or mode sequence for a particular sensor?

- What parameter values should be selected for the operation of sensors?

The simplest job of sensor management is to choose the optimal sensor parameter values given one or more sensors with respect to a given task, see for example, the chapter in [134]. This is also called active perception, where sensors are to be configured optimally for a specific purpose. More general problems of (multi-)sensor management are, however, related to decisions about what sensors to use and for which purposes, as well as when and where to use them. Widely acknowledged is the fact that it is not realistic to continually observe everything in the environment, and therefore selective perception becomes necessary, requiring the sensor management system to decide when to sense what, and with which sensors. Typical temporal complexities, which must be accommodated in the sensor management process, were discussed in [125].

3.4.2 Role in information fusion

Sensor management merits incorporation in the information fusion processes. Although terminology has not yet fully stabilized, it is generally acknowledged that information fusion is a collective concept comprising situation assessment (level

2), threat or impact assessment (level 3) and process refinement (level 4) in the so–called JDL model of data fusion [136]. As pointed out in [133], in addition to intelligence interpretation, information fusion should be equipped with techniques for proactive or reactive planning and management of collection resources such as sensors and sensor platforms, in order to make the best use of these assets with respect to identified intelligence requirements. Sensor management, aiming at improving data fusion performance by controlling sensor behavior, plays the role of level 4 functions in the JDL model.

Sensor management indeed provides information feedback from data fusion results to sensor operations [115], [128]. The representation of the data fusion process as a feedback closed-loop structure is depicted in Figure 3.23, where the sensor manager on level 4 uses the information from levels $0 \longrightarrow 3$ to plan future sensor actions. The feedback is intended to improve the data collection process with expected benefits of earlier detection, improved tracking, and more reliable identification, or to confirm what might be tactically inferred from previously gathered evidence. Timeliness is a necessary requirement on the feedback management of sensors for fast adaptation to environment changes. That is to say, a prompt decision on sensor functions has to be made before the development of the tactical situation has made such a decision obsolete.

As a categorization of process refinement, Steinberg and Bowman [132] classified responses of resources (including sensors) as reflexive, feature-based, entity relation based, context-sensitive, cost-sensitive, and reaction- sensitive, in terms of input data/information types. This categorization is viewed as an expansion of Dasarathy's model [117] in which data fusion functions are subdivided considering merely data, features, and objects as possible input/output types.

3.4.3 Architecture classes

The architecture of a multi-sensor management system is closely related to the form of data fusion unit. Typically there are three alternatives for a system structure, namely,

1. **Centralized:** In a centralized paradigm the data fusion unit is treated as central mechanism. It collects information from all the different platforms and sensors, and decides which jobs must be accomplished by the individual sensors.

   All commands sent from the fusion center to the respective sensors must be accepted and followed with the proper sensor actions.

2. **Decentralized:** In a decentralized system data are fused locally with a set of local agents rather than by a central unit. In this case, every sensor or plat-

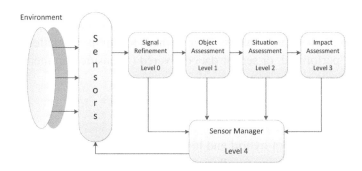

Figure 3.23: Feedback connection in a data fusion process.

form can be viewed as an intelligent asset having some degree of autonomy in decision-making. Sensor coordination is achieved based on communication in the network of agents, in which sensors share locally fused information and cooperate with each other. Durrant-Whyte and Stevens [118] stated that decentralized data fusion exhibits many attractive properties by being:

- scalable in structure without being constrained by centralized computational bottlenecks or communication bandwidth limitations;

- survivable in the face of on-line loss of sensing nodes and to dynamic changes of the network;

- modular in the design and implementation of fusion nodes.

However, the effect of redundant information is a serious problem that may arise in decentralized data fusion networks [116]. It is not possible, within most filtering frameworks, to combine information pieces from multiple sources unless they are independent or have known cross-covariance [121]. Moreover, without any common communication facility, data exchange in such a network must be carried out strictly on a node-to-node basis. A delay between the sender and receiver could result in transient inconsistencies of the global state among different parts of the network, causing degradation of overall performance [120].

3. **Hierarchical:** This can be regarded as a mixture of centralized and decentralized architectures. In a hierarchical system there are usually several levels of hierarchy in which the top level functions as the global fusion center and the lowest level consists of several local fusion centers [129]. Every

local fusion node is responsible for management of a sensor subset. The partitioning of the whole sensor assembly into different groups can be realized based on either sensors' geographical locations or platforms, sensor functions performed, or sensor data delivered (to ensure commensurate data from the same sensor group).

3.4.4   Hybrid and hierarchical architectures

Two interesting instances of hybrid and hierarchical sensor management architectures are given in the following for illustration.

The macro/micro architecture proposed by [115] can be classified as a two-level hierarchical system. It consists of a macro sensor manager playing a central role and a set of micro sensor managers residing with respective sensors. The macro sensor manager is in charge of high level strategic decisions about how to best utilize the available sensing resources to achieve the mission objectives. The micro sensor manager schedules the tactics of a particular sensor to best carry out the requests from the macro manager. Thus it is clear that every managed sensor needs its own micro manager.

Another hybrid distributed and hierarchical approach was suggested in [131] for sensor-rich environments exemplified by an aircraft health and usage monitoring system. The main idea is to distribute the management function across system functional or physical boundaries with global oversight of mission goals and information requests. One such model for management of numerous sensors is shown in Figure 3.24, see [131]. At the top of the model is the *mission manager* tasked with converting mission goals to information needs, which are then mapped by the *information instantiator* into a set of measurement patterns in accordance with those needs. The role of the *meta-manager* is to enable natural subdivision of a single manager into a set of mostly independent *local resource managers*, each being responsible for a particular sensor subset. Occasionally, these local managers need to be coordinated by the meta-manager if there is a request for information which cannot be satisfied by a single sensor suite. A major difficulty in implementing this hybrid architecture of sensor management lies with the meta-manager. It is not yet obvious how to best translate global functional needs into a set of local resource managers and how to coordinate the disparate local managers distributed across functional or physical boundaries.

3.4.5   Classification of related problems

Multi-sensor management is a broad concept referring to a set of distinct issues of planning and control of sensor resource usage to enhance multi-sensor data

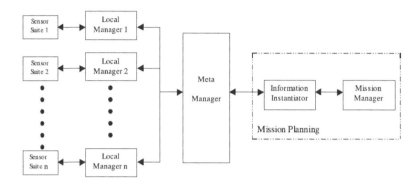

Figure 3.24: A distributed and hierarchical sensor management model.

fusion performance. Various aspects of this area have been discussed in chapters in the open literature. Generally, these problems fall into three main categories, i.e., sensor deployment, sensor behavior assignment, and sensor coordination.

1. *Sensor deployment:* Sensor deployment is a critical issue for intelligence collection in an uncertain dynamic environment. It concerns making decisions about when, where, and how many sensing resources need to be deployed in reaction to the state of the world and its changes. In some situations, it should be beneficial to proactively deploy sensing resources according to a predicted situation development tendency in order to get prepared to observe an event which is likely to happen in the upcoming period.

   Sensor placement [130] needs special attention in sensor deployment. It consists of positioning multiple sensors simultaneously in optimal or near optimal locations to support surveillance tasks when necessary. Typically it is desired to locate sensors within a particular region determined by tactical situations to optimize a certain criterion usually expressed in terms of global detection probability, quality of tracks, etc. This problem can be formulated as one of constrained optimization of a set of parameters. It is subject to constraints due to the following factors:

   - sensors are usually restricted to specified regions due to tactical considerations;
   - critical restrictions may be imposed on relative positions of adjacent sensors to enable their mutual communication when sensors are arranged as distributed assets in a decentralized network;
   - the amount of sensing resources that can be positioned in a given period is limited due to logistical restrictions.

In simple cases, decisions on sensor placement are to be made with respect to a well-prescribed and stationary environment. As examples, we may consider such application scenarios as:

- placing radars to minimize the terrain screening effect in detection of an aircraft approaching a fixed site;

  arrangement of a network of intelligence gathering assets in a specified region to target another well-defined area.

In the above scenarios, mathematical or physical models such as terrain models, propagation models, etc. are commonly available and they are used as the basis for evaluation of sensor placement decisions. More challenging are those situations in which the environment is dynamic and sensors must repeatedly be repositioned to be able to refine and update the state estimation of moving targets in real time. Typical situations where reactive sensor placement is required are:

- submarine tracking by means of passive sonobuoys in an anti-submarine warfare scenario;
- locating moving transmitters using ESM (electronic support measures) receivers;
- tracking of tanks on land by dropping passive acoustic sensors.

2. *Sensor behavior assignment*

The basic purpose of sensor management is to adapt sensor behavior to dynamic environments. By sensor behavior assignment is meant efficient determination and planning of sensor functions and usage, according to changing situation awareness or mission requirements. Two crucial points are involved here:

(a) Decisions about the set of observation tasks (referred as system-level tasks) that the sensor system is supposed to accomplish currently or in the near future, on grounds of the current/predicted situation as well as the given mission goal;

(b) Planning and scheduling of actions of the deployed sensors to best accomplish the proposed observation tasks and their objectives.

Owing to limited sensing resources, it is prevalent in real applications that available sensors are not able to serve all of the desired tasks and achieve all their associated objectives simultaneously. Therefore a reasonable compromise between conflicting demands is sought. Intuitively, more urgent or important tasks should be given higher priority in their competition for resources. Thus a scheme is required to prioritize observation tasks. Information about task priority can be very useful in scheduling of sensor actions and for negotiation between sensors in a decentralized paradigm.

To concertize this class of problems, let us consider a scenario including a number of targets as well as multiple sensors, which are capable of focusing on different objects with different modes for target tracking and/or classification. The first step for the sensor management system should be to utilize evidence gathered to decide objects of interest and to prioritize which objects to look at in the time following. Subsequently, in the second step, different sensors, together with their modes, are allocated across the interesting objects to achieve the best situation awareness. In fact, owing to the constraints on sensor and computational resources, it is in general not possible to measure all targets of interest with all sensors in a single time interval. Also, improvement of the accuracy on one object may lead to degradation of performance on another object. What is required is a suitable compromise among different targets.

It is worth noting that although several distinct terms appear in the literature such as sensor action planning in [124], sensor selection in [119], [122], [123], as well as sensor-to-task assignment in [126], [127], these terms inherently signify the same aspect of distributing resources among observation tasks, thus they belong to the second issue of this problem class. In this chapter we present the more general concept, sensor behavior assignment, which involves not only the arrangement of operations for individual sensors but also inferences about system-level tasks and objectives to be accomplished. Actually, specification of tasks at the system level can be considered as postulating expected overall behaviors of the perception system as a whole, while planning and scheduling of the sensor actions define, the local behaviors residing with specific sensors. Dynamic information associated with time-varying utility and availability serves here as the basis for decision making about sensor behaviors.

3. *Sensor coordination in a decentralized sensor network*

There are two general ways to integrate a set of sensors into a sensor network. One is the centralized paradigm, where all actions of all sensors are decided by a central mechanism. The other alternative is to treat sensors in the network as distributed intelligent agents with some degree of autonomy

[135]. In such a decentralized architecture, bi-directional communication between sensors is enabled, so that communication bottlenecks possibly existing in a centralized network can be avoided. A major research objective of decentralized sensor management is to establish cooperative behavior between sensors with no or little external supervision.

3.5   Notes

This chapter provided a review of distributed estimation techniques used for dynamical systems. Special emphasis is observed for sensor networks over the last few years. The distributed Kalman filter proposed in [515] is applied on a second order dynamical system in a scenario of ten sensor nodes. The sensor nodes try to estimate the states of the dynamical system with embedded consensus filters. The results show that the distributed estimation algorithm approximates the central Kalman filter. It is concluded that the distributed estimation techniques for distributed dynamical systems require further extensive research.

A distributed estimation approach was considered in WSN using the extended information filter for target tracking and a cost function based on the GDOP for sensor selection. The convergence properties of a distributed estimator for heterogeneous WSNs are addressed. Accordingly, the ASS algorithm is presented which incorporates the absolute error between the measured and predicted observations in selection of the number of active sensors as well as the best topology for the least possible tracking error. The number of active sensors was adaptively determined based on the absolute local innovations vector. Simulation results revealed that the ASS with SS achieves the similar performance as the ASS with GNS.

3.6   Proposed Topics

1. An interesting research topic that arises naturally in wireless sensor networks concerns the combined sensor placement and sensor selection in multi-sensor and multi-target applications. Building up an analytical model that describes this problem would definitely smooth out several standing issues.

2. Adaptive signal filtering with respect to sensor placement is considered a very attractive research topic. An adaptive mechanism is definitely desirable to decide which filter to use based on the situation and the mission goal.

3. It has been reported that with sensor collaboration, potentially powerful wireless sensor networks (WSNs), comprising a large number of geographically distributed nodes characterized by low power constraints and limited

computation capability, can be constructed in principle to monitor and control environments. A major problem which arises is that *bandwidth is limited*, necessitating the estimator to be formed using quantized versions of the original observations. In this setup, quantization becomes an integral part of the estimation process. It is therefore of great interest to study the area of bandwidth-constrained distributed mean location parameter estimation in additive white Gaussian noise (AWGN). One objective would be to seek maximum-likelihood estimators (MLEs) and benchmark their variances with the CramerRao lower bound (CRLB) that, at least asymptotically, is achieved by the MLE.

Chapter 4

Distributed Kalman Filtering

The problem of distributed Kalman filtering (DKF) for sensor networks is one of the most fundamental distributed estimation problems for scalable sensor fusion. This chapter addresses the DKF problem by reducing it to two separate dynamic consensus problems in terms of weighted measurements and inverse-covariance matrices. These two data fusion problems are solved is a distributed way using low-pass and band-pass consensus filters. Consensus filters are distributed algorithms that allow calculation of average-consensus of time-varying signals. It is shown that a central Kalman filter for sensor networks can be decomposed into n micro-Kalman filters with inputs that are provided by two types of consensus filters. This network of micro-Kalman filters collectively are capable of providing an estimate of the state of the process (under observation) that is identical to the estimate obtained by a central Kalman filter, given that all the nodes agree on two central sums. Later, we demonstrate that our consensus filters can approximate these sums, and that gives an approximate distributed Kalman filtering algorithm.

4.1   Introduction

In a hi-tech environment, a strict surveillance unit is required for an appropriate supervision. It often utilizes a group of distributed sensors which provide information about the local targets. Compared to the centralized Kalman filtering (CKF), which can be used in mission critical scenarios, where every local sensor is important with its local information, the distributed fusion architecture has many advantages. There is no second thought that in certain scenarios, a centralized Kalman filter plays a major role, and it involves minimum information loss. A general structure for the Distributed Kalman Filter (DKF) can be seen in Figure 4.1.

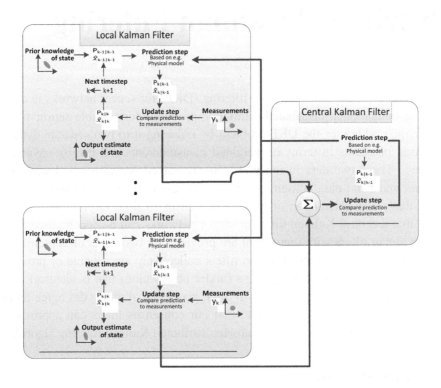

Figure 4.1: A general structure of distributed Kalman filter.

The distributed system architecture, on the whole, is very powerful since it allows the design of the individual units or components to be much simpler, while not compromising too much on the performance. Additional benefits include increased robustness to component loss, increased flexibility in that the components can be reconfigured for many different tasks, and so on. However, the design of such systems challenges various problems of assumptions, handling, and fusing the architecture of such systems. Our purpose is to provide a bibliographic survey of DKF and its architectures, comprised of distribution, fusion, filtering and estimation.

4.2 Distributed Kalman Filtering Methods

The subject of distributed Kalman filtering can be introduced through alternative viewpoints. In this section, we restrict attention to two aspects:

- Different methods that would eventually promote, better filtering approach,

- Various scenarios of applications will be considered.

Other aspects will be discussed in other sections of the book.

4.2.1 Different methods

Under uncertain observations, a method which includes a measurement with a false alarm probability is considered as a special case in [606], and randomly variant dynamic systems with multiple models are considered in [228]. Optimal centralized and distributed fusers are algebraically equivalent in this case [229]. Looking at mode estimation in power systems, a trust-based distributed Kalman filtering approach to estimate the modes of power systems is presented in [230]. Using the standard Kalman filter locally together with a consensus step in order to ensure that the local estimates agree is shown in [231]. A frequency-domain characterization of the distributed estimator's steady-state performance is presented in [40]. A version of extended Kalman filtering to globally optimal Kalman filtering for the dynamic systems with finite-time correlated noises is shown in [234]. Distributed Kalman-type processing schemes essentially make use of the fact that the sensor measurements do not enter into the update equation for the estimation error covariance matrices, that is, covariance matrices of all the sensors calculated at each individual sensor site without any further need of communication are presented in [235]. Also, in distributed fusion Kalman filtering, a weighted covariance approach is reported in [236]. Distributed Kalman filtering fusion with passive packet loss or initiative intermittent communications from local estimators to a fusion center while the process noise exists, is presented in [237]. For each Kalman update, an infinite number of consensus steps as demonstrated [238, 239]. For each Kalman update, state estimates additionally exchanged, are presented in [231]. When only the estimates at each Kalman update over-head are exchanged, the results are reported in [267]. Analysis of the number of messages used to exchange between successive updates in a distributed Kalman filter is documented in [240]. Global optimality of distributed Kalman filtering fusion exactly equal to the corresponding centralized optimal Kalman filtering fusion, is shown in [241]. A parallel and distributed state estimation structure is developed in the form of a hierarchical estimation structure is specified in [257]. A computational procedure to transform a hierarchical Kalman filter into a partially decentralized estimation structure is presented in [242]. An optimally distributed Kalman filter based on a-priori determination of measurements is given in [243].

Estimation of sparsely connected, large scale systems is reported in [244] and an n-th order with multiple sensors presentation is shown in [245]. Data-fusion over arbitrary communication networks is shown in [246]. Iterative consensus protocols are provided in [247]. Using bipartite fusion graphs, the issue of how DKF is performed is the subject of [248]. Local average consensus algorithms for DKF are shown in [249]. Consensus strategies for DKF are reported in [250]. Semidefinite programming based consensus iterations, developed for DKF, are shown in [251]. Converge speed of consensus strategies, is given in [252]. Distributed Kalman filtering, with a focus on limiting the required communication bandwidth, is shown in [266]. Distributed Kalman-type processing schemes, which provide optimal track-to-track fusion results at arbitrarily chosen instants of time, are developed in [253]. Distributed architecture of track-to-track fusion for computing the fused estimate from multiple filters tracking a maneuvering target with the simplified maximum likelihood estimator, are presented in [254]. The original batch form of the Maximum Likelihood (ML) estimator, is developed in [255] and a modified probabilistic neural network is shown in [256].

Remark 4.2.1 *In [237], an ℓ-sensor distributed dynamic system is described by:*

$$x_{k+1} = \phi_k x_k + v_k, k = 0, 1, ... \tag{4.1}$$

$$y_k^i = H_k^i x_k + w_k^i, i = 1, ..., \ell \tag{4.2}$$

where ϕ_k is a matrix of order $r \times r$, x_k, $v_k \in \mathcal{R}^r$, $H_i^k \in \mathcal{R}^{N_i \times r}$, y_k^i, $w_i^k \in \mathcal{R}^{N_i}$. The process noise v_k and measurement noise w_i^k are both zero-mean random variables independent of each other temporally but w_i^k and w_j^k may be cross-correlated for $i \neq j$ at the same time instant k.

To compare performances between the centralized and distributed filtering fusion, the stacked measurement equation is written as:

$$y_k = H_k x_k + w_k \tag{4.3}$$

where

$$y_k = (y_k^{1^t}, ..., y_k^{\ell^t})^t, H_k = (H_k^{1^t}, ..., H_k^{\ell^t})^t,$$
$$w_k = (w_k^{1^t}, ..., w_k^{\ell^t})^t \tag{4.4}$$

and the covariance of the noise w_k is given by:

$$Cov(w_k) = R_k, R_k^i = Cov(w_k^i), \quad i = 1, ..., \ell \tag{4.5}$$

where R_k and R_k^i are both invertible for all i. According to the standard results of Kalman filtering, the local Kalman filtering at the i-th sensor is expressed as:

$$\widehat{K}_k^i = \widehat{P}_{k/k}^i H_k^{i^t} \widehat{R}_k^{i-1} \tag{4.6}$$

$$\widehat{x}_{k/k}^i = \widehat{x}_{k/k-1}^i + \widehat{K}_k^i (y_k^i - H_k^i \widehat{x}_{k/k-1}^i) \tag{4.7}$$

$$\widehat{P}_{k/k}^i = \widehat{P}_{k/k-1}^i - \widehat{K}_k^i H_k \widehat{P}_{k/k-1}^i \tag{4.8}$$

$$\tag{4.9}$$

where the covariance of filtering error can be stated as:

$$\widehat{P}_{k/k}^{i-1} = \widehat{P}_{k/k-1}^{i-1} + H_k^{i^t} \widehat{R}_k^{i-1} H_k^i \tag{4.10}$$

with

$$\widehat{x}_{k/k-1}^i = \widehat{\Phi}_k \widehat{x}_{k-1/k-1}^i,$$
$$\widehat{P}_{k/k}^i = E[(\widehat{x}_{k/k}^i - \widehat{x}_k)(\widehat{x}_{k/k-1}^i - \widehat{x}_k)^t]$$
$$\widehat{P}_{k/k-1}^i = E[(\widehat{x}_{k/k-1}^i - \widehat{x}_k)(\widehat{x}_{k/k-1}^i - \widehat{x}_k)^t]$$

Similarly, the centralized Kalman filtering with all sensor data is given by:

$$\widehat{K}_k = \widehat{P}_{k/k} H_k^t \widehat{R}_k^{-1} \tag{4.11}$$

$$\widehat{x}_{k/k} = \widehat{x}_{k/k-1} + \widehat{K}_k(y_k - H_k \widehat{x}_{k/k-1}) \tag{4.12}$$

$$\widehat{P}_{k/k} = \widehat{P}_{k/k-1} - \widehat{K}_k H_k \widehat{P}_{k/k-1} \tag{4.13}$$

$$\tag{4.14}$$

where, the covariance of filtering error can be described as:

$$\widehat{P}_{k/k}^{-1} = \widehat{P}_{k/k-1}^{-1} + H_k^{t} \widehat{R}_k^{-1} H_k \tag{4.15}$$

with

$$\widehat{x}_{k/k-1} = \widehat{\Phi}_k \widehat{x}_{k-1/k-1},$$
$$\widehat{P}_{k/k} = E[(\widehat{x}_{k/k} - \widehat{x}_k)(\widehat{x}_{k/k-1} - \widehat{x}_k)^t]$$
$$\widehat{P}_{k/k-1} = E[(\widehat{x}_{k/k-1} - \widehat{x}_k)(\widehat{x}_{k/k-1} - \widehat{x}_k)^t]$$

It is quite clear when the sensor noises are cross-dependent that

$$H_k^t \widehat{R}_k^{-1} H_k = \sum_{i=1}^{l} H_k^{it} \widehat{R}_k^{i-1} H_k^i$$

Likewise, the centralized filtering and error matrix could be explicitly expressed in terms of the local filtering and error matrices as follows:

$$\widehat{P}_{k/k}^{-1} = \widehat{P}_{k/k-1}^{-1} + \sum_{i=1}^{l} (\widehat{P}_{k/k}^{i-1} - \widehat{P}_{k/k-1}^{i-1}) \tag{4.16}$$

and

$$\begin{aligned} \widehat{P}_{k/k}^{-1} \widehat{x}_{k/k} &= \widehat{P}_{k/k-1}^{-1} \\ &+ \sum_{i=1}^{l} (\widehat{P}_{k/k}^{i-1} \widehat{x}_{k/k}^{i} - \widehat{P}_{k/k-1}^{i-1} \widehat{x}_{k/k-1}^{i}) \end{aligned} \tag{4.17}$$

Also,

$$H_k^{i'} \widehat{R}_k^{i-1} y_k^i = \widehat{P}_{k/k}^{i-1} \widehat{x}_{k/k}^{i} - \widehat{P}_{k/k-1}^{i-1} \widehat{x}_{k/k-1}^{i} \tag{4.18}$$

In what follows, we are going to deal with the practical situation in which the local sensors may fail to send their estimates to the fusion center. In this case, the measurement equation of the corresponding centralized multi-sensor system has to be modified, that is, the original multiple individual observations should be stacked as a modified single observation.

4.2.2   Pattern of applications

A large amount of research has been carried out in the framework of modified filters. Multi-sensor networks are developed that are amenable to parallel processing in [258]. Then, a two-sensor fusion filter system has been applied in [259], followed by federated square root filter in [260]. Fusion filters are developed for linear time-invariant (LTI) systems with correlated noises and multi-channel ARMA signals, respectively in [261] and [262]. Fusion de-convolution estimators for the input of white noise are worked out in [263]-[264]. Distributed Kalman filtering for cooperative localization is re-formulated as a parameter estimation problem in [265]. DKF techniques for multi-agent localization is dealt with in [266, 267]. Collaborative processing of information, and gathering scientific data from spatially distributed sources is described in [268]. Particle filter implementations using Gaussian approximations are documented in [286]. Channel estimation method based on the recent methodology of distributed compressed sensing (DCS) and frequency domain Kalman filter is worked out in [287]. Algorithms for distributed Kalman filtering, where global information about the state covariances is required in order to compute the estimates are shown in [266]. The synthesis of a distributed algorithm to compute weighted least squares estimates with sensor measurements correlated is presented in [288]. Distributive and efficient computation of linear minimum mean square error (MMSE) for the multi-user detection problem is presented in [289]. A statistical approach for calculating the exact PDF approximated by a well-placed Extended Kalman Filter is presented in [290]. A distributed object tracking system which employs a cluster-based Kalman filter in a network of wireless cameras is presented in [269]. A distributed recursive mean-square error (MSE) optimal quantizer-estimator based on the quantized observations is presented in [291] [292]. Designing a communication access protocol for wireless sensor networks tailored to converge rapidly to the desired estimate and provides scalable error performance is presented in [293], [294]. A decentralized versions of the Kalman filter are presented in [40]. A novel distributed filtering/smoothing approach, flexible to trade-off estimation delay for MSE reduction while enhancing robustness, is presented in [270]. In distributed estimation agents, where a bank of local Kalman filters is embedded into each sensor and a diagnosis decision is performed by a distributed hypothesis testing consensus method is presented in [271]. The state estimation of dynamical stochastic processes based on severely quantized observations is reported in [303, 295]. A scheme for approximate DKF that is based on reaching an average-consensus is presented in [40]

In the multi-sensor random parameter matrices case [606], sometimes, even if the original sensor noises are mutually independent, the sensor noises of the converted system are still cross-correlated. Hence, such multi-sensor system seems not satisfying the conditions for the distributed Kalman filtering fusion given in [607]-[608]. It was proved that when the sensor noises or the random measurement matrices of the original system are correlated across sensors, the sensor noises of the converted system are cross-correlated. Even if so, similarly with [232], centralized random parameter matrices Kalman filtering, where the fusion center can receive all sensor measurements, can still be expressed by a linear combination of the local estimates. Therefore, the performance of the distributed filtering fusion is the same as that of the centralized fusion under the assumption that the expectations of all sensor measurement matrices are of full row rank. When there is no feedback from the fusion center to local sensors, a distributed Kalman filtering fusion formula under a mild condition is presented as [273]. A rigorous performance analysis for Kalman filtering fusion with feedback is presented in [274].

Low-power DKF based on a fast polynomial filter is shown in [275]. Consensus problems and their special cases are reported in [276]. DKF for sparse large-scale systems monitored by sensor networks is treated in [277]. DKFs to estimate actuator faults for deep space formation flying satellites are developed in [278]. An internal model average consensus estimator for distributed Kalman filtering is worked out in [279]. Distributed "Kriged" Kalman filtering is addressed in [280]. The behavior of the distributed Kalman filter that varies smoothly from a centralized Kalman filter to a local Kalman filter with an average consensus update is presented in [40]. Both track fusion formulas with feedback and without feedback are analyzed in [281]. A decoupled distributed Kalman fuser presented by using Kalman filtering method and white noise estimation theory is shown in [282]. Decomposition of a linear process model into a cascade of simpler subsystems is given in [283]. Distributed fusion steady-state Kalman filtering by using the modern time series analysis method is shown in [284]. Distributed Kalman filtering with weighted covariance is reported in [285]. The work of [228] shows that this result can be applied to Kalman filtering with uncertain observations, as well as randomly variant dynamic systems with multiple models.

Under some regularity conditions as shown in [607], in particular the assumption of independent sensor noises, an optimal Kalman filtering fusion was proposed which is proved to be equivalent to the centralized Kalman filtering using all sensor measurements; therefore such fusion is globally optimal. In the multi-sensor random parameter matrices case, sometimes, even if the original sensor noises are mutually independent, the sensor noises of the converted system are still cross-correlated. Hence, such a multi-sensor system seems not to satisfy the conditions for the distributed Kalman filtering fusion given in [607].

4.2.3   Diffusion-based filtering

Diffusion-based distributed expected maximization (EM) algorithm for Gaussian mixtures is shown in [296]. Diffusion-based Kalman filtering and smoothing algorithm is shown in [297]. Distributed EM algorithm over sensor networks, consensus filter used to diffuse local sufficient statistics to neighbors and estimate global sufficient statistics in each node is shown in [301]. Consensus filter diffusion of local sufficient statistics over the entire network through communication with neighbor nodes is presented in [302]. Distributed Kalman filtering proposed in the context of diffusion estimation is treated in [298]. Distributed Kalman filtering proposed in the context of average consensus [299][300].

Remark 4.2.2 *In [296], a diffusion scheme of EM (DEM) algorithm for Gaussian mixtures in Wireless Sensor Networks (WSNs) is proposed. At each iteration, the time-varying communication network is modeled as a random graph. A diffusion-step (D-step) is implemented between the E-step and the M-step. In the E-step,*

*sensor nodes compute the local statistics by using local observation data and pa-
rameters estimated at the last iteration. In the D-step, each node exchanges local
information with only its current neighbors, and updates the local statistics with
the exchanged information. In the M-step, the sensor nodes compute the estimation of parameter using the updated local statistics by the D-step at this iteration.
Compared with the existing distributed EM algorithms, the proposed approach can
extensively save communication for each sensor node while maintaining the estimation performance. Different from the linear estimation methods such as the
least-squares and the least-mean squares estimation algorithms, each iteration of
an EM algorithm is a nonlinear transform of measurements. The steady-state performance of the proposed DEM algorithm can not be analyzed in a linear way.
Instead, we show that the DEM algorithm can be considered as a stochastic approximation method to find the maximum likelihood estimation for Gaussian Mixtures. In this regard, we have in mind a network of M sensor nodes, each of
which has N_m data observations $\{y_{m,n}\}$, $m = 1, 2, ..., M$, $n = 1, 2, ..., N_m$.
These observations are drawn from K Gaussian mixtures with mixture probabilities $\alpha_1, ..., \alpha_k$.*

$$y_{m,n} \sim \sum_{j=1}^{K} \alpha_j.N(\mu_j, \Sigma_j) \tag{4.19}$$

*where $N(\mu, \Sigma)$ denote the Gaussian density function with mean μ and covariance
Σ. Let $z \in \{1, 2, ..., K\}$ denote the missing data where Gaussian y comes from.*

4.2.4 Multi-sensor data fusion systems

Sensor noises of converted systems cross-correlated but independent of the original system are covered in [607]-[608]. Sensor noises of converted system's cross-
correlated, and also correlated with the original system are treated in [606]. Centralized fusion center, expressed by a linear combination of the local estimates is
presented in [232]. As treated in [233], algorithms without centralized fusion center tend to be highly resilient to lose one or more sensing nodes. Discrete smoothing fusion with ARMA signals is presented in [603]. Linear minimum variance
(LMV) with an information fusion filter is developed in [604, 605]. Deconvolution
estimation of an ARMA signal with multiple sensors is presented in [304]. Fusion
criterion weighted by scalars is proposed in [305]. Functional equivalence of two
measurement fusion methods is provided in [306]. A centralized filter where data
processed/communicated centrally is discussed in [307]. New performance bounds
for sensor fusion with model uncertainty are developed in [307]. All prior fusion
results with asynchronous measurements is provided in [316]. A unified fusion

model and unified batch fusion rules are presented in [315, 314]. Unified rules by examples are found in [313]. Computing formulation for cross-covariance of the local estimation are presented in [312]. Conditions for centralized and distributed fusers to be identical are developed in [311]. Relationships among the various fusion rules are given in [310]. Optimal rules for each sensor to compress its measurements are considered in [309]. Various issues unique to fusion for dynamic systems are developed in [308]. Bayesian framework for adaptive quantization, fusion-center feedback, and estimation of a spatial random field and its parameters are treated in [317]. A framework for alternates to quartile quantizer, and fusion centers is provided in [318].

Diagonal weighting matrices are presented in [319]. Different fusion rates for the different states are contained in [320]. Optimal distributed estimation fusion in the linear minimum variance (LMV) estimation is presented in [321]. Median fusion and information fusion, not based on weighted sums of local estimates, are presented in [322]. Distributed filtering algorithms, optimal in mean square sense linear combinations of the matrix or scalar weights with derivations are developed in [323]. A closed-form analytical solution of steady fused covariance of information matrix fusion with an arbitrary number of sensors derived is developed in [324]. Focus on various issues unique to fusion for dynamic systems, presenting a general data model for discretized asynchronous multi-sensor systems, are treated in [325]. Recursive BLUE fusion without prior information is worked out in [326]. Statistical interval estimation fusion is contained in [327]. Fused estimate communicated to a central node to be used for some task is presented in [328]. An optimal distributed estimation fusion algorithm with the transformed data is proposed in [329], which is actually equivalent to the centralized estimation fusion. A state estimation fusion algorithm, optimal in the sense of maximum a posterior (MAP) is developed in [330]. A corresponding distributed fusion problem, proposed based on a unified data model for linear unbiased estimator is presented in [331]. An algorithm that fuses one step predictions at both the fusion center and all current sensor estimates is given in [353]. In a multi-sensor linear dynamic system, several efficient algorithms of centralized sensor fusion, distributed sensor fusion, and multi-algorithm fusion to minimize the Euclidean estimation error of the state vector are documented in [354].

Derivation of an approximation technique for arbitrary probability densities, providing the same distributive fusion structure as the linear information filter is presented in [332]. Multi-sensor distributed fusion filters based on three weighted algorithms, applied to the systems with uncertain observations and correlated noises are detailed in [333, 334]. Multi-sensor distributed fusion in state estimation fields, and easy fault detection, isolation and more reliability are developed in [334, 335, 336]. Centralized fusion Kalman filtering algorithm, obtained by combining all measurement data is developed in [337]. The design of a general and optimal asynchronous recursive fusion estimator for a kind of multi-sensor asynchronous sampling system is presented in [355]. To assure the validity of data fusion, a centralized trust rating system is presented in [338]. White noise filter weighted by scalars based on Kalman predictor is developed in [339]. A white noise deconvolution estimators are described in [340]. Optimal information fusion distributed Kalman smoother given for discrete time multi-channel auto-regressive moving average (ARMA) signals with correlated noise are presented in [341]. Optimal dimensional reduction of sensor data by using the matrix decomposition, pseudoinverse, and eigenvalue techniques is contained in [342]. Multi-sensor Information fusion distributed Kalman filter and applications is presented in [343]. Multisensor data fusion approaches to resolve problem of obtaining a joint state-vector estimate being better than the individual sensor-based estimates is documented in [344, 345, 346].

A distributed reduced-order fusion Kalman filter (DRFKF) is treated in [347]. A fusion algorithm based on multi-sensor systems and a distributed multi-sensor data fusion algorithm based on Kalman filtering are presented in [348]. Track fusion formulas with feedback are, like the track fusion without feedback contained in [281]. The optimal DKF fusion algorithms for the case with feedback and cross-uncorrelated sensor measurement noises are presented in [349]. General optimal linear fusion is worked out in [350]. Information fusion in distributed sensor networks is shown in [351]. Multi-scale recursive estimation, data fusion, and regularization are proposed in [352].

Remark 4.2.3 *In [603], using estimators of white measurement noise, an optimal information fusion distributed Kalman smoother is given for multichannel ARMA signals with correlated noise. Work on the ARMA signal and information fusion is also done in [604] and [605]. Basically it has a three-layer fusion structure with fault tolerant and robust properties. The first fusion layer and the second fusion layer both have nested parallel structures to determine the prediction error crosscovariance of the state and the smoothing error cross-covariance of the ARMA signal between any two faultless sensors at each time step. And the third fusion layer is the fusion center used to determine the optimal matrix weights and obtain the optimal fusion distributed smoother for ARMA signals. The computation formula of the smoothing error cross-covariance matrix between any two sensors is given for white measurement noise. The computation formula of smoothing error cross-covariance matrix between any two sensors is given for white measurement noise.*

The discrete time multi-channel ARMA signal system considered here with L

sensors is:

$$B(q^{-1})s(t) = C(q^{-1})w(t) \tag{4.20}$$
$$y_i(t) = s(t) + v_i(t), \; i = 1, \; ...., \; L \tag{4.21}$$

where $s(t) \in \Re^m$ is the signal to estimate, $y_i(t) \in \Re^m$ is the measurement of the ith sensor, $w(t) \in \Re^r$ is the process noise, $v_i(t) \in \Re^m$ is the measurement noise of the ith sensor, L is the number of sensors, and $B(q^{-1})$, $C(q^{-1})$ are polynomial matrices having the form

$$X(q^{-1}) = X_0 + X_1(q^{-1}) + \; ..... + X_{n_x} q^{-n_x}$$

where the argument q^{-1} is the back shift operator, that is, $q^{-1}x(t) = x(t - 1)$, X_i, $i = 0, 1, \; ,....., \; n_x$ are the coefficient matrices, the degree of $X(q^{-1})$ is denoted by n_x.

In the multi-sensor random parameter matrices case, sometimes, even if the original sensor noises are mutually independent, the sensor noises of the converted system are still cross-correlated. Hence, such a multi-sensor system seems not to satisfy the conditions for the distributed Kalman filtering fusion as given in [607, 608]. In [606], it was proved that when the sensor noises or the random measurement matrices of the original system are correlated across sensors, the sensor noises of the converted system are cross-correlated. Even if so, similar to, [232], centralized random parameter matrices Kalman filtering, where the fusion center can receive all the sensor measurements, can still be expressed by a linear combination of the local estimates. Therefore, the performance of the distributed filtering fusion is the same as that of the centralized fusion under the assumption that the expectations of all sensor measurement matrices are of full row rank. Numerical examples are given which support our analysis and show significant performance loss when ignoring the randomness of the parameter matrices. The following discrete time dynamic system is considered:

$$x_{k+1} = F_k x_k + v_k \tag{4.22}$$
$$y_k = H_k x_k + \omega_k, k = 0, 1, 2, 3, ... \tag{4.23}$$

where $x_k \in \Re^r$ is the system state, $y_k \in \Re^N$ is the measurement matrix, $v_k \in \Re^r$ is the process noise, and $\omega_k \in \Re^N$ is the measurement noise. The subscript k is the time index. $F_k \in \Re^{r \times r}$ and $H_k \in \Re^{N \times r}$ are random matrices.

4.2.5 Distributed particle filtering

Consensus-based distributed implementation of the unscented particle filter is shown in [371]. Particle filtering transformation into continuous representations is presented in [356]. Consensus-based, distributed implementation of the unscented particle filter is shown in [371]. Particle filter implementations using Gaussian approximations for the local posteriors are proposed in [357, 358]. A novel framework for delay-tolerant particle filtering, with delayed (out-of-sequence) measurements is treated in [359]. An approach that stores sets of particles for the last l time steps, where ℓ is the predetermined maximum delay, is reported in [360]. Markov chain Monte Carlo (MCMC) smoothing step for (out-of-sequence) measurements is presented in [361]. Approximate OOSM particle filter based on retrodiction (predicting backward) is given in [372]. Recent advances in particle smoothing, storage-efficient particle filters are documented in [364]. A number of heuristic metrics to estimate the utility of delayed measurements is proposed in [365] and a threshold based procedure to discard uninformative delayed measurements, calculating their informativeness, is reported in [363]. Optimal estimation using quantized innovations, with application to distributed estimation over sensor networks using Kalman-like particle filter is the subject of [366]. SOI-Particle-Filter (SOI-PF) derived to enhance the performance of the distributed estimation procedure is presented in [367]. The problem of tracking a moving target in a multi-sensor environment using distributed particle filters (DPFs) is described in [368]. An optimal fusion method, introduced to fuse the collected GMMs with different numbers of components, is presented in [369]. Two distributed particle filters to estimate and track the moving targets in a wireless sensor network are provided in [369]. Updating the complete particle filter on each individual sensor node is given in [370]. Out-of-sequence measurement processing for tracking ground targets using particle filters is presented in [372]. A comparison of the KF and particle filter based out-of-sequence measurement filtering algorithms is documented in [373].

4.2.6 Self-tuning based filtering

Multi-sensor systems with unknown model parameters and noise variances, using the information matrix approach, with self-tuning distributed state fusion information filters are presented in [374]. Self-tuning distributed state fusion leads to Kalman filtering with weighted covariance approach as reported in [375]. A self-tuning decoupled fusion Kalman predictor is proposed in [376] and a self-tuning weighted measurement Kalman filter is included in [377]. Multi-sensor systems with unknown noise variances, and a new self-tuning weighted measurement fusion Kalman filter is presented in [378], which has asymptotic global optimality. Weighted self-tuning state fusion filters are given in [379, 380]. A sign of innovation-particle filter (SOI-PF) improves the tracking performance when the target moves according to a linear and a Gaussian model as presented in [381]. Efficiency of the SOI-PF in a nonlinear and a non Gaussian case, considering a jump-state Markov model for the target trajectory is derived in [382]. A self-tuning information fusion reduced-order Kalman predictor with a two-stage fusion structure based on linear minimum variance is reported in [272]. An optimal self-tuning smoother is proposed in [383]. A new convergence analysis method for the self-tuning Kalman predictor is presented in [384]. A self-tuning measurement system using the correlation method can be viewed as the least-squares (LS) fused estimator and found in [385]. Self-tuning filtering for systems with unknown models and/or noise variances is presented in [386]-[387]. Self-tuning distributed state fusion Kalman estimators are reported in [388][389] Self-tuning distributed (weighed) measurement fusion Kalman filters are shown in [390, 391, 392].

Remark 4.2.4 *For a self-tuning decoupled fusion Kalman predictor, the following multi-sensor linear discrete time-invariant stochastic system is considered in the chapter [393]:*

$$
\begin{aligned}
x(t+1) &= \Phi x(t) + \Gamma w(t) & (4.24)\\
y_i(t) &= H_i x(t) + v_i(t)\,,\ i = 1,\ \ldots\ldots,\ L & (4.25)
\end{aligned}
$$

where $x(t) \in \Re^n, y_i(t) \in \Re^{m_i}$, $w(t) \in \Re^r$ and $v_i(t) \in \Re^{m_i}$ are the state, measurement, process and measurement noises of the ith sensor subsystem, respectively, and Φ, Γ and H_i are constant matrices with compatible dimensions.

4.3   Information Flow

Consider a sensor network with n sensors that are interconnected via an overlay network G (that is a connected undirected graph as shown in Figure 4.2).

   This section describes the so-called information form of the Kalman filter (IKF) according to [41, 42]. Let us describe the model of a process (e.g., a physical phenomenon or a moving object) and the sensing model of the IKF as follows:

$$
\begin{aligned}
x_{k+1} &= A_k x_k + B_k w_k; \quad x_0 \\
z_k &= H_k x_k + v_k & (4.26)
\end{aligned}
$$

where $z_k \in \Re^{np}$ represents the vector of p-dimensional measurements obtained via n sensors, w_k and v_k are white Gaussian noise (WGN), and $x_0 \in \Re^m$ denotes the initial state of the process that is a Gaussian random variable. Here is the information regarding the statistics of these variables:

$$
E(w_k w_l^T) = Q_k \delta_{kl},\ E(v_k v_l^T) = R_k \delta_{kl} \qquad (4.27)
$$

$$
x_0 = N(\bar{x}_0, P_0). \qquad (4.28)
$$

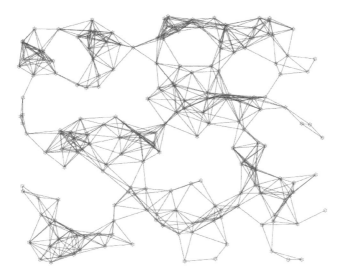

Figure 4.2: A sensor network with n = 200 nodes and l = 1074 links.

Given the measurements $Z_k = \{z_0, z_1, \ldots, z_k\}$, the state estimates can be expressed as

$$\hat{x}_k = E(x_k|Z_k), \bar{x}_k = E(x_k|Z_{k-1}), \tag{4.29}$$

$$P_k = \sum_{k|k-1}, M_k = \sum_{k|k} \tag{4.30}$$

where $\sum_{k|k-1}$ and $\sum_{k|k-1}$ denote the state covariance matrices, and their inverses are known as the information matrices. Note that $\sum_{0|-1} = P_0$. Here are the Kalman filter iterations in the information form:

$$M_k^{-1} = P_k^{-1} + H_k' \Re_k^{-1} H_k \tag{4.31}$$

$$K_k = M_k H_k' \Re_k^{-1} \tag{4.32}$$

$$\hat{x}_k = \bar{x}_k + K_k(z_k - H_k \bar{x}_k) \tag{4.33}$$

$$P_{k+1} = A_k M_k A_k' + B_k Q_k B_k' \tag{4.34}$$

$$\bar{x}_{k+1} = A_k \hat{x}_k \tag{4.35}$$

4.3.1 Micro-Kalman filters

Our first objective is to show how the information form of a central Kalman filter for a sensor network observing a process of dimension m with an n_p-dimensional measurement vector z_k can be equivalently expressed in consensus form using n micro-Kalman filters (μKF) with p-dimensional measurement vectors which are embedded in each sensor so that the network of micro-Kalman filters collectively, in a distributed way, calculate the same state estimate $\hat{x}$ obtained via application of a central Kalman filter located at a sink node (e.g., for a moving object in a plane $p = 2, m = 4$ and $n \gg 1$).

Let us assume that there are n sensors with $p \times m$ measurement matrices H_i and sensing model:

$$z_i(k) = H_i x(k) + v_i(k)$$

Thus, defining the central measurement, observation noise, and observation matrix as

$$
\begin{aligned}
z_c &= col(z_1, z_2, \ldots, z_n), \quad v_c = col(v_1, \ldots, v_n), & (4.36) \\
H_c &= [H_1; H_2; \ldots; H_n], & (4.37)
\end{aligned}
$$

where H_c is a column block matrix. We get

$$z_c(k) = H_c x(k) + v_c(k) \qquad (4.38)$$

where the subscript "c" means "central". Let

$$R_c = diag(R_1, R_2, \ldots, R_n)$$

denote the covariance of v_c (i.e. we assume v_i's are uncorrelated). The iteration numbers are dropped when ever no confusions occur. We have

$$M = (P + H_c^{'}\Re_c^{-1}H_c)^{-1}, \quad K_c = MH_c^{'}\Re_c^{-1}.$$

Thus, the state propagation equation can be expressed as

$$
\begin{aligned}
\hat{x} &= \bar{x} + K_c(z_c - H_c\bar{x}) \\
&= x + M(H_c^{'}\Re_c^{-1}z_c - H_c^{'}\Re_c^{-1}H_c\bar{x}) \qquad (4.39)
\end{aligned}
$$

Defining the following $m \times m$ average inverse-covariance matrix

$$S = \frac{1}{n}H_c^{'}\Re_c^{-1}H_c = \frac{1}{n}\sum_{i=1}^{n}H_i^{'}\Re_i^{-1}H_i \qquad (4.40)$$

and the m-vector of average measurements

$$y_i = H_i^{'}\Re_i^{-1}z_i, y = \frac{1}{n}\sum_{i=1}^{n}y_i, \qquad (4.41)$$

one gets the Kalman state update equation of a μKF as

$$x = \bar{x} + M_\mu(y - S\bar{x}) \qquad (4.42)$$

with a micro-Kalman gain of $M_\mu = nM$, measurement consensus y, and inverse-covariance consensus value of S. The expression for M_μ can be stated as follows:

$$M_\mu = nM = ((nP)^{-1} + S)^{-1}. \qquad (4.43)$$

Denoting $P_\mu = nP$ and $Q_\mu = nQ$, we obtain an update equation of dimension $m \times m$ for a μKF:

$$P_\mu^+ = AM_\mu A' + BQ_\mu B' \qquad (4.44)$$

Based on the above argument, we have the following decomposition theorem for Kalman filtering in sensor networks:

Theorem 4.3.1 *(distributed Kalman filter) Consider a sensor network with n sensors and topology G that is a connected graph observing a process of dimension m using $p \leq m$ sensor measurements. Assume the nodes of the network solve two consensus problems that allow them to calculate average inverse-covariance S and average measurements y at every iteration k. Then, every node of the network can calculate the state estimate $\hat{x}$ at iteration k using the update equations of its micro-Kalman filter (or μKF iterations)*

$$
\begin{align}
M_\mu &= (P_\mu^{-1} + S)^{-1}, \qquad && (4.45) \\
\hat{x} &= \bar{x} + M_\mu(y - S\bar{x}), \qquad && (4.46) \\
P_\mu^+ &= AM_\mu A' + BQ_\mu B', \qquad && (4.47) \\
\bar{x}^+ &= A\hat{x}. \qquad && (4.48)
\end{align}
$$

This gives an estimate identical to the one obtained via a central Kalman filter.

Remark 4.3.1 *The gain M_μ of the micro-Kalman filter has $O(m^2)$ elements, whereas the Kalman gain K of the central Kalman filter has $O(m^2 n)$ elements. Thus, the calculations of the central KF require manipulation of large matrices which is not computationally feasible.*

Remark 4.3.2 *We assume all nodes know n or solve a consensus problem to calculate n. This is necessary for calculation of $Q_\mu = nQ$.*

Considering that both S and y are time-varying quantities, one needs to solve two dynamic consensus problems that allow asymptotic tracking of the values of $S(k)$ and $y(k)$ [42]. The nature of these two dynamic consensus problems differ in nature. Consensus in $y(k)$ requires sensor fusion for noisy measurements y_i that can be solved using a newly found distributed low-pass consensus filter [41]. The consensus regarding the inverse-covariance matrices for calculation of S requires a band-pass consensus filter that will be described in the next section. Neither problem can be solved using a high-pass consensus filter alone.

Based on the results in [41], the nodes of a network that uses a consensus filter only reach an ϵ-consensus (for non-static cases), meaning that all agents reach a state that is in a closed-ball of radius $\epsilon\ell1$ around the group decision value [41]. This means that practically every node calculates its approximate consensus values $\hat{S}_i$ and $\hat{y}_i$ and that all belong to small neighborhoods around S and y, respectively. This gives the following state and covariance update equations for the ith μKF:

$$M_i = (P_i^{-1} + \hat{S}_i)^{-1}, \tag{4.49}$$

$$\hat{x} = \bar{x} + M_i(\hat{y}_i - \hat{S}_i\bar{x}), \tag{4.50}$$

$$P_i^+ = AM_iA' + BQ_\mu B', \tag{4.51}$$

$$\bar{x}^+ = A\hat{x}, \tag{4.52}$$

with $P_i = nP$. This is the perturbed version of the exact iterations of the μKF equation in Theorem 4.3.1. The convergence analysis of the collective dynamics of the perturbed μKF equations is the subject of future research.

4.3.2 Frequency-type consensus filters

We wish to emphasize that Theorem 4.3.1 does not amount to the solution of the DKF problem. So far, we have only managed to show that if two dynamic consensus problems in S and y are solved, then a distributed algorithm for Kalman filtering in sensor networks exists. The crucial part of solving the DKF problem is solving its required dynamic consensus problems which have been addressed in [41].

We state the distributed algorithms for three consensus filters: a low-pass filter, a high-pass filter, and a resulting band-pass filter. Let us denote the adjacency and Laplacian matrix [10] of G by A and $L = diag(A1) - A$, respectively

- **Low-Pass Consensus Filter** (CF_{lp}, [24]): Let q_i denote the m-dimensional state of node i and u_i denote the m-dimensional input of node i. Then, the following dynamic consensus algorithm

$$\dot{q}_i = \sum_{j \in N_i}(q_j - q_i) + \sum_{j \in N_i \cup \{i\}}(u_j - q_i) \tag{4.53}$$

that can be equivalently expressed as

$$\dot{q} = -\hat{L}q - \hat{L}u + (I_n + \hat{A})(u - x) \tag{4.54}$$

with $q = col(q_1, \ldots, q_n), \hat{A} = A \otimes I_m$ and $\hat{L} = L \otimes I_m$ gives a low-pass consensus filter with a MIMO transfer function

$$H_{lp}(s) = (s+1)I_n + \hat{A} + \hat{L})^{-1}(I_n + \hat{A}) \qquad (4.55)$$

from input u to output x. This filter is used for fusion of the measurements that calculate, $\hat{y}_i$ by applying the algorithm to $H_i' \Re_i^{-1} z_i$ as the input of node i.

- **High-Pass Consensus Filter** (CF_{hp}, [24, 29]): Let p_i denote the m dimensional state of node i, and u_i denote the m-dimensional input of node i. Then, the following dynamic consensus algorithm

$$\dot{p}_i = \sum_{j \in N_i} (p_j - p_i) + \dot{u}_i \qquad (4.56)$$

that can be equivalently expressed as

$$\dot{e} = -\hat{L}_e - \hat{L}u_i \qquad (4.57)$$
$$p = e + u \qquad (4.58)$$

with $\hat{L} = L \otimes I_m$. This gives a high-pass consensus filter with an improper MIMO transfer function

$$H_{hp}(s) = (sI_n + \hat{L})^{-1}s \qquad (4.59)$$

from input u to output x that becomes I_n as $s \to 1$. This filter apparently propagates high-frequency noise and by itself is inadequate for sensor fusion.

- **Band-Pass Consensus Filter** (CF_{bp}): This distributed filter can be defined as

$$H_{bp}(s) = H_{lp}(s)H_{hp}(s) \qquad (4.60)$$

that can be equivalently stated in the form of a dynamic consensus algorithm

$$\dot{e}_i = -\hat{L}e_i - \hat{L}u_i, \qquad (4.61)$$
$$p_i = e_i + u_i, \qquad (4.62)$$
$$\dot{q}_i = \sum_{j \in N_i} (q_j - q_i) + \sum_{j \in N_i \cup \{i\}} (p_j - q_i) \qquad (4.63)$$

with a state $(e_i, q_i) \in \Re^{2m}$, input u_i, and output q_i. This filter is used for inverse-covariance consensus that calculates $\hat{S}_i$ column-wise for node i by applying the filter on columns of $H_i' \Re_i^{-1} H_i$ as the inputs of node i. The matrix version of this filter can take $H_i' \Re_i^{-1} H_i$ as the input.

Figure 4.3 shows the architecture of each node of the sensor network for distributed Kalman filtering. Note that consensus filtering is performed with the same frequency as Kalman filtering. This is a unique feature that completely distinguishes our algorithm with some related work in [30, 33].

4.3.3 Simulation example 1

In this section, we use consensus filters jointly with the update equation of the micro-Kalman filter of each node to obtain an estimate of the position of a moving object in $\Re^2$ that (approximately) goes in circles. The output matrix is $H_i = I_2$ and the state of the process dynamics is 2-dimensional corresponding to the continuous-time system

$$\dot{x} = A_0 x + B_0 w$$

with

$$A_0 = \begin{bmatrix} 0 & -1 \\ 1 & 0 \end{bmatrix}, B_0 = I_2$$

The network has $n = 200$ sensors with a topology shown in Figure 4.2. We use the following data:

$$R_i = 100(i^{\frac{1}{2}})I_2, Q = 25, P_0 = I_2, x_0 = (15, -10)'.$$

with a step-time of $T = 0.02$ (sec). Figs 4.4 and 4.5 and show the estimate obtained by nodes $i = 100, 25$. Apparently, the distributed and central Kalman filters provide almost identical estimates. Of course, the difference is in scalability of the DKF. In Figure 4.6, the consecutive snapshots of estimates of all the nodes are shown. The estimates appear as a cohesive set of particles that move around the location of the object.

4.3.4 Simulation example 2

The importance of distributed Kalman filtering (DKF) for sensor networks was discussed. We addressed the DKF problem by reducing it to two separate dynamic consensus problems in terms of weighted measurements and inverse-covariance matrices that can be viewed as two data fusion problems with different natures. Both data fusion problems were solved in a distributed way using consensus filters. Consensus filters are distributed algorithms that allow calculation of average-consensus of time-varying signals. We employed a low-pass consensus filter for fusion of the measurements and a band-pass consensus filter for fusion of the

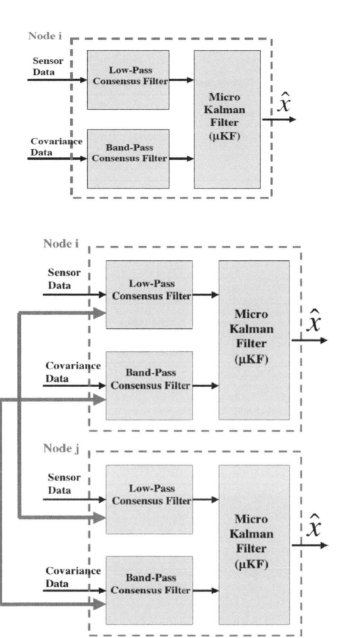

Figure 4.3: Node and network architecture for distributed Kalman filtering: (top) architecture of consensus filters and μKF of a node and (bottom) communication patterns between low-pass/band-pass consensus filters of neighboring nodes.

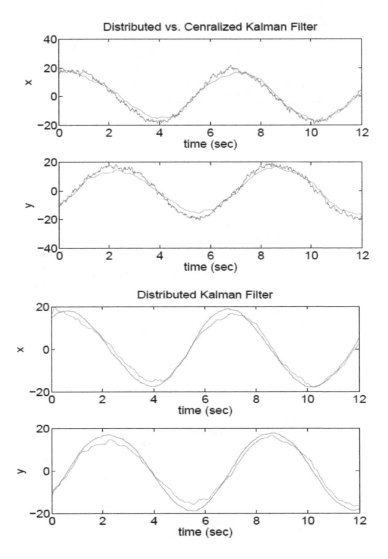

Figure 4.4: Distributed position estimation for a moving object by node i = 100: (top) DKF vs. KF and (bottom) Distributed Kalman filter estimate vs. the actual position of the object.

inverse-covariance matrices. Note that the stability properties of consensus filters is discussed in a companion Chapter [24]. We established that a central Kalman filter for sensor networks can be decomposed into n micro-Kalman filters with inputs that are provided by two consensus filters. This network of micro-Kalman filters was able to collaboratively provide an estimate of the state of the observed process. This estimate is identical to the estimate obtained by a central Kalman filter given that all nodes agree on two central sums. Consensus filters can approximate these sums and that gives an approximate distributed Kalman filtering algorithm for sensor networks. Computational and communication architecture of the algorithm was discussed. Simulation results are presented for a sensor network with 200 nodes and 1074 links.

4.4 Consensus Algorithms in Sensor Networked Systems

This section is concerned with the average-consensus algorithm for the case of n measurements of noisy signals obtained from n sensors in the form of a distributed low-pass filter called the Consensus Filter. The role of this consensus filter is to perform distributed fusion of sensor measurements that is necessary for implementation of a scalable Kalman filtering scheme. It will be shown that consensus filters can be also used independently for distributed sensor fusion.

4.4.1 Basics of graph theory

Hereafter, we provide basic information pertaining to algebraic graph theory. The reader is referred to [411] for a rigorous exposition and to the Appendix for a short account. Let $G = (V, E)$ be a graph with a nonnegative adjacency matrix $A = [a_{ij}]$ that specifies the interconnection topology of a network of dynamic systems, sensors, or agents. The set of nodes is denoted by $V = \{1, \ldots, n\}$. For complex networks, we refer to $|V|$ and $|E|$ as the scale and size of the network, respectively. Let $N_i = \{i \in V : a_{ij} \neq 0\}$ denote the set of neighbors of node i and $J_i = N_i \cup \{i\}$ denote the set of inclusive neighbors of node i. A consensus algorithm can be expressed in the form of a linear system

$$\dot{x}_i(t) = \sum_{j \in N_i} a_{ij}(x_j(t) - x_i(t)), \quad x(0) = c \in \Re^n. \tag{4.64}$$

Given a connected network G, all the solutions of system (4.64) converge to an aligned state $x^* = (\mu, \ldots, \mu)^T$ with identical elements equal to $\mu = \bar{x}(0) = \frac{1}{n}\sum_i c_i$. This explains why the term "average-consensus" is used to refer to the

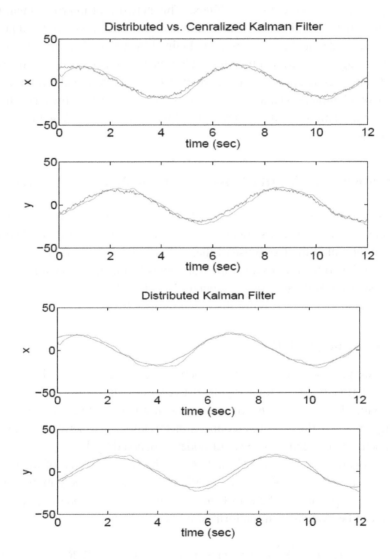

Figure 4.5: Distributed position estimation for a moving object by node i = 25: (top) DKF vs. KF and (bottom) Distributed Kalman filter estimate vs. the actual position of the object.

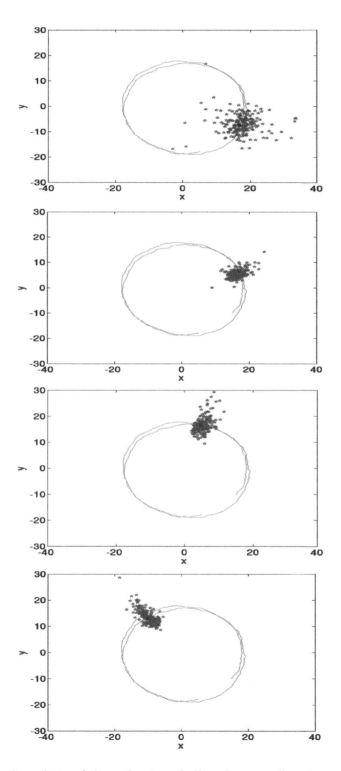

Figure 4.6: Snapshots of the estimates of all nodes regarding the position of a moving object.

distributed algorithm in (4.64). In a more compact form, system (4.64) can be expressed as

$$\dot{x} = -Lx, \tag{4.65}$$

where L is the Laplacian matrix [411] of graph G and is defined as

$$L = \Delta - A \tag{4.66}$$

where $\Delta = diag(A\mathbf{1})$ is the degree matrix of G with diagonal elements $d_i = \sum_j a_{ij}$. Here, $\mathbf{1} = (1, \ldots, 1)^T \in \Re^n$ denotes the vector of ones that is always a right eigenvector of L corresponding to $\lambda_1 = 0$ (i.e. $L\mathbf{1} = 0$). The second smallest eigenvalue λ_2 of L determines the speed of convergence of the algorithm.

4.4.2   Consensus algorithms

Consider a sensor network of size n with information flow * G. Assume each sensor is measuring a signal $s(t)$ that is corrupted by noise v_i that is a zero-mean white Gaussian noise (WGN). Thus, the sensing model of the network is

$$u_i(t) = r(t) + v_i(t), i = 1, \ldots, n \tag{4.67}$$

or $u(t) = r(t)\mathbf{1} + v(t)$. Let R_i denote the covariance matrix of v_i for all i.

The objective here is to design the dynamics of a distributed low-pass filter with state $x = (x_1, \ldots, x_n)^T \in \Re^n$ that takes u as the input and $y = x$ as the output with the property that asymptotically all nodes of the network reach an $\epsilon-$ consensus regarding the value of signal $r(t)$ in all time t. By ϵ-consensus, we mean there is a ball of radius ϵ that contains the state of all nodes (i.e., approximate agreement).

In most applications, $r(t)$ is a low-to-medium frequency signal and $v(t)$ is a high-frequency noise. Thus, the consensus filter must act as a low-pass filter.

The following dynamic consensus algorithm is proposed

$$\dot{x}_i(t) = \sum_{j \in N_i} a_{ij}(x_j(t) - x_i(t)) + \sum_{j \in J_i} a_{ij}(u_j(t) - x_i(t)), \tag{4.68}$$

as a candidate for a distributed low-pass consensus filter. The remainder of the chapter is devoted to establishing the properties of this distributed filter. Note that the algorithm in (4.68) only requires communication among neighboring nodes of the network and thus is a distributed algorithm.

---

*Keep in mind that the information flow in a sensor network might (or might not) be the same as the overlay network (i.e., communication network).

Remark 4.4.1 *In discrete-time, the dynamic consensus algorithm in (4.68) can be stated as follows:*

$$x_i^+ = x_i + \delta[\sum_{j \in N_i} a_{ij}(x_j - x_i) + \sum_{j \in J_i} a_{ij}(u_j - x_i)], \qquad (4.69)$$

where x_i is the current state of node i, x_i^+ is the next state, and δ is the step-size of iterations. We will conduct all of our analysis in continuous-time.

Proposition 4.4.1 *The distributed algorithm in (4.68) gives a consensus filter with the following collective dynamics*

$$\dot{x} = -(I_n + \Delta + L)x + (I_n + A)u \qquad (4.70)$$

that is an LTI system with specification $A = -(I + \Delta + L)$, $B = I_n + A$, $C = I_n$ and a proper MIMO transfer function.

Proof 4.4.1 *First, let us rewrite the system in (3.9) as*

$$
\begin{aligned}
\dot{x}_i &= \sum_{j \in N_i} a_{ij}(x_j - x_i) + \sum_{j \in J_i} a_{ij}(u_j - u_i + u_i - x_i), \\
&= \sum_{j \in N_i} a_{ij}(x_j - x_i) + \sum_{j \in N_i} a_{ij}(u_j - u_i) \\
&+ |J_i|(u_i - x_i).
\end{aligned}
$$

Noting that $|J_i| = 1 + d_i$, from the definition of graph Laplacian, we get

$$
\begin{aligned}
\dot{x} &= -Lx - Lu + (I_n + \Delta)(u - x), \\
&= -(I_n + \Delta + L)x + (I_n + \Delta - L)u
\end{aligned}
$$

But $\Delta - L = A$ and therefore $\dot{x} = Ax + Bu$, $y = Cx$ with matrices that are defined in the question.

The transfer function of the consensus filter is given by

$$H(s) = [sI_n + (I_n + \Delta + L)]^{-1}(I_n + A) \qquad (4.71)$$

Applying the Gershgorin theorem to matrix $A = I_n + 2\Delta + A$ guarantees that all poles of $H(s)$ are strictly negative and fall within the interval $[-(1 + d_{\min}), -(1 + 3d_{\max})]$ with $d_{\max} = \max_i d_i$ and $d_{\min} = \min_i d_i$, that is, $1 + d_{\min} \leq \lambda_i(A) \leq (1 + 3d_{\max})$ for all i. This immediately implies the following stability property of the consensus filter.

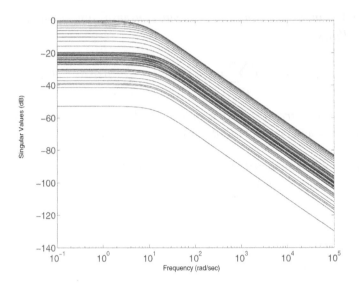

Figure 4.7: The singular value plots of the low-pass consensus filter for a regular network.

Corollary 4.4.1 *The consensus filter in (4.71) is a distributed stable low-pass filter.*

Proof 4.4.2 *Apparently, all the poles of $H(s)$ are strictly negative and thus the filter is stable. On the other hand, $H(s)$ is a proper MIMO transfer function satisfying $\lim_{s \to \infty} H(s) = 0$ which means it is a low-pass filter.*

Figure 4.7 shows the singular value plots of the low-pass consensus filter (or CF_{lp}) for a regular network with $n = 100$ nodes and degree $k = 6$.

Remark 4.4.2 *The following dynamic consensus algorithm*

$$\dot{x} = -Lx + \dot{u}(t)$$

gives a high-pass consensus filter (CF_{hp}) that is useful for distributed data fusion applications with low-noise data.

It remains to establish that all the nodes asymptotically can reach an ϵ-consensus regarding $r(t)$.

Proposition 4.4.2 *Let $r(t)$ be a signal with a uniformly bounded rate $|\dot{r}(t)| \leq \nu$. Then, $x^*(t) = r(t)1$ is a globally asymptotically ϵ-stable equilibrium of the dynamics of the consensus filter given by*

$$\dot{x} = -Lx - Lu + (I_n + \Delta)(u - x) \tag{4.72}$$

with input $u = r(t)\mathbf{1}$ and

$$\epsilon = \frac{\nu\sqrt{n}(1 + d_{\max})\lambda_{\max}^{\frac{1}{2}}(A)}{\lambda_{\min}^{\frac{1}{2}}(A)} \tag{4.73}$$

Proof 4.4.3 *Given the input $u = r(t)\mathbf{1}$, the dynamics of the system in (4.72) reduces to*

$$\dot{x} = -Lx + (I_n + \Delta)(r(t)\mathbf{1} - x) \tag{4.74}$$

with an obvious equilibrium at $x = r(t)\mathbf{1}$ that is an aligned state with elements that are identical to the signal $r(t)$. This is due to the fact that $L\mathbf{1} = 0$. Defining the error variable $\eta = x - r(t)\mathbf{1}$ gives

$$\dot{\eta} = -A\eta + \dot{r}(t)\mathbf{1} \tag{4.75}$$

where $A = I_n + \Delta + L$ is a positive definite matrix with the property that

$$1 + d_{\min} \leq \lambda_{\min}(A) \leq \lambda_{\max}(A) \leq 1 + 3d_{\max}. \tag{4.76}$$

Let us define the Lyapunov function $\varphi(\eta) = \frac{1}{2}\eta^T A\eta$ for the perturbed linear system in (4.75). We have

$$\begin{aligned}
\dot{\varphi} &= -\|A\eta\|^2 + \dot{r}(t)(\mathbf{1}^T A\eta) \\
&\leq -\lambda_{\min}^2(A)\|\eta\|^2 + \nu\sqrt{n}(1 + d_{\max})\|\eta\|.
\end{aligned}$$

This is because

$$\mathbf{1}^T A = \mathbf{1}^T + \mathbf{1}^T\Delta = (1 + d_1, 1 + d_2, \ldots, 1 + d_n),$$

and thus

$$|\mathbf{1}^T A\eta| \leq [\sum_i (1 + d_i)^2]^{\frac{1}{2}}\|\eta\| \leq \sqrt{n}(1 + d_{\max})\|\eta\|.$$

As a result, one obtains

$$\begin{aligned}
\dot{\varphi}(\eta) &\leq -\left(\lambda_{\min}(A)\|\eta\| - \frac{\nu\sqrt{n}(1 + d_{\max})}{2\lambda_{\min}(A)}\right)^2 \\
&\quad + \left(\frac{\nu\sqrt{n}(1 + d_{\max})}{2\lambda_{\min}(A)}\right)^2
\end{aligned}$$

Let B_ρ be a closed ball centered at $\eta = 0$ with radius

$$\rho \frac{\nu\sqrt{n}(1 + d_{\max})}{\lambda^2_{\min}(A)} \tag{4.77}$$

and let $\Omega_c = \{\eta : \varphi(\eta) \leq c\}$ be a level-set of the Lyapunov function $\varphi(\eta)$ with $c = \frac{1}{2}\lambda_{\max}(A)\rho^2$. Then, B_ρ is contained in Ω_c because

$$\|\eta\| \leq \rho \Rightarrow \varphi(\eta) = \frac{1}{2}\eta^T A\eta \leq \frac{1}{2}\lambda_{\max}(A)\rho^2 = c,$$

and thus $\eta \in \Omega_c$. As a result, any solution of (4.75) starting in $\Re^n$ $Omega_c$ satisfies $\dot\varphi < 0$. Thus, it enters Ω_c in some finite time and remains in Ω_c thereafter (that is, Ω_c is an invariant level-set). This guarantees global asymptotic ϵ-stability of $\eta = 0$ with a radius $\epsilon = \rho\lambda_{\max}(A)/\lambda_{\min}(A)$. To show this, note that

$$\frac{1}{2}\lambda_{\min}(A)\|\eta\|^2 \leq \varphi(\eta) \leq \frac{1}{2}\lambda_{\max}(A)\rho^2 \tag{4.78}$$

Thus, the solutions enter the region

$$\|\eta\| \leq \rho\sqrt{\frac{\lambda_{\max}(A)}{\lambda_{\min}(A)}}$$

which implies the radius of ϵ-stability is

$$\epsilon = \frac{\nu(1 + d_{\max})}{\lambda^2_{\min}(A)}\sqrt{\frac{n\lambda_{\max}(A)}{\lambda_{\min}(A)}} \tag{4.79}$$

Of course, ϵ-stability of $\eta = 0$ implies ϵ-tracking of $r(t)$ by every node of the network (i.e. ϵ-consensus is asymptotically reached).

The following result describes the occurrence of a critical phenomenon in regular complex networks.

Proposition 4.4.3 *Consider a regular network G of degree k. Let $r(t)$ be a signal with a finite rate $|\dot r| \leq \nu$. Then, the dynamics of the consensus filter in the form*

$$\dot x = -Lx - Lu + (I + \Delta)(u - x) \tag{4.80}$$

satisfies the following properties:

- *The mean $\mu(t) = Px(t)$ of the state of all nodes is the output of a scalar low-pass filter*

$$\dot{\mu} = (k+1)(\bar{u}(t) - \mu) \tag{4.81}$$

  with an input $\bar{u}(t) = r(t) + w(t)$ and a zero-mean noise $w(t) = \frac{1}{n}\sum_i v_i(t)$.

- *Assume the network node degree $k = \beta n^\gamma$ is exponentially scale-dependent. Then, there exists a critical exponent $\gamma_c = \frac{1}{2}$ such that for all $\gamma > \gamma_c$ (or networks with more than $O(n^{1.5})$ links), the radius of ϵ-tracking vanishes as the scale n becomes infinity large for any arbitrary ν, β(ϵ is defined in Proposition 4.4.2).*

Proof 4.4.4 *Part i) follows from the fact that $\mu = \frac{1}{n}(1^T x)$ and $1^T L = 0$. Moreover, for regular networks with degree $k, I_n + \Delta = (k+1)I_n$. To show **part ii)**, note that for a regular network with degree $k, d_{\max} = d_{\min} = k = \beta n^\gamma$ and $\lambda_{\max}(A) = \lambda_{\min}(A) = 1 + k$ (the least conservative upper bound on ϵ is attained by a regular network). Hence, the expression for ϵ greatly simplifies as*

$$\epsilon = \frac{\nu\sqrt{n}}{1+k}\nu\sqrt{n}1 + \beta n^\gamma \tag{4.82}$$

Thus, for all $\gamma > \gamma_c = \frac{1}{2}, \epsilon_n \to 0$ as $n \to \infty$ regardless of the values of $\beta, \nu < \infty$. In other words, ϵ_n-tracking of $r(t)$ is achieved asymptotically by every node with a vanishing ϵ for large-scale regular networks of a size (i.e. $nk/2$) greater than $O(n^{1.5})$.

Remark 4.4.3 *The white noise $w(t) = \frac{1}{n}\sum_i v_i(t)$ has a covariance matrix $\frac{1}{n}\bar{R}$ that is n times smaller than the average covariance $\bar{R} = \frac{1}{n}\sum_i R_i$ of all (uncorrelated) $v_i's$. For a large-scale network, $w(t)$ can possibly become multiple orders of magnitude weaker than all the $v_i's$.*

Corollary 4.4.2 *(Scale-uncertainty principle) A regular complex network with density $\sigma = (2|E| + n)/n^{1.5}$ and tracking uncertainty $\varepsilon = \epsilon/\nu$ that runs the dynamic consensus algorithm in (3.9) satisfies the following uncertainty principle*

$$(network density) \times (tracking uncertainty) = 1, \tag{4.83}$$

or $\sigma \times \varepsilon = 1$.

Proof 4.4.5 *The proof follows from (4.82) and the identity $2|E| := \sum_i d_i = nk$.*

4.4.3   Simulation example 3

Defining the performance of tracking as $1/\varepsilon$, we get the following trade-off between tracking performance and network density:

$$(network density) \propto (tracking performance).$$

The most common application is to track a signal that has a single, or multiple, sinusoidal components.

Consider the case of a signal $r(t) = a\sin(\omega t)$ with $a, \omega > 0$ that is being measured by every sensor in a sensor network. This signal could possibly represent the x-coordinate of the position of a moving object that goes in circles. The main question of interest is how large must the sensor network be? This is important for the purpose of tracking $r(t)$ within a tube of radius $\epsilon \leq \delta a$ (e.g. $\delta = 0.1$).

Notice that $\nu = a\omega$ and therefore the tracking uncertainty satisfies. To guarantee $\epsilon \leq \delta a$, we must have $\varepsilon = \epsilon/\nu \leq \delta/\omega$. Using the uncertainty principle, $\sigma \times \varepsilon = 1$ and thus $\omega \leq \delta \times \sigma$.

For a network with $n = 1000$ nodes and weighted degree $k = \beta n^\gamma$ with $\beta = 10, \gamma = 0.6 > \gamma_c$ (all weights of the graph are in $\{0, \beta\}$), we get $k = 631$ and $\omega \leq 2$ (rad/sec) for $\epsilon = 0.1a$ accuracy. This is a relatively conservative bound and in practice the network is capable of tracking much faster signals with only 100 nodes. Finding a less conservative uncertainty principle is a real challenge.

One cannot arbitrarily increase β because based on the low-pass filter with state μ, this is equivalent to using a high-gain observer for $\bar{u}$ that amplifies noise.

4.4.4   Simulation example 4

In this section, we present simulation results for sensor networks with two type of topologies: a) a regular network of degree $k = 6$ and b) a random network obtained as a spatially induced graph from $n = 400$ points with coordinates $\{q_i\}i \in V$ that are distributed uniformly at random in an $n \times n$ square region with a set of neighbors $N_i = \{q_j : \|q_i - q_i\| < \rho_0\}$ and a radio range of $\rho_0 = 2\sqrt{n}$. These networks are shown in Fig 4.8. Networks (a) and (b), shown in Figure 4.8, have an average-degree of 6 and 7.1, respectively. Apparently, the random network is irregular.

We use the following three test signals

$$\begin{aligned}
r_1(t) &= \sin(2t); \\
r_2(t) &= \sin(t) + \sin(2t + 3) + \sin(5t + 4), \\
r_3(t) &= \sin(10t).
\end{aligned}$$

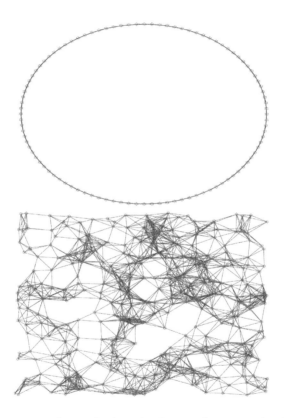

Figure 4.8: Sensor network topologies: (top) a regular network with $n = 100$ and degree $k = 6$ and (bottom) a random network with $n = 400$ and 2833 links.

For r_1 and r_2, we set the covariance matrix to $R_i = 0.3$ for all nodes and for r_3, $R_i = 0.6$ for all i.

Figure 4.9 demonstrates sensor fusion using a low-pass consensus filter with a regular network topology for sensor measurements $r_1(t) + v_i(t)$ obtained from $n = 100$ nodes. The fused measurements Figure 4.9 (b) have a covariance that is almost 100 times smaller than the covariance of the sensor data.

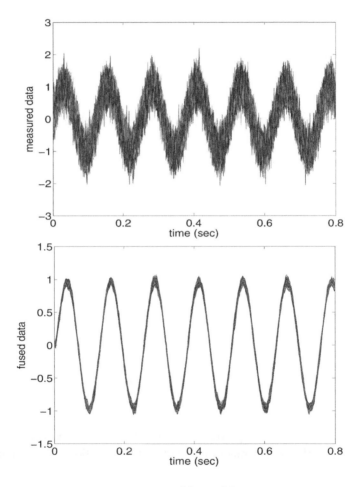

Figure 4.9: (top) sensor measurements $r_1(t) + v_i(t)$ and (bottom) fused sensor data via a low-pass consensus filter in a regular network.

Similarly, Figure 4.10 demonstrates sensor fusion using a distributed low-pass consensus filter for sensor data $r_2(t) + v_i(t)$ obtained from $n = 100$ nodes. Again, the network topology is regular. All nodes are apparently capable of tracking $r_3(t)$ within a radius of uncertainty that is determined by $|\dot{r}_3|$ and the noise covariance R_i.

To demonstrate tracking capabilities of larger networks, we consider tracking $r_3(t)$ that is 5 times faster than $r_1(t)$ using a consensus filter in a network with random topology.

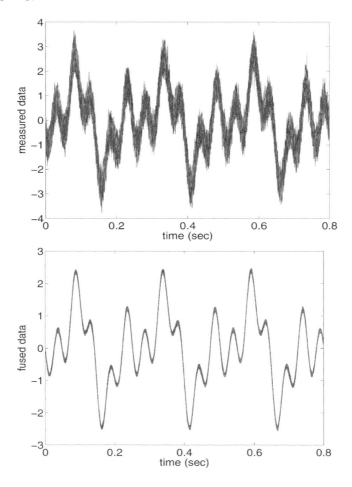

Figure 4.10: (top) sensor measurements r2(t) + vi(t) and (bottom) fused sensor data via a low-pass consensus filter with a regular network topology.

4.5 Application of Kalman Filter Estimation

It is well known that the throughput performance of the IEEE 802.11 distributed coordination function (DCF) is very sensitive to the number n of competing stations. The objective of this section is threefold:

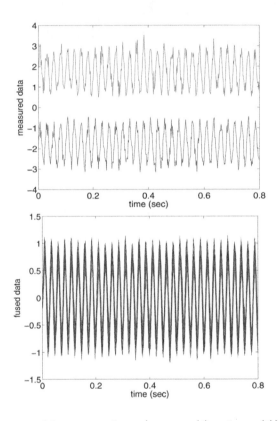

Figure 4.11: Upper and lower envelops $(\max_i u_i(t), \min_i u_i(t))$ of sensor measurements $r_3(t) + v_i(t)$ (top) and (bottom) fused measurements (i.e. states x_i) after consensus filtering in a sensor network with randomly distributed nodes.

1. We show that n can be expressed as a function of the collision probability encountered on the channel; hence, it can be estimated based on run-time measurements.

2. We show that the estimation of n, based on exponential smoothing of the measured collision probability (specifically, an ARMA filter), results in a biased estimation, with poor performance in terms of accuracy/tracking trade-offs.

3. We propose a methodology to estimate n, based on an extended Kalman filter coupled with a change detection mechanism.

This approach indicates both high accuracy as well as prompt reactivity to changes in the network occupancy status. Numerical results show that, although devised in the assumption of saturated terminals, our proposed approach results are effective also in non saturated conditions, and specifically in tracking the average number of competing terminals.

4.5.1 Preliminaries

IEEE 802.11 [422] employs DCF (Distributed Coordination Function) as the primary mechanism to access the medium. DCF is a random access scheme, based on the Carrier Sense Multiple Access with Collision Avoidance $(CSMA/CA)$ protocol and binary exponential back-off. Several performance evaluation studies of the IEEE 802.11 DCF [423], [424], [425], [426] show that performance is very sensitive to the number of stations competing on the channel, especially when the Basic Access mode is employed. Specifically, performance strongly depends on the number n of "competing" stations, i.e., the number of terminals that are simultaneously trying to send a packet on the shared medium. This information cannot be retrieved from the protocol operation. On one side, DCF does not rely on a privileged station to control the access to the channel. But even considering the existence of an Access Point (AP), as in Infrastructured 802.11 Networks, the information available at the AP is limited to the number of associated stations, a number which may be very different from the number of competing stations, i.e., stations that are actually in the process of transmitting packets.

The ability to acquire knowledge of n leads to several implications. It has been shown [427], [428] that, in order to maximize the system performance, the back-off window should be made to depend upon n. While, in the standard IEEE 802.11 protocol [422], the back-off parameters were hard-wired in the PHY layer, the idea of adaptively setting the back-off window has been recently taken into consideration in the activities of the 802.11e working group.

Indeed, the knowledge of n has several possible practical implications also in currently deployed 802.11 networks. The 802.11 standard is designed to allow both Basic Access and RTS/CTS access modes to coexist. The standard suggests that the RTS/CTS access mode should be chosen when the packet payload exceeds a given RTS threshold. However, it has been shown [425] that the RTS threshold which maximizes the system throughput is not a constant value, but significantly depends on the number n of competing stations. Specifically, as the number of stations in the network increases, the Basic Access becomes ineffective. It results to switch to the RTS/CTS mode even in the presence of short packets. Clearly, this operation requires each station to be capable of estimating n.

A second application scenario of emerging importance occurs when Infrastructured 802.11 networks are arranged in a cellular-like pattern, to provide wireless access in confined high-populated terrestrial areas, called "hot spots", such as convention centers, malls, university campuses, residential areas, etc. It appears that, in the very recent months, 802.11 is becoming a complementary (or even an alternative) access infrastructure to 3G systems, thus offering new perspectives and market shares for emerging wireless Internet providers. In this cellular-like 802.11 scenario, the estimated knowledge of traffic load and number of terminals sharing an 802.11 cell might effectively drive load-balancing and handover algorithms to achieve better network resource utilization.

In this section, we propose an efficient technique to estimate the number of competing stations in an 802.11 network. Our technique is based on an Extended Kalman filter approach, coupled with a change detection mechanism to capture variations in the number of competing terminals in the network. The estimation methodology builds on the existence of a mathematical relationship between the number of competing stations and the packet collision probability encountered on the shared medium. Such a relation is a straightforward, although originally unforeseen, extension of the analysis carried out in [425]. It is obtained in the assumption of terminals in saturation conditions (i.e. always having a packet waiting for transmission), and in the assumption of ideal channel conditions, i.e., no packet corruption and no hidden terminals and capture [429], [430]. Since this relationship is independent of the access mode adopted, it is suited for application to any DCF access mode scenario, including hybrid Basic-RTS/CTS operations.

While the extension of the work to account for non ideal channel conditions is left for future research activity, we will show that the proposed estimation mechanisms also apply to the non-saturated regime. Specifically, in such conditions, our estimation mechanism allows us to determine the average number of competing terminals (rather than the total number of terminals, as in the saturated regime).

Table 4.1: Performance parameters of the IEEE 802.11 protocol

PHY	Slot Time (σ)	CW_{min}	CW_{max}
FHSS	$50\mu s$	16	1024
DSSS	$20\mu s$	32	1024
IR	$8\mu s$	64	1024

4.5.2 802.11 Distributed coordination function

The IEEE 802.11 distributed coordination function (DCF) is briefly summarized as follows. A station with a new packet to transmit monitors the channel activity. If the channel is idle for a period of time equal to a distributed interFrame space (DIFS), the station transmits. Otherwise, if the channel is sensed busy (either immediately or during the DIFS), the station persists to monitor the channel until is measured idle for a DIFS. At this point, the station generates a random back-off interval before transmitting (this is the collision avoidance (CA) feature of the protocol), to minimize the probability of collision with packets being transmitted by other stations. In addition, to avoid channel capture, a station must wait for a random back-off time between two consecutive new packet transmissions, even if the medium is sensed idle in the DIFS time.

For efficiency reasons, DCF employs a discrete-time back-off scale. The time immediately following an idle DIFS is slotted and a station is allowed to transmit only at the beginning of each Slot Time.

DCF adopts an exponential back-off scheme. At each packet transmission, the back-off time is uniformly chosen in the range $(0, w - 1)$. The value w is called "contention window", and depends on the number of transmissions failed for the packet. At the first transmission attempt, w is set equal to a value CW_{min}, called the minimum contention window. After each unsuccessful transmission, w is doubled, up to a maximum value $CW_{max} = 2^m CW_{min}$. The values CW_{min} and CW_{max} reported in the final version of the standard [422] are PHY-specific and are summarized in Table 4.5.1.

The back-off time counter is decremented as long as the channel is sensed idle, "frozen" when a transmission is detected on the channel, and reactivated when the channel is sensed idle again for more than a DIFS. The station transmits when the back-off time reaches 0.

Figure 4.12 illustrates this operation. Two stations A and B share the same wireless channel. At the end of the packet transmission, station A waits for a DIFS and then chooses a back-off time equal to 9, before transmitting the next packet. We assume that the first packet of station B arrives at the time indicated with an arrow in the figure. After a DIFS, the packet is transmitted. Note that the transmission of

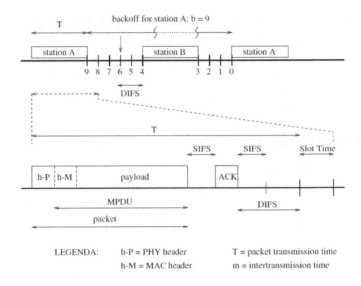

Figure 4.12: Example of basic access mechanism.

packet B occurs during the Slot Time corresponding to a back-off value, for station A, equal to 4. As a consequence of the channel being sensed busy, the back-off time is frozen to its value 4, and the back-off counter decrements again only when the channel is sensed idle for a DIFS.

Since the CSMA/CA does not rely on the capability of the stations to detect a collision by hearing their own transmission, a positive acknowledgment (ACK) is transmitted by the destination station to signal the successful packet reception. The ACK is immediately transmitted at the end of the packet, after a period of time called short interFrame space (SIFS). As the SIFS (plus the propagation delay) is shorter than a DIFS, no other station is able to detect the channel idle for a DIFS until the end of the ACK. If the transmitting station does not receive the ACK within a specified ACK Timeout, or it detects the transmission of a different packet on the channel, it reschedules the packet transmission according to the given back-off rules.

The above described two-way handshaking technique for the packet transmission is called "basic access" mechanism. DCF defines an additional four-way handshaking technique to be optionally used for a packet transmission. This mechanism, known by the name RTS/CTS, is shown in Figure 4.13. A station that wants to transmit a packet waits until the channel is sensed idle for a DIFS, follows the back-off rules explained above, and then, instead of the packet, preliminarily transmits a special short frame called "request to send" (RTS). When the receiving

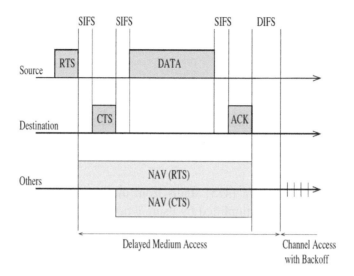

Figure 4.13: RTS/CTS access mechanism.

station detects an RTS frame, it responds, after a SIFS, with a Clear To Send (CTS) frame. The transmitting station is allowed to transmit its packet only if the CTS frame is correctly received.

The RTS/CTS mechanism provides two fundamental advantages in terms of system performance. First, the RTS/CTS mechanism standardized in 802.11 has been specifically designed to combat the so-called problem of "Hidden Terminals", which occurs when pairs of mobile stations are unable to hear each other. In fact, the frames RTS and CTS explicitly carry, in their payload, the information of the length of the packet to be transmitted. This information can be read by any listening station, which is then able to update a Network Allocation Vector (NAV) containing the information of the period of time in which the channel will remain busy. Therefore, when a station is hidden from either the transmitting or the receiving station, by detecting just one frame among the RTS and CTS frames, it can suitably delay further transmission, and thus avoid collision.

Second, the RTS/CTS is proven to effectively increase, in most cases, the throughput performance even in ideal channel conditions. When two colliding stations employ the RTS/CTS mechanism, collision occurs only on the RTS frames, and it is detected early by the transmitting stations by the lack of CTS responses. Since, after the lack of CTS reception, packets are no longer transmitted, the duration of a collision is considerably reduced, especially when long packets are involved. The price to pay is a slightly increased transmission overhead (i.e., the

RTS/CTS frame exchange) in the case of successful transmissions. A detailed performance discussion can be found in [425].

4.6   Estimating the Competing Stations

In this section, we show that, starting from the model proposed in [425], it is an immediate task to seek deriving a formula that explicitly relates the number of competing stations with a performance criteria that can measure the run-time by each station.

The analysis proposed in [425] stems from the observation that the modeling of DCF can be greatly simplified by using a non-uniform discrete time scale, where each slot corresponds either to an empty slot (thus lasting a slot-time σ), or to a transmission or collision slot (e.g., slot 4 in figure 4.12), where the slot duration corresponds to the (random) duration of a transmission or collision. This approach allows to derive results independent of the access mode considered (Basic, RTS/CTS or a combination of the two), since the access mode employed only affects the duration of the busy slots.

Following [425], we consider a scenario composed of a fixed number n of contending stations, each operating in saturation conditions, i.e., whose transmission queue always contains at least one packet ready for transmission. Channel conditions are ideal: no hidden terminals and no packet corruption is considered. For convenience, the exponential back-off parameters are expressed as W and m, where $W = CW_{\min}$ and $CW_{\max} = 2^m CW_{\min}$, i.e. $m = log_2(CW_{\max}/CW_{\min})$.

Let p be the probability (called conditional collision probability) that a packet being transmitted on the channel collides, and let τ be the probability that a station transmits in a randomly chosen slot time. In the fundamental assumption that, regardless of the number of retransmissions suffered, the probability p is constant and independent at each transmission attempt, it has been shown in [425] that:

1. the probability can be expressed as a function of p as:

$$\tau = \frac{2(1-2p)}{(1-2p)(W+1) + pW(1-(2p)^m)} \tag{4.84}$$

2. the probability p can be expressed as a function of τ and n as:

$$p = 1 - (1-\tau)^{n-1} \tag{4.85}$$

Substituting , as expressed by (4.84), into (4.85), and solving the equation with respect to n, we obtain:

$$n = f(p) = 1 + \frac{log(1-p)}{log\left(1 - \frac{2(1-2p)}{(1-2p)(W+1)+pW(1-(2p)^m)}\right)} \tag{4.86}$$

This equation is of fundamental importance for the subsequent analysis. Effectively, it provides an explicit expression of n versus the conditional collision probability p, and the (known and constant) back-off parameters m and W. Since the conditional collision probability p can be independently measured by each station by simply monitoring the channel activity, it follows that each station can estimate the number of competing stations.

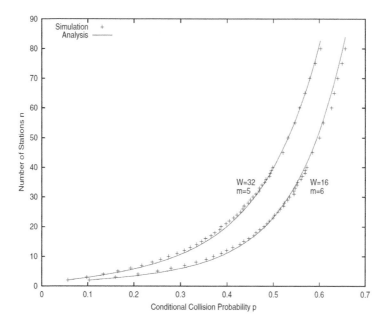

Figure 4.14: Number of stations versus conditional collision probability.

Specifically, each individual station can efficiently measure p as follows. We recall that the conditional collision probability p is defined as the probability that a packet transmitted by the considered station fails. This happens if, in the slot time selected for transmission, another transmission occurs. It might appear that the estimation of p requires each station to count the number of failed transmission and divide such a number for the total transmission attempts. However, it is imperative to understand that a much more efficient procedure is to monitor all the slot times (thus significantly increasing the number of samples), regardless of the fact that a transmission attempt is performed. Since in each busy slot an eventual packet transmission would have failed, the conditional collision probability can be obtained by counting the number of collisions experienced, C_{coll}, as well as the number of observed busy slots, C_{busy}, and dividing this sum by the total number

Table 4.2: Packet format and parameter values

packet payload	8184 bits
MAC header	272 bits
ACK length	112 bits + PHY header
PHY header	128 bits
Channel Bit Rate	1 Mbit/s
Propagation Delay	$1\ \mu s$
RxTx_Turnaround_Time	$20\ \mu$ s
Busy_Detect_Time	$29\ \mu$ s
SIFS	$28\ \mu s$
DIFS	$130\ \mu s$
ACK_Timeout	$300\ \mu s$
Slot Time (σ)	$50\ \mu s$

B of observed slots on which the measurement is taken, i.e.:

$$p = \frac{C_{busy} + C_{coll}}{B} \tag{4.87}$$

The agreement of formula (4.86) with respect to simulation results is shown in figure 4.14. This figure plots the number of contending stations n versus the conditional collision probability p, for two different sets of back-off parameters corresponding to the two different physical layer specifications (table I): FHSS, characterized by $W = 16$ and $m = 6$, and DSSS characterized by $W = 32, m = 5$. Lines represent the analytical relation given in (4.86), while symbols provide simulation results[†]. Each simulation point has been obtained considering a constant number n of stations, each in saturation conditions, and measuring the resulting conditional collision probability p. The values of the parameters used in the simulation program are summarized in Table II. The packet size has been set to a constant value. No MSDU fragmentation occurs, so that each MSDU corresponds exactly to an MPDU. Each MPDU is composed of a payload, a MAC header, and a PHY header, whose sizes, shown in table II, are those defined in [422], except for the payload length that we have chosen, equal to about half of the maximum value defined in the standard.

[†]Simulation results have been obtained using a custom-made object-oriented event-driven simulator software written in C++. The simulation program reproduces all the details of DCF, as defined in [422]. Simulation results have been obtained using the basic access procedure described earlier. It is important to note that the same curve would be obtained with RTS/CTS. Each simulation run lasts 1000 seconds. To reach saturation conditions, the offered load has been set greater than the per-MS throughput, and a 10 second warmup time has been added at the beginning of the simulation.

The figure shows that the agreement between simulation results (symbols) and analytical results (lines) is remarkable: the difference between simulation and analysis never exceeds 3%.

4.6.1   ARMA filter estimation

To provide a run-time adaptive estimation of n, it is sufficient to define a convenient run-time estimation algorithm, so that (depending on the specific application in mind) each station or Access Point, on the basis of channel monitoring, can independently evaluate the time-varying number of competing stations in the network.

In general, run-time estimation is provided by simple mechanisms, such as AR (Auto Regressive) or ARMA (Auto Regressive Moving Average) filters. In particular, we have evaluated the effectiveness of the following estimator:

$$\begin{cases} \hat{p}(t+1) = \alpha\hat{p}(t) + \frac{(1-\alpha)}{q}\sum_{i=0}^{q-1} C_{t-i} \\ \hat{n}(t+1) = f(\hat{p}(t+1)) \end{cases} \tag{4.88}$$

In this equation, $\hat{p}(t)$ is an ARMA smoothing of the conditional collision probability p. The number of competing stations is estimated from $\hat{p}(t)$ by using the non linear function $f()$ given in (4.86). The estimation $\hat{p}(t)$ is built upon the computation of the number of busy/idle slots encountered on the channel. Specifically, C_{t-i}, with $i = 0, \ldots, q-1$ are the last q slot samples. C_i is equal to 0 if in the i-th slot, either the station does not transmit and sees an empty slot, or the station transmits with success. Conversely, C_i is equal to 1 if the channel is sensed busy during the i-th slot, or the station transmits without success [‡]. We prefer an ARMA filter rather than a more traditional AR filter (i.e. $q = 1$), since the moving average taken on the last q samples better smooths the fast time scale fluctuations due to the 0/1 gross quantization of each input sample C_{t-i}.

Figure 4.15 shows the temporal behavior of the running estimate for a reference station. In the figure, two simulation plots of 400 seconds each are reported (the simulation is restarted at time 400s with different parameters). The initial number of stations in the network is set to 10, and this number is doubled after 200 seconds, to simulate an abrupt change in the network state. In the figure, the leftmost plot (seconds 0 to 400) shows the case of $\alpha = 0.999$, while the rightmost plot shows the case of $\alpha = 0.995$. In both cases, $q = 10$ has been used in the filter (3.9). Note that the value q is used to smooth the measurements that are then used to feed the exponentially weighted average, and it only marginally affects the filter performance. Conversely, the selection of a suitable value α, i.e., the filter memory,

---

[‡]Note that, according to (4.86), $f(1)$ is not defined. However, since $\hat{p}(t)$ can be equal to 1 only asymptotically, our estimation rule $\hat{n}(t) = f(\hat{p}(t))$ can be practically applied.

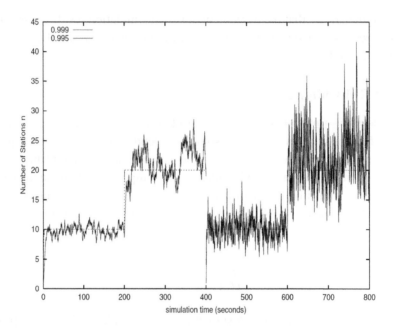

Figure 4.15: Run-time ARMA estimate of the number of contending stations n — $\alpha = 0.999$ and $\alpha = 0.995$.

determines the tradeoff between the estimation accuracy and the response time in the case of changes in the number of competing stations.

From the analysis of Figure 4.15, a number of interesting remarks can be drawn. First, we see that, for both values α considered, the estimate rapidly adapts to sudden changes in the network configuration. A closer look at the leftmost plot in Figure 4.15 in the neighborhood of time 0 and time 200 shows that the tracking performance decreases with an increasing number of stations. This behavior is motivated by the fact that, even if the value α is constant, the time constant of the filter (i.e., the filter memory), when expressed in seconds, is given by $\mathbf{E}[slottimeduration]/1\alpha$.

However, Figure 4.15 shows that the accuracy of the estimation degrades as the number of stations increases. This phenomenon is due to the slope of the curve shown in Figure 4.14, which plots $n = f(p)$ as given by (4.86). As the number of stations increases, the slope increases, too. This implies that the errors in the collision probability are amplified in the evaluation of the number of contending stations. Moreover, we see that, due to the non linearity of the relation $n = f(p)$, the estimation $\hat{n}$ is biased, as it is $f(\mathbf{E}[\hat{p}]) \neq \mathbf{E}[f(\hat{p})]$. Specifically, as clearly shown by the rightmost plot of Figure 4.15, the average estimated value $\hat{n}$ is greater than the real value n.

This fact is clearly shown in Figure 4.16, which plots the probability distributions $P_p(\hat{p})$ and $P_n(\hat{n})$ of both the collision probability estimate and the resulting network occupancy estimate. The plots have been obtained for the case of $n = 20$ contending stations, for $q = 10$, and for both values α considered in the previous Figure 4.15. The x-axis is graduated in terms of percent deviation from the nominal value $p = f^{-1}(n = 20)$. The spread of the P_p distributions depend on the filter parameters α and q. The little bias from the value 0 is due to the small mismatch between the analytical relation (3.7) and the simulation results. While the distributions P_p are almost symmetric, their images P_n through the non linear function $n = f(p)$ given in (4.86), are very distorted. The distortion is more and more evident as the spread of the $\hat{p}$ distribution increases (i.e., as the α coefficient decreases). Summarizing, the considered ARMA estimation approach is biased, and an unbiased estimate of n as function of the p estimates is possible only asymptotically, if the p estimates are very accurate, i.e., the filter memory is set to a very large value.

4.6.2 Extended Kalman filter estimation

In the previous section, we have shown that simple ARMA filtering approach results are unsatisfactory in terms of accuracy/ tracking ability tradeoff. We now show that a significantly better performance can be achieved by using an extended Kalman filtering technique. The rationale for using a Kalman filter approach is threefold. First, it allows us to adaptively tune the filter memory (say, the factor α in the ARMA filter) to faster track variations in the network occupancy status. Second, it allows us to significantly improve the accuracy of the estimation, by exploiting additional information available in the model (i.e., state updating laws, and variance of the measurement of p), whereas this information cannot be included in an elementary AR or ARMA approach. Third, the resulting complexity is comparable with that of an elementary ARMA filter, i.e., the use of this filtering technique does not have practical drawbacks.

4.6.3 Discrete state model

Let us first focus on the definition of the temporal steps at which the estimation is updated. Time is discretized in steps of B slot-times, where B is a constant predefined value. Within each time step k, the considered MS (or AP) provides a measure p_k of the conditional collision probability based on equation (4.87), and

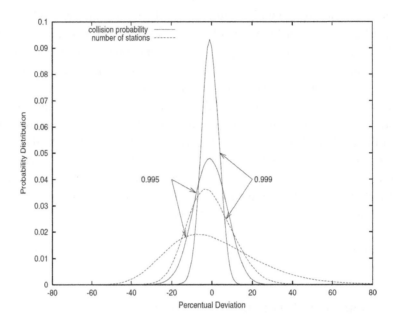

Figure 4.16: Probability density function of the estimates $\hat{p}$ and $\hat{n}$.

rewritten here, for convenience, as

$$p_k = \frac{1}{B} \sum_{i=(k-1)B}^{kB-1} C_i \qquad (4.89)$$

where, as explained in the previous section, for slot-time i, $C_i = 0$ if the slot-time is empty or the station transmits with success, while $C_i = 1$ if the slot-time is busy or the station transmits without success. p being the real (unknown) conditional collision probability suffered on the radio channel, then, for every slot-time i, $Prob(C_i = 1) = p$ and $Prob(C_i = 0) = 1 - p$. Therefore, p_k is a random variable with binomial distribution:

$$Prob\left(p_k = \frac{b}{B}\right) = \binom{B}{b} p^b (1-p)^{B-b} \quad b \in (0, B) \qquad (4.90)$$

The mean value and variance of the measure p_k are obviously $E[p_k] = p$ and $Var[p_k] = p(1-p)/B$

To devise a Kalman filter estimation technique, we need to provide a state model which consists of:

   i) a state updating law for the system under consideration, and

   ii) a measurement model, that is, the relationship between state and measures.

In our case, the system state is trivially represented by the number n_k of stations in the network at discrete time k. In most generality, the network state evolves as

$$n_k = n_{k-1} + w_k \tag{4.91}$$

where the number of stations n_k in the system at time k is given by the number of stations at time $k-1$ plus a random variable w_k (in Kalman filtering terms, a noise hereafter referred to as state noise) which accounts for stations that have activated and/or terminated in the last time interval. The suitable statistical characterization of the (non stationary) state noise w_k is a key issue in the Kalman filter design, and it will be discussed in section V-C. At the moment, just note that we won't assume any model for the arrival/departure of stations, from which the properties of w_k might be derived.

Regarding the measurement model, it is the measure of the conditional collision probability p that each station can carry out via the samples p_k obtained as in equation (3.10). If, at time k, there are n_k stations in the system, then, the conditional collision probability can be obtained as $h(n_k)$, where h is the inverse function of (3.7). Note that such inversion is possible, since the direct function is monotone. We can the thus rewrite p_k as:

$$p_k = f^{-1}(n_k) + v_k = h(n_k) + v_k \tag{4.92}$$

where, based on the previous consideration, v_k is a binomial random variable with zero mean and variance:

$$Var[v_k] = \frac{h(n_k)[1 - h(n_k)]}{B} \tag{4.93}$$

Observe that (4.91) and (4.92) thus provide a complete description of the state model for the system under consideration.

4.6.4 Extended Kalman filter

Once the state model described by (4.91) and (4.92) is given, the definition of the Extended Kalman filter is a straightforward application of basic theory, see [431]. Let $\hat{n}_{k-1}$ be the network state estimated at time instant $k-1$ and P_{k-1} be the corresponding error variance. According to the particularly simple structure of equation (4.91), at each step, the one-step state prediction is equal to the previous state estimate. Hence, the estimate n_k of the number of stations at time k is computed from the estimate at the time instant $k-1$ by the relation:

$$\hat{n}_k = \hat{n}_{k-1} + K_k z_k \tag{4.94}$$

In this relation, z_k is the innovation supported by the k-th measure, given by

$$z_k = p_k - h(\hat{n}_{k-1}) \tag{4.95}$$

where we have applied also in this case the property that the one-step state prediction is equal to the previous state estimate. In equation (3.15) K_k is the Kalman gain, given by

$$K_k = \frac{(P_{k-1} + Q_k)h_k}{(P_{k-1} + Q_k)h_k^2 + R_k} \tag{4.96}$$

In the above Kalman gain computation equation, the following symbolism has been adopted:

- Q_k is the variance of the random variable w_k, i.e., the state noise introduced in the state updating law (4.91). The values adopted for Q_k will be discussed shortly.

- R_k is the variance of the measure p_k, i.e., with reference to equation (4.92), it represents the estimated variance of the random variable v_k, obtained from equation (4.93) by replacing the actual state n_k with $\hat{n}_{k-1}$ (being the estimate coincident with the one-step predicted state itself). Summarizing:

$$R_k = \frac{h(\hat{n}_{k-1})(1 - h(\hat{n}_{k-1}))}{B}$$

- h_k is the sensitivity of the measurement, linearized around the state estimation $\hat{n}_{k-1}$. Coefficient h_k is computed by taking the derivative

$$h_k = \left. \frac{\partial h(n)}{\partial n} \right|_{n=\hat{n}_{k-1}}$$

Finally, the error variance of the new estimate is also recursively computed as:

$$P_k = (1 - K_k h_k)(P_{k-1} + Q_k) \tag{4.97}$$

Regarding the initial conditions, quick convergence is guaranteed when the initial error variance P_0 is set to a large value (in our numerical results, we have used $P_0 = 100$). In these conditions, the initial estimate of the state is not relevant, and can be set to any value (we used $n_0 = 1$).

4.6.5 Selection of state noise statistics

In order to complete the design of the Extended Kalman filter, it remains to specify the statistics of the state noise process w_k used in the state update equation (4.91). In several applications of the Kalman filter, it is generally assumed that w_k is a stationary process with a given constant variance Q. The tuning of the Kalman filter is then performed by appropriately selecting this variance.

However, this approach is quite simplistic, as it leads again to the issue (discussed for the ARMA case) of trading estimation accuracy with tracking ability. In fact, high values for Q allow to quickly react to state changes, but imply a reduced accuracy in the estimation, i.e. high error variance P_k. Conversely, low Q values give accurate estimates in stationary conditions, but have very slow transient phases when abrupt state variations occur.

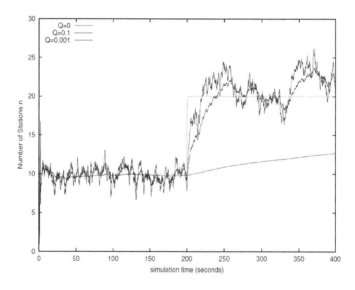

Figure 4.17: Tracking ability of the filter.

The unsatisfactory performance of such a constant noise variance approach is quantitatively shown in Figure 4.17, which compares the performance of the Kalman filter for three different values Q, namely $Q = 0$, $Q = 0.1$ and $Q = 0.001$. The simulation scenario is the same as that adopted for Figure 4.15: initially, 10 competing stations are considered, while, at time 200 seconds, an additional 10 stations abruptly activate. The figure shows that the only case in which the transient time is kept in the order of just a few seconds is the case $Q = 0.001$. However, the price to pay is the occurrence of non marginal fluctuations (order of 20% error) in the estimation. The case of $Q = 0$ is also interesting. When the number of terminals is constant, the estimation is impressively accurate. However, in such a case, the resulting state update relation of (4.91) forces the estimation to remain, in practice, stuck to the initial value, which is a serious problem when, as in the test case shown in the figure, the number of stations varies.

The above considerations, combined with the flexibility of the state model (4.91) which does not require w_k to be a stationary process, suggest that a better approach consists of selecting a time-varying noise variance Q_k. In particular, Q_k should be set to 0, or at least to a very small value, when the number of competing stations in the network appears to remain constant. Conversely, if it appears that the network state has changed, it should be set to a possibly large value Q : it is sufficient to set $Q_k = Q$ for just a single time instant k, as the sudden increase in the noise variance is propagated in the estimation error variance P_k through equation (4.97). In the next section, we show how this operation can be automated.

4.6.6   Change detection

To automate the selection of the state noise variance Q_k, we have associated to the Kalman filter a second "change detection" filter, devised to estimate whether a state change has occurred. A change in the network state can be detected by analyzing the innovation process z_k, defined in equation (4.95). If the network occupancy state is constant, the innovation process z_k is a white process with zero mean. Conversely, if the network state changes, the process z_k will move away from its zero mean value.

Among the several available statistical tests, we have implemented a change detection filter based on the CUSUM (CUmulative SUMmary) test [432], which is very intuitive to derive and very effective in terms of simplicity and performance. The CUSUM test is based on two filtered versions g_k^+ and g_k^- of the innovation process z_k. For convenience, we use a related process s_k which represents the innovation process normalized with respect to its standard deviation. Note that the rationale behind using the normalized innovation process s_k instead of the innovation process z_k is that, in this manner, the design parameters v and h of the CUSUM test can be kept constant. If the innovation sequence z_k were used, then the CUSUM test parameters should have been configured to explicitly depend on the estimated network state and on the related error variance, that is:

$$s_k = \frac{z_k}{\sqrt{(P_{k-1} + Q_k)h_k^2 + R_k}} \tag{4.98}$$

The samples g_k^+ and g_k^- are constructed from the input values s_k as follows:

$$\begin{cases} g_k^+ = \max(0, g_{k-1}^+ + s_k - v) \\ g_k^- = \min(0, g_{k-1}^- - s_k + v) \end{cases} \tag{4.99}$$

In these equations, v, called "drift parameter," is a filter design parameter. The smaller v is, the more sensible the test results are to fluctuations of the process s_k. As initial conditions, $g_0^+ = 0$ and $g_0^- = 0$. If a change in the network state occurs, the magnitude of one between g_k^+ or g_k^- tends to increase unlimited. For example, suppose that a new station activates. Then, the collision probability predicted by the current state estimate $\hat{n}_k$ results lower, on average, than the measured one. Therefore, the mean value of the normalized innovation s_k (equations (4.98) and (4.95)) becomes positive, and the process g_k^+ starts diverging when such a mean value becomes greater than v. Conversely, divergence occurs for g_k^- when there is a station departure. Hence, an additional CUSUM test parameter, called the "alarm threshold" h is defined, i.e., the change detection filter sends an alarm when $g_k^+ > h$ or $g_k^- < -h$. After an alarm, both sequences g_k^+ and g_k^- are restored to the value zero. The greater the value h, the lower the probability that a false alarm is detected, but the longer is the time to detect a state change. In this chapter, we have set the parameters h and v to be constant values; indeed, we recall that techniques are available [432] to automate the setting of these test design parameters (e.g. to match a given false-alarm rate and change the detection delay).

Alarms coming from the CUSUM test are used to adaptively set the variance Q_k of the noise w_k:

- when the change detection filter does not detect a state change (i.e. no alarm arrives at time k), $Q_k = 0$. This allows to use the new measure p_k (more precisely, the innovation z_k) to increase the accuracy of the former estimation.

- Conversely, upon an alarm generated at time k, the value, Q_k is set (for the instant of time k only) to a sufficiently large constant value Q (as discussed in the next section, this parameter marginally affects the estimator performance). This represents a noise impulse in the state update equation (3.12), which allows the Kalman filter to "move away" from the former estimate and therefore to rapidly converge to a new estimate.

4.6.7 Performance evaluation

The effectiveness of the proposed Kalman filter estimate is demonstrated in Figure 4.90. The configuration parameters for the change detection filter are $v = 0.5$ and $h = 10$. Upon a change detection alarm, the state noise variance is set to $Q = 5$, while it is set to 0 when no alarms occur. In this figure, we have simulated a scenario in which the number of stations in the network increases in steps (1, 2, 3, 5, 10, 25 and 15 stations). Although unrealistic, this scenario allows us to prove that our proposed estimation technique is able to track abrupt variations in the network state, while keeping a very high level of accuracy in the estimation. The alarms coming from the change detection filter are also reported in the figure, as small impulses on the x-axis. They demonstrate that the parameter value Q is not very critical in terms of filter performance. In fact, as shown in the step from 10 to 25 stations (simulation time 350s) and from 25 to 15 stations (simulation time 450s), it may happen that the value $Q = 5$ adopted is too small to allow the Kalman filter to capture a large variation in the network state. Indeed, this is not a critical problem, as the change detection filter will eventually send a second alarm after a few seconds (so that the Kalman filter convergence to the state change results is always possible). For comparison purposes, Figure 4.18 reports results for two ARMA filters with $\alpha = 0.9995$ and $\alpha = 0.999$. Both the ARMA filters are not satisfactory: the first in terms of tracking ability, the second in terms of estimation accuracy.

Our model, as well as all simulation results up to now, has been developed in the assumption of saturated conditions, i.e. all stations in the network are assumed to always have a packet to transmit in their transmission buffer. Figure 4.19 shows the behavior of the proposed estimation technique when the terminals are not in saturated conditions. The Kalman filter parameters are the same as those used in the previous figure, while the ARMA filter has a memory factor $\alpha = 0.999$. In particular, Figure 4.19 reports a simulation run for a network scenario of 20 stations. Packets arrive to each station according to a Poisson process, whose rate is initially set to be lower than the saturation throughput. The arrival rate is subsequently increased so that, at the end of the simulation time, all stations are in saturation conditions.

In the non saturated regime, the number of terminals attempting to transmit a packet (i.e. the number of competing terminals) shows fairly large and fast fluctuations, as highlighted by the dashed plot in Figure 4.19. Neither our Kalman filter, nor an ARMA filter, are able to follow these fast fluctuations. Instead, they both appear to estimate the average number of competing stations. In other words, the proposed estimation technique, devised in saturation assumption, appears to apply also to non saturated conditions, provided that the estimation target becomes the average number of competing stations (rather than the total number of stations in saturation conditions).

By comparing the Kalman estimator with the ARMA one, we see that our proposed mechanism appears to provide a smoother estimation in both saturated and non saturated regimes, although the different level of accuracy can be fully appreciated only in the saturated regime.

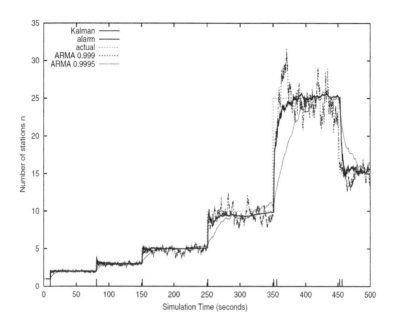

Figure 4.18: Kalman filter estimation in saturated network conditions.

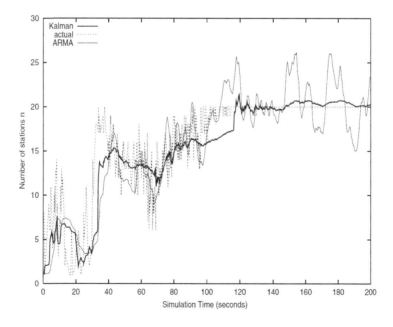

Figure 4.19: Kalman filter estimation in non saturated network conditions.

4.7   Notes

This chapter has introduced consensus filters as a tool for distributed sensor fusion in sensor networks. The consensus filter is a dynamic version of the average-consensus algorithm that has been extensively used for sensor fusion as well as other applications that involve networked dynamic systems and collaborative decision making. It was mentioned that based on a new scalable Kalman filtering scheme, a crucial part of the solution is to estimate the average of n signals in a distributed way. It was shown that consensus filters effectively solve this dynamic average-consensus problem. This distributed filter acts as a low-pass filter induced by the information flow in the sensor network. In addition, ϵ-tracking properties of consensus filters for sensor fusion were analyzed in detail. The byproduct of this analysis was a novel type critical phenomenon in complex networks that relates the size of the sensor network to its capability to track relatively fast signals. This limitation was characterized as a tracking uncertainty principle. Simulation results for large regular and random sensor network were presented.

Then we discussed the problem of estimating the number of competing stations in an 802.11 distributed coordination function Wireless LAN. The basis for the estimation is provided by a numerically accurate closed form expression that relates the number of competing stations to the probability of a collision seen by a packet being transmitted on the channel by a selected station. By independently monitoring the transmissions eventually occurring within each slot-time, each station is in the condition to estimate, though this relation, the number of competing terminals.

We then evaluated the effectiveness of a practical estimation technique based on ARMA filtering. We found that such an approach has several drawbacks. First, as expected for all ARMA filters, estimation accuracy is traded off with the tracking ability of the filter. Moreover, we have shown that, in our specific problem, the ARMA estimation results are biased, due to the strong non linearity of the expression that relates the number of terminals to the collision probability.

Hence, we have proposed a better approach, based on an extended Kalman filter estimate, in conjunction with a change-detection mechanism devised to track variations in the network state and accordingly feed the Kalman filter.

4.8   Proposed Topics

1. Develop an adaptive medium access control algorithm for minimizing the power consumption while guaranteeing reliability and delay constraints of the IEEE 802.11 protocol. Take into account retry limits, acknowledgment (ACK) and an unsaturated traffic regime. Study the robustness of the

protocol to possible errors during the estimation process on the number of devices and traffic load.

2. Generate a quantitative comparison between the various distributed Kalman filtering algorithms in terms of a generic system model and arbitrary wireless sensor network architecture with respect to some performance criteria that you choose for a typical application.

3. It is recommended to re-study the problem of estimation of the number of competing stations to account for non-ideal channel conditions. How can one show that the proposed estimation mechanisms in the foregoing section still apply to the non-saturated regime? What would be the merits of the estimation mechanism in this case?

Chapter 5

Expectation Maximization

The data-fusion detection and isolation (DFFDI) process becomes more potentially challenging if the faulty component of the system causes partial loss of data. In this chapter, we present an iterative approach to DFFDI that is capable of recovering the model and detecting the fault pertaining to that particular cause of the model loss. The method developed is an expectation-maximization (EM) based on forward-backward Kalman filtering. We test the method on a rotational drive-based electro-hydraulic system using various fault scenarios. It is established that the method developed retrieves the critical information about presence or absence of a fault from partial data-model with minimum time-delay, and provides accurate unfolding-in-time of the finer details of the fault, thereby completing the picture of fault detection and an estimation of the system under test. This in turn is completed by the fault diagnostic model for fault isolation. The experimental results obtained indicate that the method developed is capable of correctly identifying various faults, and then estimating the lost information.

5.1 General Considerations

Table 5.1 contains the variables are used throughout this chapter.

Data-fusion fault detection and isolation (DFFDI) has always been the subject of considerable interest in the process industry where the whole model structure for the plant is usually not available. The intensity of importance of this subject is the ever increasing requirement of the reliable operation of control systems, which are, in most cases, subject to a number of faults either, in the internal closed loops or from environmental factors. The data generated from the assumed model are compared with measured data from the physical system to create residuals that relate to specific faults. Faults encountered can be of many types, starting from a faulty

169

Table 5.1: Nomenclature

Symbols	Function
$I(t)$	input current
τ_v	servo-valve time constant
$A_v(t)$	servo-valve opening area
K_x	servo-valve area constant
K_v	servo-valve torque motor constant
$Q_1(t)$	flow from the servo-valve
$Q_2(t)$	flow to the servo-valve
$P_L(t)$	load pressure difference
P_s	source pressure
C_d	flow discharge coefficient
ρ	fluid mass density
$P_1(t), P_2(t)$	rotational drive chambers pressure
β	fluid bulk modulus
D_m	actuator: volumetric disp. parameter
C_L	leakage coefficient
$\theta(t)$	angular displacement
$sign$	change in direction of actuator motion
J	moment of inertia
B	viscous damping coefficient
T_L	load torque
C_{0_i}	latent variable

sensor in the production line to a broken transducer or a burnt out coil not transforming the assigned accurate information. Once system faults have occurred, they can cause unrecoverable losses and result in unacceptable environmental pollution, etc. Occasionally, the occurrence of a minor fault has resulted in disastrous effects. With an accurate process model and under appropriate assumptions, it is possible to accomplish fault detection and isolation (FDI) for specific fault structures.

Generally, fault diagnostic methods can be divided into two categories:

- Model-based fault diagnostics, and

- Data-fusion fault diagnostics, including knowledge-based Fault diagnostics.

Model-based fault diagnostic methods are generally dependent on the mathematical models of the process developed either from first principles or from identification of the system. The data extracted from the model is then compared with measured data from the physical system to create residuals that relate to specific faults. With an accurate process model and under appropriate assumptions, it is possible to accomplish fault diagnostics for specific fault structures (see, for example, [450, 46, 49, 52]). Data-fusion methods, on the other hand, rely on process measurements in order to perform fault diagnostics. Analyzing process measurements gives the location and direction of the system trajectory in the state space. The databases contain a great amount of redundant information and must be processed by means of suitable algorithms, most of which belong to the great area of data mining. The strategy presented in this work is an alternative methodology to process large databases and to design appropriate monitoring systems integrated with fault treatment approaches. Tools such as adaptive principle component analysis, fuzzy logic, and neural networks have also demonstrated their capability on treating important data-fusion systems. It is then possible, particularly for linear process systems, to extract information about the fault by comparing the location and/or direction of the system trajectory in the state space with past faulty behavior, for example, [56]. Several methods have been developed that manipulate the measured data to reduce their dimension and extract information from the data with respect to actuator/sensor faults using principle component analysis (PCA) or partial least squares (PLS) techniques (for example, [51, 55, 54]. These methods reduce the dimensionality of the data by eliminating directions in the state space with low common-cause variance. Other methods have been developed that consider the contribution of particular states to the overall shift from normal operation [51]. Some data-fusion methods take advantage of PCA to find correlations within the data [45]. Work has also been done to group data-fusion on process structure or process distinct time scales as in multi-block or multi-scale PCA. While many of these methods have been successful in achieving fault detection, fault isolation remains a difficult task, particularly for non-linear processes where historical data under faulty operation are insufficient to discriminate between faults. For a comprehensive review of model-based and data-fusion FDI methods, the reader may refer to [58, 57].

In this chapter, a data-fusion fault diagnostics scheme is developed using expectation maximization as on the data-fusion Kalman filter. The salient feature is that it does not rely on prior knowledge and mathematical information about the system under consideration. We construct an iterative approach to data-fusion fault detection and isolation (DBFDI) that is capable of recovering the model and detecting the fault pertaining to that particular cause of the model loss. It is essentially an expectation-maximization (EM) based on forward-backward Kalman filtering. We test the method on a rotational drive-based electro-hydraulic system using various fault scenarios. It is shown that the method developed retrieves the critical information about presence or absence of a fault from a partial data-model with minimum time-delay and provides accurate unfolding-in-time of the finer details of the fault, thereby completing the picture of fault detection and an estimation of the system under test. In turn, this is completed by the fault diagnostic model for fault isolation. The experimental results obtained indicate that the method developed is capable of correctly identifying various faults, and then estimating the lost information.

5.2   Data-Fusion Fault Diagnostics Scheme

To construct an effective fault diagnostics scheme, we have assumed that various faults in the system have been successfully monitored, estimated and protected through tolerance by the encapsulation of the expectation maximization algorithm and diagnostic model scheme. Fault tolerant control systems are designed to achieve high reliability and survivability of the dynamic systems and processes. The fault tolerant scheme can work in various steps. The fault tolerant scheme has the following general possible steps [53, 47]:

• Fault modeling of the system comprising sensor and actuator faults.

• Fault detection and estimation using forward-backward Kalman filter-based expectation maximization.

• Fault isolation using diagnostics model.

Consider a linear-time discrete model of the system:

$$
\begin{aligned}
x(k+1) &= A_d x(k) + B_d u(k) \\
y(k) &= C_d x(k) + D_d u(k)
\end{aligned}
\tag{5.1}
$$

where A_d, B_d, C_d, and D_d are the matrices of the discrete-time system of appropriate dimensions.

5.2.1 Modeling with sensor and actuator faults

During the system operation, faults or failures may affect the sensors, the actuators, or the system components. These faults can occur as additive or multiplicative faults due to a malfunction or equipment aging. For fault detection and identification (FDI), a distinction is usually made between additive and multiplicative faults. The faults affecting a system are often represented by a variation of system parameters. Thus, in the presence of a fault, the system model can be written as:

$$\begin{aligned} x_f(k+1) &= A_f x_f(k) + B_f u_f(k) \\ y_f(k+1) &= C_f x_f(k) \end{aligned} \tag{5.2}$$

where the new matrices of the faulty system are defined by:

$$A_f = A + \delta A; B_f = B + \delta B; C_f = C + \delta C; \tag{5.3}$$

where δA, δB, and δC correspond to the deviation of the system parameters with respect to the nominal values. However, when a fault occurs in the system, it is very difficult to get these new matrices on-line. Process monitoring is necessary to ensure effectiveness of the process control, and consequently a safe and a profitable plant operation. As presented in the next paragraph, the effect of actuator and sensor faults can also be represented as an additional unknown input vector acting on the dynamics of the system or on the measurements. The effect of actuator and sensor faults can also be represented using an unknown input vector $f_j \in \Re^l$, $j = a$ (for actuators), s (for sensors) acting on the dynamics of the system or on the measurements.

5.2.2 Actuator faults

It is important to note that an actuator fault corresponds to the variation of the global control input U applied to the system, and not only to u:

$$U_f = \Gamma U + U_{f0} \tag{5.4}$$

where
- U is the global control input applied to the system.
- U_f is the global faulty control input.
- u is the variation of the control input around the operating point

$$U_0, (u = U - U_0, u_f = U_f - U_0$$

- U_{f0} corresponds to the effect of an additive actuator fault.
- ΓU represents the effect of a multiplicative actuator fault with $\Gamma = diag(\alpha)$.

$$\alpha = [\alpha_1....\alpha_i....\alpha_m]^T \tag{5.5}$$
$$U_{f0} = [U_{f01}....U_{f0i}....U_{f0m}]^T \tag{5.6}$$

The i_{th} actuator is faulty if $a_i \neq 1$ or $u_{fo} \neq 0$. In the presence of an actuator fault, the linearized system 5.1 can be given by:

$$x(k+1) = Ax(k) + B(\Gamma U(k) + U_{f0} - U_0) \tag{5.7}$$
$$y(k) = Cx(k) \tag{5.8}$$

The previous equation can be re-written as follows:

$$x(k+1) = Ax(k) + B(\Gamma - 1)U(k) + U_{f0}$$
$$y(k) = Cx(k) \tag{5.9}$$

By defining $f_a(k)$ as an unknown input vector corresponding to actuator faults (5.2), an equation can be represented as follows:

$$x(k+1) = Ax(k) + Bu(k) + F_a f_a k$$
$$y(k) = Cx(k) \tag{5.10}$$

where F_a=B, and f_a= $(\Gamma - 1) U + U_{f0}$. If the ith actuator is declared to be faulty, then F_a corresponds to the ith column of matrix B and f_a corresponds to the magnitude of the fault affecting this actuator.

5.2.3 Sensor faults

In similar way, considering f_s as an unknown input illustrating the presence of a sensor fault, the linear faulty system will be represented by:

$$x(k+1) = Ax(k) + Bu(k) + F_s f_s k$$
$$y(k) = Cx(k) \tag{5.11}$$

The state-space representation of a system that may be affected by an actuator and/or sensor fault is:

$$x(k+1) = Ax(k) + Bu(k) + F_a f_a k$$
$$y(k) = Cx(k) + F_s f_s k \tag{5.12}$$

where matrices F_a and F_s are assumed to be known and f_a and f_s correspond to the magnitude of the actuator and the sensor faults, respectively. The magnitude and time occurrence of the faults are assumed to be completely unknown. In the presence of sensor and actuator faults, a system can also be represented by the unified general formulation:

$$
\begin{aligned}
x(k+1) &= Ax(k) + Bu(k) + F_x f(k) \\
y(k) &= Cx(k) + F_y f(k)
\end{aligned}
\tag{5.13}
$$

where $f = [f_a^T \quad f_s^T]^T \in \mathcal{R}^v$ ($v = m + q$) is a common representation of sensor and actuator faults. $F_x \in \mathcal{R}^{n \times v}$ and $F_y \in \mathcal{R}^{q \times v}$ are respectively the actuator and sensor fault matrices with $F_x = [B \quad 0_{n \times q}]$ and $F_y = [B \quad 0_q]$. The objective is to isolate faults. This is achieved by generating residuals sensitive to certain faults and insensitive to others, commonly called structured residuals. The fault vector f in (5.13) can be split into two parts. The first part contains the "d" faults to be isolated $f_0 \in \Re^d$. In the second part, the other "$v - d$" faults are gathered in a vector $f^* \in \Re^{v-d}$. Then, the system can be written by the following equations:

$$
\begin{aligned}
x(k+1) &= Ax(k) + Bu(k) + F_x^0 f^0(k) + F_x^* f^*(k) \\
y(k) &= Cx(k) + F_y f(k) + F_y^0 f^0(k) + F_y^* f^*(k)
\end{aligned}
\tag{5.14}
$$

Matrices F_x^0, F_x^*, F_y^0, and F_y^*, assumed to be known, characterize the distribution matrices of f^* and f^0 acting directly on the system dynamics and on the measurements respectively. In case of an ith actuator fault, the system can be represented according to 5.14 by:

$$
\begin{aligned}
x(k+1) &= Ax(k) + Bu(k) + B_i f^0(k) + [\bar{B}_i \quad 0_{n \times q}] f^*(k) \\
y(k) &= Cx(k) + [0_{q \times (p-1)} \quad I_q] f^*(k)
\end{aligned}
\tag{5.15}
$$

where B_i is the ith column of matrix B and $\bar{B}_i$ is matrix B without the ith column. Similarly, for a jth sensor fault, the system is described as follows:

$$
\begin{aligned}
x(k+1) &= Ax(k) + Bu(k) + [B \quad 0_{n \times (q-1)}] f^*(k) \\
y(k) &= Cx(k) + E_j f^0(k) + [0_{q \times p} \quad \bar{E}_j] f^*(k)
\end{aligned}
\tag{5.16}
$$

where $E_j = [0...1...0]^t$ represents the jth sensor fault effect on the output vector and $\bar{E}_j$ is the identity matrix without the jth column.

5.2.4 The expected maximization algorithm

In what follows, we look at the situation where joint fault detection and data estimation is considered. By and large, the expected maximization (EM) algorithm is a method for finding maximum likelihood estimates of parameters in statistical models, where the model depends upon unobserved latent variables. In our case, the latent variable is C_{0_i}. The EM has two steps:

- The $E-step$ is obtained with respect to the underlying unknown variables conditioned on the observations and

- the $M-step$ provides a new estimation of the parameters (or when some of the data are missing).

With reference to [43], the formulation has been applied for the fault diagnosis case, where the EM is being used for fault detection and data estimation. In the ideal case, we will be estimating fault α_i using the measurement equation, for maximizing the corresponding log likelihood function:

$$\Upsilon_{eh_i} = C_{0_i}\alpha_{i-1} + v_i \tag{5.17}$$

where Υ_{eh_i} presents the electro-hydraulic profile, C_{0_i} represents the measurement matrix perturbed by the fault which cause some data loss which is to be recovered/estimated, α_{i-1} represents the leakage profile (a type of fault), and Υ_i represents the noise assumed to be Gaussian. For example, when the system is obeying the input–output relation (so that in $\ell\ n\ p(\Upsilon_{eh_i}/C_{0_i}, \alpha_i) = -\|\Upsilon_i - C_{0_i}\alpha_i\|^2_{\sigma_n^{-2}}$ up to some additive constant), (so that $\ell n\ p(\alpha_i) = -\|\alpha_i\|^2_{\Pi^{-1}}$) in this case, the maximum a posterior MAP estimate is given by:

$$\hat{\alpha}_i^{MAP} = argmin_{\alpha_i}[\|\Upsilon_i - C_{0_i}\alpha_i\|^2_{\sigma_n^{-2}} + \|\alpha_i\|^2_{\Pi^{-1}}] \tag{5.18}$$

Considering the case of monitoring the fault detection, however, the input C_{0_i} is not observable, as we have different scenarios for the profiles of faults depending upon the potency of the fault considered. Thus, we use the expectation-maximization algorithm, and maximize instead an average form of the log-likelihood function. Thus, the E-step for the expected maximization algorithm for the example given above when starting from an initial estimate, $\hat{\alpha}_i^0$ to estimate $\hat{\alpha}_i$ is calculated iteratively, with the estimate at the $j-th$ iteration given by:

$$\hat{\alpha}_i^{MAP} = arg\max_{\alpha_i}[E_{\alpha_i/\Upsilon_{eh_i}^{j-1}}\ell n\ p(\Upsilon_{eh_i}/C_{0_i},\alpha_i)+lnp(\alpha_i)] \tag{5.19}$$

Likewise, the M-step for the example given above will be as follows:

$$\hat{\alpha}_i^j = arg\min_{\alpha_i}[\|\Upsilon_i - E[C_{0_i}]\alpha_i\|^2_{\sigma_n^{-2}}$$
$$+ \|\alpha_i\|^2_{cov[i^*]+\|\alpha_i\|^2_{\Pi^{-1}}}] \tag{5.20}$$

Where the two moments of C_{0_i} are taken given the output Υ_i and the most recent flow/height of water estimate, $\hat{\alpha}_i^{j-1}$. We now derive the EM algorithm for the time variant case.

Consider the system expressed earlier, essentially described by the state-space model:

$$\underline{\alpha}_{i+1} = F\underline{\alpha}_i + G\underline{u}_i \qquad (5.21)$$

$$\Upsilon_{eh_i} = C_{0_i}\alpha_{i-1} + \nu_i \qquad (5.22)$$

we can obtain the *maximum a posterior* estimate by maximizing the log-likelihood as:

$$L = \ell n\ (\Upsilon_{eh_0}^t/C_0^t, \underline{\alpha}_0^t) + \ell n\ p(\underline{\alpha}_0^t) \qquad (5.23)$$

where T is the sampling time. Now, for describing the terms of likelihood, consider the two equations of the state–space model (16) and (17). Considering (5.22), we can express the first term of likelihood as:

$$
\begin{aligned}
\ell n\ p(\Upsilon_{eh_0}^t/C_0^T, \underline{\alpha}_0^T) &= \sum_{i=0}^{T} \ell n\ p(\Upsilon_{eh_i}/C_i, \underline{\alpha}_i) \\
&= -\sum_{i=0}^{T} \|\Upsilon_{eh_i} - C_i\underline{\alpha}_i\|_{1/\sigma_n^2}^2 \qquad (5.24)
\end{aligned}
$$

Similarly, considering (5.21), we can express the second term of likelihood as:

$$
\begin{aligned}
\ell n\ p(\alpha_0^t) &= \sum_{i=1}^{T} \ell n\ p(\underline{\alpha}_i, \alpha_{i-1}) + \ell n\ p(\underline{\alpha}_0) \\
&= -\sum_{k=1}^{T} \|\underline{\alpha}_k - F\alpha_{k-1}\|_{1/\sigma_n^2 GG*}^2 - \|\underline{\alpha}_0\|_{\pi_0^{-1}}^2 \qquad (5.25)
\end{aligned}
$$

Considering these two expressions 5.24 and 5.25, we get:

$$
\begin{aligned}
L &= -\sum_{i=0}^{T} \|\Upsilon_{eh_i} - C_i\underline{\alpha}_i\|_{1/\sigma_n^2}^2 \\
&\quad - \sum_{k=1}^{T} \|\underline{\alpha}_k - F\alpha_{k-1}\|_{1/\sigma_n^2 GG*}^2 - \|\underline{\alpha}_0\|_{\pi_0^{-1}}^2 \qquad (5.26)
\end{aligned}
$$

Now, the forward-backward Kalman scheme is implemented to get the input and output sequences.

• *Forward run:*

Starting from the initial condition $P_{01-1} = var(\Upsilon_{eh})$ and $\underline{\alpha}_{01-1}$ and for $i = 1, ..., T$, calculate

$$R_{e,i} = \sigma_n^2 I_{N+P} + C_{0_{leak}} P_{i/i-1} C_{0_{leak}} \tag{5.27}$$

$$K_{f,i} = P_{i/i-1} C_{0_{leak}}{}^* R_{e,i}^{-1} \tag{5.28}$$

$$\hat{\alpha}_i = (I_{N+P} - K_{f,i}, C_{0_{leak}})\hat{\alpha}_{i-1} + K_{f,i} * Y_i \tag{5.29}$$

$$\hat{\alpha}_{i+1/i} = F\underline{\alpha}_i \tag{5.30}$$

$$P_{i+1,i} = F_i(P_{i/i-1} - K_{f,i}R_{e,i}, K_{f,i}^*)F^* + \frac{1}{\sigma_n^2}GG^* \tag{5.31}$$

• *Backward run:*

Starting from $\lambda_{T+1/T} = 0$ and for $i = T, T - 1, ..., 0$, calculate

$$\begin{aligned} \lambda_{i/T} &= I_{P+N} - C_{0_{leak}}^* K_{f,i}^* F_i^* \lambda_{T+1/T} \\ &+ C_{0_{leak}} R_{e,i}^{-1}(y_i - C_{0_{leak}}\hat{\alpha}_{i-1}) \end{aligned} \tag{5.32}$$

$$\hat{\alpha}_{i/T} = \hat{\alpha}_{i/i-1} + P_{i/i-1}\lambda_{i/T} \tag{5.33}$$

The desired estimate is $\alpha_{i/T}$:

The forward-backward Kalman derives the MAP estimate of the system impulse response. In the forward step, the filer obtains the MAP estimate. Our aim, however, is to obtain the MAP estimate of $\underline{\alpha}_i$ given the whole sequence C_i^0 . The backward step adds the contribution of C_{i+1}^t to the MAP estimate of $\underline{\alpha}_i$.

5.2.5   Initial system estimation

The purpose is to get the observation C_0 from α, in order to get Υ.

We can obtain the initial system estimation from the measurement equation of (12). We can do this by implementing the forward–backward Kalman filter to the state-space model with substitution of $C_{0_i} \to C_{0_i I_P}$ and $\Upsilon_i \to \Upsilon_{ehi,I_P}$:

$$\underline{\alpha}_{i+1} = F\underline{\alpha}_i + G\underline{u}_i \tag{5.34}$$

$$\Upsilon_{height,I_P} = C_{0_i I_P}\alpha_{i-1} + \nu_{i/p} \tag{5.35}$$

5.2.6 Computing the input moments

Input moments can be computed using the application of Bayes rules for evaluating the pdf of the function of the system:

Applying the Bayes rule:

$$
\begin{aligned}
f(C_{0_i}(l)/Y_i(l), \alpha_i(l)) &= \frac{f(C_{0_i}(l)/Yi(l), \alpha_i(l))}{f(Y(l)_i/\alpha_i(l))} \\[2mm]
&= \frac{f(C_{0_i}, Yi/\alpha_i)}{\sum_{C_{0I}=A1}^{A_M} f(C_{0_i}, Yi/\alpha_i)} \\[2mm]
&= \frac{f(Yi/C_{0_i}, \alpha_l)f(C_{0l}, \alpha_l)}{\sum_{C_{0I}=A1}^{A_M} f(Yi/C_{0_i}, \alpha_l)f(C_{0l}, \alpha_l)} \\[2mm]
&= \frac{e^{\frac{-|Y_i - \alpha C_{0_i}|^2}{\sigma_n^2}}}{\sum_{j=1}^{M} e^{\frac{-|Y_i - \alpha A_j|^2}{\sigma_n^2}}}
\end{aligned}
\tag{5.36}
$$

where we have dropped the dependence on l. We have used the fact here that C_{0_l} is drawn from the alphabet $A = A_1, A_2, A_3, A_4$ and A is not fixed here. There are four $A's$ considered for the expectation of the fault scenarios where each A is showing a particular fault scenario; that is, for finding the expected value at each instant, the expectation of the four elements will be taken. We can use this to show that:

First moment:

$$
E[C_{0_{fault}}(l)/Y_i(l), \alpha_i(l)] = \frac{\sum_{j=1}^{M} A_j e^{\frac{-|Y_i - \alpha C_{0_i}|^2}{\sigma_n^2}}}{\sum_{j=1}^{M} e^{\frac{-|Y_i(l) - \alpha_i(l)A_j|^2}{\sigma_n^2}}}
$$

Second moment:

$$
E[C_{0_{fault}}(l)/Y_i(l), \alpha_i(l)] = \frac{\sum_{j=1}^{M} |A_j|^2 e^{\frac{-|Y_i - \alpha C_{0_i}|^2}{\sigma_n^2}}}{\sum_{j=1}^{M} e^{\frac{-|Y_i(l) - \alpha_i(l)A_j|^2}{\sigma_n^2}}}
$$

Thus, the EM-Based FB Kalman algorithm has been shown for implementation of fault detection and estimation.

5.3   Fault Isolation

The approach employed for fault isolation is based on a diagnostic model, which directly relates the diagnostic parameters to the input and output. The diagnostic parameters are identified off line by performing a number of experiments. The diagnostic model relating the reference input r the diagnostic parameter γ and the residual $e(k)$, is given by:

$$e(k) = y(k) - y^0(k) = \sum_{i=1}^{q} \psi^T(k-1)\theta_i^{(1)}\Delta\gamma_i + \nu(k) \tag{5.37}$$

where, $\Delta\gamma_i = \gamma - \gamma_i^0$ is the perturbation in γ ; $y^0(k)$ and γ_i^0 are the fault-free (nominal) output and parameter, respectively, $\theta_i^{(1)} = \frac{\delta\theta}{\delta\gamma_i}$, and ψ is the data vector formed of the past outputs and past reference inputs. The gradient $\theta_i^{(1)}$ is estimated by performing a number of off line experiments which consist of perturbing the diagnostic parameters, one at a time. The input-output data from all the perturbed parameter experiments is then used to identify the gradients $\theta_i^{(1)}$. The outcome can be seen in the form of the cross-spectral density between the faulty data and fault-free data.

Remark 5.3.1 *Power spectral density function shows the strength of the variations (energy) as a function of frequency. In other words, it shows at which frequencies variations are strong and at which frequencies variations are weak. It is a very useful tool if you want to identify oscillatory signals in our time series data and want to know their amplitude. For example, we have a chemical plant, and we have hydraulic drives operating to drive the fluid in the pipe network, and some of them have motors inside to pull the fluid with pressure. You detect unwanted vibrations from somewhere. You might be able to get a clue to locate offending machines by looking at power spectral density which would give you the frequencies of the vibrations. When we have two sets of time series data at hand and we want to know the relationships between them, we compute coherency function and some other functions computed from cross spectral density function of two time series data and power spectral density functions of both time series data. In this chapter, we have two time series data of fault and fault-free case respectively. This property of power spectral density helps to treat the isolation case in a better way.*

Remark 5.3.2 *Cross-correlation is a measure of similarity of two waveforms as a function of a time-lag applied to one of them, whereas in autocorrelation, there will always be a peak at a lag of zero, unless the signal is a trivial zero signal. This property of cross-correlation helps to capture the fault signatures more coherently.*

5.3.1 System description

The electro-hydraulic system for this study is a rotational hydraulic drive at the LITP (Laboratoire d'Intégration des technologies de production) of the University of Québec École de technologie supérieure (ÉTS). The setup is generic and allows for simple extension of the results herewith to other electro-hydraulic systems, for example, double-acting cylinders. Referring to the functional diagram in Figure 5.1, a DC electric motor drives a pump, which delivers oil at a constant supply pressure from the oil tank to each component of the system. The oil is used for the operation of the hydraulic actuator and is returned through the servo-valve to the oil tank at atmospheric pressure. An accumulator and a relief valve are used to maintain a constant supply pressure from the output of the pump. The electro-hydraulic system includes two Moog Series 73 servo-valves which control the movement of the rotary actuator and the load torque of the system. These servo-valves are operated by voltage signals generated by an Opal-RT real-time digital control system.

The actuator and load are both hydraulic motors connected by a common shaft. One servo-valve regulates the flow of hydraulic fluid to the actuator and the other regulates the flow to the load. The actuator operates in a closed-loop while the load operates open-loop, with the load torque being proportional to the command voltage to the load servo-valve. While the actuator and load chosen for this study are rotary drives, the exact same setup could be used with a linear actuator and load, and thus, they are represented as generic components in Figure 5.1. The test setup includes three sensors, two Noshok Series 200 pressure sensors with a 0.50V output corresponding to a range of 20.7MPa (3000 PSI) that measure the pressure in the two chambers of the rotational drive, as well as a tachometer to measure the angular velocity of the drive. In order to reduce the number of sensors used (a common preference for commercial application), angular displacement is obtained by numerically integrating the angular velocity measurement.

Figure 5.2 shows the layout of the system and the Opal-RT RT-LAB digital control system. The RT-LAB system consists of a real-time target and a host PC. The real-time target runs a dedicated commercial real-time operating system (QNX), reads sensor signals using an analog-to-digital (A/D) conversion board, and generates output voltage signals for the servo-valves using a digital-to-analog (D/A) conversion board. The host PC is used to generate code for the target using MAT-LAB/Simulink and Opal-RT–RT-LAB software * and also to monitor the system. Controller parameters can also be adjusted on the fly from the host in RT-LAB.

---

*OPAL-RT is the world leader in the development of PC/FPGA Based Real-Time Digital Simulators, Hardware-In-the-Loop (HIL) testing equipment and Rapid Control Prototyping (RCP) systems. RT-LAB is Professional-Real Time Digital Simulation Software.

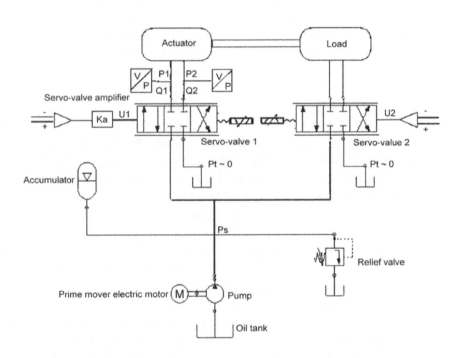

Figure 5.1: Functional diagram.

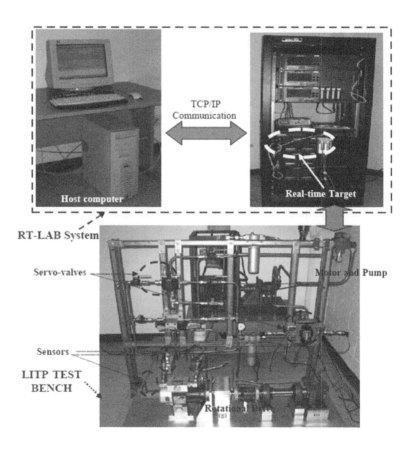

Figure 5.2: Physical layout.

5.3.2   Fault model for rotational hydraulic drive

In general, a rotational hydraulic drive system is a drive or transmission system that uses pressurized hydraulic fluid to drive hydraulic machinery. The rotational hydraulic drive may experience various faults that reduce the performance and reliability. These can occur in components such as the pipe system, sensors, actuators, controllers, communication system elements and the actual platform. Since in the rotational hydraulic drive achieving good driving control is essential, this also requires the flow system of the hydraulic fluid to be working appropriately. Also, an excessive torque load can result in affecting the control of hydraulic drive. With these factors in mind, the faults considered in this study are those that cause leakage faults and controller faults.

A mathematical model of the system described is now developed based on the approach in [48] and [50]. First, the servo-valves are modeled using the following assumptions:

1. The servo-valves are matched and symmetric.
2. The internal leakage inside the servo-valve can be neglected.

The dynamic equation for the servo-valve spool movement is given as [48] and [50].

$$\tau_v(\frac{dA_v(t)}{dt}) + A_v(t) = K_x.K_v.I(t), \tag{5.38}$$

where t denotes time, $I(t)$ is the command input current, τ_v is the servo-valve time constant, $A_v(t)$ is the servo-valve opening area with the sign dependent on flow direction, K_x is the servo-valve area constant, and K_v is the servo-valve torque motor constant. $A_v(t)$ is said to have a positive sign when the servo-valve directs the flow such that the supply drives $P_1(t)$ and $P_2(t)$ drives the fluid to the tank. The reverse configuration is represented using a negative sign for $A_v(t)$, although the actual servo-valve opening area is always a positive number.

Let $Q_1(t)$ represent the flow from the servo-valve and let $Q_2(t)$ represent the flow to the servo-valve. Then,

$$Q_1(t) = Q_2(t) = C_d A_v(t)\sqrt{\frac{P_s - P_L(t)}{\rho}} \tag{5.39}$$

where $P_L(t)$ is the load pressure difference, P_s is the source pressure, C_d is the flow discharge coefficient and ρ is the fluid mass density. $P_L(t)$ and P_s are given by $P_L(t) = P_1(t) - P_2(t)$ and $P_s = P_1(t) + P_2(t)$ with $P_1(t)$ and $P_2(t)$ denoting the pressure in the two chambers of the rotational drive.

The fluid dynamic equation of the actuator, considering the compressibility of oil and internal leakage is given by:

$$\frac{V}{2\beta}\dot{P}_L(t) = C_d A_v(t)\sqrt{\frac{P_s - P_L(t)sign(A_v t}{\rho}}$$
$$- D_m\dot{\theta}(t) - C_L P_L(t), \tag{5.40}$$

where V is the oil volume under compression in one chamber of the actuator, β is the fluid bulk modulus, D_m is the volumetric displacement parameter of the actuator, C_L is the leakage coefficient, $\theta(t)$ is the angular displacement, and $sign$ is the sign function, which accounts for the change in direction of motion of the actuator.

Neglecting friction, the torque-acceleration equation of the actuator is given by:

$$J\ddot{\theta}(t) = D_m(P_1(t) - P_2(t)) - B\dot{\theta}(t) - T_L \tag{5.41}$$

where J is the moment of inertia, B is the viscous damping coefficient, and T_L is the load torque. The variables $\dot{\theta}(t)$, $P_L(t)$, $A_v(t)$ are now normalized by dividing them by their respective maximum values denoted by ω_{max}; P_s and $A_v^{max} = K_x.K_v$. I_{max} to reduce numerical errors while performing simulation and real-time computations. The $sign$ function is approximated by the sigmoid function defined as:

$$sigm(x) = \frac{1 - e^{-ax}}{1 + e^{-ax}} \tag{5.42}$$

where $a > 0$, to address the non-differentiable nature of the sign function.

5.3.3   Fault scenarios

Fault scenarios are created by using the rotational hydraulic drive in the simulation program. In these scenarios leakage fault and controller fault are being considered.

• Scenario I: Leakage fault

In this scenario, while the system is working in real time, a leakage fault is being introduced in the hydraulic fluid flow linked to the servo-valve of the system. The leakage fault is considered as $\omega_h C_{L_{leakage}} x_3(t)$ in state 3.

• Scenario II: Controller fault

In this scenario, while the system is working in real time and getting the input for driving the dynamics of the system, a fault has been introduced by increasing the torque load in the hydraulic drive, then affecting the controller, $-\frac{\omega_h}{\alpha} t_{L_{fault}}$ is considered in state 2 of the system.

Using (5.38)-(5.41) and the fault scenarios, the fault model of the system can be represented in state-space form as:

$$\dot{x}_1(t) = \omega_{max} x_2(t) \tag{5.43}$$

$$\dot{x}_2(t) = -\gamma \frac{\omega_h}{\alpha} x_2(t) + \frac{\omega_h}{\alpha} x_3(t) -$$
$$\frac{\omega_h}{\alpha} t_L - \frac{\omega_h}{\alpha} t_{L_{fault}} \tag{5.44}$$

$$\dot{x}_3(t) = -\alpha \omega_h x_2(t) - \omega_h C_L x_3(t)$$
$$+ \alpha \omega_h x_4(t) \sqrt{1 - x_3(t) sigm(x_4(t))}$$
$$- \omega_h C_{L_{leakage}} x_3(t) \tag{5.45}$$

$$\dot{x}_4(t) = -\frac{1}{\tau_v} x_4(t) + \frac{i(t)}{\tau_v} \tag{5.46}$$

where

$$x_1(t) = \theta(t), \ x_2(t) = \frac{\dot{\theta}(t)}{\omega_{max}},$$

$$x_3(t) = \frac{P_L(t)}{P_s}, \ x_4(t) = \frac{A_v(t)}{A_v^{max}},$$

$$u_1(t) = i(t) = \frac{I(t)}{I_{max}}, \ u_2 = t_L = \frac{T_L}{P_s D_m},$$

$$\gamma = \frac{B\omega_{max}}{P_s D_m},$$

$$\omega_h = \sqrt{\frac{2\beta D_m^2}{JV}},$$

$$\alpha = \frac{(C_d A_v^{max} \sqrt{P_s \rho}) J \omega_h}{P_s D_m^2},$$

$$c_L = \frac{J C_L \omega_h}{D_m^2}$$

and $C_{L_{leakage}}$ is the leakage fault considered in state 3, $t_{L_{fault}}$ is the controller fault in the form of torque load in state 2.

Using the sign convention for $A_v(t)$ and the definition of $x_3(t)$, it follows that $0 \leq x_3(t) sigm(x_4(t)) \leq 1$. It is also noted here that $0 \leq x_3(t) sigm(x_4(t)) \leq 1$, because $P_1(t)$ and $P_2(t)$ are both positive and the condition $x_3(t) sigm(x_4(t)) = 1$ implies that $P_1(t) = P_s$ and $P_2(t) = 0$ or $P_2(t) = P_s$ and $P_1(t) = 0$, indicating zero pressure drop across the open ports of the servo-valve and thus, no flow to or from the actuator, a situation that would occur if the rotational motion of the drive is impeded.

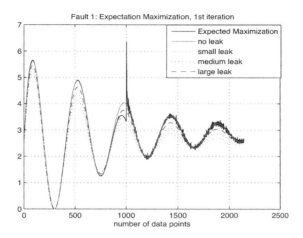

Figure 5.3: Implementation results on the leakage profile at iteration 1.

5.4 EM Algorithm Implementation

The evaluation of the proposed scheme will be made on the the electro-hydraulic system. The following sections show the detailed implementation and simulation of the proposed scheme. In what follows, we present simulation results for the proposed fault diagnostics scheme covering the fault detection and estimation. The tasks of our EM-Based Forward Backward Kalman scheme have been executed here with an increasing precision accompanied with a more detailed fault picture by increasing the number of iterations. Two sets of faults have been considered here i.e., the leakage fault in state 3 and a controller fault. Firstly, the data collected from the plant has been initialized and the parameters have been optimized which comprises the pre-processing and normalization of the data. Then, the EM Based Forward Backward Kalman is implemented with an iterative process giving not only the recovery of the correct data, but also detecting the correct fault profile.

5.4.1 Leakage fault

The EM Based FB Kalman scheme has been followed and employed here for the leakage fault to get a final profile of the lost data and the fault detection. It has been shown that the estimated profile at iteration 3 (Figure 5.5) and Iteration 4 (Fig 5.6) is performing better in following the original output as compared to the initial iterations, thus pointing clearly to the fault detection and recovery of the correct data.

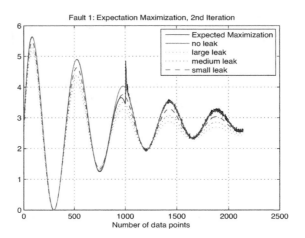

Figure 5.4: Fault 1: Implementation results on the leakage profile at iteration 2.

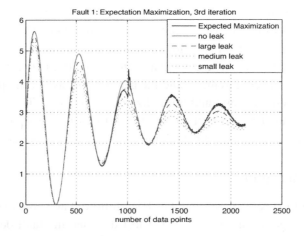

Figure 5.5: Implementation results on the leakage profile at iteration 3.

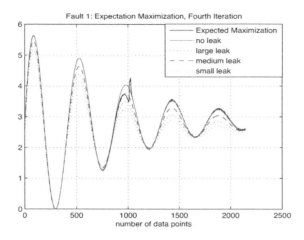

Figure 5.6: Implementation results on the leakage profile at iteration 4.

Further, the information is fused in the fault-diagnostic model, and the results show that on the scale of a number of observations, we can judge the isolation of the fault by cross spectral density, as can be seen in Figure 5.14, Figure 5.15 and Figure 5.16.

5.4.2   Controller fault

The EM Based FB Kalman scheme has been followed and employed here for the controller torque fault, to get a final profile of the lost data and the fault detection. It has been shown that the estimated profile at iteration 3 (Figure 5.12) and Iteration 4 (Fig 5.13) is performing better in following the original output as compared to the initial iterations, thus pointing clearly to the fault detection and recovery of the correct data.

Further, the information is fused in the fault-diagnostic model, and the results show that on the scale of a number of observations, we can judge the isolation of the fault by cross spectral density. as can be seen in Figure 5.14, Figure 5.15 and Figure 5.16.

5.5   Notes

This chapter has presented a general approach to integrating data-fusion fault detection and estimation with a fault isolation scheme. The proposed scheme has

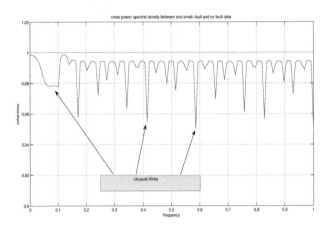

Figure 5.7: Cross-power spectral density between small fault and no fault data.

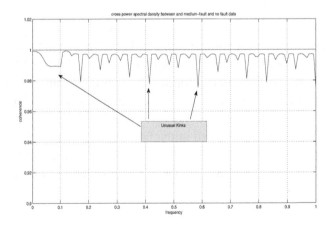

Figure 5.8: Cross-power spectral density between medium fault and no fault data.

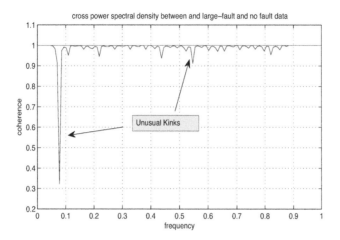

Figure 5.9: Cross-power spectral density between large fault and no fault data.

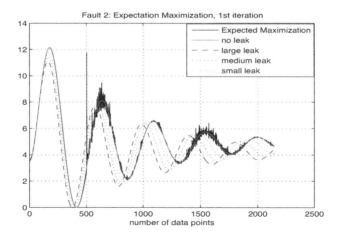

Figure 5.10: Implementation results on the controller fault at iteration 1.

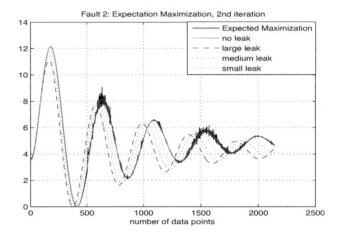

Figure 5.11: Fault 2: Implementation results on the controller fault at iteration 2.

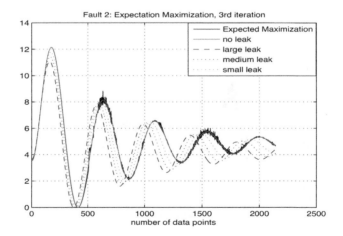

Figure 5.12: Implementation results on the controller fault at iteration 3.

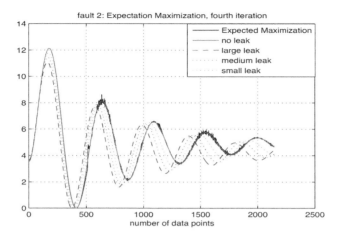

Figure 5.13: Implementation results on the controller fault at iteration 4.

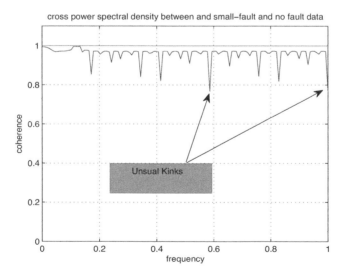

Figure 5.14: Cross-power spectral density between small fault and no fault data.

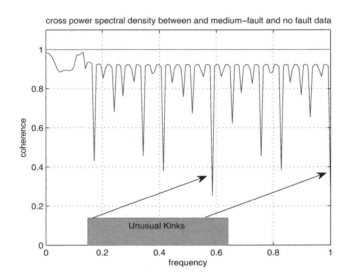

Figure 5.15: Cross-power spectral density between medium fault and no fault data.

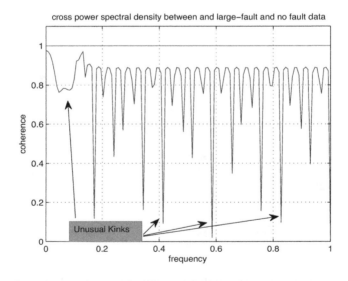

Figure 5.16: Cross-power spectral density between large fault and no fault data.

been developed based on an expectation maximization (EM) algorithm and diagnostics model. The proposed scheme can function when information about the system faults, and the structure and dynamics of the underlying data generation mechanism is inaccessible, incomplete, or partially missing. The EM approach has been motivated by its articulation on a forward-backward Kalman filter thereby initiating the picture of fault detection and estimation. This picture is then completed by a fault diagnostic isolation scheme. The proposed scheme has been evaluated on an electro-hydraulic system, thus ensuring the effectiveness of the approach. The major contribution of the chapter is the integration of iterative expectation maximization-based approach boiled on forward-backward Kalman filter for loss data-recovery, with a fault detection and fault diagnostic model for fault isolation to achieve both accuracy and reliability of a data-fusion FDI scheme.

5.6 Proposed Topics

1. It is known that one consequence of the EM approach is the increased storage and latency requirements of the forward-backward Kalman filter. Suggest a distributed/parallel algorithm to perform the required computations to help in processing multiple OFDM symbols simultaneously. One get benefits for the work of [43].

2. From the literature, it is known that an expectation maximization algorithm provides a simple, easy-to-implement and efficient tool for learning parameters of a model. An interesting research topic would be the outcome of an EM to build up an algorithm for control purposes. Show that the combined algorithms render an effective and practical technique for controlling industrial plants.

3. Develop an a modified expectation maximization algorithm that will guarantee an improved convergence speed and/or computational structure. By looking at models of classes of industrial processes, some hints might help in building the developed algorithm. Provide an assessment of the performance of the algorithm versus the complexity, as well as the hardware requirements.

4. An interesting extension to the results in this chapter would be the estimation of the parameters of a mixture density model. It is suggested to develop a decentralized expectation-maximization (EM) algorithm to estimate these parameters for use in distributed learning tasks performed with data collected at spatially deployed wireless sensors. It is suggested to associate an E-step in the iterative scheme to local information available to individual sensors,

while during the M-step sensors exchange information only with their one hop neighbors to reach consensus and eventually percolate the global information needed to estimate the wanted parameters across the wireless sensor network (WSN).

5. In a variety of well-known data mining tasks in distributed environments such as clustering, anomaly detection, target tracking, and density estimation, to name a few, it is considered desirable to have a distributed expectation maximization algorithm for learning parameters of Gaussian mixture models (GMM) in large peer-to-peer (P2P) environments. Design such an algorithm in the two-step approach: In the monitoring phase, the algorithm checks if the model "quality" is acceptable by using an efficient local algorithm. This is then used as a feedback loop to sample data from the network and rebuild the GMM when it is outdated. Simulate the developed algorithm to validate the theoretical results.

Chapter 6

Wireless Estimation Methods

This chapter is devoted to the development of wireless estimation methods as distinct applications of wireless sensor networks. It is known that the usage of WSNs for state-estimation has recently gained increasing attention due to its cost effectiveness and feasibility. One of the major challenges of state-estimation via WSNs is the distribution of the centralized state-estimator among the nodes in the network. Significant emphasis has been on developing non-centralized state-estimators considering communication, processing–demand and estimation-error.

6.1 Partitioned Kalman Filters

State-estimation is a widely used technique in monitoring and control applications. The method requires that all process-measurements are sent to a central system which estimates the global state-vector of the process. The interest in using WSNs to retrieve the measurements has recently grown [394], due to improved performance and feasibility in new application areas. However, for WSNs consisting of a large amount of nodes a central state-estimator becomes impracticable due to high processing demand and energy consumption. As a result, the distribution of the centralized Kalman filter, in which each node estimates its own state-vector, has become a challenging and active research area.

6.1.1 Introduction

Within this research area two different directions can be noticed. In one direction each node estimates the global state vector, and a central system is used to fuse the information of all the nodes together. In the second direction the central estimation is absent. Instead, each node estimates a part of the global state-vector using information from other nodes in its local region, preferably its direct neighbors. Articles that describe these distributed Kalman filters (DKFs) are [399]–[404].

The purpose of this section is to provide a critical overview of existing partitioned (non-centralized) Kalman filters (PKF), which would help in choosing a particular method for a particular application. For each method we present its characteristics, algorithm and amount of decentralization in terms of processing demand and communication requirements per node. Finally all methods are assessed in a benchmark problem on their performance in estimation, communication, and robustness to data loss or node break down.

6.1.2 Centralized Kalman filter

Suppose a WSN is used in combination with a centralized Kalman filter to estimate the states of a global process. All the nodes send their measurements to one system where the centralized Kalman filter estimates the global state-vector. The measurements of the kth sample instant are combined in the measurement-vector $y[k]$ with the measurement-noise $v[k]$. The global state-vector of the process is defined as $x[k]$ with process-noise $w[k]$. With this, the discretized process-model becomes:

$$
\begin{aligned}
x[k] &= Ax[k-1] + w[k-1], \\
y[k] &= Cx[k] + v[k]
\end{aligned}
\tag{6.1}
$$

The probability density functions (PDF) of both $w[k]$ and $v[k]$ are described by a Gaussian-distribution, that is,

$$
\begin{aligned}
\mathbf{E}(w[k]) &= 0 \ \ and \ \ \mathbf{E}(w[k]w^T[k]) = Q[k], \\
\mathbf{E}(v[k]) &= 0 \ \ and \ \ \mathbf{E}(v[k]v^T[k]) = R[k]
\end{aligned}
\tag{6.2}
$$

The standard (centralized) Kalman filter estimates the global state-vector $\hat{x}[k]$ and the global error-covariance matrix $P[k]$. Let $\mathbf{E}(\alpha)$ represent the expectation of the stochastic variable α. Then, $\hat{x}[k]$ and $P[k]$ are defined as:

$$
\hat{x}[k] = \mathbf{E}(x[k]), \quad P[k] = \mathbf{E}((x[k] - \hat{x}[k])(x[k] - \hat{x}[k])^T).
\tag{6.3}
$$

The centralized Kalman filter consists of two stages that are performed at each sample instant k:

- the "prediction-step"

- the "measurement-update"

First the prediction-step computes the predicted state-vector $\hat{x}[k|k-1]$ and error-covariance $P[k|k-1]$. Second, the measurement-update calculates the estimated state-vector $\hat{x}[k|k]$ and error-covariance $P[k|k]$. The centralized Kalman filter, with initial values $\hat{x}[0|0] = x_0$ and $P[0|0] = P_0$, is formally described by the following set of equations:

Prediction – step
$$\hat{x}[k|k-1] = A\hat{x}[k-1|k-1],$$
$$P[k|k-1] = AP[k-1|k-1]A^T + Q[k-1],$$
Measurement – update
$$K[k] = P[k|k-1]C^T(CP[k|k-1]C^T + R[k])^{-1},$$
$$\hat{x}[k|k] = \hat{x}[k|k-1] + K[k](y[k] - C\hat{x}[k|k-1]),$$
$$P[k|k] = (I - K[k]C)P[k|k-1] \tag{6.4}$$

For large scale WSNs the centralized implementation of (6.4) results in high processing demand, communication requirements and energy consumption, which prevents the usage of a centralized Kalman filter. To overcome this issue, a number of methodologies to implement the Kalman filter in a distributed fashion were designed. However, until now there has been no comparison or evaluation of the results obtained in this direction. The purpose of this chapter is to provide a critical overview of existing methods for designing non-centralized Kalman filters.

Before explaining the different methods of this overview in detail, we present three assumptions. If not indicated otherwise, these assumptions hold for the presented method.

1. The existence of a WSN consisting of N nodes is assumed in which each node i has its own measurement-vector y_i with corresponding measurement-noise v_i. The global measurement-vector y, observation-matrix C and (6.1) are rewritten as follows:

$$y_i[k] = C_i x[k] + v_i[k] \Rightarrow \begin{cases} y = (y_1, y_2, \ldots, y_N)^T \\ C = (C_1, C_2, \ldots, C_N)^T. \end{cases} \tag{6.5}$$

2. The measurement-noises of two different nodes are uncorrelated, i.e. $R_{(i,j)} = E(v_i v_j^T) = 0$, if $i \neq j$. Resulting in an R-matrix of the form:

$$R = blockdiag(R_{(1,1)}, R_{(2,2)}, \ldots, R_{(N,N)}). \tag{6.6}$$

3. All nodes j that are directly connected to a node i are collected in the set N_i, which also includes the node i. This means that if node j is connected to node i, then $j \in N_i$. Usually, N_i is contains only the direct neighbors of node i. However, it is also possible that N_i contains other nodes besides the direct neighbors and in the case of global communication $N_i = N$. This will be made clear for each estimation method.

6.1.3   Parallel information filter

In what follows, a parallel implementation of the Kalman filter [526] is described. Each node i has its own Kalman filter calculating the global state-estimates $\hat{x}_i$ and P_i of node i using only its measurement-vector y_i. In the algorithm an information-matrix I_i and an information-vector i_i are computed from the y_i and $R_{(i,i)}$. Each node sends its state-estimates to a central system which calculates the global state-estimates of the whole WSN, i.e. $\hat{x}$ and P.

The sets of equations of the parallel information filter(PIF) for node i are:

node i prediction-step
$$\hat{x}_i[k|k-1] = A\hat{x}_i[k-1|k-1], \tag{6.7a}$$
$$P_i[k|k-1] = AP_i[k-1|k-1]A^T + Q[k-1],$$

node i information-update
$$I_i[k] = C_i^T R_{(i,i)}^{-1}[k]C_i, \quad i_i[k] = C_i^T R_{(i,i)}^{-1}[k]y_i[k], \tag{6.7b}$$

node i measurement-update
$$P_i^{-1}[k|k] = P_i^{-1}[k|k-1] + I_i[k], \tag{6.7c}$$
$$\hat{x}_i[k|k] = P_i[k|k](P_i^{-1}[k|k-1]\hat{x}_i[k|k-1] + i_i[k]).$$

The global state-estimates $\hat{x}$ and P are calculated taking the covariance intersection into account [405]:

$$\alpha_i[k] = \frac{(tr(P_i[k|k]))^{-1}}{\sum_{i=1}^{N}(tr(P_i[k|k]))^{-1}}, \quad P^{-1}[k] = \sum_{i=1}^{N}\alpha_i[k]P_i^{-1}[k|k],$$

$$\hat{x}[k] = \sum_{i=1}^{N}\alpha_i[k]P[k]P_i^{-1}[k|k]\hat{x}_i[k|k]. \tag{6.8}$$

The calculation of $\hat{x}[k]$ and $P[k]$ is done in a central system, which can be located in one node only or even in every node. A drawback of this method is that every node estimates a global state-vector leading to a high processing-demand. A second drawback is global communication, for every node needs to send information to at least one central system. This method was improved in the decentralized information filter presented in the next section.

6.1.4 Decentralized information filter

In [399] the decentralized information filter (DIF) was proposed to overcome some drawbacks of the PIF. Again each node i has its own global state-estimates $\hat{x}_i$ and P_i. However, the central estimation is decentralized among the nodes and a node i is only connected to its neighboring nodes in N_i. These nodes exchange their information-matrix I_i and information-vector i_i. Meaning that node i receives I_j and i_j from the nodes j with $j \in N_i, j \neq i$. The received I_j and i_j are added to I_i and i_i respectively. The sets of equations of the DIF for node i are:

node i prediction-step
$$\hat{x}_i[k|k-1] = A\hat{x}_i[k-1|k-1], \tag{6.9a}$$
$$P_i[k|k-1] = AP_i[k-1|k-1]A^T + Q[k-1],$$

node i information-update
$$I_i[k] = C_i^T R_{(i,i)}^{-1}[k]C_i, \quad i_i[k] = C_i^T R_{(i,i)}^{-1}[k]y_i[k], \tag{6.9b}$$

local measurement-update
$$P_i^{-1}[k|k] = P_i^{-1}[k|k-1] + \sum_{j \in N_i} I_j[k], \tag{6.9c}$$
$$\hat{x}_i[k|k] = P_i[k|k](P_i^{-1}[k|k-1]\hat{x}_i[k|k-1] + \sum_{j \in N_i} i_j[k]).$$

An important aspect of this DKF is that if node i is connected to all other nodes and assumptions (6.5) and (6.6) are valid, its state-estimates $\hat{x}_i$ and P_i are exactly the same as the estimates of a centralized Kalman filter. An advantage is that only local communication is required. A drawback however, is that each node estimates the global state-vector.

6.1.5 Hierarchical Kalman filter

In [400]–[402] decoupled hierarchical Kalman filters (DHKFs) are presented. The common feature of this method is that the global state-vector x and the process-model are divided in N parts. Each node estimates one of the N parts and exchanges its state-estimates with all the other nodes in the WSN. The process-model is described as:

$$
\begin{pmatrix} x_1[k] \\ \vdots \\ x_N[k] \end{pmatrix} = A \begin{pmatrix} x_1[k-1] \\ \vdots \\ x_N[k-1] \end{pmatrix} + \begin{pmatrix} w_1[k-1] \\ \vdots \\ w_N[k-1] \end{pmatrix},
$$

$$
\begin{pmatrix} y_1[k] \\ \vdots \\ y_N[k] \end{pmatrix} = C \begin{pmatrix} x_1[k] \\ \vdots \\ x_N[k] \end{pmatrix} + \begin{pmatrix} v_1[k] \\ \vdots \\ v_N[k] \end{pmatrix}, \tag{6.10}
$$

where

$$
A = \begin{pmatrix} A_{(1,1)} & \cdots & A_{(1,N)} \\ \vdots & \ddots & \vdots \\ A_{(N,1)} & \cdots & A_{(N,N)} \end{pmatrix}, C = \begin{pmatrix} C_{(1,1)} & \cdots & C_{(1,N)} \\ \vdots & \ddots & \vdots \\ C_{(N,1)} & \cdots & C_{(N,N)} \end{pmatrix}.
$$

Just as R, also the matrices Q and P are both assumed to be block-diagonal matrices. Therefore we define $Q_i = E(w_i w_i^T)$ and $P_i = E((x_i - \hat{x}_i)(x_i - \hat{x}_i)^T)$. Node i estimates $\hat{x}_i[k]$ and $P_i[k]$. The algorithm for each node i is:

node i prediction-step

$$
\hat{x}_i[k|k-1] = \sum_{j=1}^{N} A(i,j)\hat{x}_j[k-1|k-1], \tag{6.11a}
$$

$$
P_i[k|k-1] = \sum_{j=1}^{N} (A_{(i,j)} P_j[k-1|k-1] A_{(i,j)}^T) + Q_i[k-1],
$$

node i measurement-update

$$
K_i[k] = P_i[k|k-1] C_{(i,i)}^T (\sum_{j=1}^{N} (C_{(i,j)} P_j[k|k-1] C_{(i,j)}^T) + R_{(i,i)}[k])^{-1},
$$

$$
\hat{x}_i[k|k] = \hat{x}_i[k|k-1] + K_i[k](y_i[k] - \sum_{j=1}^{N} C_{(i,j)}\hat{x}_j[k|k-1]),
$$

$$
P_i[k|k] = (I - K_i[k]C_{(i,i)})P_i[k|k-1]. \tag{6.11b}
$$

Notice that this method is better compared to the PIF and DIF in terms of processing-demand and the amount of data transfer required. A drawback however, is that global communication is still required.

6.1.6 Distributed Kalman filter with weighted averaging

In previous methods each node sends a vector with its corresponding covariance-matrix to the other nodes, i.e. ii with I_i or $\hat{x}_i$ with P_i. In the distributed Kalman filter with weighted averaging (DKF-WA) [403] a node i only sends its state-vector, without a covariance-matrix, to its neighboring nodes in the set N_i. The weighted average of all received state-vectors forms the node's estimated global state-vector $\hat{x}_i$. One remark should be made: in this case the matrix R is not necessarily block-diagonal, i.e. $R_{(i,j)} \neq 0, \forall j \in N_i$.

The algorithm of the DKF-WA is divided into an on-line and an off-line part. In the on-line part each node has its own estimate of the global state-vector $\hat{x}_i$ which is partly calculated using the equations of the centralized Kalman filter. In this method a node i has a fixed, pre-calculated Kalman gain K_i. After the measurement-update the nodes exchange their estimated state-vector. A node i receives the state-vectors $\hat{x}_j (j \in N_i)$ which are then weighted with a fixed, pre-calculated matrix $W_{(i,j)}$. The weighted average is chosen as the new estimated global state-vector of node i, that is, $\hat{x}_i$. The on-line algorithm is:

node i prediction-step (on-line)

$$\hat{x}_i[k|k-1] = A\hat{x}_i[k-1|k-1], \tag{6.12a}$$

node i measurement-update (on-line)

$$\hat{x}_i[k|k] = \hat{x}_i[k|k-1] + K_i(y_i[k] - C_i\hat{x}_i[k|k-1]), \tag{6.12b}$$

local weighted average (on-line)

$$\hat{x}_i[k|k] = \sum_{j \in N_i} W_{(i,j)}\hat{x}_j[k|k]. \tag{6.12c}$$

Next, we explain the off-line algorithm which is used to calculate K_i and $W_{(i,j)}$. For that, the error-covariance between the estimated global state-vectors of node i and j is:

$$P_{(i,j)}[k] = E((x[k] - \hat{x}_i[k])(x[k] - \hat{x}_j[k])^T). \tag{6.13}$$

The off-line algorithm uses the same stages for $P_{(i,j)}$ as the on-line algorithm for $\hat{x}_i$ in (6.12).

First the "prediction-step" (6.12a) and "measurement-update" (6.12b) of a node i are given to calculate $P_{(i,j)}[k|k-1]$ and $P_{(i,j)}[k|k]$, with $j \in N_i$:

node i prediction-step (off-line)

$$P_{(i,j)}[k|k-1] = AP_{(i,j)}[k-1|k-1]A^T + Q[k-1], \tag{6.14a}$$

node i measurement-update (off-line)

$$K_i[k] = P_{(i,i)}[k|k-1]C_i^T(C_iP_{(i,i)}[k|k-1]C_i^T + R_{(i,i)}[k])^{-1},$$

$$P_{(i,j)}[k|k] = (I - K_i[k]C_i)P_{(i,j)}[k|k-1](I - K_j[k]C_j)^T$$

$$+K_i[k]R_{(i,j)}K_j^T[k]. \tag{6.14b}$$

Notice $R_{(i,j)}$ and the calculation of K_i in (6.14b). The next step is calculating $W_{(i,j)}$ of the weighted average as in (6.12c). To keep the state-estimation unbiased the following constraint is introduced:

$$\sum_{j \in N_i} W_{(i,j)}[k] = I_{n \times n}. \tag{6.15}$$

From (6.12c) and (6.15) we can derive:

$$\begin{aligned}
x[k] - \hat{x}_i[k|k] &= \sum_{j \in N_i} W_{(i,j)}x[k] - \sum_{j \in N_i} W_{(i,j)}\hat{x}_j[k|k], \\
&= \sum_{j \in N_i} W_{(i,j)}(x[k] - \hat{x}_j[k|k]). \tag{6.16}
\end{aligned}$$

Using (6.13) the weighted average of $P_{(i,j)}[k|k]$ results in:

$$P_{(i,j)}[k|k] = \sum_{p \in N_i} \sum_{q \in N_j} W_{(i,p)}[k]P_{(p,q)}[k|k]W_{(j,q)}^T[k]. \tag{6.17}$$

Equation (6.17) can also be written in matrix form. If $N_i = (i_1, i_2, \ldots, i_{N_i})$ and we define $\overline{W}_i = (W_{(i,i_1)}, \ldots, W_{(i,i_{N_i})})$, equation (6.17) becomes:

$$P_{(i,j)}[k|k] = \overline{W}_i \begin{pmatrix} P_{(i_1,j_1)}[k|k] & \cdots & P_{(i_1,j_{N_j})}[k|k] \\ \vdots & \ddots & \vdots \\ P_{(i_{N_i},j_1)}[k|k] & \cdots & P_{(i_{N_i},j_{N_j})}[k|k] \end{pmatrix} \overline{W}_j^T. \tag{6.18}$$

The last step in this off-line algorithm is to minimize $P_{(i,i)}[k|k]$ with respect to $\overline{W}_i = (W_{(i,i_1)}, \ldots, W_{(i,i_{N_i})})$ taking constraint (6.15) into account. For further details we refer the interested reader to [403]. The off-line algorithm runs until the values K_i and $W_{(i,j)}$ remain constant. These values are then used in the on-line algorithm.

An important aspect in the performance of this method is that each node estimates the global state-vector, but due to the fixed matrices K_i and $W_{(i,j)}$ its processing-demand remains low. It was already noticed that the DKF-WA has low communication requirements. However, it is not robust against lost data or nodes breaking down, for in that case the weighted averaging of (6.12c) will not be accurate.

6.1.7 Distributed consensus Kalman filter

In [404],[406] the distributed Kalman filter with consensus filter (DKF-CF) was proposed. In this method a node i has its own estimate of the global state-vector $\hat{x}_i$ and the node can only communicate with its neighboring nodes collected in N_i. Instead of averaging the received state-vectors, a node tries to reach consensus on them using a correction-factor ε.

Basically the algorithm of the DKF-CF adds an extra stage to the algorithm of the DIF in (6.9), i.e., the "local-consensus"- stage. Hence, every node has its own global state-estimates $\hat{x}_i$ and P_i. We define $\hat{x}_i^c$ to be the estimated global state-vector of node i before the consensus-stage. The algorithm is:

node i prediction-step
$$\hat{x}_i[k|k-1] = A\hat{x}_i[k-1|k-1], \tag{6.19a}$$
$$P_i[k|k-1] = AP_i[k-1|k-1]A^T + Q[k-1],$$

node i information-update
$$I_i[k] = C_i^T R_{(i,i)}^{-1}[k]C_i, \quad i_i[k] = C_i^T R_{(i,i)}^{-1}[k]y_i[k], \tag{6.19b}$$

local measurement-update
$$P_i^{-1}[k|k] = P_i^{-1}[k|k-1] + \sum_{j \in N_i} I_j[k], \tag{6.19c}$$
$$\hat{x}_i^c[k|k] = P_i[k|k](P_i^{-1}[k|k-1]\hat{x}_i[k|k-1] + \sum_{j \in N_i} i_j[k]),$$

local consensus
$$\hat{x}_i[k|k] = \hat{x}_i^c[k|k] + \varepsilon \sum_{j \in N_i} (\hat{x}_j^c[k|k] - \hat{x}_i^c[k|k]). \tag{6.19d}$$

Due to the "local-consensus"-stage this method requires more communication than the DIF, but it does not necessarily lead to an improved estimation-error. A drawback is that each node estimates the global state-vector, meaning high processing-demand and data transfer per node. A DKF that overcomes this problem is the distributed Kalman filter with bipartite fusion graphs.

6.1.8 Distributed Kalman filter with bipartite fusion graphs

Originally, the usage of graphs to show how sensors are related to state estimates in DKFs was employed in [407]. More recently, DKFs with bipartite fusion graphs

(DKFBFG) were presented in [408]. The method assumes that each node is connected only to its neighboring nodes collected in N_i. Furthermore, a node has its own state-estimate which is only a part of the global state-vector. This means that the global state-vector at node i, i.e. x_i^{global}, is divided into two parts: a part that is estimated, i.e. x_i, and a part that is not estimated, i.e. d_i. The vectors x_i and d_i are defined using some transformation-matrices G_i and S_i as follows:

$$\begin{pmatrix} x_i[k] \\ d_i[k] \end{pmatrix} = \begin{pmatrix} \Gamma_i \\ S_i \end{pmatrix} x_i^{global}[k]. \tag{6.20}$$

Preferably, the states of x_i are determined by taking those states of x_i^{global} that have a direct relation with the measurement-vector y_i. Meaning that Γ_i and S_i are defined by observation-matrix C_i. Assume I is the identity matrix with size equal to the number of states in x_i^{global}. If the jth column of C_i contains non-zero elements, the jth row of I is put into Γ_i. If not, the jth row of I is put into S_i. An example of C_i with its corresponding Γ_i and S_i is:

$$C_i = \begin{pmatrix} c_{11} & c_{12} & 0 & 0 & c_{15} \\ 0 & c_{22} & 0 & 0 & 0 \end{pmatrix} \Rightarrow$$

$$\Gamma_i = \begin{pmatrix} 1 & 0 & 0 & 0 & 0 \\ 0 & 1 & 0 & 0 & 0 \\ 0 & 0 & 0 & 0 & 1 \end{pmatrix}, S_i = \begin{pmatrix} 0 & 0 & 1 & 0 & 0 \\ 0 & 0 & 0 & 1 & 0 \end{pmatrix}. \tag{6.21}$$

Due to the fact that a node i estimates a part of the global state-vector, the node also has its own process-model derived from the global one. This is done by using Γ_i and S_i on the global process-model. The following matrices are defined: $A_i = \Gamma_i A \Gamma_i^T, D_i = \Gamma_i A S_i^T, H_i = C_i \Gamma_i^T$ and $w_i[k] = \Gamma_i w[k]$. With this, the process-model of node i becomes:

$$\begin{aligned} x_i[k] &= A_i x_i[k-1] + D_i d_i[k-1] + w_i[k-1], \\ y_i[k] &= H_i x[k] + v_i[k]. \end{aligned} \tag{6.22}$$

The method assumes that the state-vector x_i is estimated by node i as $\hat{x}_i$, state-vector d_i is sent by other nodes and is represented by node i as $\hat{d}_i$. What remains is the matrix $Q_i[k] = E(w_i[k]w_i^T[k])$.

Now that the characteristics of the DKF-BFG are presented, we proceed with the estimation algorithm. Notice that the algorithmic procedure is actually based on the DIF algorithm in (6.9). Each node shares its local information-matrix I_i and information-vector i_i with its neighbors in N_i. But because the state-vectors in different nodes are not necessarily equal, in contrast with the DIF, the structure of

I_i and i_i differs per node. This means that I_i cannot be added to I_j, as is the case in (6.9c). This is solved by using Γ_i and S_i as shown in the algorithm:

node i prediction-step
$$\hat{x}_i[k|k-1] = A_i\hat{x}_i[k-1|k-1] + D_i\hat{d}_i[k-1], \qquad (6.23a)$$
$$P_i[k|k-1] = A_iP_i[k-1|k-1]A_i^T + Q_i[k-1],$$

node i information-update
$$I_i[k] = H_i^T R_{(i,i)}^{-1} H_i, \quad i_i[k] = H_i^T R_{(i,i)}^{-1} y_i[k], \qquad (6.23b)$$

local measurement-update
$$P_i^{-1}[k|k] = P_i^{-1}[k|k-1] + \sum_{j\in N_i} (\Gamma_i\Gamma_j^T)I_j[k](\Gamma_i\Gamma_j^T)^T,$$
$$\hat{x}_i[k|k] = P_i[k|k]P_i^{-1}[k|k-1]\hat{x}_i[k|k-1]$$
$$+P_i[k|k] \sum_{j\in N_i} (\Gamma_i\Gamma_j^T)i_j[k](\Gamma_i\Gamma_j^T)^T. \qquad (6.23c)$$

An important issue in the performance of this method is whether the global process-model is sparse and localized so that the node's process-model can be derived without loss of generality. If this is indeed the case, its performance should be equal to the DIF. A drawback is that although only local communication is assumed in [408], it is also assumed that the states of $\hat{d}_i$ are sent by other nodes. This means that extended or even global communication may still be needed. A benefit of this method is that a node only estimates a part of the global state-vector so that its processing-demand per node is low.

6.1.9 Simulation example 1

This section assesses the partitioned Kalman filters presented in this chapter in terms of state-estimation error, communication requirements, and robustness against data loss or node break-down.

The benchmark process is the heat transfer of a bar. The bar is divided into 100 segments and the temperature T_n of each segment n is estimated. The state-vector of the global process is therefore $x = (T_1, T_2, \ldots, T_{100})^T$. The bar is heated at the 48th segment. The WSN consists of 5 nodes, placed at segment 11, 31, 51, 71 and 91. Each node measures the temperature of its own specific segment. Several of the DKFs are used to estimate the temperature at all 100 segments. A graphical description of this system is shown in Figure 6.1.

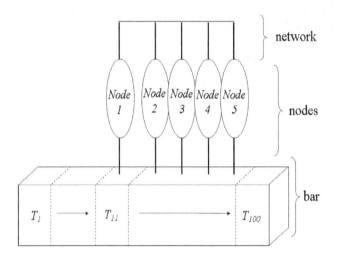

Figure 6.1: Bar with wireless sensor network.

The DKFs are first initialized. The sampling time is 10 seconds and the model runs for 10,000 seconds. The initial state-vector and error-covariance together with Q and R are the same for all methods. This concludes the design of the PIF and the DHKF. Communication is only allowed with the neighboring nodes. For example, node 3 receives from and sends data to node 2 and 4. In this way the design of the DIF is also completed. For the DKF-CF the value of e is set 0.1, which gave good simulation-results. The design of this parameter is critical, for if it's too big the estimation algorithm becomes unstable, while if it is too little the method has no improvements over the DIF algorithm. Matrices Γ_i and S_i of the DKF-BFG are constructed in such a way that node 1 estimates state 1 to 21, node 2 state 1 to 41, node 3 state 21 to 61, node 4 state 41 to 81 and node 5 state 61 to 100.

Figure 6.2 and Figure 6.3 show the real temperature of all the states together with the measurements (with noise), both at 10,000 seconds. Also the estimated states of the different methods are plotted. The estimation of a state nearest the node is plotted, i.e., the plotted states 41 to 60 were estimated by node 3. In case of Figure 6.2 no data loss was simulated. In Figure 6.3 however, we simulated a 5% loss of the communicated data-packages.

Beside state-estimation, communication is also an important aspect. Table 6.1.9 shows which variables need to be transmitted and whether they are transmitted locally (i.e., to node in N_i) or globally (i.e., to all nodes in N). The total number of sent items is shown in the fourth column. Take for example DIF; ii has

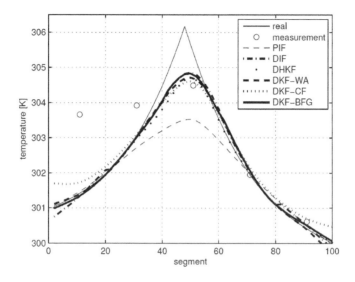

Figure 6.2: State-estimation at time 10,000 seconds without data loss.

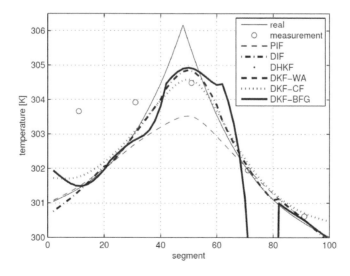

Figure 6.3: State-estimation at time 10,000 seconds with 5% data loss.

Table 6.1: Required communication.

DKF	variables	nodes	sent items per node
PIF	$\hat{x}_i[k|k] P_i[k|k]$	N	40,400
DIF	$i_i[k] I_i[k]$	N_i	10,100 or 20,200
DHKF	$\hat{x}_i[k|k - 1] P_i[k|k - 1] \hat{x}_i[k|k] P_i[k|k]$	N	3360
DKF-WA	$\hat{x}_i[k]$	N_i	200
DKF-CF	$i_i[k] I_i[k] \hat{x}_i^c[k|k]$	N_i	20,400
DKF-BFG	$i_i[k] I_i[k] / \hat{x}_i[k|k]$	N_i/N	500 or 1800 or 3440

100 items and I_i 10,000 items. Nodes 2, 3 and 4 send this data to 2 other nodes which leads to 20,200 items to be sent per node. Nodes 1 and 5 send to 1 other node, resulting in 10,100 sent items per node.

Figure 6.2 and Figure 6.3 together with Table 6.1.9 show the performance, robustness to data loss and the communication requirement, respectively, for each method.

Unfortunately, the methods that require the least data transfer, like DKFWA and DHKF, suffer the most from data loss. Note that the estimated temperature values obtained with these two methods do not even appear in Figure 6.3 (they are around 100K). Furthermore, also the DKF-BFG estimator, although it needs much less communication than the DIF estimator, in the presence of data loss is not robust, as can be observed in Figure 6.3. Overall, the least estimation error was obtained for the DIF estimator, which is also the most robust against data loss. Another aspect that can be observed is that the process-model is almost localized and sparse, as the results of the DKF-BFG closely resemble the ones obtained with the DIF, when no data loss occurs.

6.2 Wireless Networked Control System

This section is mainly concerned about the wireless networked control system (WNCS), in which wireless networks have lower installing costs and more free mobility than the wired ones. And a NCS must be wireless, as in some cases dedicated cabling for communication may not be possible. As the wireless technology is widely spread, research on WNCS has become a trend in this domain. Nevertheless, the wireless networks also have some drawbacks to overcome that can lead to the degradation of the control performance and even an unstable system, such as long time delay, data packet collision, ambient noise and interferences, and wireless channel congestion.

Other researchers investigate the WNCS on the wireless communication primarily. New network scheduling algorithms and protocols were presented to provide a reliable communication channel which could make the controller work at the optimal condition. A new protocol named COMAC was adopted in [215], which resulted in improved output Signal-to-Noise Ratio, hence reduced the number of retransmission, packet losses, and higher link reliability.

Simulations are a feasible way to test and evaluate the network and control strategies. Simulation studies will, hopefully, unravel these matters and lead to a coherent theory, best practices knowledge, and design expertise of WNCSs. The currently available simulation tools for WNCSs are few. Most of the simulators concentrate on either the network or the control part. At the moment there are only a couple of co-simulators, where both the network and control system are simulated simultaneously. The simulation tools like NS2, MATLAB and OPNET, PiccSIM and Truetime are most commonly used [216].

The main target is to apply the networked predictive control algorithm to a wireless networked control system in which the wireless network is emulated by the combination of several fading models. Channel fading is the main reason for communication errors like data dropouts, so it is more suitable to imitate a wireless network in this way. And the simulation is done within Simulink; all parts of the WNCS are implemented by the C-MEX s-function.

In the following section, we present some background and analysis of the wireless channel errors which are described by corresponding mathematical models first. Then the structure of the designed wireless control system will be shown.

6.2.1 Sources of wireless communication errors

The characteristics of wireless networks make the system more complex to analyze and degrades the performance of the controller. Here we consider three main sources of wireless communication errors: ambient wireless traffic, block fading, and multi-path fading. Ambient wireless traffic is the wireless traffic generated by neighboring nodes, while block fading and fast fading are directly related to the characteristics of the wireless channel. Wireless channel errors occur typically in bursts followed by practically error-free periods rather than occurring completely randomly [215].

Ambient Wireless Traffic

Since a wireless channel is of broadcast nature, a node's transmissions may be influenced by ambient wireless traffic. Collisions may occur while the data packets are transmitted through wireless channel. Some standard techniques have been widely utilized to solve this problem. In the Wireless Local Area Network (WLAN), IEEE 802.11 MAC uses a contention-based medium access mechanism named Distributed Coordination Function (DCF) which is responsible for avoiding collisions and resolving them when they appear as multiple wireless nodes try to transmit simultaneously.

Block Fading and Bursty Channel Errors

Block fading leads to bursty packet losses in the wireless communication and the bursty error characteristics can often be modeled by the Gilbert/Elliot model [217]. The GE model is a discrete Markov chain with two states G and B as shown in Figure 6.4. At any given time, the channel is either available or not available and named as "G (for good)" or "B (for bad)," respectively.

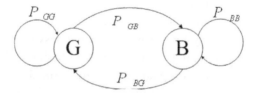

Figure 6.4: GE model of the packet loss.

where P_G and P_B are the probability of the channel error occurring in the "G" and "B" states, respectively. The next state of the channel is determined by state transmission probabilities P_{GB} and P_{BG} after each packet. Since state transition probabilities are usually very small, the channel state remains unchanged for some time after a transition imitating bursts of packet loss when the model is in the bad states, and periods of nearly error free transmission when the model is in the good state. If the transmission channel is in a steady state, the probability of state change from "G" and "B"can be formulated as follows:

$$\pi_G = \frac{P_{BG}}{P_{BG} + P_{GB}} \quad , \quad \pi_B = \frac{P_{GB}}{P_{BG} + P_{GB}}$$

The average packet loss rate in the GE model is

$$P = P_G \pi_G + P_B \pi_B \tag{6.24}$$

Multi-path Fading

Sometimes in a WNCS with numerous obstacles and no direct line of sight (LOS) between the transmitter and receiver, multi-path fading results in fluctuations of the signal amplitude because of the addition of signals arriving with different phases.

This phase difference is caused by the fact that the signals have traveled different distances along different paths. Because the phases of the arriving paths are changing rapidly, the received signal amplitude undergoes rapid fluctuation that is often modeled as a random variable with a particular distribution [218].

The most commonly used distribution for multi-path fading is the Rayleigh distribution, whose distribution density function (PDF) is given by:

$$f_{ray}(r) = \frac{r}{\sigma^2} \exp(-\frac{r^2}{2\sigma^2}), \quad r \geq 0 \tag{6.25}$$

The random variable corresponding to the signal amplitude is r. But when a strong LOS signal component also exists, the distribution is found to be Gaussian, and the PDF is given by:

$$f_{ric}(r) = \frac{r}{\sigma^2} \exp\left(\frac{-(r^2 + K^2)}{2\sigma^2}\right) I_0\left(\frac{Kr}{\sigma^2}\right), r, K \geq 0 \tag{6.26}$$

Where K is the factor that determines how strong the LOS component is relative to the rest of the multi-path signals, I_0 is the zero-order modified Bessel function of the first kind.

Many models have been proposed to generate Rayleigh fading, of which the Jakes' model is the most well-known. Jakes popularized the model based on summing sinusoids. However, it has been shown that the waveforms are correlated among themselves, that is, they have a non-zero cross-correlation except in special circumstances. The model is also deterministic (it has no random element to it once the parameters are chosen). So a modified model in [219] is employed in this chapter because of its improved statistical properties and simple software realization.

6.2.2 Structure of the WNCS

Figure 6.5 shows the structure of the wireless networked predictive control systems.

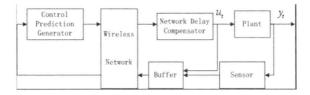

Figure 6.5: The structure of wireless networked predictive control system.

It mainly consists of three parts: control prediction generator, network delay compensator, and buffer.

The buffer is used to store historical data of the network delay compensator output u_t and the plant output y_t. All the data will be packed into one package and sent to the controller through the wireless network. The maximum length of the buffer can be preset as d, i.e., the packages $[u_t, u_{t-1}, \cdots, u_{t-d+1}]^T$ and $[y_t, y_{t-1}, \cdots, y_{t-d+1}]^T$ will be sent at time instant t. And it will update automatically in each sample interval.

The control prediction generator can do more than just calculate the control signal at time instant t. According to the information received from the buffer, the latest one is selected to generate a control prediction sequence:

$$[u_{t|t}, u_{t+1|t}, \cdots, u_{t+N-1|t}]^T$$

where N is the upper bound of round trip network delay (the sum of the delay from controller to actuator and the delay from sensor to controller). The sequence is put into one package and sent to the actuator. A most suitable control signal will be chosen from the received sequence. The network delay compensator can really compensate for data packet dropout. For example, if the data missed at time instant $t + 1$, the control signal $u_{t+1|t}$ is used as u_{t+1}.

In this wireless networked predictive control system, different drive modes are applied for different parts. The controller is event-driven which calculates the control signal sequence and sends it to the actuator after being triggered by an external event, while the sensor and the actuator are time-driven which updates their data in every sample period automatically.

6.2.3   Networked control design

The networked predictive control (NPC) strategy has been widely used in recent years because of its effectiveness in compensating the time delays and data packet dropouts induced by the network. Two policies are applied to compensate for them respectively.

Compensation for time delays

Consider the following MIMO discrete system in the state-space form:

$$\begin{aligned}
x_{t+1} &= Ax_t + Bu_t \\
y_t &= Cx_t
\end{aligned} \tag{6.27}$$

where x_t, u_t, and y_t are the state, input, and output vectors of the system, respectively. A, B and C are the constant matrices of appropriate dimensions.

Suppose the pair (A, B) is controllable, and the pair (A, C) is observable absolutely. Thus, the state-observer is designed as

$$x_{t+1|t} = Ax_{t|t-1} + Bu_t + L(y_k - Cx_{t|t-1}) \tag{6.28}$$

where the matrix L can be obtained by using observer design approaches. From the state observer described by (6.28), the state predictions from time $t - k$ (k is an integer multiple of the sampling period) to t are constructed as

$$
\begin{aligned}
x_{t-k+1|t-k} &= Ax_{t-k|t-k+1} + Bu_{t-k} \\
&\quad + L(y_{t-k} - Cx_{t-k|t-k-1}) \\
x_{t-k+2|t-k} &= Ax_{t-k+1|t-k} + Bu_{t-k+1} \\
&\quad \vdots \\
x_{t|t-k} &= Ax_{t-1|t-k} + Bu_{t-1}
\end{aligned}
\tag{6.29}
$$

which can get

$$
\begin{aligned}
x_{t|t-k} &= A^{k-1}(A - LC)x^{t-k|t-k-1} \\
&\quad + \sum_{j=1}^{k} A^{k-j}Bu_{t-k+j-1} + A^{k-1}Ly_{t-k}, \\
j &= 1, 2, \cdots, k
\end{aligned}
\tag{6.30}
$$

If there are time delay and data dropout from controller to actuator, the state prediction from time t to $t + i$ is given by

$$
\begin{aligned}
x_{t+1|t-k} &= Ax_{t|t-k} + Bu_{t|t-k} \\
x_{t+2|t-k} &= Ax_{t+1|t-k} + Bu_{t+1|t-k} \\
&\quad \vdots \\
x_{t+i|t-k} &= Ax_{t+i-1|t-k} + Bu_{t+i-1|t-k}
\end{aligned}
\tag{6.31}
$$

Particularly, the system with $u_t = Kx_{t|t-1}$ can be described as follows:

$$
\begin{aligned}
x_{t+1|t} &= (A + BK - LC)x_{t|t-1} + LCx_t \\
x_{t+1} &= Ax_t + BKx_{t|t-1}
\end{aligned}
\tag{6.32}
$$

The controller and the observer can be designed separately for the case of no network delay based on the assumption above.

When the delay i from the controller to the actuator and the delay k from the sensor to the controller both exist, the control predictions are calculated by

$$u_{t+i|t-k} = K x_{t+i|t-k} \tag{6.33}$$

where K is the state feedback matrix. Thus

$$
\begin{aligned}
x_{t+i|t-k} &= (A+BK)^i x_{t|t-k} \\
&= (A+BK)^i (A^{k-1}(A-LC)x_{t-k|t-k-1} \\
&\quad + \sum_{j=1}^{k} A^{k-j} B u_{t-k+j-1} + A^{k-1} L y_{t-k})
\end{aligned}
\tag{6.34}
$$

So the output of the NPC at time instant t is determined by

$$
\begin{aligned}
u_{t|t-k} &= K A^{k-1}(A-LC) x_{t-k|t-k-1} \\
&\quad + \sum_{j=1}^{k} A^{k-j} B u_{t-k+j-1} + K A^{k-1} L y_{t-k})
\end{aligned}
\tag{6.35}
$$

From (6.35), it is obvious that the future control predictions depend on the state estimation $x_{t-k|t-k-1}$, the past control input up to u_{t-1}, and the past output up to y_{t-k} of the system. Since there are the network delays i and k, the control input of the plant is designed as

$$u_t = u_{t|t-i-k} \tag{6.36}$$

Combining this predictive controller with a network delay compensator, a good control performance can be obtained. With respect to the analysis of the stability of closed-loop system, it is discussed in detail in [220].

Compensation for packet dropouts

Due to the specific characteristics of a wireless channel, the transmission of data packets is influenced greatly by the ambient disturbances. Packet dropouts are more likely to happen in a wireless network under imperfect radio conditions.

In addition to real packet dropouts caused by the sources in the foregoing section, quasi packet dropouts exist in WNCS. Control packets may arrive at the actuator with such long time delays that the control information they contain is of no further use. The actuator will drop the delayed control packets if newer information is available. Consequently, such packet dropouts are called quasi packet dropouts.

In this work, the control packet will be dropped and be treated as a packet dropout if it cannot be received by the deadline, which can be set to a reasonable value.

Consider the following system:

$$\begin{aligned} x_{t+1} &= Ax_t + Bu_t + \omega_t \\ y_t &= Cx_t + \nu_t \end{aligned} \tag{6.37}$$

where ω_t is a zero-mean uncorrelated white noise and Q_1 is the covariance matrix. A binary random number γ_t is defined to show whether the observed values arrived or not, whose distribution meets $P(\gamma_t = 1) = \lambda_t$, $\lambda_t \in (0,1)$. And γ_t, γ_s are independent when $t \neq s$. ν_t is the noise of output, which is defined as

$$\nu_t \sim \begin{cases} N(0, R_1), & \gamma_t = 1 \\ N(0, \sigma^2 I), & \gamma_t = 0 \end{cases}$$

The Kalman filter based on the state observer is designed as

$$\begin{aligned} x_{t|t-1} &= Ax_{t-1|t-1} + Bu_{t-1} \\ x_{t|t} &= x_{t|t-1} + \gamma_t(y_t - Cx_{t|t-1}) \\ P_{t|t} &= P_{t|t-1} - \gamma_t K_t C P_{t|t-1} \\ K_t &= P_{t|t-1}C^T(CP_{t|t-1}C^T + R)^{-1} \end{aligned} \tag{6.38}$$

where $P_{t|t-1}$ is the solution of the following Riccati equation. Thus the state prediction sequence is given by

$$\begin{aligned} x_{t+1|t} &= Ax_{t|t-1} + Bu_t + \gamma_t AK_t(y_t - Cx_{t|t-1}) \\ x_{t+2|t} &= Ax_{t+1|t} + Bu_{t+1|t} \\ &\;\;\vdots \\ x_{t+N|t} &= Ax_{t+N-1|t} + Bu_{t+N-1|t} \end{aligned} \tag{6.39}$$

If the control law is selected as

$$u_t = u_{t|t} = Lx_{t|t} \tag{6.40}$$

then the control prediction sequence can be generated as

$$u_{t+k|t} = Lx_{t+k|t}, \quad k = 0, 1, 2, \cdots, N \tag{6.41}$$

From (6.39) we can get

$$x_{t+k|t} = (A + BL)^{k-1}x_{t+1|t} \tag{6.42}$$

Further we can get

$$
\begin{aligned}
x_{t+k|t} &= (A + BL)^{k-1}[Ax_{t|t-1} + Bu_t + \gamma_t Ak_t(y_t - Cx_{t|t-1})] \\
&= (A + BL)^{k-1}[(A + BL - \gamma_t(AK_t \\
&- BLK_t)C)x_{t|t-1} + \gamma_t(AK_t - BLK_t)Cx_t]
\end{aligned}
\tag{6.43}
$$

and

$$
\begin{aligned}
u_{t+k|t} &= L(A + BL)^{k-1}[(A + BL - \gamma_t(AK_t \\
&- BLK_t)C)x_{t|t-1} + \gamma_t(AK_t - BLK_t)Cx_t]
\end{aligned}
\tag{6.44}
$$

The predictive control sequence is then put into a package and sent to the side of the controlled plant. The network delay compensator will select an optimal control signal to give it to the actuator. The simulation in the next section will prove the validity of this method discussed above.

6.2.4   Simulation example 2

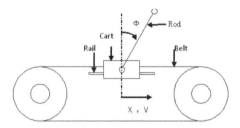

Figure 6.6: The inverted pendulum.

The control plant considered here is the inverted pendulum shown in Figure 6.6. The inverted pendulum is an open loop unstable plant whose mathematical model is nonlinear. It consists of a vertical rod that can rotate around a fixed point on a cart [221]. The corresponding rod angle from the vertical upward position is ϕ which is measured using a rotational encoder. The cart displacement x is also measured using a rotational encoder and the cart velocity is measured using a tachometer. The cart is driven by a DC electric motor controlled by a power amplifier with an input voltage u and coupled to the cart through a transmission belt. There is no sensor available to measure the angular velocity $\dot{\phi}$.

The nonlinear mathematical model can be given below where M, m are the mass of the cart and the rod respectively, l the length of rod, I the rod inertia, b the

Table 6.2: Parameters of the inverted pendulum

symbol	value	
m	0.109	Kg
M	1.096	Kg
l	0.25	m
b	0.1	$N/m/s$
I	0.0034	$Kg.m.m$
g	9.81	m/s^2

cart friction constant, and g the gravity acceleration.

$$(M + m)\ddot{x} + b\dot{x} + ml\ddot{\phi}cos\phi - ml\dot{\phi}^2sin\phi = u \tag{6.45}$$

$$-(I + ml^2)\ddot{\phi} + mgl\,sin\phi = ml\ddot{x}cos\phi \tag{6.46}$$

Choosing the state vector $x = [x, \dot{x}, \phi, \dot{\phi}]^T$ and linearizing (22) and (23) around the equilibrium point with the parameters in Table 6.2 result in the continuous time linear state space model

$$\begin{aligned} \dot{x}(t) &= Ax(t) + Bu(t) \\ y(t) &= Cx(t) \end{aligned} \tag{6.47}$$

where

$$A = \begin{bmatrix} 0 & 1 & 0 & 0 \\ 0 & 0 & 0 & 0 \\ 0 & 0 & 0 & 1 \\ 0 & 0 & 29.43 & 0 \end{bmatrix}, B = \begin{bmatrix} 0 \\ -1 \\ 0 \\ 3 \end{bmatrix}$$

$$C = \begin{bmatrix} 1 & 0 & 0 & 0 \\ 0 & 0 & 1 & 0 \end{bmatrix}$$

Using a sample interval of $T = 0.01s$ produces the equivalent discrete-time state model.

$$\begin{aligned} x_{k+1} &= \begin{bmatrix} 1 & 0.01 & 0 & 0 \\ 0 & 1 & 0 & 0 \\ 0 & 0 & 1.0015 & 0.01 \\ 0 & 0 & 0.2941 & 1.0015 \end{bmatrix} x_k \\ &+ \begin{bmatrix} -0.0001 \\ -0.01 \\ 0.0002 \\ 0.03 \end{bmatrix} u_k \end{aligned} \tag{6.48}$$

A linear-quadratic regulator (LQR) is used for the calculation of the optimal gain matrix F such that control law F minimizes the cost function $J_{LQG}(Q,R)$ with

$$Q = \begin{bmatrix} 1000 & 0 & 0 & 0 \\ 0 & 0 & 0 & 0 \\ 0 & 0 & 500 & 0 \\ 0 & 0 & 0 & 0 \end{bmatrix}, R = 1$$

We can adjust the controller such as

$$F = \begin{bmatrix} -31.6228 & -20.8404 & -77.0990 & -13.6080 \end{bmatrix} \tag{6.49}$$

The simulation tool used here is MATLAB/Simulink. We design the function blocks of the inverted pendulum wireless networked control system by defining corresponding C-MEX s-function. The control block diagram is shown in Figure 6.7.

The s-function 1 named pen_pre donates the controller with an NPC algorithm which can produce a sequence of predictive control signals. The s-function 2 called pen_com is on behalf of the network delay compensator which can choose an optimal control signal from the output of pen_pre and send to the inverted pendulum that is shown in a discrete state-space form. The s-functions 3 and 4 are the wireless networks in the forward and feedback channels respectively, but there is no difference between them in the characteristics of the networks. And the block named memory is just the buffer in Figure 6.5, which is very important to the whole control system.

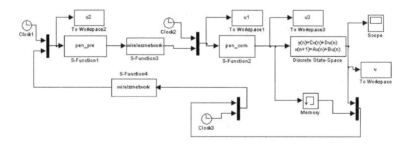

Figure 6.7: Inverted pendulum simulation diagram.

The simulation result is presented in Figure 6.8. The number k represents the number of samples. It is clear that the cart and the rod can reach the equilibrium point after an adjustment time of about 6 s, which implies that the NPC can compensate for time delay and data packet dropout effectively.

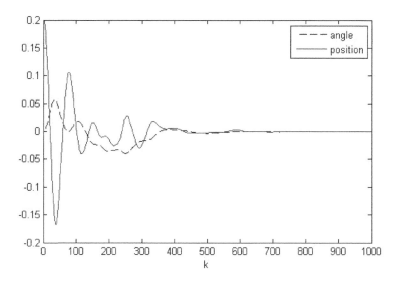

Figure 6.8: Simulation result.

6.3 Notes

In the first section, we presented an overview of different methodologies for designing partitioned (non-centralized) Kalman filters that can be used in WSNs. Each method was described and analyzed in terms of communication requirements, robustness and estimation-error. It turned out that the DKFWA requires the least communication and provides a low state-estimation error. However, it lacks robustness for its estimation error increases significantly when data is lost or nodes break down, which is usually the case in WSNs. For this reason it is not suitable for most WSNs. A method that can deal with unreliable data transfer and node loss, but still has a low state-estimation error is the DIF. It also has average requirements regarding the amount of data transfer needed compared to other methods. The amount of computations and communication per node can be decreased when the DKF-BFG is used. However, this approach is valid only for processes that have a localized and sparse structure, and assuming that there is no data-loss. Hence, the DKF-BFG is not suitable for usage in WSNs.

The next section is mainly about the application of networked predictive control in a wireless inverted pendulum control system. In order to make it different from the wired network, we consider the wireless network from the perspective of channel fading, since it is the main reason for wireless communication errors, such as network delay and data packet dropout. The mathematical models of different fading episodes that occurred in wireless communications are used to emulate a wireless network. Networked predictive control is applied to WNCS because it can produce a sequence of predictive control signals as an effective remedy to wireless network-induced problems. The control module (NPC), the network module, and the compensation module in the simulation diagram are all implemented by a C-MEX s-function. And the simulation result shows the validity of NPC.

6.4   Proposed Topics

1. An interesting topic to explore is to take mathematical models for communication into account. Meaning that both communication topology as well as the introduced errors and noises due to wireless communication links are used in the noise- and stability analysis, as described in [409].

2. Considering the results of partitioned estimators, it is desired to find new methods, suitable for WSNs, for reducing the communication and computation requirements, without losing robustness to data loss. Improving the robustness of the DKF-BFG seems to be a possible solution. Verify this.

3. From the literature, little is known about how different data-fusion link estimation methods affect the reliability, latency, stability, and energy efficiency of routing. This is an important problem because, as low power wireless sensor networks are increasingly deployed for mission critical tasks such as industrial monitoring, it is critical to ensure high reliability, low latency, and high predictability in routing. Develop a comparative study of the different methods of data-fusion in wireless link estimation and routing.

Chapter 7

Multi-Sensor Fault Estimation

In this chapter, an integrated design framework to utilize multi-sensor data fusion techniques is proposed for process fault monitoring. The multi-sensor data fusion technique is presented by frameworks of centralized and decentralized unscented Kalman filter architectures. A set of simulation studies has been conducted to demonstrate the performance of the proposed scheme on a quadruple tank system and an industrial boiler. It is established that the decentralized integrated framework retrieves more effectively the critical information about presence or absence of a fault from the dynamic model with a minimum time delay, and provides accurate unfolding-in-time of the finer details of the fault as compared to the centralized integrated framework, thus completing the overall picture of fault monitoring of the system under test. Experimental results on the quadruple tank system and industrial utility boiler, show that the proposed method is able to correctly identify various faults even when the dynamics of the systems are large.

Nomenclature

The following variables, Table 7.1, are used in this chapter.

7.1   Introduction

The problem of fault monitoring has always been an area of much importance for the research departments in the industries. And this importance becomes more prioritized when we are dealing with the non-linear systems. In this chapter, an unscented Kalman filter will be proposed in an integrated design framework to utilize multi-sensor data fusion techniques for process fault monitoring, thus completing a picture of a new automated fault detection and diagnosis system based on an enhanced unscented Kalman filter (UKF) estimator. The proposed methodology utilizes a multi-sensor data fusion (MSDF) technique to enhance the accuracy and reliability of parametric estimation in the process fault detection. This technique seeks to combine data from multiple sensors and related information to achieve improved accuracies and more specific inferences. The technique encapsulates the unscented Kalman filter in centralized and decentralized architectures for multi-sensor data fusion for improved fault monitoring. The extended Kalman filter (EKF) based centralized and decentralized fault monitoring multi-sensor data fusion was originally proposed for a continuous time stirred tank reactor problem in [447]. We use a similar approach where an unscented Kalman filter is used to detect the presence or absence of a fault. The proposed scheme has then been successfully evaluated on a quadruple tank system and an industrial utility boiler, thus corroborating the theory underpinning it.

We will summarize below the related work that have been conducted in this area of performance monitoring of plants: model based schemes for fault detection, model-free schemes for fault detection, and probabilistic models with fault detection.

7.1.1 Model-based schemes

The model-based approach is popular for developing fault diagnosis and isolation techniques [448]. It mainly consists of generating residuals by computing the difference between the measured output from the system and the estimated output obtained from the state system estimator used (like the Kalman filter). Any departure from zero of the residuals indicates a fault has likely occurred [449]. However, these methods are developed mainly for linear systems, assuming that a precise mathematical model of the system is available. This assumption, however, may be difficult to satisfy in practice, especially as engineering systems in general are nonlinear and are becoming more complex [450].

7.1.2 Model-free schemes

For model-free approaches, only the availability of a large amount of historical process data is assumed. There are different ways in which this data can be transformed and presented as a priori knowledge to a diagnostic system. This is known as the feature extraction process from the process history data, and it is done to facilitate later diagnosis [451]. This extraction process can proceed mainly as either a quantitative or a qualitative feature extraction process. Quantitative feature extraction can be either statistical or non-statistical. Model-free techniques such as neural networks, fuzzy logic, and genetic algorithms are used to develop models for FDI techniques. These models not only can represent a wide class of nonlinear systems with arbitrary accuracy, they can also be trained from data. Among these techniques, neural networks are well recognized for their ability to approximate nonlinear functions and for their learning ability [452]. For these reasons, they have been used as models to generate residuals for fault detection [453]. However, it is very difficult to isolate faults with these networks as they are black boxes in nature. Further, it is also desirable that a fault diagnostic system should be able to incorporate the experience of the operators [454]. Fuzzy reasoning allows symbolic generalization of numerical data by fuzzy rules and supports the direct integration of the experience of the operators in the decision making process of fault detection and isolation (FDI) in order to achieve more reliable fault diagnoses [455]. A rule-based expert system for fault diagnosis in a cracker unit is described in [455]. Optimization algorithms such as a genetic algorithm (GA) and particle swarm optimization (PSO) that simulate biological processes to solve search and optimization problems are also implemented to have a better pictorial view of fault detection and even classification.

7.1.3 Probabilistic schemes

Bayesian belief networks (BBN) provide a probabilistic approach to consider the cause-and-effect relation between process variables. There have been a few attempts to apply Bayesian belief networks for fault detection and diagnosis. [456] has worked in probabilistic sensor fault detection and identification. [457] proposed an approach to present a BBN model in the form of a set of nonlinear equations and constraints that should be solved for the unknown probabilities. As an inference tool, [458] used genetic algorithm for fault diagnosis in a Bayesian belief network representing a fluid catalytic cracking process. In [459] the learning capability of Bayesian belief networks is used to incorporate process data in an adaptable fault diagnosis strategy. Bayesian belief networks are also used to perform fault diagnosis and detection (FDD) for discrete events like walking [460]. A probabilistic approach with application to bearing fault-detection is also implemented in [461].

7.2 Problem Statement

Fault is an undesirable factor in any process control industry. It affects the efficiency of the system operation and reduces economic benefits to the industry. The early detection and diagnosis of faults in mission critical systems becomes highly crucial for preventing failure of equipment, loss of productivity and profits, management of assets and reduction of shutdowns.

To have an effective fault diagnosis approach to highly non-linear systems, we have assumed various faults in the system which have been successfully monitored and estimated through the encapsulation of the unscented Kalman filter (UKF) in various architectures of multi-data fusion technique. Figure 7.1 shows the proposed scheme has been introduced here by showing the implementation plan of fault monitoring using an unscented Kalman filter.

7.3   Improved Multi-Sensor Data Fusion Technique

The multi-sensor data fusion method has received major attention for various industrial applications. Data fusion techniques combine data from multiple sensors which are installed in the plant, and related information from associated databases, to achieve improved accuracies and more specific inferences that could not be achieved by the use of a single sensor alone. Specifically, in mission critical systems, where timely information is of immense importance, precision and accuracy achieved through multi-sensor data fusion technique can be very handy.

For a particular industrial process application, there might be plenty of associated sensor measurements located at different operational levels and having various accuracy and reliability specifications. One of the key issues in developing a multi-sensor data fusion system is the question of how the multi-sensor measurements can be fused or combined to overcome the uncertainty associated with individual data sources to obtain an accurate joint estimate of the system state vector. There exist various approaches to resolve this multi-sensor data fusion problem, of which the Kalman filter is one of the most significant and applicable solutions.

7.3.1   Unscented Kalman filter

The unscented Kalman filter (UKF) essentially addresses the approximation issues of the extended Kalman filter (EKF) [462]. The state distribution is again represented by Gaussian random variables (GRV), but is now specified using a minimal set of carefully chosen sample points that completely capture the true mean and covariance of the GRV. When propagated through a true nonlinear system, it captures the posterior mean and covariance accurately to the second order (Taylor Series Expansion) for any nonlinearity. The structure of the UKF is elaborated by unscented transformation.

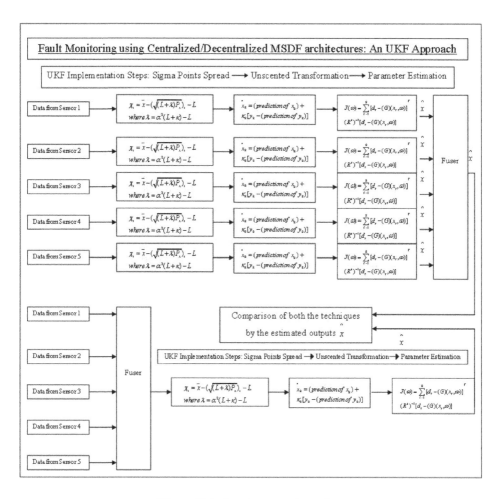

Figure 7.1: Implementation plan.

Remark 7.3.1 *Consider a state-space model given by:*

$$\begin{aligned}
\mathbf{x}_t &= f(\mathbf{x}_t - 1) + \epsilon, \quad \mathbf{x}_t \in \mathbf{R}^m \\
\mathbf{y}_t &= g(\mathbf{x}_t) + \nu, \quad \mathbf{y}_t \in \mathbf{R}^D
\end{aligned} \qquad (7.1)$$

Here, the system noise $\epsilon \sim N(0, \Sigma_\epsilon)$ and the measurement noise $\nu \sim N(0, \Sigma_\nu)$ noise are both Gaussian. The EKF linearizes f and g, at the current estimate of $\mathbf{x}_t$ and treats the system as a non-stationary linear system even though it is not. The UKF propagates several estimates of $\mathbf{x}_t$ through f and g and reconstructs a Gaussian distribution assuming the propagated values came from a linear system. Moreover, in nonlinear processes, when we are using EKF, the pdf is propagated through a linear approximation of the system around the operating point at each time instant. In doing so, the EKF needs the Jacobian matrices which may be difficult to obtain for higher order systems, especially in the case of time-critical applications. Further, the linear approximation of the system at a given time instant may introduce errors in the state which may lead the state to diverge over time. In other words, the linear approximation may not be appropriate for some systems. Also in the EKF algorithm, during the time-update (prediction) step, the mean is propagated through the nonlinear function, in other words, this introduces an error since in general $\bar{y} \neq g(\bar{\mathbf{x}})$. Whereas, in case of the UKF, during the time-update step, all the sigma points are propagated through the nonlinear function which makes the UKF a better and more effective nonlinear approximation. The UKF principle is simple and easy to implement as it does not require the calculation of Jacobian at each time step. The UKF is accurate up to second order moments in the pdf propagation whereas the EKF is accurate up to first order moment [467].

7.3.2 Unscented transformation

The unscented transformation (UT) is a method for calculating the statistics of a random variable which undergoes a nonlinear transformation [462]. Consider propagating a random variable $\mathbf{x}$ (dimension L) through a nonlinear function, $y = f(\mathbf{x})$. Assume $\mathbf{x}$ has mean $\bar{\mathbf{x}}$ and covariance $\mathbf{P_x}$. To calculate the statistics of y, we form a matrix χ of $2L + 1$ sigma vectors χ_i according to the following:

$$\begin{aligned}
\chi_0 &= \bar{\mathbf{x}} & (7.2) \\
\chi_i &= \bar{\mathbf{x}} + (\sqrt{(L + \lambda)\mathbf{P_x}})_i, i = 1, ...L & (7.3) \\
\chi_i &= \bar{\mathbf{x}} - (\sqrt{(L + \lambda)\mathbf{P_x}})_i - L, i = L + 1, ...2L & (7.4)
\end{aligned}$$

where $\lambda = \alpha^2(L + \kappa) - L$ is a scaling parameter. The constant α determines the spread of the sigma points around $\bar{\mathbf{x}}$, and is usually set to a small positive value (e.g., $1 \leq \alpha \leq 10^{-4}$). The constant κ is a secondary scaling parameter, which is usually set to $3 - L$, and β is used to incorporate prior knowledge, it is a tunable parameter of the distribution of $\mathbf{x}$ (for Gaussian distributions, $\beta = 2$) is optimal, used in equation (7.8). $(\sqrt{(L + \lambda)\mathbf{P_x}})_i$ is the ith column of the matrix square root (that is, lower-triangular Cholesky factorization). These sigma vectors are propagated through the nonlinear function

$$\xi_i = f(\chi_i), i = 0, ..., 2L \tag{7.5}$$

and the mean and covariance for y are approximated using a weighted sample mean and covariance of the posterior sigma points,

$$\bar{y} \approx \sum_{i=0}^{2L} W_i^m \xi_i, \ \mathbf{P}_y \approx \sum_{i=0}^{2L} W_i^c (\xi_i - \bar{y})(\xi_i - \bar{y})^T \tag{7.6}$$

with weights W_i given by:

$$W_0^{(m)} = \frac{\lambda}{L + \lambda} \tag{7.7}$$

$$W_0^{(c)} = \frac{\lambda}{L + \lambda} + 1 - \alpha^2 + \beta \tag{7.8}$$

$$W_i^{(m)} = W_i^{(c)} = \frac{1}{2(L + \lambda)}, i = 1, ..., 2L \tag{7.9}$$

Note that this method differs substantially from general Monte Carlo sampling methods which require orders of magnitude more sample points in an attempt to propagate an accurate (possibly non-Gaussian) distribution of the state. The deceptively simple approach taken with the UT results in approximations that are accurate to the third order for Gaussian inputs for all nonlinearities. For non-Gaussian inputs, approximations are accurate to at least the second order, with the accuracy of the third- and higher order moments being determined by the choice of α and β.

In view of the foregoing, the unscented Kalman Filter (UKF) is a straightforward extension of the UT to the recursive estimation in the following equation:

$$\hat{\mathbf{x}}_k = \mathbf{x}_{k predicted} + \kappa_k [y_k - y_{k predicted}] \tag{7.10}$$

where the state random variable (RV) is redefined as the concentration of the original state and noise variables:

$$\mathbf{x}_k^a = \begin{bmatrix} \mathbf{x}_k^T & \nu_k^T & \mathbf{n}_k^T \end{bmatrix} \tag{7.11}$$

The unscented transformation sigma points selection scheme, equation (7.2-7.4), is applied to this new augmented state RV to calculate the corresponding sigma matrix, χ_k^a. The UKF equations are given as follows. Note that no explicit calculations of the Jacobian or Hessian are necessary to implement this algorithm.

- Initialize with:

$$\hat{\mathbf{x}}_0 = \mathsf{E}[\mathbf{x}_0], \ \mathbf{P}_0 = \mathsf{E}[(\mathbf{x}_0 - \hat{\mathbf{x}})(\mathbf{x}_0 - \hat{\mathbf{x}})^T] \tag{7.12}$$

$$\hat{\mathbf{x}}_0^a = \mathsf{E}[\mathbf{x}^a] = [\hat{\mathbf{x}}_0^T \ \ 0 \ \ 0]^T \tag{7.13}$$

For $k\epsilon \ [1,....,\infty]$

- Calculate the sigma points:

$$\chi_{k-1}^a = [\hat{\mathbf{x}}_{k-1}^a \ \ \hat{\mathbf{x}}_{k-1}^a + \gamma\sqrt{P_{k-1}^a} \ \ \hat{\mathbf{x}}_{k-1}^a -$$

$$\gamma\sqrt{\mathbf{P}_{k-1}^a}] \tag{7.14}$$

- The time-update equations are:

$$\chi_{k|k-1}^x = \mathbf{F}(\chi_{k-1}^x, \mathbf{u}_{k-1}, \chi_{k-1}^\nu) \tag{7.15}$$

$$\hat{\mathbf{x}}_k^- = \sum_{i=0}^{2L} W_i^m \chi_{i,k|k-1}^x \tag{7.16}$$

$$\mathbf{P}_k^- = \sum_{i=0}^{2L} W_i^c (\chi_{i,k|k-1}^x - \hat{\mathbf{x}}_k^-)$$

$$(\chi_{i,k|k-1}^x - \hat{\mathbf{x}}_k^-)^T \tag{7.17}$$

$$\xi_{k|k-1} = \mathbf{H}(\chi_{k|k-1}^x, \chi_{k-1}^n) \tag{7.18}$$

$$\hat{y}_k^- = \sum_{i=0}^{2L} W_i^m \chi_{i,k|k-1}^Y \tag{7.19}$$

• The measurement-update equations are:

$$\mathbf{P}_{\bar{y}_k \bar{y}_k} = \sum_{i=0}^{2L} W_i^c (\xi_{i,k|k-1} - \hat{y}_k^-)$$
$$(\xi_{i,k|k-1} - \hat{y}_k^-)^T \tag{7.20}$$

$$\mathbf{P}_{x_k y_k} = \sum_{i=0}^{2L} W_i^c (\chi_{i,k|k-1} - \hat{\mathbf{x}}_k^-)$$
$$(Y_{i,k|k-1} - \hat{y}_k^-)^T \tag{7.21}$$

$$\kappa_k = \mathbf{P}_{x_k y_k} \mathbf{P}_{\bar{y}_k \bar{y}_k}^{-1} \tag{7.22}$$

$$\hat{\mathbf{x}}_k = \hat{\mathbf{x}}_k^- + \kappa_k (y_k - \hat{y}_k^-) \tag{7.23}$$

$$\mathbf{P}_k = \mathbf{P}_k^- - \kappa_k \mathbf{P}_{\bar{y}_k \bar{y}_k} \kappa_k^T \tag{7.24}$$

where

$$\mathbf{x}^a = \begin{bmatrix} \mathbf{x}^T & \nu^T & \mathbf{n}^T \end{bmatrix}^T$$
$$\chi^a = \begin{bmatrix} (\chi^x)^T & (\chi^v)^T & (\chi^n)^T]^T \end{bmatrix}^T$$
$$\gamma = \sqrt{L + \lambda}$$

Note that λ is the composite scaling parameter, L is the dimension of the augmented state, $\mathbf{R}^\nu$ is the process-noise covariance, $\mathbf{R}^n$ is the measurement-noise covariance, and W_i are the weights.

Next, we consider parameter estimation. It basically involves learning a nonlinear mapping $\mathbf{y}_k = \mathbf{G}(\mathbf{x}_k, \mathbf{w})$, where $\mathbf{w}$ corresponds to the set of known parameters. $\mathbf{G}(.)$ may be a neural network or another parametrized function. The extended Kalman filter (EKF) may be used to estimate the parameters by writing a new state-space representation

$$\mathbf{w}_{k+1} = \mathbf{w}_k + \mathbf{r}_k$$
$$\mathbf{d}_k = \mathbf{G}(x_k, w_k) + e_k \tag{7.25}$$

where $\mathbf{w}_k$ corresponds to a stationary process with an identity state transition matrix, driven by noise $\mathbf{r}_k$. The desired output $\mathbf{d}_k$ corresponds to a non-linear observation on $\mathbf{w}_k$.

From the optimized perspective, the following prediction error cost is minimized:

$$J(\mathbf{w}) = \sum_{i=1}^{k} [\mathbf{d}_t - \mathbf{G}(\mathbf{x}_t, \mathbf{w})]^T (\mathbf{R}^e)^{-1} [\mathbf{d}_t - \mathbf{x}\mathbf{G}(x_t, w)] \tag{7.26}$$

Thus, if the "noise" covariance $\mathbf{R}^e$ is a constant diagonal matrix, then, in fact, it cancels out of the algorithm, and hence can be set arbitrarily(e.g., $\mathbf{R}^e = 0.5I$). Alternatively, $\mathbf{R}^e$ can be set to specify a weighted MSE cost. The innovations covariance $\mathbf{E}[\mathbf{r}_k \quad \mathbf{r}_k^T] = \mathbf{R}_k^r$, on the other hand, affects the convergence rate and tracking performance. Roughly speaking, the larger the covariance, the more quickly older data is discarded. There are several options on how to choose $\mathbf{R}_k^r$.

1) Set $\mathbf{R}_k^r$ to an arbitrary "fixed" diagonal value, which may then be "annealed" towards zero as training continues.

2) Set $\mathbf{R}_k^r = (\lambda_{RLS}^{-1} - 1)\mathbf{P}_{w_i}$, where $\lambda_{RLS} \in (0,1]$ is often referred to as the "forgetting factor." This provides for an approximate exponentially decaying weighting on past data. Note that λ_{RLS} should not be confused with λ used for sigma-point calculation.

3) Set

$$
\begin{aligned}
\mathbf{R}_k^r &= (1 - \alpha_{RM})\mathbf{R}_r^{k-1} + \alpha_{RM}\mathbf{K}_k^w[\mathbf{d}_k - \mathbf{G}(\mathbf{x}_k, \hat{\mathbf{w}})] \\
&\times [\mathbf{d}_k - \mathbf{G}(\mathbf{x}_k, \hat{\mathbf{w}})]^T (\mathbf{K}_k^w)^T
\end{aligned}
$$

which is a Robbins-Monroe stochastic approximation scheme for estimating the innovations. The method assumes that the covariance of the Kalman update model is consistent with the actual update model. Typically, $\mathbf{R}_k^r$ is also constrained to be a diagonal matrix, which implies an independence assumption on the parameters. Note that a similar update may also be used for $\mathbf{R}_k^e$.

7.3.3 Multi-sensor integration architectures

Multi-sensor data fusion can be done at a variety of levels from the raw data or observation level to the feature/state vector level and the decision level. This idea can lead to utilization of different possible configurations or architectures to integrate the data from disparate sensors in an industrial plant to extract the desired monitoring information. Using Kalman filtering as the data fusion algorithm, multiple sensors can be integrated in two key architecture scenarios called the centralized method and the decentralized or distributed method. These methods have been widely studied over the last decade [468] and [469].

7.3.4 Centralized integration method

In the centralized integration method (CIM), all the raw data from different sensors is sent to a single location to be fused. This architecture is sometimes called the measurement fusion integration method [468] and [469], in which observations or sensor measurements are directly fused to obtain a global or combined measurement data matrix H^*. Then, it uses a single Kalman filter to estimate the global state vector based upon the fused measurement. Although this conventional method provides high fusion accuracy to the estimation problem, the large number of states may require high processing data rates that cannot be maintained in practical real time applications. Another disadvantage of this method is the lack of robustness in case of failure of a sensor or the central filter itself. For these reasons, parallel structures can often provide improved failure detection and correction, enhance redundancy management, and decrease costs for multi-sensor system integration.

7.3.5   Decentralized integration method

As such, there has recently been considerable interest shown in the distributed integration method in which the filtering process is divided between some local Kalman filters working in parallel to obtain individual sensor-based state estimates and one master filter combining these local estimates to yield an improved global state estimate. This architecture is sometimes called the state-vector fusion integration method [468] and [469]. The advantages of this method are higher robustness due to parallel implementation of fusion nodes and a lower computation load and communication cost at each fusion node. It is also applicable in modular systems where different process sensors can be provided as separate units. On the other hand, distributed fusion is conceptually a lot more complex and is likely to require higher bandwidth compared with centralized fusion.

In this chapter, the architectures are being implemented using the unscented Kalman filter (UKF) as shown in Figure 8.2 and Figure 8.3 respectively. The UKF addresses this problem by using a deterministic sampling approach. The state distribution is again approximated by a Gaussian Random Variable (GRV), but is now represented using a minimal set of carefully chosen sample points. These sample points completely capture the true mean and covariance of the GRV, and when propagated through the true nonlinear system, capture the posterior mean and covariance accurately to second order (Taylor Series Expansion) for any nonlinearity.

7.4   Simulation Results

The developed scheme was evaluated and simulated on the following systems:
- A quadruple tank system, and
- An industrial utility boiler.

The detailed implementation and simulation are now presented:

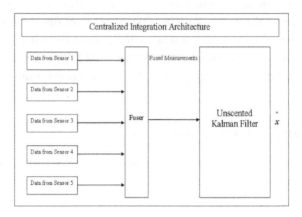

Figure 7.2: Centralized integration method using Unscented Kalman Filter.

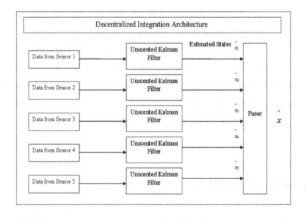

Figure 7.3: Decentralized integration method using Unscented Kalman Filter.

7.4.1 An interconnected-tank process model

The process is called an interconnected-tank process and consists of four interconnected water tanks and two pumps. Its manipulated variables are voltages to the pumps and the controlled variables are the water levels in the two lower tanks. The quadruple tank system presents a multi-input-multi-output (MIMO) system. This system is a real-life control problem prototyped to experiment on, to try to solve in the most efficient way; since it deals with multiple variables, it gives a reflection of the large systems in industry.

The schematic description of the four tank system can be visualized by Figure 7.4. The system has two control inputs (pump throughput) which can be manipulated to control the water level in the tanks. The two pumps are used to transfer water from a sump into four overhead tanks. The two tanks at the upper level drain freely into the two tanks at the bottom level and the liquid levels in these bottom two tanks are measured by pressure sensors. The piping system is designed such that each pump affects the liquid levels of both measured tanks. A portion of the flow from one pump is directed into one of the lower level tanks and the rest is directed to the overhead tank that drains into the other lower level tank. By adjusting the bypass valves of the system, the proportion of the water pumped into different tanks can be changed to adjust the degree of interaction between the pump throughput and the water levels. The output of each pump is split into two, using a three-way valve. Thus each pump output goes to two tanks, one lower and another upper, diagonally opposite, and the ratio of the split up is controlled by the position of the valve. Because of the large water distribution load, the pumps have been supplied with 12 V each. The mathematical modeling of the quadruple tank process can be obtained by using Bernoulli's law [470]. The constants are denoted in Table 5.1. A nonlinear mathematical model of the four-tank model is derived based on mass balances and Bernoulli's law. Mass balance for one of the tanks is:

$$A\frac{dh}{dt} = q_{out} - q_{in} \tag{7.27}$$

where A denotes the cross section of the tank, h, q_{in} and q_{out} denote the water level, the inflow, and the outflow of the tank, respectively. In order to establish a relationship between output and height Bernoulli's law is used. It states that

$$q_{out} = a\sqrt{2gh} \tag{7.28}$$

where a is the cross section of the outlet hole (cm^2) and g is the acceleration due to gravity. A common multiplying factor for an orifice of the type being used in this system is coefficient of discharge k. We can therefore rewrite Bernoulli's equation as:

$$q_{out} = ak\sqrt{2gh} \tag{7.29}$$

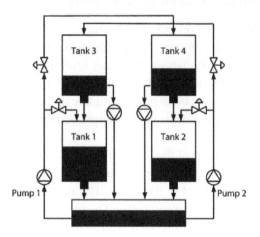

Figure 7.4: Schematic diagram of an interconnected-tank process.

The flow through each pump is split so that a proportion of the total flow travels to each corresponding tank. This can be adjusted via one of the two valves shown in Figure 7.4. Assuming that the flow generated is proportional to the voltage applied to each pump, (change) v, and that q_T and q_B are the flows going to the top and bottom tanks, respectively, we are able to come up with the following relationships. $q_B = \gamma k \nu$, $q_T = (1 - \gamma) k \nu$ where $\gamma \in [0, 1]$.

Combining all the equations for the interconnected four-tank system we obtain the physical system. A fault model can then be constructed by adding extra holes to each tank. The mathematical model of the faulty quadruple tank system can be given as:

$$
\begin{aligned}
\frac{dh_1}{dt} &= -\frac{a_1}{A_1}\sqrt{2gh_1} + \frac{a_3}{A_1}\sqrt{2gh_3} + \frac{\gamma_1 k_1}{A_1}\nu_1 + \frac{d}{A_1} \\
&\quad - \frac{a_{leak1}}{A_1}\sqrt{2gh_1} \\
\frac{dh_2}{dt} &= -\frac{a_2}{A_2}\sqrt{2gh_2} + \frac{a_4}{A_2}\sqrt{2gh_4} + \frac{\gamma_2 k_2}{A_2}\nu_2 - \frac{d}{A_2} \\
&\quad - \frac{a_{leak2}}{A_2}\sqrt{2gh_2} \\
\frac{dh_3}{dt} &= -\frac{a_3}{A_3}\sqrt{2gh_3} + \frac{(1-\gamma_2)k_2}{A_3}\nu_2 \\
&\quad - \frac{a_{leak3}}{A_3}\sqrt{2gh_3} \\
\frac{dh_4}{dt} &= -\frac{a_4}{A_4}\sqrt{2gh_4} + \frac{(1-\gamma_1)k_1}{A_4}\nu_1 \\
&\quad - \frac{a_{leak4}}{A_4}\sqrt{2gh_4} \\
\frac{d\nu_1}{dt} &= -\frac{\nu_1}{\tau_1} + \frac{1}{\tau_1}u_1 \\
\frac{d\nu_2}{dt} &= -\frac{\nu_2}{\tau_2} + \frac{2}{\tau_2}u_2
\end{aligned}
\tag{7.30}
$$

Two fault scenarios are created by using the quadruple tank system in the simulation program. In these scenarios incipient single and multiple tank faults (i.e., leakages) are created by changing some system parameters manually during the simulation at certain times. The system inputs, outputs and/or some states are corrupted by Gaussian noise with zero mean and a standard deviation of 0.1.

- Scenario I: Leakage fault in tank 1

  In this scenario, while the system is working in real time, a single incipient fault (i.e., tank 1 leakage percentage), is created by changing the parameter a_{leak1} to $0.81 cm^2$ (i.e., the value $0.81 is 30\%$ of the cross-section of the outlet hole of the tank 1) in the quadruple tank at 350 seconds.

- Scenario II: Leakage fault in tank 2 and 3 In this scenario, while the system is working in real time, multiple incipient faults (i.e., tank 2 and 3 leakage percentages) are created by changing the parameter a_{leak2} to $1.62\ cm^2$, a_{leak3} to $0.54\ cm^2$ (i.e., the value 1.62 is 60 percent of the cross-section of the outlet holes of the tank 2, and 0.54 is 20 percent of the cross-section of the outlet holes of the tank 3) in the quadruple tank at 350 seconds.

The general implementation structure of an unscented Kalman filter on one of the states of the quadruple tank systems can be described as follows:

Step 1   Initializing the parameters, and defining the conditions according to the system:

$\mathbf{n} = 6$ (where $\mathbf{n}$ is the number of states)

$\mathbf{q} = 0$ (where $\mathbf{q}$ shows the standard deviation of process)

$\mathbf{r} = 1.9$ (where $\mathbf{r}$ shows the standard deviation of measurement)

$\mathbf{Q} = \mathbf{q}^2 \times eye(\mathbf{n})$ (where $\mathbf{Q}$ represents the covariance of process)

$\mathbf{R} = \mathbf{r}^2$ (where $\mathbf{R}$ represents the covariance of measurement)

Step 2   Defining the states and measurement equation of the system:

$\mathbf{f} = @(\mathbf{x})[\mathbf{x}(1); \mathbf{x}(2); \mathbf{x}(3); \mathbf{x}(4); \mathbf{x}(5); \mathbf{x}(6)]$ (where $\mathbf{f}$ represents the nonlinear state equations)

$\mathbf{h} = @(\mathbf{x})\mathbf{x}(1)$; (where $\mathbf{h}$ represents the measurement equation)

$\mathbf{s} = [0; 0; 1; 0; 1; 1]$; (where $\mathbf{s}$ defines the initial state

$\mathbf{x} = \mathbf{s} + \mathbf{q} \times randn(6, 1)$; (where $\mathbf{x}$ defines the initial state with noise)

$\mathbf{P} = eye(\mathbf{n})$; (where $\mathbf{P}$ defines the initial state covariance)

N (where N presents the total dynamic steps)

Step 3   Upgrading the estimated parameter under observation: $for k = 1 : N$

$\mathbf{x}_1(1) = initializing$, $\mathbf{x}_2(1) = initializing$

$\mathbf{x}_3(1) = initializing$, $\mathbf{x}_4(1) = initializing$

$\mathbf{x}_5(1) = initializing$, $\mathbf{x}_6(1) = initializing$;

$\mathbf{H} = [\mathbf{x}_1(k); \mathbf{x}_2(k); \mathbf{x}_3(k); \mathbf{x}_4(k); \mathbf{x}_5(k); \mathbf{x}_6(k)]$

$\mathbf{z} = \mathbf{h}(s) + \mathbf{r} \times randn$; (measurements)

$sV(:, k) = \mathbf{s}$; (save actual state)

$zV(:, k) = \mathbf{z}$; (save measurement)

Then injecting the things in the following equation (function).

$[\mathbf{x}, \mathbf{P}] = ukf(\mathbf{x}, \mathbf{P}, hmeas, \mathbf{z}, \mathbf{Q}, \mathbf{R}, \mathbf{h})$

There will be three functions completing the process of unscented Kalman filter.

Function 1: The Unscented Kalman filter   Function 1 is defined as:

$function[\mathbf{x}, \mathbf{P}] = ukf(\mathbf{x}, \mathbf{P}, hmeas, \mathbf{z}, \mathbf{Q}, \mathbf{R}, \mathbf{h})$

(where [**x**,**P**] = ukf(f,x,P,h,z,Q,R))

returns state estimate, **x** and state covariance, **P**)

We note for a nonlinear dynamic system (for simplicity, noises are assumed as additive)

$\mathbf{x}_{k+1} = \mathbf{f}(\mathbf{x}_k) + \mathbf{w}_k$

$\mathbf{z}_k = \mathbf{h}(\mathbf{x}_k) + \nu_k$

(where $\mathbf{w} \sim N(0,\mathbf{Q})$ meaning $\mathbf{w}$ is Gaussian noise with covariance $\mathbf{Q}$)

(where $\nu \sim N(0,\mathbf{R})$ meaning ν is Gaussian noise with covariance $\mathbf{R}$)

(Inputs: $\mathbf{f}$: function handle for $f(\mathbf{x})$

$\mathbf{x}$: "a priori" state estimate

$\mathbf{P}$: "a priori" estimated state covariance

$\mathbf{h}$: function handle for $h(\mathbf{x})$

$\mathbf{z}$: current measurement

$\mathbf{Q}$: process noise covariance

$\mathbf{R}$: measurement noise covariance

Output: $\mathbf{x}$: "a posterior" state estimate

$\mathbf{P}$: "a posterior" state covariance

$\mathbf{L} = \text{numel}(\mathbf{x})$; (where $\mathbf{L}$ presents the number of states)

$\mathbf{m} = \text{numel}(\mathbf{z})$; (where $\mathbf{m}$ presents the number of measurements)

$\alpha = 1e^{-3}$; (default, tunable)

$k_i = 3\text{-}\,\mathbf{L}$; (default, tunable)

$\beta = 2$; (default, tunable)

$\lambda = \alpha^2 \times (\mathbf{L}+k_i\text{-}\mathbf{L})$; (scaling factor)

$c = \mathbf{L} + \lambda$; (scaling factor)

$\mathbf{W}_m = (\frac{\lambda}{c} + \frac{0.5}{c}+\text{zeros}(1,2 \times \mathbf{L}))$; (weights for means)

$\mathbf{W}_c = \mathbf{W}_m$;

$\mathbf{W}_c(1) = \mathbf{W}_c(1) + (1 - alpha^2 + \beta)$; weights for covariance

$c = \sqrt{c}$;

$\mathbf{P}_1 = \mathbf{P} + \mathbf{Q}$;

$X = \text{sigmas}(\mathbf{x}\,,\mathbf{P}\,,c)$; (sigma points around $\mathbf{x}$)

$[\mathbf{x}_1, X1, \mathbf{P}_1, X2]=\text{ut}(\text{fstate},\, X\,,\, \mathbf{W}_m,\, \mathbf{W}_c,\, \mathbf{L},\, \mathbf{Q})$; (unscented transformation of process)

$X2 = X - \mathbf{x}\,(:,\text{ones}(1,2 \times \mathbf{L}+1))$;

$[\mathbf{z}_1, Z1, \mathbf{P}_2, Z2]=\text{ut}(\text{hmeas}\,,\, X,\, \mathbf{W}_m,\, \mathbf{W}_c,\, \mathbf{m},\, \mathbf{R},\, \mathbf{h})$; (unscented transformation of measurements)

$\text{P}12 = X2 \times \text{diag}(\mathbf{W}_c) \times Z2'$; (transformed cross-covariance)

$\mathbf{R} = \text{chol}(\text{P}2)$; (where chol presents the Cholesky factorization)

$\text{K}=(\frac{\text{P}12/\mathbf{R}}{\mathbf{R}'})$; (Filter gain)

$\text{K}=\mathbf{P}12 \times inv(\mathbf{P}2)$;

$\mathbf{x}=\mathbf{x}+\text{K} \times (\mathbf{z}\text{-}\mathbf{z}1)$; (state update)

$\mathbf{P}=\mathbf{P}1\text{-}\text{K} \times \mathbf{P}12'$; (covariance update)

Function 2: Unscented Transformation   Function 2 is defined as:

$function[y, Y, \mathbf{P}, Y1] = ut(\mathbf{F}, X, \mathbf{W}_m, \mathbf{W}_c, n, \mathbf{R}, \mathbf{h})$

Unscented Transformation

Input:

f: nonlinear map

X: sigma points

$\mathbf{W}_m$: weights for mean

$\mathbf{W}_c$: weights for covariance

n: number of outputs

$\mathbf{R}$: additive covariance

Output:

y: transformed mean

Y: transformed sampling points

$\mathbf{P}$: transformed covariance

$Y1$: transformed deviations

L=size(X,2);

y=zeros(n,1);

Y=zeros(n,L);

x=[**h**];

for k=1:**L**

Y(k)=**x**(**h**);

y=y+$\mathbf{W}_m$(k)$\times$ Y(k);

end

Y1=Y-y;

P=Y1 $\times$ diag($\mathbf{W}_c$) $\times$ Y1'+**R**;

Function 3: Sigma Points   Function 3 is defined as as:

function $X = sigmas(\mathbf{x}, \mathbf{P}, c)$

Sigma points around reference point

Inputs:

x: reference point

P: covariance c: coefficient

Output:

X: Sigma points

$A = c * chol(\mathbf{P})'$;

$Y = \mathbf{x}(:, ones(1, numel(\mathbf{x})));$

$X = [\mathbf{x} Y + AY - A];$

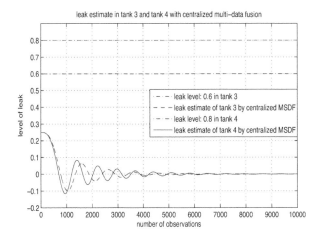

Figure 7.5: Quadruple tank system: Leak Estimate of tank 3 and tank 4 with centralized UKF MSDF approach.

A series of simulation runs was conducted on the quadruple tank system to evaluate and compare the effectiveness of the multi-sensor decentralized and centralized integration approaches based on the unscented Kalman filter (UKF) data fusion algorithm. To perform a different set of experiment the same fault scenarios have been used as defined.

The simulation results of the unscented Kalman filter embedded in the centralized structure of the multi-sensor data fusion technique are depicted in Figure 7.5, from which it is evident that the centralized structure was able to estimate the fault but there was a considerable offset in the estimation.

Next, the simulation results of the unscented Kalman filter embedded in the decentralized structure of a multi-sensor data fusion technique are depicted in Figure 7.6 and Figure 7.7, from which it is shown that with an increasing precision accompanied with a more detailed fault picture, the decentralized structure was able to estimate the fault in a much better way compared to the centralized architecture.

We now look at the drift detection. A fault may occur in any phase and in any part of the plant. Critical faults, not detected in time, can lead to adverse effects. In the sequel, the drift detection of the faults using the unscented Kalman filter is clarified. It is seen from Figure 7.8 that the fault is so incipient that apart from in the beginning, the level of water is achieving the same height. Thus, drift detection can give us a better picture for the fault scenario as shown in Figure 7.9. The kinks showing the middle of the height achievement can alarm the engineer about some unusual practice going on in the process.

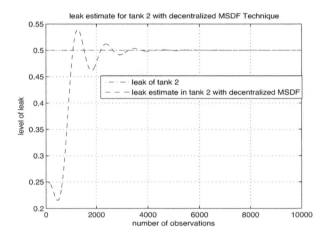

Figure 7.6: Quadruple tank system: Leak estimate of tank 2 with a decentralized UKF MSDF approach.

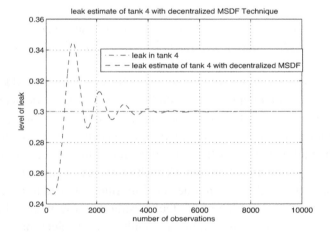

Figure 7.7: Quadruple tank system: Leak estimate of tank 4 with a decentralized UKF MSDF approach.

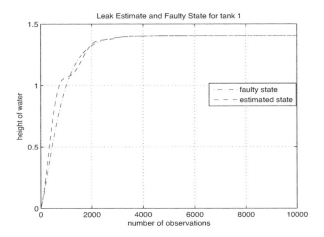

Figure 7.8: Quadruple tank system: Leak estimate and fault estimate of tank 1.

7.4.2 Utility boiler

The utility boilers in Syncrude Canada are water tube drum boilers. Since steam is used for generating electricity and process applications, the demand for steam is variable. The control objective of the co-generation system is to track steam demand while maintaining the steam pressure and the steam temperature of the header at their respective set-points. In the system, the principal input variables are u_1, feed water flow rate (kg/s); u_2, fuel flow rate (kg/s); and u_3, attemperator spray flow rate (kg/s); the states are x_1, fluid density, x_2, drum pressure, x_3, water flow input, x_4, fuel flow input, x_5, and the spray flow input. The principal output variables are y_1, drum level (m); y_2, drum pressure kPa; and y_3, steam temperature C^0 [471]. The schematic diagram of the utility boiler can be seen in Figure 7.10. Treatment of the different components is carried out as follows

1. Steam flow dynamics:

   Steam flow plays an important role in the drum-boiler dynamics. Steam flow from the drum to the header, through the super heaters, is assumed to be a function of the pressure drop from the drum to the header. We use a modified form of Bernoulli's law to represent flow versus pressure, with friction [473]. This expression is written as:

   $$q_s = K\sqrt{x_2^2 - P_{header}^2} \qquad (7.31)$$

   where q_s is the steam mass flow rate, K is a constant, and x_2 and P_{header} are the upstream and downstream pressures, respectively. The constant K is

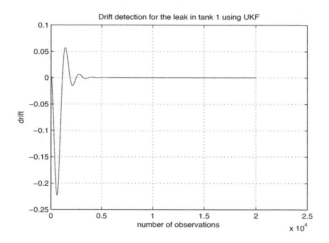

Figure 7.9: Quadruple tank system: drift detection for the leak in tank 1.

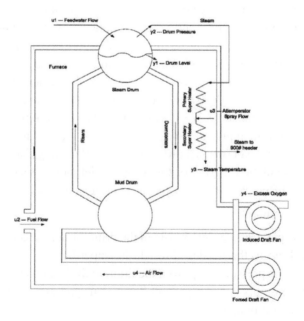

Figure 7.10: Schematic diagram of an industrial utility boiler.

chosen to produce agreement between measured flow and the pressure drop at a reference condition. Because, for the real system $P_{header}= 6306(kPa)$, by measuring the steam flow and drum pressure in the real system, the value of K is identified and the steam flow in the system can be modeled as:

$$q_s = 0.03\sqrt{x_2^2 - 6306^2} \tag{7.32}$$

2. Drum pressure dynamics:

To model the pressure dynamics, first step identification is done to observe the behavior of the system. By applying step inputs to the three different inputs at different operating points, we observe that for a step increase in the feed-water and fuel flow, the system behaves like a first order system with the same time constant. By applying a step to spray flow input, the system behaves like a first order system with a different time constant. The dynamics for the drum pressure is chosen as follows:

$$\dot{x}_2 = (c_1 x_2 + c_2)q_s + c_3 u_1 + c_4 u_2 \tag{7.33}$$

$$y_2 = x_2 \tag{7.34}$$

Finally, the dynamics of the drum pressure can be modeled as:

$$\begin{aligned} \dot{x}_2 &= (-1.8506 \times 10^{-7} x_2 - 0.0024)\sqrt{x_2^2 - (6306)^2} \\ &\quad -0.0404 u_1 + 3.025 u_2 \\ y_2(t) &= x_2(t) + p_0 \end{aligned} \tag{7.35}$$

where p_0=8.0715, p_0=-0.6449 and p_0=-6.8555 for low, normal, and high loads, respectively. At the three operating points, the initial conditions are x_{2_0} =6523.6, x_{2_0} =6711.5 and x_{2_0} =6887.9 for low, normal, and high loads, respectively.

3. Drum level dynamics:

Identification of the water level dynamics is a difficult task. Applying step inputs to the inputs separately, shows that the level dynamics is unstable. By increasing the water flow rate, the level increases and by increasing the fuel flow, the level decreases. Three inputs, water flow, fuel flow, and steam flow affect the drum water level. Letting x_1, and V_T denote the fluid density and total volume of the system, then we have

$$\dot{x}_1 = \frac{u_1 - q_s}{V_T} \tag{7.36}$$

where V_T=155.1411. By doing several experiments, it was observed that the dynamics of the drum level can be given by:

$$y_1 = c_5 x_1 + c_6 q_5 + c_7 u_5 + c_8 u_2 + c_9 \qquad (7.37)$$

The constants $c_i, i = 5, ..., 9$ should be identified from the plant data. The initial values of x_1 at the three operating points are given by $x_{1_0} = 678.15$, $x_{1_0} = 667.1$, and $x_{1_0} = 654.628$ for low, normal and high load, respectively.

4. Steam temperature:

   In the utility boiler, the steam temperature must be kept at a certain level to avoid overheating of the super-heaters. To identify a model for steam temperature, first step identification is used. By applying a step to the water flow input, steam temperature increases and the steam temperature dynamics behave like a first order system. Applying a step to the fuel flow input, the steam temperature increases and the system behaves like a second order system. Applying a step to the spray flow input, steam temperature decreases and the system behaves like a first order system. Then, a third order system is selected for the steam temperature model. This step identification gives an initial guess for local time constants and gains. By considering steam flow as input and applying input PRBS at the three operating points, local linear models for the steam temperature dynamics are defined. Combining the local linear models, the following nonlinear model is identified for all three operating points with a good fitness.

$$
\begin{aligned}
\dot{x}_3(t) &= (-0.0211\sqrt{x_2^2 - (6306)^2} + x_4 - 0.0010967 u_1 \\
&\quad +0.0475 u_2 + 3.1846 u_3 \qquad\qquad\qquad (7.38)\\
\dot{x}_4(t) &= 0.0015\sqrt{x_2^2 - (6306)^2} + x_5 + 0.001 u_1 \\
&\quad +0.32 u_2 - 2.9461 u_3 \qquad\qquad\qquad\quad (7.39)\\
\dot{x}_5(t) &= -1.278 \times 10^{-3}\sqrt{x_2^2 - (6306)^2} - 0.00025831 \\
&\quad x_3 - 0.29747\, x_4 - 0.8787621548\, x_5 \\
&\quad 0.00082 u_1 - 0.2652778\, u_2 + 2.491\, u_3 \qquad (7.40)\\
y_3 &= x_3 + T_0 \qquad\qquad\qquad\qquad\qquad\qquad\quad (7.41)
\end{aligned}
$$

where T_0=443.3579, T_0=446.4321, and T_0=441.9055 for the low load, normal load, and high load, respectively. At three operating points, we have for the low load,

$$x_{3_0} = 42.2529, x_{4_0} = 3.454, x_{5_0} = 3.45082$$

for the normal load,

$$x_{3_0} = 49.0917, x_{4_0} = 2.9012, x_{5_0} = 2.9862$$

and for the high loads

$$x_{3_0} = 43.3588, x_{4_0} = -0.1347, x_{5_0} = -0.2509.$$

Combining the results achieved so far, the identified model for the utility boiler can be obtained. In addition, the following limit constraints exist for the three control variables:

$$0 \leq u_1 \leq 120, \ 0 \leq u_2 \leq 7 \tag{7.42}$$
$$0 \leq u_3 \leq 10 \tag{7.43}$$

5. Fault model for utility boiler:

A fault model for the utility boiler will be developed. To construct this model extra holes are added to each tank. The mathematical model of the faulty utility boiler can be given as follows, where the faults of steam pressure are there in states 4 and 5:

$$\dot{x}_1(t) = \frac{u_1 - 0.03\sqrt{x_2^2 - (6306)^2}}{155.1411} \tag{7.44}$$

$$\dot{x}_2(t) = (-1.8506 \times 10^{-7}x_2 - 0.0024)\sqrt{x_2^2 - (6306)^2}$$
$$-0.0404u_1 + 3.025u_2 \tag{7.45}$$

$$\dot{x}_3(t) = -0.0211\sqrt{x_2^2 - (6306)^2} + x_4 - 0.0010967u_1$$
$$+0.0475u_2 + 3.1846u_3 \tag{7.46}$$

$$\dot{x}_4(t) = 0.0015\sqrt{x_2^2 - (6306)^2} + x_5 - 0.001u_1$$
$$+0.32u_2 - 2.9461u_3$$
$$+(a_{st\ pr})\sqrt{x_2^2 - (6306)^2} \tag{7.47}$$

$$\dot{x}_5(t) = -1.278 \times 10^{-3}\sqrt{x_2^2 - (6306)^2}$$
$$-0.00025831\ x_3 - 0.29747\ x_4$$
$$-0.8787621548\ x_5 - 0.00082\ u_1 - 0.2652778$$
$$u_2 + 2.491\ u_3$$
$$+(a_{st\ pr})\sqrt{x_2^2 - (6306)^2} \tag{7.48}$$

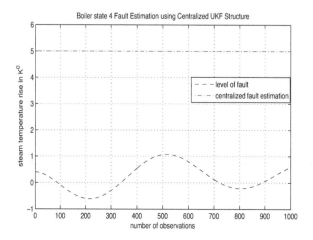

Figure 7.11: Utility boiler: estimate of the state 4 using centralized UKF.

Two fault scenarios are created by using the utility boiler in the simulation program. In these scenarios, steam pressure is added there in states 4 and 5 resulting in a more uncontrolled non-linear system. In what follows, we present the simulation results for UKF with both centralized and decentralized multi-sensor data fusion methods.

For assuring the gist of UKF-based fault estimation and monitoring, a series of experiments was also performed on the industrial utility boiler system to evaluate and compare the effectiveness of the multi-sensor decentralized and centralized integration approaches based on the unscented Kalman filter data fusion algorithm. A series of simulation runs was performed with a fault in the fourth state 4 in the boiler in form of the increased steam temperature. The simulation results of the unscented Kalman filter embedded in the centralized structure of the multi-sensor data fusion technique are depicted in Figure 7.11, from which there is a considerable offset in the estimation. The results of the unscented Kalman filter embedded in the decentralized structure of the multi-sensor data fusion technique can be seen in Figure 7.12. The comparison of both centralized and decentralized schemes with the fault estimation is depicted in Figure 7.13.   In utility boiler, several faults may occur in any part of the boiler. Critical faults not detected in time, can lead to adverse effects. This section shows the drift detection of the faults using an unscented Kalman filter.

In Figure 7.14, the estimated parameter and the fault parameter are shown, from which it is seen that there is difference between them despite the same pattern they are following. By drift detection, as shown in Figure 7.15, we can see

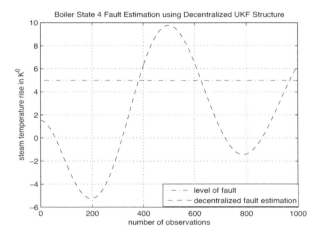

Figure 7.12: Utility Boiler: estimate of state 4 with decentralized UKF.

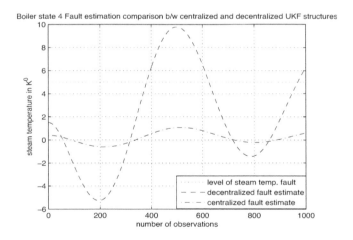

Figure 7.13: Utility Boiler: estimate comparisons.

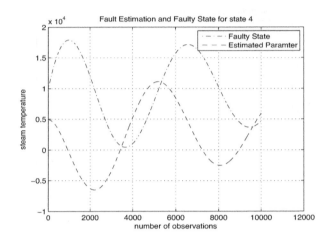

Figure 7.14: Utility boiler: estimated and fault parameters.

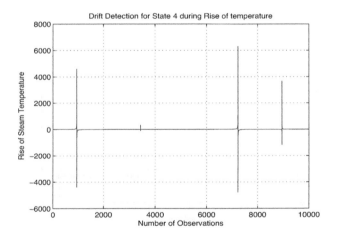

Figure 7.15: Utility boiler: drift detection.

the prominent kinks in the profile of the faulty parameter estimation, thus giving sufficient signs for a necessary action.

Remark 7.4.1 *In the simulation section, a multi-sensor data fusion technique with centralized and decentralized structures of UKF is implemented on the quadruple tank system under different leakage fault scenarios and on the utility boiler with an uncontrolled steam pressure fault scenario. It has been shown that in the cases of both physical systems i.e., the quadruple tank system and the utility boiler, the decentralized structure shows better results than the centralized structure, followed by the drift detection which has been made effectively showing prominent kinks of fault. The effectiveness of the decentralized structure in the case of the utility boiler was less compared to the quadruple tank system, because of the large fault of steam pressure introduced in state 4 and state 5 respectively.*

7.5 Notes

In this chapter, a complete fault estimation and monitoring scheme has been developed by integrating the techniques of multi-sensor data fusion, and centralized and decentralized architectures for providing quality reports and ensuring reliable fault detection, and it has been made effective by using a deterministic sampling approach known as the unscented Kalman filter, thus capturing the posterior mean and covariance accurately, to complete the overall diagnostic picture. It has been demonstrated that this approach can be used for reliable detection of incipient faults, which in turn leads to an efficient and cost-effective preventive maintenance scheme. The major contributions of the chapter are the implementation of the integrated centralized and decentralized architectures of multi-sensor data fusion, implemented thoroughly on a quadruple tank system and an industrial utility boiler to achieve both accuracy and reliability of the fault monitoring schemes.

7.6 Proposed Topics

1. An integral part of multi-sensor fault estimation is the necessity to design models of faults according to each fault candidate. Such design requires time, computational resources, and a large history of data. It is desired to look for systematic design of models of faulty behaviors, which retains the merits of the available methods and benefits from the correlation of diagnosis performances with the selection of estimated outputs for residuals design.

2. Expanding on the algorithms developed in this chapter, the analysis of cumulative residuals in the residual space and the covering of the residual space

to select appropriate residuals provides an interesting topic that deserves further research work. A deeper interpretation of the distance as a probability or likelihood is worth investigation.

3. Wireless sensor networks have become a new information collection and monitoring solution for a variety of applications. Faults occurring at sensor nodes are common due to the sensor device itself and the harsh environment where the sensor nodes are deployed. In order to ensure the network quality of service it is necessary for the WSN to be able to detect the faults and take actions to avoid further degradation of the service. An important research topic is to design a distributed detection algorithm that is capable of locating the faulty sensors in wireless sensor networks. It would be beneficial to examine the implementation complexity of the algorithm and the probability of correct diagnoses in the existence of large fault sets.

Table 7.1: Nomenclature

Symbols	Function
$\bar{x}, P_x$	mean, covariance
$2L + 1$	Sigma vectors in unscented Kalman filter (UKF)
α	Spread of the sigma points around $\bar{x}$
κ, λ	Scaling parameters
β	incorporate prior knowledge of the distrb. of $\mathbf{x}$
L	Dimension of the augmented state
R^v	Process-noise covariance
R^n	Measurement-noise covariance
W_i	weights
$\mathbf{w}_k$	Stationary process with identity state transition matrix
r_k	Noise
d_k	desired output
w_k	nonlinear observation
λ_{RLS}	forgetting factor
h_i	Level of water in tank i
a_i	Area of water flowing out from tank i
A_i	Area of tank i
γ_1	tank 1 and tank 4 water diverting ration
γ_2	tank 2 and tank 3 water diverting ration
k_1, k_2	Gains of Pumps 1 and 2
ν_1	Manipulated input 1 (pump 1)
ν_2	Manipulated input 2 (pump 2)
g	Gravitational constant
a_{leak1}	Leak in pipe of tank 1
a_{leak2}	Leak in pipe of tank 2
a_{leak3}	Leak in pipe of tank 3
a_{leak4}	Leak in pipe of tank 4
q_{in}, q_{out}	inflow and outflow
u_1	feed water flow rate (kg/s)
u_2	fuel flow rate (kg/s)
u_3	at temperature spray flow rate (kg/s)
y_1	drum level (m)
y_2	drum pressure kPa
y_3	steam temperature C^0
x_2	Upstream pressure
P_{header}	Downstream pressure
x_1	fluid density of the system
V_T	Total volume of the system

Chapter 8

Multi-Sensor Data Fusion

Data fusion is a multi-disciplinary research field with a wide range of potential applications in areas such as defense, robotics, automation and intelligent system design, and pattern recognition. This chapter provides an analytical review of recent developments in the multi-sensor data fusion domain. The different sections will present a comprehensive review of the data fusion ingredients, exploring its conceptualizations, benefits, and challenging aspects, as well as existing methodologies.

8.1 Overview

Multi-sensor data fusion is a technology to enable combining information from several sources in order to form a unified picture. Data fusion systems are now widely used in various areas such as sensor networks, robotics, video and image processing, and intelligent system design, to name a few. Data fusion is a wide ranging subject and many terminologies have been used interchangeably. These terminologies and ad hoc methods in a variety of scientific, engineering, management, and many other publications, show the fact that the same concept has been studied repeatedly. The focus of this chapter is on multi-sensor data fusion. Thus, throughout this chapter the terms data fusion and multi-sensor data fusion are used interchangeably.

The data fusion research community has achieved substantial advances, especially in recent years. Nevertheless, realizing a perfect emulation of the data fusion capacity of the human brain is still far from accomplished.

8.1.1   Multi-sensor data fusion

Many definitions for data fusion exist in the literature. Joint Directors of Laboratories (JDL) [585] defines data fusion as a multilevel, multifaceted process handling the automatic detection, association, correlation, estimation, and combination of data and information from several sources. Klein [586] generalizes this definition, stating that data can be provided either by a single source or by multiple sources. Both definitions are general and can be applied in different fields including remote sensing. In [587], the authors present a review and discussion of many data fusion definitions. Based on the identified strengths and weaknesses of previous work, a principled definition of information fusion is proposed as: *Information fusion is the study of efficient methods for automatically or semi-automatically transforming information from different sources and different points in time into a representation that provides effective support for human or automated decision making.* Data fusion is a multidisciplinary research area borrowing ideas from many diverse fields such as signal processing, information theory, statistical estimation and inference, and artificial intelligence.

Generally speaking, performing data fusion has several advantages [588], [583]. These advantages mainly involve enhancements in data authenticity or availability. Examples of the former are improved detection, confidence, and reliability, as well as reduction in data ambiguity, while extending spatial and temporal coverage belong to the latter category of benefits. Data fusion can also provide specific benefits for some application contexts. For example, wireless sensor networks are often composed of a large number of sensor nodes, hence posing a new scalability challenge caused by potential collisions and transmissions of redundant data. Regarding energy restrictions, communication should be reduced to increase the lifetime of the sensor nodes. When data fusion is performed during the routing process, that is, sensor data is fused and only the result is forwarded, the number of messages is reduced, collisions are avoided, and energy is saved.

Various conceptualizations of the fusion process exist in the literature. The most common and popular conceptualization of fusion systems is the JDL model [585]. The JDL classification is based on the input data and produced outputs, and originated from the military domain. The original JDL model considers the fusion process in four increasing levels of abstraction, namely, object, situation, impact, and process refinement. Despite its popularity, the JDL model has many shortcomings, such as being too restrictive and especially tuned to military applications, which have been the subject of several extension proposals [589], [590] attempting to alleviate them. The JDL formalization is focused on data (input/output) rather than processing. An alternative is Dasarathys framework [591] that views the fusion system, from a software engineering perspective, as a data flow characterized by input/output as well as functionalities (processes). Another general conceptualization of fusion is the work of Goodman et al. [592], which is based on the notion of random sets. The distinctive aspects of this framework are its ability to combine decision uncertainties with the decisions themselves, as well as presenting a fully generic scheme of uncertainty representation. One of the most recent and abstract fusion frameworks is proposed by Kokar et al. [593]. This formalization is based on category theory and is claimed to be sufficiently general to capture all kinds of fusion, including data fusion, feature fusion, decision fusion, and fusion of the relational information. It can be considered as the first step toward development of a formal theory of fusion. The major novelty of this work is the ability to express all aspects of multi-source information processing, that is, both data and processing. Furthermore, it allows for consistent combination of the processing elements (algorithms) with measurable and provable performance. Such formalization of fusion paves the way for the application of formal methods to standardized and automatic development of fusion systems.

8.1.2 Challenging problems

There are a number of issues that make data fusion a challenging task [582]. The majority of these issues arise from the data to be fused, imperfection and diversity of the sensor technologies, and the nature of the application environment as following:

- *Data imperfection:* data provided by sensors is always affected by some level of impreciseness as well as uncertainty in the measurements. Data fusion algorithms should be able to express such imperfections effectively, and to exploit the data redundancy to reduce their effects.

- *Outliers and spurious data:* the uncertainties in sensors arise not only from the impreciseness and noise in the measurements, but are also caused by the ambiguities and inconsistencies present in the environment, and from the inability to distinguish between them [594]. Data fusion algorithms should be able to exploit the redundant data to alleviate such effects.

- *Conflicting data:* fusion of such data can be problematic especially when the fusion system is based on evidential belief reasoning and Dampster's rule of combination [595]. To avoid producing counter-intuitive results, any data fusion algorithm must treat highly conflicting data with special care.

- *Data modality:* sensor networks may collect qualitatively similar (homogeneous) or different (heterogeneous) data such as auditory, visual, and tactile measurements of a phenomenon. Both cases must be handled by a data fusion scheme.

- *Data correlation:* this issue is particularly important and common in distributed fusion settings, e.g. wireless sensor networks, as for example some sensor nodes are likely to be exposed to the same external noise biasing their measurements. If such data dependencies are not accounted for, the fusion algorithm may suffer from over/under confidence in the results.

- *Data alignment/registration:* sensor data must be transformed from each sensor's local frame into a common frame before fusion occurs. Such an alignment problem is often referred to as sensor registration and deals with the calibration error induced by individual sensor nodes. Data registration is of critical importance to the successful deployment of fusion systems in practice.

- *Data association:* multi-target tracking problems introduce major complexity to the fusion system compared to the single-target tracking case [596]. One of these new difficulties is the data association problem, which may come in two forms: measurement-to-track and track-to-track association. The former refers to the problem of identifying from which target, if any, each measurement originated, while the latter deals with distinguishing and combining tracks which are estimating the state of the same real-world target [584].

- *Processing framework:* data fusion processing can be performed in a centralized or decentralized manner. The latter is usually preferable in wireless sensor networks, as it allows each sensor node to process locally collected data. This is much more efficient compared to the communication burden required by a centralized approach, when all measurements have to be sent to a central processing node for fusion.

- *Operational timing:* the area covered by sensors may span a vast environment composed of different aspects varying in different rates. Also, in the case of homogeneous sensors, the operation frequency of the sensors may be different. A well-designed data fusion method should incorporate multiple time scales in order to deal with such timing variations in data. In distributed fusion settings, different parts of the data may traverse different routes before reaching the fusion center, which may cause out-of-sequence arrival of data. This issue needs to be handled properly, especially in real-time applications, to avoid potential performance degradation.

- *Static vs. dynamic phenomena:* the phenomenon under observation may be time-invariant or varying with time. In the latter case, it may be necessary for the data fusion algorithm to incorporate a recent history of measurements into the fusion process [597]. In particular, data freshness, that is, how quickly data sources capture changes and update accordingly, plays a vital role in the validity of fusion results. For instance in some recent work [598], the authors performed a probabilistic analysis of the recent history of measurement updates to ensure the freshness of input data, and to improve the efficiency of the data fusion process.

- *Data dimensionality:* the measurement data could be preprocessed, either locally at each of the sensor nodes or globally at the fusion center to be compressed into lower dimensional data, assuming a certain level of compression loss is allowed. This pre-processing stage is beneficial, as it enables saving on the communication bandwidth and power required for transmitting data, in the case of local pre-processing [599], or limiting the computational load of the central fusion node, in the case of global pre-processing [600].

8.1.3 Multi-sensor data fusion approaches

Different multi-sensor data fusion techniques have been proposed with different characteristics, capabilities, and limitations. A data-centric taxonomy is discussed in [539] to show how these techniques differ in their ability to handle different imperfection aspects of the data. This section summarizes the most commonly used approaches to multi-sensor data fusion.

1. *Probabilistic fusion:* Probabilistic methods rely on the probability distribution/density functions to express data uncertainty. At the core of these methods lies the Bayes estimator, which enables fusion of pieces of data, hence, the name Bayesian fusion [540]. More details are provided in the next section.

2. *Evidential belief reasoning:* Dempster-Shafer (D-S) theory introduces the notion of assigning beliefs and plausibility to possible measurement hypotheses along with the required combination rule to fuse them. It can be considered as a generalization to the Bayesian theory that deals with probability mass functions. Unlike the Bayesian Inference, the D-S theory allows each source to contribute information in different levels of detail [540].

3. *Fusion and fuzzy reasoning:* Fuzzy set theory is another theoretical reasoning scheme for dealing with imperfect data. Due to being a powerful theory

to represent vague data, fuzzy set theory is particularly useful to represent and fuse vague data produced by human experts in a linguistic fashion [540].

4. *Possibility theory:* Possibility theory is based on fuzzy set theory but was mainly designed to represent incomplete rather than vague data. The possibility theory's treatment of imperfect data is similar in spirit to probability and D-S evidence theory with a different quantification approach [541].

5. *Rough set-based fusion:* Rough set is a theory of imperfect data developed by Pawlak [542] to represent imprecise data, ignoring uncertainty at different granularity levels.This theory enables dealing with data granularity.

6. *Random set theoretic fusion:* The most notable work on promoting random finite set theory (RFS) as a unified fusion framework has been done by Mahler in [543]. Compared to other alternative approaches of dealing with data imperfection, RFS theory appears to provide the highest level of flexibility in dealing with complex data while still operating within the popular and well-studied framework of Bayesian inference. RFS is a very attractive solution to the fusion of complex soft/hard data that is supplied in disparate forms and may have several imperfection aspects [544].

7. *Hybrid fusion approaches:* The main idea behind the development of hybrid fusion algorithms is that different fusion methods such as fuzzy reasoning, D-S evidence theory, and probabilistic fusion should not be competing, as they approach data fusion from different (possibly complementary) perspectives [539].

8.1.4   Multi-sensor algorithms

Multi-sensor data fusion can be performed at four different processing levels, according to the stage at which the fusion takes place: signal level, pixel level, feature level, and decision level. Figure 8.1 illustrates of the concept of the four different fusion levels [535]:

1. *Signal level fusion:* In signal-based fusion, signals from different sensors are combined to create a new signal with a better signal-to noise ratio than the original signals.

2. *Pixel level fusion:* Pixel-based fusion is performed on a pixel-by-pixel basis. It generates a fused image in which information associated with each pixel is determined from a set of pixels in source images to improve the performance of image processing tasks such as segmentation.

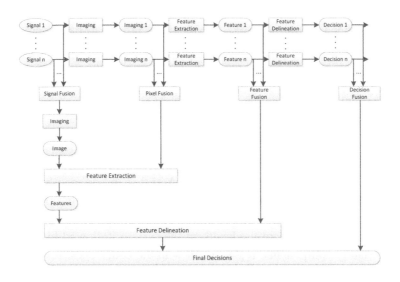

Figure 8.1: A categorization of the fusion algorithms.

3. *Feature level fusion:* Feature-based fusion at the feature level requires an extraction of objects recognized in the various data sources. It requires the extraction of salient features which depend on their environment, such as pixel intensities, edges, or textures. These similar features from input images are fused.

4. *Decision-level fusion:* consists of merging information at a higher level of abstraction, and combines the results from multiple algorithms to yield a final fused decision. Input images are processed individually for information extraction. The obtained information is then combined by applying decision rules to reinforce common interpretation.

We provide a brief account of some of the algorithms. One standard algorithm is the principal component analysis (PCA) transform. It basically converts inter-correlated multi-spectral (MS) bands into a new set of uncorrelated components. Multi-resolution or multi-scale methods, such as pyramid transformation, have been adopted for data fusion since the early 1980s [536]. The pyramid-based image fusion methods, including the Laplacian pyramid transform, were all developed from the Gaussian pyramid transform, and have been modified and widely used, and substituted by the wavelet transform methods to some extent in recent years [537]. Recently, wavelet transforms has provided a framework in which an image is decomposed, with each level corresponding to a coarser resolution band. Moreover, artificial neural networks (ANNs) have proven to be a more powerful and self-adaptive method of pattern recognition, compared to traditional linear and simple nonlinear analyses. For a recent detailed account, the reader is referred to [538].

8.2   Fault Monitoring

The problem of fault monitoring has always been an area of much importance for the research departments in the industries. And this importance becomes more prioritized when we are dealing with the nonlinear systems. Monitoring of uncommon behavior of the plants and detecting the unprecedented changes in system are the essential steps to maintain the health of the system, followed by covering the removal of faulty components, replacement with the better ones, restructuring the system architecture, and thus improving the overall system reliability. However, with the increasing complexity of the modern nonlinear systems, process engineers are facing tough challenges to understanding and troubleshooting possible system problems [433]. Further recent applications on these challenges are found in [434]–[440]. Therefore due to the large system structure, highly efficient fault monitoring methods have become a valuable asset for the life of the systems.

8.2.1   Introduction

In industry, failures can be classified as *sudden* or *incipient*. Sudden failures are often the simplest form to diagnose since they usually have a dramatic impact on performance, which can be detected at a number of downstream sensors. When a blast occurs, the impact can be felt everywhere in the vicinity. However, incipient or gradual failures are difficult to detect since they manifest themselves as a slow degradation in performance which can only be detected over time. The techniques which are based on identifying a model and using performance or error metrics to detect failure will find it difficult to identify a failure which occurs at a rate slower (often, much slower) than the model drifts under normal conditions. It should be noted here that there is another class of failure which has largely been ignored in the fault detection and isolation (FDI) literature is a 'pre-existing' failure. This is because a model of correct operation is impossible to identify simply by looking at a unit after it has already failed, and so most practical fault detection techniques are simply inapplicable to this type of fault. Some recent chapters on the same or similar topic have been published in [441],[442] where modified Kalman filters have been used for various applications such as green house climate control, chaos systems, types of estimation problems, etc. Most recent approaches for fault monitoring and detection handling induction motors and machines can be found in [443]–[446].

In this chapter, a UKF has been proposed in an integrated design framework to utilize MSDF techniques for process fault monitoring, thus completing a picture of a new automated fault detection and diagnosis system based on an enhanced UKF estimator. The proposed methodology utilizes an MSDF technique to enhance the accuracy and reliability of parametric estimation in the process fault detection. This technique seeks to combine data from multiple sensors and related information to achieve improved accuracies and more specific inferences.

It should be noted here that only one sensor is employed for each state of the system. Each of these sensors collects data from its respective state and fuses the information, followed by the implementation of UKF on the whole in the case of centralized MSDF, and individual state data implementation of UKF followed by fusion of the UKF-based each state data in the case of decentralized MSDF. The technique encapsulates the UKF in centralized and decentralized architectures for MSDF, promoting an improved fault monitoring. The extended Kalman filter (EKF) based centralized and decentralized fault monitoring MSDF was originally proposed for a continuous time stirred tank reactor problem in [447]. We use a similar approach where a UKF is used to detect the presence or absence of a fault. It should be noted that to have an effective fault diagnosis of highly nonlinear systems, we have assumed various faults in the system. The proposed scheme has then been successfully evaluated on a QTS and an IUB, thus corroborating the theory underpinning it.

8.2.2 Problem formulation

In this section, we maintained a hat over a variable which indicates that an estimate of the variable, e.g. $\hat{x}$ is an estimate of x, and the time index k appears as a function of time. Assume that a process is monitored by N different sensors, described by the following general nonlinear process and measurement models in a discrete time state-space framework:

$$
\begin{aligned}
x(k) &= f(x(k-1), u(k-1), d(k-1)) + w(k-1) \\
z_i(k) &= h_i(x(k)) + \nu_i(k); \quad i = 1, ...N
\end{aligned}
\tag{8.1}
$$

where $f(.)$ and $h_i(.)$ are the known nonlinear functions, representing the state transition model and the measurement model, respectively; $x(k) \in \Re^{n_x}$ is the process state-vector, $u(k) \in \Re^{n_u}$ denotes the manipulated process variables, $d(k) \in \Re^{n_d}$ represents the process faults modeled by the process disturbances, $z_i(k) \in \Re^{n_{zi}}$ are the measured variables obtained from the N installed sensors, $w(k)$ and $v_i(k)$ indicate the stochastic process and measurement disturbances modeled by zero-mean white Gaussian noises with covariance matrices $Q(k)$ and $R_i(k)$ respectively.

8.2.3   Discrete time UKF

In most practical applications of interest, the process and/or measurement dynamic models are described by nonlinear equations, represented in the system (8.1). This means that the nonlinear behavior can affect the process operation at least through its own process dynamics or measurement equation. In such cases, the standard Kalman filter algorithm is often unsuitable to estimate the process states using its linearized time-invariant state-space model at the desired process nominal operating point. UKF gives a simple and effective remedy to overcome such a nonlinear estimation problem. Its basic idea is to locally linearized the nonlinear functions, described by system (8.1), at each sampling time instant around the most recent process condition estimate. This allows the Kalman filter to be applied to the following linearized time varying model:

$$
\begin{aligned}
x(k) &= A(k)x(k-1) + B_u(k)u(k-1) + B_d(k) \\
      &\quad\; d(k-1) + w(k-1) \\
z_i(k) &= H_i(k)x(k) + \nu_i(k); \quad i = 1, ..., N
\end{aligned}
\tag{8.2}
$$

where the state transition matrix $A(k)$, the input matrices $B_u(k)$ and $B_d(k)$, and the observation matrix $H_i(k)$ are the Jacobian matrices which are evaluated at the most recent process operating condition in real-time rather than the process fixed nominal values as:

$$
\begin{aligned}
A(k) &= \left.\frac{\partial f}{\partial x}\right|_{\hat{x}(k)}, \; B_u(k) = \left.\frac{\partial f}{\partial u}\right|_{u(k)} \\
B_d(k) &= \left.\frac{\partial f}{\partial d}\right|_{\hat{d}(k)}, \; H_i(k) = \left.\frac{\partial h_i}{\partial x}\right|_{\hat{x}(k)}, \quad i = 1, ..., 10
\end{aligned}
\tag{8.3}
$$

Remark 8.2.1 *It should be observed that the UKF essentially addresses the approximation issues of the extended Kalman filter (EKF) [462]–[464]. The basic difference between the EKF and the UKF stems from the manner in which Gaussian random variables (GRV) are presented through system dynamics. In the EKF, the state distribution is approximated by GRV, which is then propagated analytically through the first-order linearization of the non-linear system. This can introduce large errors in the true posterior mean covariance of the transformed GRV, which may lead to sub-optimal performance and sometimes divergence of the filter.*

The UKF addresses this problem by using a deterministic sampling approach. The state distribution is again approximated by a GRV, but is now represented using a minimal set of carefully chosen sample points. These sample points completely capture the true mean and covariance of the GRV, and when propagated through the true *non-linear system, capture the posterior mean and covariance accurately to the second order (Taylor series expansion) for any nonlinearity. The EKF, in contrast, only achieves first-order accuracy.*

In classical control, the disturbance variable $d(k)$ is treated as a known input with distinct entry in the process state-space model. This distinction between state and disturbance as non-manipulated variables, however, is not justified from the monitoring perspective using the estimation procedure. Therefore, a new augmented state variable vector $x^*(k) = [d^T(k) \quad x^T(k)]^T$ is developed by considering the process disturbances or faults as additional state variables. To implement this view, *the process faults are assumed to be random state variables governed by the following stochastic auto-regressive (AR) model equation:*

$$d(k) = d(k-1) + w_d(k-1) \tag{8.4}$$

This assumption changes the linearized model formulations in system (8.2) to the following augmented state-space model:

$$
\begin{aligned}
x^*(k) &= A^*(k)x^*(k-1) + B_u^*(k)u(k-1) + w^*(k-1) \\
z_i(k) &= H_i^*(k)x^*(k) + \nu_i(k); \quad i = 1, ..., N
\end{aligned}
\tag{8.5}
$$

Noting that

$$A^*(k) = \begin{bmatrix} I^{n_d \times n_d} & 0^{n_d \times n_x} \\ B_d(k)^{n_x \times n_d} & A(k)^{n_x \times n_x} \end{bmatrix} \tag{8.6}$$

$$B_u^*(k) = \begin{bmatrix} 0^{n_d \times n_u} & B_u(k)^{n_x \times n_u} \end{bmatrix}^T \tag{8.7}$$

$$H_i^*(k) = \begin{bmatrix} 0^{1 \times n_d} & H_i(k)^{1 \times n_x} \end{bmatrix} \tag{8.8}$$

$$w^*(k-1) = \begin{bmatrix} w_d(k-1)^{n_d \times 1} & w(k-1)^{n_x \times 1} \end{bmatrix}^T \tag{8.9}$$

Assumption 8.2.1 *There exists a known positive constant L_0 such that for any norm bounded $x_1(k)$, $x_2(k) \in \Re^n$, the following inequality holds:*

$$\|f(u(k), z(k), x_1(k)) - f(u(k), z(k), x_2(k))\| \leq L_0 \|x_1(k) - x_2(k)\| \tag{8.10}$$

Assumption 8.2.2 $H[sI - (A - KH)]^{-1}B$ *is strictly positive real, where $K \in \Re^{n \times r}$ is chosen such that $A - KH$ is stable.*

Remark 8.2.2 *For a given positive definite matrix $Q > 0 \in \Re^{n \times n}$, there exist matrices $P = P^T > 0 \in \Re^{n \times n}$ and a scalar $\boldsymbol{R}$ such that:*

$$
\begin{aligned}
(A - KH)^T P + P(A - KH) &= -Q \\
PB &= H^T R
\end{aligned}
\tag{8.11}
$$

To detect the fault, the following is constructed:

$$
\begin{aligned}
\hat{x}(k) &= A\hat{x}(k) + f(u(k), z(k)) + B\xi_H f(u(k), z(k), \\
&\quad \hat{x}(k)) + K(z(k) - \hat{z}(k)) \tag{8.12} \\
\hat{z}(k) &= H\hat{x}(k) \tag{8.13}
\end{aligned}
$$

where $\hat{x}(k) \in \Re^n$ is the state estimate, the input is $u \in \Re^m$, and the measurement variable is $z \in \Re^r$. The pair (A, H) is observable. The nonlinear term $g(u(k), z(k))$ depends on $u(k)$ and $z(k)$ which are directly available.
The $f(u(k), z(k), x(k)) \in \Re^r$ is a nonlinear vector function of $u(k)$, $z(k)$ and $x(k)$. The $\xi(k) \in \boldsymbol{R}$ is a parameter which changes unexpectedly when a fault occurs. Since it has been assumed that the pair (A, H) is observable, a gain matrix K can be selected such that $A - KH$ is a stable matrix. We define:

$$
e_x(k) = x(t) - \hat{x}(k), \quad e_z(k) = z(k) - \hat{z}(k)
\tag{8.14}
$$

Then, the error equations can be given by:

$$
\begin{aligned}
e_x(k+1) &= (A - KH)e_x(k) + B[\xi(k)f(u(k), z(k), \\
&\quad x(k)) - \xi_H f(u(k), z(k), \hat{x}(k))], \tag{8.15} \\
e_z(k) &= He_x(k) \tag{8.16}
\end{aligned}
$$

The convergence of the above filter is guaranteed by the following theorem:

Theorem 8.2.1 *Under Assumption (8.2.2), the filter is asymptotically convergent* when no fault occurs $(\xi(k) = \xi_H)$, that is, $\lim_{k \to \infty} e_z(k) = 0$.*

Proof 8.2.1 *Consider the following Lyapunov function:*

$$
V(e(k)) = e_x^T(k) P e_x(k)
\tag{8.17}
$$

---

*The convergence and optimality property of Kalman filter-based algorithms is discussed in [465]–[466].

where P is given by the system (8.11), and Q is chosen such that $\rho_1 = \lambda_{min}(Q) - 2\|C\|.|R|\xi_H L_0 > 0$. Along the trajectory of the fault-free system (8.15), the corresponding Lyapunov difference along the trajectories $e(k)$ is:

$$
\begin{aligned}
\Delta V &= E\{V(e(k+1)|e_k, p_k)\} - V(e(k)) \\
&= E\{e^T(k+1)P_i e(k+1)\} - e^T(k)P_i e(k) \\
&= (A_e e_x + B_L u_e)^T P (A_e e_x + B_L u_e) \\
&- e_x^T(k) P e_x(k) \\
&= e^T(k)[(P(A - KH) + (A - KH)^T P) \\
&+ PB\xi_H[f(u(k), z(k), x(k)) \\
&- f(u(k), z(k), \hat{x}(k))]]e(k)
\end{aligned}
\tag{8.18}
$$

From (8.2.1) and system (8.11), one can further obtain that

$$
\begin{aligned}
\Delta V &\leq -e_x^T(k)Q e_x(k) + 2\|e_z(k)\|.|R|\xi_H L_0\|e_x(k)\| \\
&\leq -\rho_1\|e_x\|^2 < 0
\end{aligned}
\tag{8.19}
$$

Thus, $\lim_{k\to\infty} e_x(k) = 0$ and $\lim_{k\to\infty} e_z(k) = 0$. This completes the proof.

8.2.4 Unscented procedure

The structure of the UKF is elaborated by an unscented procedure for calculating the statistics of a random variable which undergoes a nonlinear transformation [464]. Observe that this method differs substantially from general Monte Carlo sampling methods which require orders of magnitude more sample points in an attempt to propagate an accurate (possibly non-Gaussian) distribution of the state. Consider propagating a random state x of dimension L through a nonlinear function, $z=f(x)$, where x is the augmented state of the system. Assume x has mean $\bar{x}$ and covariance P_x. To calculate the statistics of z, we form a matrix $\mathcal{X}$ of $2L + 1$ sigma vectors $\mathcal{X}_i$ according to:

$$
\begin{aligned}
\mathcal{X}_0 &= \bar{x}, \\
\mathcal{X}_i &= \bar{x} + (\sqrt{(L + \lambda)P_x})_i, \; i = 1, ..L \\
\mathcal{X}_i &= \bar{x} - (\sqrt{(L + \lambda)P_x})_i - L, \; i = L + 1, ..2L
\end{aligned}
\tag{8.20}
$$

where $\lambda = \alpha^2(L + \kappa) - L$ is a composite scaling parameter. The constant α determines the spread of the sigma points around $\bar{x}$, and is usually set to a small positive value ($1 \leq \alpha \leq 10^{-4}$). The constant κ is a secondary scaling parameter, which is usually set to $3 - L$, and β is used to incorporate prior knowledge of the distribution of x (for Gaussian distributions, $\beta = 2$) is optimal). $(\sqrt{(L+\lambda)P_x})_i$ is the i-th column of the matrix square root (i.e., lower-triangular Cholesky factorization). These sigma vectors are propagated through the nonlinear function $\mathcal{Z}_i = f(\mathcal{X}_i), i = 0, ...2L$. Now the mean and covariance for z are approximated[†] using a weighted sample mean and covariance of the posterior sigma points and weights W_i:

$$\bar{z} \approx \sum_{i=0}^{2L} W_i^m \mathcal{Z}_i,$$

$$P_z \approx \sum_{i=0}^{2L} W_i^c (\mathcal{Z}_i - \bar{z})(\mathcal{Z}_i - \bar{z})^T,$$

$$W_0^m = \frac{\lambda}{L+\lambda},$$

$$W_0^c = \frac{\lambda}{L+\lambda} + 1 - \alpha^2 + \beta,$$

$$W_i^m = W_i^c = \frac{1}{2(L+\lambda)}, i = 1, ...2L \qquad (8.21)$$

In view of the foregoing, the UKF is an extension of the UT to the following recursive estimation:

$$\hat{x}_k = x_{k_{prediction}} + \kappa_k \left[z_k - z_{k_{prediction}} \right] \qquad (8.22)$$

where the state random variables (RV) are re-defined as the concentration of the original state and noise variables: $x_k^a = [x_k^T \quad v_k^T \quad n_k^T]^T$. The UT sigma point selection scheme is then applied to this new augmented state RV to calculate the corresponding sigma matrix, $\mathcal{X}_k^a$. The UKF equations are given below. Note that no explicit calculations of Jacobian or Hessian are necessary to implement this algorithm. Initialize with :

$$\hat{x}_0 = \mathbf{E}[x_0],$$
$$P_0 = \mathbf{E}[(x_0 - \hat{x}_0)(x_0 - \hat{x}_0)^T],$$
$$\hat{x}_0^a = \mathbf{E}[x^a] = [\hat{x}_0^T \quad 0 \quad 0]^T. \qquad (8.23)$$

For $k \in [1, ..., \infty]$, calculate the sigma points as:

$$\mathcal{X}_{k-1}^a = [\hat{x}_{k-1}^a \quad \hat{x}_{k-1}^a + \gamma\sqrt{P_{k-1}^a} \quad \hat{x}_{k-1}^a - \gamma\sqrt{P_{k-1}^a}] \qquad (8.24)$$

[†]The deceptively simple approach taken with the UT results in approximations that are accurate to the third order for Gaussian inputs for all nonlinearities. For non-Gaussian inputs, approximations are accurate to at least the second order, with the accuracy of the third- and higher order moments being determined by the choice of α and β.

The UKF time-update equations are as follows:

$$\mathcal{X}^x_{k|k-1} = \mathsf{F}(\mathcal{X}^x_{k-1}, \quad u_{k-1}, \quad \mathcal{X}^\nu_{k-1}), \tag{8.25}$$

$$\hat{x}^-_k = \sum_{i=0}^{2L} W^m_i \mathcal{X}^x_{i,k|k-1}, \tag{8.26}$$

$$P^-_k = \sum_{i=0}^{2L} W^c_i (\mathcal{X}^x_{i,k|k-1} - \hat{x}^-_k)(\mathcal{X}^x_{i,k|k-1} - \hat{x}^-_k)^T, \tag{8.27}$$

$$\mathcal{Z}_{k|k-1} = \mathsf{H}(\mathcal{X}^x_{k|k-1}, \mathcal{X}^n_{k-1}), \tag{8.28}$$

$$\hat{z}^-_k = \sum_{i=0}^{2L} W^m_i \mathcal{Z}_{i,k|k-1} \tag{8.29}$$

where the system dynamical model F and H are assumed to be known and P^-_k and $\hat{x}^-_k$ are the prior covariance matrix and prior state estimate respectively.

The UKF measurement-update equations are:

$$
\begin{aligned}
P_{\bar{z}_k \bar{z}_k} &= \sum_{i=0}^{2L} W^c_i (\mathcal{Z}_{i,k|k-1} - \hat{z}^-_k)(\mathcal{Z}_{i,k|k-1} - \hat{z}^-_k)^T, \\
P_{x_k z_k} &= \sum_{i=0}^{2L} W^c_i (\mathcal{X}_{i,k|k-1} - \hat{x}^-_k)(\mathcal{Z}_{i,k|k-1} - \hat{y}^-_k)^T, \\
\kappa_k &= P_{x_k z_k} P^{-1}_{\bar{z}_k \bar{z}_k}, \\
\hat{x}_k &= \hat{x}^-_k + \kappa_k(z_k - \hat{z}^-_k), \\
P_k &= P^-_k - \kappa_k P_{\bar{z}_k \bar{z}_k} \kappa^T_k
\end{aligned}
\tag{8.30}
$$

where $x^a = [x^T \ v^T \ n^T]^T$, $\mathcal{X}^a = [(\mathcal{X}^x)^T \ (\mathcal{X}^v)^T \ (\mathcal{X}^n)^T]^T$ and $\gamma = \sqrt{L + \lambda}$. In addition, λ is the composite scaling parameter, L is the dimension of the augmented state, $\Re^v$ is the process-noise covariance, $\mathbf{R}^n$ is the measurement-noise covariance, and W_i are the weights.

Initializing with

$$\hat{x}_0 = \mathbf{E}[x_0], \quad P_0 = \mathbf{E}[(x_0 - \hat{x}_0)(x_0 - \hat{x}_0)^T], \quad \forall k \in [1, ...\infty]$$

calculate the sigma points as:

$$\mathcal{X}_{k-1} = [\hat{x}_{k-1} \ \hat{x}_{k-1} + \gamma \sqrt{P_{k-1}} \ \hat{x}_{k-1} - \gamma \sqrt{P_{k-1}}] \tag{8.31}$$

Note that (8.25)–(8.31) are for the augmented sigma points. Here we augment the sigma points with additional points derived from the matrix square root of the process noise covariance. Now, the time-update equations are:

$$\mathcal{X}^*_{k|k-1} = \mathsf{F}(\mathcal{X}_{k-1}, u_{k-1}), \tag{8.32}$$

$$\hat{x}^-_k = \sum_{i=0}^{2L} W^m_i \mathcal{X}^*_{i,k|k-1}, \tag{8.33}$$

$$\mathcal{X}^*_{k|k-1} = [\mathcal{X}^*_{k|k-1} \ \mathcal{X}^*_{0,k|k-1} + \gamma\sqrt{\Re^v}$$
$$\mathcal{X}^*_{0,k|k-1} - \gamma\sqrt{\Re^v}], \tag{8.34}$$

$$P^-_k = \sum_{i=0}^{2L} W^c_i (\mathcal{X}^*_{i,k|k-1} - \hat{x}^-_k)(\mathcal{X}^*_{i,k|k-1} - \hat{x}^-_k)^T + \Re^v \tag{8.35}$$

$$\mathcal{Z}_{k|k-1} = \mathsf{H}(\mathcal{X}_{k|k-1}), \tag{8.36}$$

$$\hat{z}^-_k = \sum_{i=0}^{2L} W^m_i \mathcal{X}_{i,k|k-1} \tag{8.37}$$

And the measurement-update equations are:

$$P_{\bar{z}_k \bar{z}_k} = \sum_{i=0}^{2L} W^c_i (\mathcal{Z}_{i,k|k-1} - \hat{z}^-_k)(\mathcal{Z}_{i,k|k-1} - \hat{z}^-_k)^T + \Re^n,$$

$$P_{x_k z_k} = \sum_{i=0}^{2L} W^c_i (\mathcal{X}_{i,k|k-1} - \hat{x}^-_k)(\mathcal{Z}_{i,k|k-1} - \hat{z}^-_k)^T,$$

$$\kappa_k = P_{x_k z_k} P^{-1}_{\bar{z}_k \bar{z}_k}, \ \hat{x}_k = \hat{x}^-_k + \kappa_k(z_k - \hat{z}^-_k),$$

$$P_k = P^-_k - \kappa_k P_{\bar{z}_k \bar{z}_k} \kappa^T_k \tag{8.38}$$

8.2.5   Parameter estimation

Parameter estimation involves learning a nonlinear mapping $\mathbf{z}_k = \mathsf{H}(x_k, w)$, where w corresponds to the set of unknown parameters [464]. $\mathsf{H}(.)$ may be a neural network or another parametrized function. Consider a state-space model given by $x_t = f(x_t - 1) + \epsilon$, $x_t \in R^m$ and $z_t = h(x_t) + \nu$, $z_t \in \mathbf{R}^D$ where the system noise $\epsilon \sim N(0, \Sigma_\epsilon)$ and the measurement noise $\nu \sim N(0, \Sigma_\nu)$ are both Gaussian. The EKF linearizes f and h at the current estimate of x_t and treats the system as a non-stationary linear system even though it is not. Then the pdf is propagated through a linear approximation of the system around the operating point at each time instant. In doing so, the EKF needs the Jacobian matrices which may be difficult to obtain for higher order systems, especially in the case of time-critical applications. Further, the linear approximation of the system at a given time instant may introduce errors in the state which may lead the state to diverge over time. In other words, the linear approximation may not be appropriate for some systems. Also in an EKF algorithm, during the time-update (prediction) step, the mean is propagated through the nonlinear function, in other words, this introduces an error since in general $\bar{z} \neq h(\bar{x})$. Whereas, the UKF propagates several estimates of x_t through f and h, and reconstructs a Gaussian distribution assuming the propagated values came from a linear system. During the time-update step, all the sigma points are propagated through the nonlinear function which makes the UKF a better and more effective nonlinear approximation. The UKF principle is simple and easy to implement as it does not require the calculation of Jacobian at each time step. The UKF is accurate up to second order moments in the pdf propagation whereas the EKF is accurate up to first order moment [467]. The EKF may be used to estimate the parameters by writing a new state-space representation

$$
\begin{aligned}
w_{k+1} &= w_k + r_k & \text{(8.39)} \\
d_k &= \mathsf{H}(x_k, w_k) + e_k & \text{(8.40)}
\end{aligned}
$$

where w_k corresponds to a stationary process with an identity state transition matrix, driven by noise r_k. The desired output d_k corresponds to a non-linear observation on w_k. From the optimized perspective, the following prediction error cost is minimized:

$$
J(w) = \sum_{i=1}^{k} [d_t - \mathsf{H}(x_t, w)]^T (\mathbf{R}^e)^{-1} [d_t - x\mathsf{H}(x_t, w)] \tag{8.41}
$$

Thus, if the noise covariance $\mathbf{R}^e$ is a constant diagonal matrix, then, in fact, it cancels out of the algorithm, and hence can be set arbitrarily (e.g., $\mathbf{R}^e = 0.5I$). Alternatively, $\mathbf{R}^e$ can be set to specify a weighted mean square error (MSE) cost. The innovations covariance $\mathbf{E}[r_k \quad r_k^T] = \mathbf{R}_k^r$, on the other hand, affects the convergence rate and tracking performance. Roughly speaking, the larger the covariance, the more quickly older data is discarded. There are several options on how to choose $\mathbf{R}_k^r$.

• Set $\mathbf{R}_k^r$ to an arbitrary fixed diagonal value, which may then be annealed toward zero as training continues.

• Set $R_k^r = (\lambda_{RLS}^{-1} - 1)P_{w_i}$, where $\lambda_{RLS} \in (0,1]$ is the forgetting factor. This provides for an approximate exponentially decaying weighting on past data. Note that λ_{RLS} should not be confused with λ used for sigma-point calculation.

- Set $\Re_k^r = (1 - \alpha_{RM})\mathbf{R}_r^{k-1} + \alpha_{RM} K_k^w \left[d_k - \mathsf{H}(x_k, \hat{w}) \right] \times \left[d_k - \mathsf{H}(x_k, \hat{w}) \right]^T$ $(K_k^w)^T$ which is a Robbins-Monroe stochastic approximation scheme for estimating the innovations. The method assumes that the covariance of the Kalman update model is consistent with the actual update model. Typically, $\mathbf{R}_k^r$ is also constrained to be a diagonal matrix, which implies an independence assumption on the parameters. Note that a similar update may also be used for $\mathbf{R}_k^e$.

8.2.6   Improved MSDF techniques

The MSDF method has received major attention for various industrial applications. MSDF can be done at a variety of levels from the raw data or observation level to the feature/state vector level and the decision level. This idea can lead to utilization of different possible configurations or architectures to integrate the data from disparate sensors in an industrial plant to extract the desired monitoring information. Using Kalman filtering as the data fusion algorithm, multiple sensors can be integrated in two key architecture scenarios called the centralized method and the decentralized or distributed method. These methods have been widely studied over the last decade [468] and [469].

1. *Centralized integration method:* In the centralized integration method, all the raw data from different sensors is sent to a single location to be fused. This architecture is sometimes called the measurement fusion integration method [468] and [469], in which observations or sensor measurements are directly fused to obtain a global or combined measurement data matrix H^T. Then, it uses a single Kalman filter to estimate the global state vector based upon the fused measurement. Although this conventional method provides high fusion accuracy to the estimation problem, the large number of states may require high processing data rates that cannot be maintained in practical real time applications. Another disadvantage of this method is the lack of robustness in case of failure in a sensor or the central filter itself. For these reasons, parallel structures can often provide improved failure detection and correction, enhance redundancy management, and decrease costs for multi-sensor system integration. This method integrates the sensor measurement information as follows:

$$
\begin{aligned}
z(k) &= [z_1(k) \dots z_N(k)]^T, \quad H(k) = [H_1(k) \dots H_N(k)]^T \\
R(k) &= diag[R_1(k) \dots R_N(k)]
\end{aligned}
$$

   where $R_j(k)$ is the covariance matrix.

2. *Decentralized integration method:* As such, there has recently been considerable interest shown in the distributed integration method in which the filtering process is divided between some local Kalman filters working in parallel to obtain individual sensor-based state estimates and one master filter combining these local estimates to yield an improved global state estimate. This architecture is sometimes called as the state-vector fusion integration method [468] and [469]. The advantages of this method are higher robustness due to parallel implementation of fusion nodes and lower computation load and communication cost at each fusion node. It is also applicable in modular systems where different process sensors can be provided as separate units. On the other hand, distributed fusion is conceptually a lot more complex, and is likely to require higher bandwidth compared with centralized fusion. This method integrates the sensor measurement information as follows:

$$z(k) = [\sum_{j=1}^{N} R_j^{-1}(k)]^{-1} \sum_{j=1}^{N} R_j^{-1}(k) z_j(k),$$

$$H(k) = [\sum_{j=1}^{N} R_j^{-1}(k)]^{-1} \sum_{j=1}^{N} R_j^{-1}(k) H_j(k),$$

$$R(k) = [\sum_{j=1}^{N} R_j^{-1}(k)]^{-1}$$

The architectures are being implemented using the UKF as shown in Figure 8.2 and Figure 8.3 respectively.

3. *Fusion methods*

It is well-known that MSDF plays a crucial role in providing improved probability of detection, extended spatial and temporal coverage, reduced ambiguity, and improved system reliability and robustness. Among the various techniques available for MSDF, the Kalman filtering-based approach is used for the present case, as it proves to be an efficient recursive algorithm suitable for real-time application using digital computers. Among different approaches for Kalman Filter-based sensor fusion, two commonly employed techniques are:

   (a) *State-vector fusion.* It uses covariance of the filtered output of individual noisy sensor data to obtain an improved joint state estimate.

   (b) *Measurement fusion.* It directly fuses the sensor measurements to obtain a weighted or combined measurement, and then uses a single Kalman filter to obtain the final state estimate based on the fused measurement

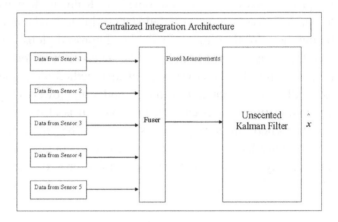

Figure 8.2: Centralized integration method using UKF.

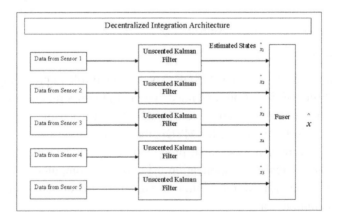

Figure 8.3: Decentralized integration method using UKF.

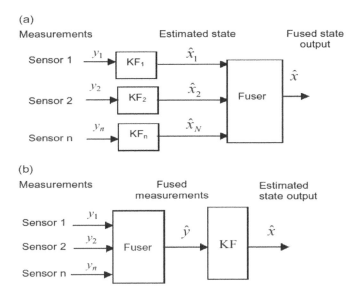

Figure 8.4: (a) State-vector fusion, (b) Measurement fusion.

The two philosophies are enumerated in Figure 8.4. Indeed, both systems have their own merits and demerits. The measurement fusion method, which combines multi-sensor data using a minimum mean square error estimate, requires that the sensors should have identical measurement matrices. Although the measurement fusion method provides better overall estimation performance, state-vector fusion has lower computational cost and possesses the advantage of parallel implementation and fault tolerance. Judicious trade-off between computational complexity, computational time, and numerical accuracy has to be made for the selection of an algorithm for practical application.

4. Typical cases Multi-sensor fusion and integration is a rapidly evolving research area and requires interdisciplinary knowledge in control theory, signal processing, and artificial intelligence. Intelligent systems equipped with multiple sensors can interact with and operate in an unstructured environment without the complete control of a human operator. Due to the fact that the system is operating in a totally unknown environment, a system may lack of sufficient knowledge concerning the state of the outside world. In this regard, sensors can allow a system to learn the state of the world as needed and to continuously update its own model of the world. A typical case of

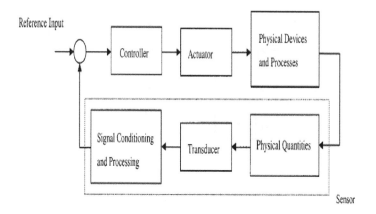

Figure 8.5: Role of sensors in mechatronic systems.

mechatronic systems is depicted in Figure 8.5, where a complete package of a sensor is indicated in the dashed line.

In transportation systems, such as automatic train control systems, intelligent vehicle and highway systems, GPS-based vehicle navigation, and aircraft landing tracking systems, multi-sensor fusion techniques are utilized to increase reliability, safety, and efficiency. In [474], sensor fusion for train speed and position measurement using different combination of global positioning by satellite (GPS), inertia navigation systems (INS), tachometers, Doppler radar, was designed and implemented, see Figure 8.7.

8.3   Notes

In this chapter, a complete fault estimation and monitoring scheme has been developed by integrating the techniques of MSDF and centralized and decentralized architectures for providing quality reports and ensuring reliable fault detection, and it has been made effective by using a deterministic sampling approach known as the UKF, thus capturing the posterior mean and covariance accurately, completing the overall diagnostic picture. It has been demonstrated that this approach can be used for the reliable detection of incipient faults, which in turn leads to an efficient and cost-effective preventive maintenance scheme. Some aspects of multi-sensor data fusion in target tracking are outlined.

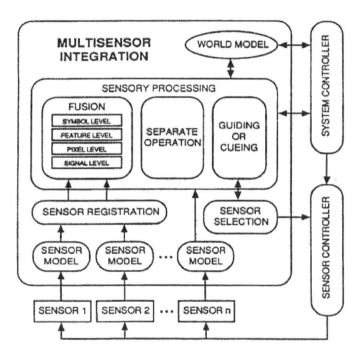

Figure 8.6: Functional diagram of multi-sensor fusion and integration.

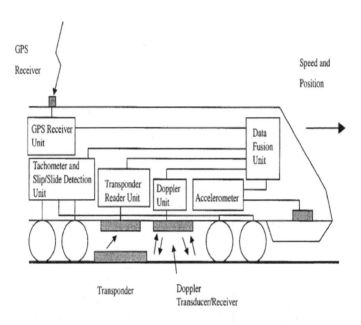

Figure 8.7: Integration of different sensors in train position and speed measurement [474].

8.4   Proposed Topics

1. Expanding on the foregoing work, consider a discrete time-varying linear stochastic control system with ℓ sensors. Under the standard assumptions that are often utilized, it is desired to design a multi-sensor optimal information fusion decentralized Kalman filter with a two-layer fusion structure for the case of multiple sensors and correlated noises. Demonstrate the merits and demerits of the proposed design.

2. Consider a typical multi-sensor environment where several sensors, in general, dissimilar observe the same dynamic system, where each sensor is attached to a local processor. An important issue in such an integrated network system is that of time-alignment. Let us assume that the integrated network consists of several sites whereby each site has more than one sensor closely located. Under the assumption of independence of the measurement noise processes at each sensor, a natural candidate is a two-level Kalman filter structure with parallel processing capabilities. Provide a complete analysis of this structure and evaluate its computational load.

3. Decentralized data fusion (DDF) methods were initially motivated by the insight that the information or canonical form of the conventional Kalman filter data fusion algorithm could be implemented by simply adding information contributions from observations as shown below for a system with n sensors,

$$\hat{y}(k|k) = \hat{y}(k|k-1) + \sum_{i=1}^{N} H_i(k) R_i^{-1}(k) z_i(k),$$

$$Y(k|k) = Y(k|k-1) + \sum_{i=1}^{N} H_i(k) R_i^{-1}(k) H_i^t(k).$$

As these (vector and matrix) additions are commutative, the update or data fusion process can be optimally distributed among a network of sensors. A structure of a DDF sensor node is shown in Figure 8.8. The sensor is modeled directly in the form of a likelihood function.

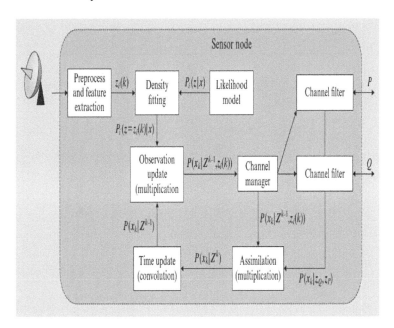

Figure 8.8: Structure of a decentralized data fusion node.

Investigate the properties of practical implementation of such a structure.

Chapter 9

Approximate Distributed Estimation

In this chapter, we propose an approximate distributed estimation within a distributed networked control formalism. This is made possible by using a Bayesian-based forward-backward (FB) system with generalized versions of a Kalman filter. The analytical treatment is presented for cases with complete, incomplete, or no prior information with bounds and then followed by estimation fusion for all three cases. The proposed scheme is validated on a rotational drive-based electro-hydraulic system and the ensuing results ensured the effectiveness of the scheme underpinning it.

9.1 Introduction

Distributed and decentralized estimations have been the point of attraction in the past with a large associated literature. The classic work of Rao and Durrant-Whyte [545] presents an approach to decentralized Kalman filtering which accomplishes globally optimal performance in the case where all sensors can communicate with all other sensors. Further, this design fails gracefully as individual sensors are removed from the network, due to its distributed design. When tackling the distributed structure, problems encountered regarding fusion of the data coming from various sensors of the plant or network. Data fusion techniques combine data from multiple sensors and related information to achieve more specific inferences than could be achieved by using a single, independent sensor. Sensor noises of converted systems are cross-correlated, while original system independence is shown in [607]–[608]. Sensor noises of a converted system are cross-correlated, while an original system also correlated is presented in [606]. A centralized fusion center, expressed by a linear combination of the local estimates is pictured in [232]. No centralized fusion center, but an algorithm highly resilient to losing one or more sensing nodes is shown in [233]. Discrete smoothing fusion with ARMA Signals is shown in [603]. Linear minimum variance (LMV) with an information fusion filter is shown in [604]–[605]. Much attention has been devoted to multi-sensor data fusion for both military and civilian applications. For civilian applications, the monitoring of manufacturing processes, robotics, medical applications/environmental monitoring. In military applications, target recognition, guidance for autonomous vehicles, and battle field surveillance are important.

The estimation problem has also been dealt with by using consensus algorithms. Consensus problems [609] and their special cases have been the subject of intensive studies by several researchers [610], [611], [612], [613], [614], [615], [616] in the context of formation control, self-alignment, and flocking [617] in networked dynamic systems.

In distributed estimation and fusion, Kalman filtering is a fundamental tool, and it is an essential element used to provide functionality particularly in sensor networks. An in-depth comparison between the distributed Kalman filter and the existing decentralized sensor fusion algorithms, both with and without fusion centers, are presented in [618], [619], [620], [621].

In this chapter, we have derived an approximate distributed estimation for different prior cases for dynamic systems, with the help of the Bayesian-based FB Kalman filter; the estimation is derived on a distributed networked control system [602]. Then, to reduce the time complexity, upper bound and lower bound methods for time complexity reduction have been derived on all three cases of prior knowledge. After achieving a distributed structure, we have used a data fusion technique to consider it for a distributed structure. The proposed scheme is then validated on a rotational drive-based electro-hydraulic system, where different types of faults were introduced, and then different fault profile data is considered for the evaluation of the proposed scheme.

9.2   Problem Formulation

Consider a distributed control system as in [602] consisting of N agents, in which there is no communication loss. The discrete-time linear dynamic model of the agent j can be described as:

$$x_j(k+1) \quad = \quad \sum_{i=1}^{N} A_{ij}x_i(k) + G_j w_j(k) \tag{9.1}$$

where $k \in \mathbf{Z}^+$, $x_j(k) \in \mathbf{R}_{n_x}$ is the state of the agent j at time k, $w_j(k) \in R_{n_w}$ is a white noise process, $A_{ij} \in \mathbf{R}_{n_x \times n_x}$, and $G_j \in R_{n_x \times n_w}$. Hence, the state of the agent j is governed by the previous states of all N agents. We can also consider $A_{ij} x_i(k)$ as a control input from the agent i to the agent j for $i \neq j$.

Now consider a distributed networked control system (DNCS), in which agents communicate with each other over a lossy communication channel, e.g., a wireless channel. We assume an erasure channel between a pair of agents. At each time k, a packet sent by the agent i is correctly received by the agent j with probability p_{ij}. We form a communication matrix $P_{com} = [p_{ij}]$. Let $Z_{ij}(k) \in \{0, 1\}$ be a Bernoulli random variable, such that $Z_{ij}(k) = 1$ if a packet sent by the agent i is correctly received by the agent j at time k, otherwise, $Z_{ij}(k) = 0$. Since there is no communication loss within an agent, $p_{ii} = 1$ and $Z_{ii}(k) = 1$ for all i and k. For each (i, j) pair, $\{Z_{ij}(k)\}$ are i.i.d. (independent identically distributed) random variables such that $P(Z_{ij}(k) = 1) = p_{ij}$ for all k; and $Z_{ij}(k)$ are independent from $Z_{lm}(k)$ for $l \neq i$ or $m \neq j$. Then we can write the dynamic model of the agent j under lossy links as:

$$x_j(k+1) = \sum_{i=1}^{N} Z_{ij}(k) A_{ij} x_i(k) + G_j w_j(k) \qquad (9.2)$$

Let $x(k) = [x_1(k)^T, ..., x_N(k)^T]^T$ and $w(k) = [w_1(k)^T, ..., w_N(k)^T]^T$, where y^T is a transpose of y. Let $\bar{A}_{ij}$ be a $N_{n_x} \times N_{n_x}$ block matrix. The entries of $\bar{A}_{ij}$ are all zeros except the $(j, i) - th$ block is A_{ij}. For example, when $N = 2$.

$$\bar{A}_{12} = \begin{bmatrix} 0_{nx} & 0_{nx} \\ A_{12} & 0_{nx} \end{bmatrix}$$

where 0_{nx} is a $n_x \times n_x$ zero matrix. Then the discrete-time linear dynamic model of the DNCS with lossy links can be represented as the following:

$$x(k+1) = (\sum_{i=1}^{N} \sum_{j=1}^{N} Z_{ij}(k) \bar{A}_{ij}) x(k) + G w(k) \qquad (9.3)$$

where G is a block diagonal matrix of $G_1, ..., G_N$. For notational convenience, we introduce a new index $n \in 1, ..., N_2$ such that ij is indexed by $n = N(i-1) + j$. With this new index n, the dynamic model (9.3) can be rewritten as:

$$x(k+1) = (\sum_{n=1}^{N^2} Z_n(k) \bar{A}_n) x(k) + G w(k) \qquad (9.4)$$

By letting $A(k) = (\sum_{n=1}^{N^2} Z_n(k)\bar{A}_n)$ we see that (9.4) is a time-varying linear dynamic model:

$$x(k+1) \quad = \quad A(k)x(k) + Gw(k) \tag{9.5}$$

Until now we have assumed that $\bar{A}_n$ is fixed for each n. Now suppose a more general case where the matrix A is time-varying and its values are determined by the communication link configuration $Z(k) = [Z_1(k), \ldots, Z_{N^2}(k)]^T$. Hence, A is a function of $Z(k)$ and this general case can be described as:

$$x(k+1) \quad = \quad A(Z(k))x(k) + Gw(k) \tag{9.6}$$

The dynamic model (9.6) or (9.4) is a special case of the linear hybrid model or a jump linear system [623] since $A(k)$ takes an element from a set of a finite number of matrices. We will call the dynamic model (9.4) as the "simple" DNCS dynamic model and (9.6) as the "general" DNCS dynamic model.

In the following sections, we will derive Kalman filter fusion with cases of prior information, and their modifications which can bound the covariance matrices [602]. The Bayesian-based FB Kalman filter is expressed below, where the simple Bayesian-based optimal Kalman filter is expressed in [601].

Forward Run: For $(k = 0;\ k < T;\ +k)$

$$R_{e,i} = R_i + H_k P_{k+1/k} H_k^* \tag{9.7}$$

$$\widehat{K}_{f,i} = F_{k+1/k} \widehat{P}_{k+1/k} H_k^T (H_k \widehat{P}_{k/k-1} H_k^T + R_{e,i}^{-1}) \tag{9.8}$$

$$\widehat{x}_{k/k}^{MAP} = \widehat{x}_{k+1/k} + \widehat{K}_{f,i}(y_k - H_k \widehat{x}_{k+1/k}) \tag{9.9}$$

$$\widehat{x}_{k+1/k} = F_k \widehat{x}_{k+1/k} \tag{9.10}$$

$$\widehat{P}_{k+1/k} = F_{k+1/k} P_{k+1/k} F_{k+1/k}^T + G_i Q G_i^*$$

$$-\widehat{K}_{p,i} R_{e,i} \widehat{K}_{p,i}^* \tag{9.11}$$

$$\widehat{P}_{k/k} = \widehat{P}_{k+1/k} - F_{k/k+1} \widehat{K}_k H_k \widehat{P}_{k+1/k} \tag{9.12}$$

Backward Run: For $(k = T - 1; t \geq 0; -k)$

$$\widehat{J}_{k-1/T} = \widehat{P}_{k-1/T} F_k^T \widehat{P}_{k-1/T}^{-1} \tag{9.13}$$

$$\widehat{x}_{k-1/T} = \widehat{x}_{k-1/k-1}^i + \widehat{J}_{k-1}(\widehat{x}_{k-1/T} - \widehat{x}_{k-1/k}) \tag{9.14}$$

$$\widehat{P}_{k-1/T} = \widehat{P}_{k-1/k-1}$$

$$+\widehat{J}_{k-1}(\widehat{J}_{k-1/T} - \widehat{P}_{k-1}/k)J_{k-1}' \tag{9.15}$$

9.3 Fusion with Complete Prior Information

In this section, a generalized version of Kalman filter is presented with complete prior information. Consider the generalized DNCS dynamic model (9.6) where $w(k)$ is a Gaussian noise with zero mean and covariance Q, and measurement model (9.16) where $y(k) \in R^{n_y}$ is a measurement at time t, $C \in \mathbf{R}^{n_y \times N_{n_x}}$ and $\nu(k)$ is a Gaussian noise with zero mean and covariance k.

$$y(k) = Cx(k) + \nu(k) \tag{9.16}$$

The following theorem presents the Bayesian-based FB Kalman filter with complete prior information:

Theorem 9.3.1

$Forward\ Run$: For $(k = 0;\ k < T;\ +k)$

$$\widehat{x}_{k/k} = F_k \bar{x}_k + K_{p,k}[y_i - H_k \bar{x}_{k+1/k} - \bar{\nu}] \tag{9.17}$$

$$\widehat{x}_{k+1/k} = F_k \widehat{x}_{k+1/k} + K_{p,k}\nu_k \tag{9.18}$$

$$\widehat{R}_{e,k} = R_k + H_k P_{k+1/k} H_k^* + HC_{xv} + (HC_{xv})' \tag{9.19}$$

$$K_k = (F_k P_{k+1/k} H^* + G_k S_k)(H_k P_{k/k} H_k^* + R_{e,k})^{-1} \tag{9.20}$$

$$\widehat{P}_{k+1/k} = F_k P_{k+1/k} F_k^* + GQ_i G^*$$
$$-F_{k+1/k}K_{p,k}R_{e,k}K_{p,k}^* \tag{9.21}$$

$$\widehat{P}_{k/k} = F_k P_{k+1/k} F_k^* - K_k H_k P_{k+1/k} \tag{9.22}$$

$Backward\ Run$: For $(k = 0;\ k < T;\ +k)$

$$\widehat{J}_{k-1/T} = \widehat{P}_{k-1/T} F_k^T \widehat{P}_{k-1/T}^{-1} \tag{9.23}$$

$$\widehat{x}_{k-1/T} = \widehat{x}_{k-1/k-1}^i + \widehat{J}_{k-1}(\widehat{x}_{k-1/T} - \widehat{x}_{k-1/k}) \tag{9.24}$$

$$\widehat{P}_{k-1/T} = \widehat{P}_{k-1/k-1}$$
$$+\widehat{J}_{k-1}(\widehat{J}_{k-1/T} - \widehat{P}_{k-1}/k)J'_{k-1} \tag{9.25}$$

The error covariance and the gain matrices have the following alternative forms:

$$P = FP_{k+1/k+1}F' + KR_{e,k}K' - FPK' - (FBK')' \tag{9.26}$$

$$K = (F_k P_{k+1/k} H^* + P_{k/k})(KR_{e,k}K + HP_{k/k})^{-1} \tag{9.27}$$

Proof 9.3.1 *For linear estimation of x using data y with linear model $y = Hx + \nu$, the prior information consists of $\bar{x}$ and $\bar{\nu}$, and $C_x = cov(x)$, $C_v = cov(v)$, and C_{xv}*

$= cov(x, v)$. *When we talk about prior information, we mean prior information about x, that is $\bar{x}$, C_x, and $C_{x,v}$.*

For a dynamic case, as in Kalman filter,

$$
\begin{aligned}
\hat{x}_{k/k} &= \mathbf{E}^*[x_k|y^k] = [\bar{x}_k|y^k] \\
&= \bar{x}_k + C_{x_k}y^k C^+ y^k (y^k - \bar{y}^k), \ \bar{x}_k = \mathbf{E}[x_k] \\
P_{k/k} &= \text{MSE}(\hat{x}_{k/k}) =^\triangle \mathbf{E}[(x_k - \hat{x}_{k/k})(x_k - \hat{x}_{k/k})'] \\
&= C x_k - C x_k y^k C^+_{y^k} C'_{x_k} y^k
\end{aligned}
$$

With few exceptions, however, it is unrealistic since its computational burden increases rapidly with time (a method for decreasing time computation complexity is applied in the next section using modified Kalman filter functions of upper bound and lower bound).

$$
\begin{aligned}
\hat{x}_{k/k} &= E^*[x_k|y^k] = \mathbf{E}^*[x_k|y_k, y^{k-1}] = \hat{x}_{k/k-1} + K_k \bar{y}_{k/k-1} \\
P_{k/k} &= \text{MSE}(\hat{x}_{k/k}) = \text{MSE}(\hat{x}_{k/k-1}) - K_k C \bar{y}_{k/k-1} K'_k
\end{aligned}
$$

Let $A = P_{k/k}$ and $F_k = \zeta$. Equation (9.27) follows from the following:

$$
\begin{aligned}
&\quad (\zeta PH' + A)(C + HA)^{-1} \\
&= \{\zeta[C_x - (C_xH' + A)(HC_xH' + C + HA + (HA)')^{-1} \\
&\quad \cdot (C_xH' + A)']H' + A\}(C + HA)^{-1} \\
&= (\zeta C_x + H' + A)[I - (HC_xH' + C + HA + (HA)')^{-1} \\
&\quad \cdot (HC_xH' + (HA)')](C + HA)^{-1} \\
&= (\zeta C_xH' + A)(HC_xH' + C + HA + (HA)')^{-1} \\
&\quad \cdot (C + HA)(C + HA)^{-1} \\
&= (\zeta C_xH' + A)(C_y + HA)^{-1}
\end{aligned}
$$

9.3.1 Modified Kalman filter-I

Based on general DNCS dynamic model (9.6), where $Z(k)$ is independent from $Z(t)$ for $t \neq k$, we derive an optimal linear filter. The following terms are defined to describe the modified Bayesian-Based FB Kalman filter.

$$
\begin{aligned}
\widehat{x}_{k/k} &= \mathbf{E}[x(k)|y_k] \\
P(k|k) &= \mathbf{E}[e(k)e(k)^T|y_k] \\
\widehat{x}(k+1|k) &= \mathbf{E}[x(k+1)|y_k] \\
P(k+1|k) &= \mathbf{E}[e(k+1|k)e(k+1|k)^T|y_k] \\
J(k-1|T) &= \mathbf{E}[J(k-1|T)|P_{k/k}] \\
\widehat{x}(k-1|T) &= \mathbf{E}[e(k-1|T)|y_k] \\
P(k-1|T) &= \mathbf{E}[e(k-1|T)e(k-1|T)^T|y_k]
\end{aligned}
\tag{9.28}
$$

where $y_k = \{y(t) : 0 \le t \le k\}$, $e(k|k) = x(k) - \widehat{x}(k|k)$, and $e(k+1|k) = x(k+1) - \widehat{x}(k+1|k)$.

Suppose that we have estimates $\widehat{x}(k|k)$ and $P(k|k)$ from time k. At time $k+1$, a new measurement $y(k+1)$ is received and our goal is to estimate $\widehat{x}(k+1|k+1)$ and $P(k+1|k+1)$ from $\widehat{x}(k|k)$, $P(k|k)$ and $y(k+1)$. First, we compute $\widehat{x}(k+1|k)$ and $P(k+1|k)$.

$$
\begin{aligned}
\widehat{x}(k+1|k) &= \mathbf{E}[x(k+1)|y_k] \\
&= \mathbf{E}[A(Z)x(k) + G\omega(k)|y_k] \\
&= \widehat{A}\widehat{x}(k|k)
\end{aligned}
\tag{9.29}
$$

where

$$
\widehat{A} = \sum_{z \in Z} p_z A(z)
\tag{9.30}
$$

is the expected value of $A(Z)$. Here $p_z = P(Z = z)$, and $\mathcal{Z}$ is a set of all possible communication link configurations.

The prediction covariance can be computed as:

$$
\begin{aligned}
P(k+1|k) &= \mathbf{E}[e(k+1|k)e(k+1|k)^T|y_k] \\
&= GQG^T + \sum_{z \in Z} p_z A(z)P(k|k)A(z)^T \\
&\quad - K_{p,k}R_{e,k}K_{p,k}^* + \sum_{z \in Z} p_z A(z)\widehat{x}(k|k)\widehat{x}(k|k)^T \\
&\quad \times (A(z) - \widehat{A})^T
\end{aligned}
\tag{9.31}
$$

Given $\widehat{x}(k+1|k)$ and $P(k+1|k)$, $\widehat{x}(k+1|k+1)$ and $P(k+1|k+1)$ are computed as in the standard Kalman filter (See Eqn. (9.32) and (9.33)).

$$\hat{x}(k+1|k+1) = F_k\hat{x}(k+1|k) + K(k+1)(y(k+1)$$
$$-H\hat{x}(k+1|k)) - \nu_i \tag{9.32}$$
$$P(k+1|k+1) = F_kP(k+1|k)F_k^*$$
$$-F_{k/k-1}K_k(k+1)HP(k+1|k) \tag{9.33}$$

where $K(k+1) = (FPk+1|kH^T + GS)(HPk|kH^T + R)^{-1}$.

The modified KF proposed earlier for the general DNCS is an optimal linear filter but the time complexity of the algorithm can be exponential in N since the size of $\mathcal{Z}$ is $O(2^{N(N-1)})$ in the worst case, i.e., when all agents can communicate with each other. In this section, we describe two approximate Kalman filtering methods for the general DNCS dynamic model (6) which are more computationally efficient than the modified KF by avoiding the enumeration over $\mathcal{Z}$. Since the computation of $P(k+1|k)$ is the only time-consuming process, we propose two filtering methods which can bound $P(k+1|k)$. We use the notation $A > 0$ if A is a positive definite matrix and $A \geq 0$ if A is a positive semi-definite matrix.

9.3.2 Lower-bound KF-I

The lower-bound KF (lb-KF) is the same as the modified KF described in Section 0.3.1, except we approximate $P(k+1|k)$ by $\underline{P}(k+1|k)$ and $P(k|k)$ by $\underline{P}(k|k)$. The covariances are updated as:

$$\underline{P}(k+1|k) = \hat{A}\underline{P}(k|k)\hat{A}^T + GQG^T$$
$$-\underline{K}_{p,k}\underline{R}_{e,k}\underline{K}_{p,k} \tag{9.34}$$
$$\underline{P}(k+1|k+1) = F_k\underline{P}(k+1|k)$$
$$-F_{k/k-1}\underline{K}(k+1)H_k\underline{P}(k+1|k) \tag{9.35}$$

where $\hat{A}$ is the expected value of $A(Z)$ and $\underline{K}(k+1) = F_{k+1/k}\underline{P}(k+1|k)H^T$ $(H_k\underline{P}(k+1|k)H_k^* + R)^{-1}$. Notice that $\hat{A}$ can be computed in advance and the lb-KF avoids the enumeration over $\mathcal{Z}$.

Lemma 9.3.1 *If $\underline{P}(k|k) \leq P(k|k)$, then $\underline{P}(k+1|k) \leq P(k+1|k)$.*

Proof 9.3.2 *Using (9.31), we have*

$$P(k+1|k) - \underline{P}(k+1|k) = \mathbf{E}[A(Z)P(k|k)A(Z)^T]$$
$$+ \mathbf{E}[A(Z)\hat{x}(k|k)\hat{x}(k|k)^TA(Z)^T]$$
$$- \hat{A}\hat{x}(k|k)\hat{x}(k|k)^T\hat{A}^T - \hat{A}\underline{P}(k|k)\hat{A}^T$$
$$- \underline{K}_{p,k}\underline{R}_{e,k}\underline{K}_{p,k} + K_{p,k}R_{e,k}K_{p,k}$$
$$= P_1 + P_2 \tag{9.36}$$

where $P_1 = \mathbf{E}[A(Z)P(k|k)A(Z)^T] - \widehat{A}\underline{P}(k|k)\widehat{A}^T - \underline{K}_{p,k}\underline{R}_{e,k}\underline{K}_{p,k}$ and $P_2 = \mathbf{E}[A(Z)\widehat{x}(k|k)\widehat{x}(k|k)^T A(Z)^T] - \widehat{A}\widehat{x}(k|k)\widehat{x}(k|k)^T\widehat{A}^T + K_{p,k}R_{e,k}K_{p,k}$.

If $P_1 \geq 0$ and $P_2 \geq 0$, then $P(k+1|k) - \underline{P}(k+1|k) \geq 0$

$$
\begin{aligned}
P_1 &= \mathbf{E}[A(Z)P(k|k)A(Z)^T] - \widehat{A}\underline{P}(k|k)\widehat{A}^T - \underline{K}_{p,k}\underline{R}_{e,k}\underline{K}_{p,k}^* \\
&\quad - \widehat{A}P(k|k)\widehat{A}^T + \widehat{A}P(k|k)\widehat{A}^T \\
&= \mathbf{E}[A(Z)P(k|k)A(Z)^T] - \widehat{A}P(k|k)\widehat{A}^T \\
&\quad + \widehat{A}(P(k|k) - \underline{P}(k|k))\widehat{A}^T - \underline{K}_{p,k}\underline{R}_{e,k}\underline{K}_{p,k}^* \quad\quad (9.37)
\end{aligned}
$$

Since $P(k|k)$ is a symmetric matrix, $P(k|k)$ can be decomposed into $P(k|k) = U_1 D_1 U_1^T$, where U_1 is a unitary matrix and D_1 is a diagonal matrix. Hence,

$$
\begin{aligned}
P_1 &= \mathbf{E}[(A(Z)U_1 D_1^{1/2})(A(Z)U_1 D_1^{1/2})^T] \\
&\quad - \mathbf{E}[(A(Z)U_1 D_1^{1/2})]\mathbf{E}[(A(Z)U_1 D_1^{1/2})]^T \\
&\quad + \widehat{A}(P(k|k) - \underline{P}(k|k))\widehat{A}^T - \underline{K}_{p,k}\underline{R}_{e,k}\underline{K}_{p,k}^* \\
&= Cov[(A(Z)U_1 D_1^{1/2}] + \widehat{A}(P(k|k) - \underline{P}(k|k))\widehat{A}^T \\
&\quad - \underline{K}_{p,k}\underline{R}_{e,k}\underline{K}_{p,k} \quad\quad (9.38)
\end{aligned}
$$

where $Cov[H]$ denotes the covariance matrix of H. Since a covariance matrix is positive definite and $P(k|k) - \underline{P}(k|k) \geq 0$ by assumption, $P_1 \geq 0$. P_2 is a covariance matrix since $\widehat{x}(k|k)\widehat{x}(k|k)^T$ is symmetric, hence $P_2 \geq 0$.

Lemma 9.3.2 If $\underline{P}(k+1|k) \leq P(k+1|k)$, then $\underline{P}(k+1|k+1) \leq P(k+1|k+1)$.

Proof 9.3.3 Here, we will use a matrix inversion lemma which says that $(A + UCV)^{-1} = A^{-1} - A^{-1}U(C^{-1} + VA^{-1}U)^{-1}VA^{-1}$ where A, U, C and V all denote matrices of the correct size. Applying the matrix inversion lemma to (9.33), we have $P(k+1|k+1) = (P(k+1|k)^{-1} + C^T R^{-1} C)^{-1}$. Let $P = P(k+1|k)$ and $\underline{P} = \underline{P}(k+1|k)$. Then

$$
\begin{aligned}
P &\geq \underline{P} \\
P^{-1} &\leq \underline{P}^{-1} \\
P^{-1} + C^T R^{-1} C &\leq \underline{P}^{-1} + C^T R^{-1} C \\
(P^{-1} + C^T R^{-1} C)^{-1} &\geq (\underline{P}^{-1} + C^T R^{-1} C)^{-1} \\
P(k+1|k+1) &\geq \underline{P}(k+1|k+1) \quad\quad (9.39)
\end{aligned}
$$

Finally, using Lemma 9.3.1, Lemma 9.3.2, and the induction hypothesis, we have the following theorem showing that the lb-KF maintains the state error covariance which is upper-bounded by the state error covariance of the modified KF.

Theorem 9.3.2 *If the lb-KF starts with an initial covariance $\underline{P}(0|0)$, such that $\underline{P}(0|0) \leq P(0|0)$, then $\underline{P}(k|k) \leq P(k|k)$ for all $k \geq 0$.*

9.3.3 Upper-bound KF-I

Similar to the lb-KF, the upper-bound KF (ub-KF) approximates $P(k+1|k)$ by $\overline{P}(k+1|k)$ and $P(k|k)$ by $\overline{P}(k|k)$. Let

$$\lambda_{max} = \lambda_{max}(\overline{P}(k|k)) + \lambda_{max}(\widehat{x}(k|k)\widehat{x}(k|k)^T),$$

where $\lambda_{max}(S)$ denotes the maximum eigenvalue of S. The covariances are updated as following:

$$
\begin{aligned}
\overline{P}(k+1|k) &= \lambda_{max}\mathbf{E}[A(Z)A(Z)^T] - \overline{K}_p\overline{R}_{e,k}\overline{K}_p^* \\
&\quad - \widehat{A}\overline{x}(k|k)\overline{x}(k|k)^T\widehat{A}^T + GQG^T & (9.40) \\
\overline{P}(k+1|k+1) &= F\overline{P}(k+1|k) \\
&\quad - F\overline{K}(k+1)H\overline{P}(k+1|k) & (9.41)
\end{aligned}
$$

where $\widehat{A}$ is the expected value of $A(Z)$ and $\overline{K}(k+1) = (F\overline{P}(k+1|k)H^T + GS)(H\overline{P}(k+1|k)H^T + R)^{-1}$. In the ub-KF, $\mathbf{E}[A(Z)A(Z)^T]$ can be computed in advance but we need to compute λ_{max} at each step of the algorithm. But if the size of Z is large, it is more efficient than the modified KF. (Notice that the computation of λ_{max} requires a polynomial number of operations in N while the size of $\mathcal{Z}$ can be exponential in N.)

Lemma 9.3.3 *If $\overline{P}(k|k) \geq P(k|k)$, then $\overline{P}(k+1|k) \geq P(k+1|k)$.*

Proof 9.3.4 *Let $M = \widehat{x}(k|k)\widehat{x}(k|k)^T$ and I be an identity matrix. Then using (9.31), we have*

$$
\begin{aligned}
\overline{P}(k|k) - P(k|k) &= \lambda_{max}\mathbf{E}[A(Z)A(Z)^T] \\
&\quad - \mathbf{E}[A(Z)P(k|k)A(Z)^T] - \mathbf{E}[A(Z)MA(Z)^T] \\
&\quad - K_pR_{e,k}K_p^* + \overline{K}_p\overline{R}_{e,k}\overline{K}_p^* \\
&= \mathbf{E}[A(Z)(\lambda_{max}(\overline{P}(k|k))I - P(k|k))A(Z)^T] \\
&\quad + \mathbf{E}[A(Z)(\lambda_{max}(M)I - M)A(Z)^T] \\
&\quad - K_pR_{e,k}K_p^* + \overline{K}_p\overline{R}_{e,k}\overline{K}_p^* & (9.42)
\end{aligned}
$$

Since, $\overline{P}(k|k) \geq P(k|k)$ and $\lambda_{max}(S)I - S \geq 0$ for any symmetric matrix S, $\overline{P}(k|k) - P(k|k) \geq 0$.

Using Lemma 9.3.3, Lemma 9.3.2, and the induction hypothesis, we obtain the following theorem. The ub-KF maintains the state error covariance which is lower-bounded by the state error covariance of the modified KF.

Theorem 9.3.3 *If the ub-KF starts with an initial covariance $\overline{P}(0|0)$, such that $\overline{P}(0|0) \geq P(0|0)$, then $\overline{P}(k|k) \geq P(k|k)$ for all $k \geq 0$.*

9.3.4 Convergence

The following theorem shows a simple condition under which the state error covariance can be unbounded.

Theorem 9.3.4 *If $(\mathbf{E}[A(Z)]^T, \mathbf{E}[A(Z)]^T C^T)$ is not stabilizable, or equivalently, $(\mathbf{E}[A(Z)], C\mathbf{E}[A(Z)])$ is not detectable, then there exists an initial covariance $P(0|0)$ such that $P(k|k)$ diverges as $k \to \infty$.*

Proof 9.3.5 *Let us consider the lb-KF. Let $\underline{P}_k = \underline{P}_{k|k}$. $\psi = GQG^T$, $\hat{A} = \mathbf{E}[A]$, and $F = -(C\hat{A}\underline{P}_k\hat{A}^T C^T + C\psi C^T + R)^{-1}(C\psi + C\hat{A}\underline{P}_k\hat{A}^T)$.*
Then based on Riccati difference equation [624], we can express $\underline{P}_{k+1}$ as:

$$
\begin{aligned}
\underline{P}_{k+1} &= \hat{A}\underline{P}_k\hat{A}^T + \psi \\
&- F^T(C\hat{A}\underline{P}_k\hat{A}^T C^T + C\psi C^T + R)F \\
&= (\hat{A}^T + \hat{A}^T C^T F)^T \underline{P}_k(\hat{A}^T + \hat{A}^T C^T F) \\
&+ F^T(C\psi C^T + R)F + \psi C^T F + F^T C\psi + \psi \quad (9.43)
\end{aligned}
$$

Hence, if $(\hat{A}^T + \hat{A}^T C^T F)$ is not a stability matrix, for some $\underline{P}_0 \leq P(0|0)$. $\underline{P}_k$ diverges as $k \to \infty$. Since the state error covariance of the lb-KF diverges and $\underline{P}(k|k) \leq P(k|k)$ for all $k \geq 0$ (Theorem 9.3.2), $P(k|k)$ diverges as $k \to \infty$. Here $P(k|k)$ can be $F_k P_{k+1/k} F_k^* - K_k H_k P_{k+1/k}$ for the "complete" prior case and $K_k H_k P_{k/k-1}$ for "without" prior and "incomplete" prior cases respectively.

9.3.5 Fusion without prior information

The Bayesian-Based FB Kalman filter rule of theorem 9.3.1 is not applicable if either there is no prior information about the quantity to be estimated, the information is incomplete (e.g., the prior covariance is not known or does not exist), or the quantity to be estimated is not random. In these cases, the estimation formulas are not clearly applicable.

The following theorem presents the Bayesian-Based FB Kalman filter for fusion without prior information:

Theorem 9.3.5

$Forward\ Run$: $For\ (k = 0;\ k < T;\ +k)$

$$\widehat{x}_{k/k} = K_{p,i}[y_i - \bar{\nu}] \tag{9.44}$$

$$\widehat{x}_{k+1/k} = F_k x_{k+1/k} - K_p H_k x_{k+1/k} + k_p y - k_p \nu \tag{9.45}$$

$$\widehat{P}_{k/k} = K_k H_k P_{k/k-1} \tag{9.46}$$

$$K_k = H_k^+[I - P_{k/k-1}((I - HH')(P_{k/k-1})$$

$$.(I - HH'))^+] \tag{9.47}$$

$$\tilde{K} = K + B'(I - HH') \tag{9.48}$$

$$P_{k+1/k} = K_{p,k} R_{e,k} K_{p,k}^* \tag{9.49}$$

$Backward\ Run$: $For\ (k = 0;\ k < T;\ +k)$

$$\widehat{J}_{k-1/T} = \widehat{P}_{k-1/T} F_k^T \widehat{P}_{k-1/T}^{-1} \tag{9.50}$$

$$\widehat{x}_{k-1/T} = \widehat{x}_{k-1/k-1}^i + \widehat{J}_{k-1}(\widehat{x}_{k-1/T} - \widehat{x}_{k-1/k}) \tag{9.51}$$

$$\widehat{P}_{k-1/T} = \widehat{P}_{k-1/k-1}$$

$$+\widehat{J}_{k-1}(\widehat{J}_{k-1/T} - \widehat{P}_{k-1}/k)J_{k-1}' \tag{9.52}$$

where B is any matrix of compatible dimensions satisfying $P_{k/k-1}^{\frac{1}{2}'}(I - HH^+)B = 0$, $P_{k/k-1}^{\frac{1}{2}}$ is any square root matrix of $P_{k/k-1}$. The optimal gain matrix $\tilde{K}$ is given uniquely by:

$$\tilde{K} = K = H^+[I - P_{k/k-1}(I - HH^+)^{\frac{1}{2}}((I - HH^+)^{\frac{1}{2}'}$$

$$P_{k/k-1}(I - HH^+)^{\frac{1}{2}})^{-1}(I - HH^+)^{\frac{1}{2}'}] \tag{9.53}$$

if and only if $[H,\ P_{k/k-1}^{\frac{1}{2}}]$ has full row rank, where $(I - HH^+)^{\frac{1}{2}}$ is a full-rank square root of T.

9.3.6   Modified Kalman filter-II

In this section, we outline the case without prior information. Similar to the case of complete prior information discussed earlier, the modification of the Kalman filter is focused toward the prediction covariance computing of that case.

Hence, the prediction covariance in the case of no prior information can be computed as following:

$$
\begin{aligned}
P(k+1|k) &= \mathbf{E}[e(k+1|k)e(k+1|k)^T|y_k] \\
&= -K_p R_{e,k} K_p^* \\
&\quad + \sum_{z \in Z} p_z A(z)\widehat{x}(k|k)\widehat{x}(k|k)^T (A(z) - \widehat{A})^T
\end{aligned}
$$

$$(9.54)$$

And here also, given $\widehat{x}(k+1|k)$ and $P(k+1|k)$, $\widehat{x}(k+1|k+1)$ and $P(k+1|k+1)$ are computed as in the standard Kalman filter.

$$
\begin{aligned}
\widehat{x}(k+1|k+1) &= K(k+1)[y(k+1) - \bar{\nu}] & (9.55) \\
P(k+1|k+1) &= K(k+1)H(k+1)P(k+1) & (9.56)
\end{aligned}
$$

where $K(k+1) = H(k+1)^+[I - P(k+1)((I - HH^T)(Pk+1)$.

Likewise in the foregoing sections, since the computation of $P(k+1|k)$ is the only time-consuming process, we propose two filtering methods which can bound $P(k+1|k)$ using the same notations.

Lower-bound KF: without prior information case

The lower-bound KF (lb-KF) is the same as the modified KF described in the foregoing section, except we approximate $P(k+1|k)$ by $\underline{P}(k+1|k)$ and $P(k|k)$ by $\underline{P}(k|k)$. The covariances are updated as following:

$$
\begin{aligned}
\underline{P}(k+1|k) &= \underline{K}(k+1)R\underline{K}^T(k+1) & (9.57) \\
\underline{P}(k+1|k+1) &= \underline{K}(k+1)H\underline{P}^T(k+1|k) & (9.58)
\end{aligned}
$$

where $\underline{K}(k+1) = H^+[I - \underline{P}(k+1|k)((I - HH^T)\underline{P}(k+1|k)$.

Lemma 9.3.4 *If $\underline{P}(k|k) \leq P(k|k)$, then $\underline{P}(k+1|k) \leq P(k+1|k)$.*

Proof 9.3.6 *Using (9.54), we have*

$$
\begin{aligned}
P(k+1|k) - \underline{P}(k+1|k) &= \mathbf{E}[A(Z)\widehat{x}(k|k)\widehat{x}(k|k)^T A(Z)^T] \\
&\quad - K_{p,k} R_{e,k} K_{p,k}^* \\
&\quad - \widehat{A}\widehat{x}(k|k)\widehat{x}(k|k)^T \widehat{A}^T \\
&\quad - \underline{K}_{p,k}\underline{R}_{e,k}\underline{K}_{p,k}^* \\
&= P_1 + P_2
\end{aligned}
$$

$$(9.59)$$

where

$$
\begin{aligned}
P_1 &= K_{p,k} R_{e,k} K_{p,k}^*, \\
P_2 &= \mathbf{E}[A(Z)\widehat{x}(k|k)\widehat{x}(k|k)^T A(Z)^T] - \widehat{A}\widehat{x}(k|k)\widehat{x}(k|k)^T \widehat{A}^T \\
&\quad - \underline{K}_{p,k} \underline{R}_{e,k} \underline{K}_{p,k}^*
\end{aligned}
$$

Since $P(k|k)$ is a symmetric matrix, $P(k|k)$ can be decomposed into $P(k|k)$ = $U_1 D_1 U_1^T$, where U_1 is a unitary matrix and D_1 is a diagonal matrix, but there is no $P(k|k)$ for P_1 here. Hence,

$$
P_1 = -K_{p,k} R_{e,k} K_{p,k}^* \tag{9.60}
$$

9.3.7   Upper-bound KF-II

Similar to the lb-KF, the upper-bound KF (ub-KF) approximates $P(k + 1|k)$ by $\overline{P}(k+1|k)$ and $P(k|k)$ by $\overline{P}(k|k)$. Let $\lambda_{max} = \lambda_{max}(\overline{P}(k|k)) + \lambda_{max}(\widehat{x}(k|k)\widehat{x}(k|k)^T)$, where $\lambda_{max}(S)$ denotes the maximum eigenvalue of S. The covariances are updated as following:

$$
\begin{aligned}
\overline{P}(k + 1|k) &= \lambda_{max}\mathbf{E}[A(Z)A(Z)^T] \\
&\quad + K_{p,k} R_{e,k} K_{p,k}^* \tag{9.61} \\
\overline{P}(k + 1|k + 1) &= \overline{K}(k + 1)H\overline{P}(k + 1|k) \tag{9.62}
\end{aligned}
$$

where $\overline{K}(k + 1) = H^+[I - \overline{P}(k + 1|k)(I - HH')(\overline{P}(k + 1|k))$. In the ub-KF, $\mathbf{E}[A(Z)A(Z)^T]$ can be computed in advance but we need to compute λ_{max} at each step of the algorithm.

Lemma 9.3.5 *If $\overline{P}(k|k) \geq P(k|k)$, then $\overline{P}(k + 1|k) \geq P(k + 1|k)$.*

Proof 9.3.7 *Let $M = \widehat{x}(k|k)\widehat{x}(k|k)^T$ and I be an identity matrix. Then using (9.54), we have (See Eqn. (9.63)).*

$$
\begin{aligned}
\overline{P}(k|k) - P(k|k) &= \lambda_{max}\mathbf{E}[A(Z)A(Z)^T] \\
&\quad - \mathbf{E}[A(Z)MA(Z)^T] - \mathbf{E}[\widehat{A}M\widehat{A}^T] \\
&= \mathbf{E}[A(Z)(\lambda_{max}(M)I - M)A(Z)^T] \\
&\quad + \mathbf{E}[\widehat{A}M\widehat{A}^T] + \overline{K}_{p,k}\overline{R}_{e,k}\overline{K}_{p,k}^* \\
&\quad - K_{p,k}R_{e,k}K_{p,k}^* \tag{9.63}
\end{aligned}
$$

Since, $\overline{P}(k|k) \geq P(k|k)$ and $\lambda_{max}(S)I - S \geq 0$ for any symmetric matrix S, $\overline{P}(k|k) - P(k|k) \succeq 0$.

Using Lemma 9.3.5, Lemma 9.3.2, and the induction hypothesis, we obtain the following theorem. The ub-KF maintains the state error covariance which is lower-bounded by the state error covariance of the modified KF.

Theorem 9.3.6 *If the ub-KF starts with an initial covariance $\overline{P}(0|0)$, such that $\overline{P}(0|0) \geq P(0|0)$, then $\overline{P}(k|k) \geq P(k|k)$ for all $k \geq 0$.*

The convergence will be the same as followed in Section 9.3.4 and in Theorem 9.3.4.

9.4 Fusion with Incomplete Prior Information

In practice, it is sometimes the case that prior information of some but not all the components of $\bar{x}$ are available. For example, tracking the positioning of a vehicle, it is easy to determine the prior position vector of the vehicle (it must be within a certain position range) with certain covariance, but not the velocity of the vehicle, i.e., at what speed it is traveling. Such an incomplete prior problem is presented in this section using the Bayesian-Based FB Kalman filter.

The following theorem presents the Bayesian-Based FB Kalman filter with incomplete prior information:

Theorem 9.4.1

$$Forward\ Run:\ For\ (k = 0;\ k < T;\ +k)$$

$$\widehat{x}_{k/k} = VK_{p,i}V_1'\bar{x} + VK_{p,i}[y_i - \bar{\nu}] \tag{9.64}$$

$$\widehat{x}_{k+1/k} = VK_{p,i}V_1'\widehat{x}_{k+1/k} + VK_{p,k}y_k - VK_{p,k}V' \tag{9.65}$$

$$\widehat{P}_{k/k} = K_k H_k P_{k/k-1} \tag{9.66}$$

$$K_k = H_k^+[I - P_{k/k-1}((I - HH')(P_{k/k-1})$$
$$.(I - HH'))^+] \tag{9.67}$$

$$P_{k+1/k} = G_i Q_i G_i^* - K_{p,k} R_{e,k} K_{p,k}^* \tag{9.68}$$

$$Backward\ Run:\ For\ (k = 0;\ k < T;\ +k)$$

$$\widehat{J}_{k-1/T} = \widehat{P}_{k-1/T} F_k^T \widehat{P}_{k-1/T}^{-1} \tag{9.69}$$

$$\widehat{x}_{k-1/T} = \widehat{x}_{k-1/k-1}^i + \widehat{J}_{k-1}(\widehat{x}_{k-1/T} - \widehat{x}_{k-1/k}) \tag{9.70}$$

$$\widehat{P}_{k-1/T} = \widehat{P}_{k-1/k-1}$$
$$+\widehat{J}_{k-1}(\widehat{J}_{k-1/T} - \widehat{P}_{k-1}/k)J_{k-1}' \tag{9.71}$$

Proof 9.4.1 *By Theorem 9.3.5, the problem can be converted to "without prior information" with H and C replaced by the $\tilde{H}$ and $\tilde{C}$ respectively, where, from the proof of Theorem 9.3.5, the quantity to be estimated is $u = V'x$, where V is an orthogonal matrix. This means that Theorem 9.3.5 is applicable now to u. Therefore, all the formulas in this theorem follow from Theorem 9.3.5 and the relationship:*

$$\hat{x} = V\hat{u}, \quad P = V\mathrm{MSE}(\hat{u})V'$$

The uniqueness result thus follows from Theorem 9.3.5.

9.4.1　Modified Kalman filter-III

In this section, we outline the case with incomplete prior information. As we discussed, in the case of incomplete prior information, the modification of the Kalman filter is focused toward the prediction covariance computing of that case.

The prediction covariance in the case of incomplete prior information can be computed as the following:

$$
\begin{aligned}
P(k+1|k) &= \mathbf{E}[e(k+1|k)e(k+1|k)^T|y_k] \\
&= GQG^T - K_p R_{e,k} K_p^* \\
&\quad + \sum_{z \in Z} p_z A(z)\hat{x}(k|k)\hat{x}(k|k)^T (A(z) - \hat{A})^T
\end{aligned}
$$

$$(9.72)$$

And here also, given $\hat{x}(k+1|k)$ and $P(k+1|k)$, $\hat{x}(k+1|k+1)$ and $P(k+1|k+1)$ are computed as in the standard Kalman filter.

$$
\begin{aligned}
\hat{x}(k+1|k+1) &= K(k+1)[y(k+1) - \bar{\nu}] & (9.73) \\
P(k+1|k+1) &= K(k+1)H(k+1)P(k+1) & (9.74)
\end{aligned}
$$

where $K(k+1) = \tilde{H}(k+1)^+[I - \tilde{P}(k+1|k)((I - \tilde{H}\tilde{H}^T)(Pk+1|k)$.

9.4.2　Approximating the Kalman filter

Likewise in the foregoing section, since the computation of $P(k+1|k)$ is the only time-consuming process, we propose two filtering methods which can bound $P(k+1|k)$ using the same notations.

9.4.3 Lower-bound KF-III

The lower-bound KF (lb-KF) is the same as the modified KF described earlier, except we approximate $P(k+1|k)$ by $\underline{P}(k+1|k)$ and $P(k|k)$ by $\underline{P}(k|k)$. The covariances are updated as the following:

$$\underline{P}(k+1|k) = GQG^T - \underline{K}_{p,k}\underline{R}_{e,k}\underline{K}^*_{p,k} \qquad (9.75)$$

$$\underline{P}(k+1|k+1) = V\underline{K}(k+1)H_k\underline{P}(k+1|k)^*V^T \qquad (9.76)$$

where $\underline{K}(k+1) = \tilde{H}_k^+[I - \tilde{\underline{P}}(k+1|k)(I - \tilde{H}\tilde{H}')(\tilde{\underline{P}}(k+1|k))$.

Lemma 9.4.1 *If $\underline{P}(k|k) \preceq P(k|k)$, then $\underline{P}(k+1|k) \preceq P(k+1|k)$.*

Proof 9.4.2 *Using (9.72), we have*

$$
\begin{aligned}
P(k+1|k) - \underline{P}(k+1|k) &= \mathbf{E}[A(Z)\widehat{x}(k|k)\widehat{x}(k|k)^T A(Z)^T] \\
&\quad - K_{p,k}R_{e,k}K^*_{p,k} \\
&\quad - \widehat{A}\widehat{x}(k|k)\widehat{x}(k|k)^T \widehat{A}^T \\
&\quad + \underline{K}_{p,k}\underline{R}_{e,k}\underline{K}^*_{p,k} \\
&= P_1 + P_2 \qquad (9.77)
\end{aligned}
$$

where

$$
\begin{aligned}
P_1 &= -K_{p,k}R_{e,k}K^*_{p,k}, \\
P_2 &= \mathbf{E}[A(Z)\widehat{x}(k|k)\widehat{x}(k|k)^T A(Z)^T] - \widehat{A}\widehat{x}(k|k)\widehat{x}(k|k)^T \widehat{A}^T \\
&\quad - \underline{K}_{p,k}\underline{R}_{e,k}\underline{K}^*_{p,k}
\end{aligned}
$$

Since $P(k|k)$ is a symmetric matrix, $P(k|k)$ can be decomposed into $P(k|k)$ $= U_1 D_1 U_1^T$, where U_1 is a unitary matrix and D_1 is a diagonal matrix, but here there is no $P(k|k)$ for P_1.

9.4.4 Upper-bound KF-III

Similar to the lb-KF, the upper-bound KF (ub-KF) approximates $P(k+1|k)$ by $\overline{P}(k+1|k)$ and $P(k|k)$ by $\overline{P}(k|k)$. Let

$$\lambda_{max} = \lambda_{max}(\overline{P}(k|k)) + \lambda_{max}(\widehat{x}(k|k)\widehat{x}(k|k)^T)$$

where $\lambda_{max}(S)$ denotes the maximum eigenvalue of S. The covariances are updated as the following:

$$
\begin{aligned}
\overline{P}(k+1|k) &= \lambda_{max}\mathbf{E}[A(Z)A(Z)^T] \\
&\quad + \overline{K}_{p,k}\overline{R}_{e,k}\overline{K}^*_{p,k} \qquad (9.78) \\
\overline{P}(k+1|k+1) &= \overline{K}(k+1)H\overline{P}(k+1|k) \qquad (9.79)
\end{aligned}
$$

where $\overline{K}(k+1) = \tilde{H}^+[I - \overline{\tilde{P}}(k+1|k)(I - \tilde{H}\tilde{H}')(\overline{\tilde{P}}(k+1|k))$. In the ub-KF, $\mathbf{E}[A(Z)A(Z)^T]$ can be computed in advance but we need to compute λ_{max} at each step of the algorithm.

Lemma 9.4.2 *If $\overline{P}(k|k) \geq P(k|k)$, then $\overline{P}(k+1|k) \geq P(k+1|k)$.*

Proof 9.4.3 *Let $M = \hat{x}(k|k)\hat{x}(k|k)^T$ and I be an identity matrix. Then using (9.72), we have*

$$
\begin{aligned}
\overline{P}(k|k) - P(k|k) &= \mathbf{E}[A(Z)(\lambda_{max}(M)I - M)A(Z)^T] \\
&+ \hat{A}M\hat{A}^T + \overline{K}_{p,k}\overline{R}_{e,k}\overline{K}_{p,k}^* \\
&- K_{p,k}R_{e,k}K_{p,k}^* \\
&+ GQG^T
\end{aligned}
\tag{9.80}
$$

Since, $\overline{P}(k|k) \geq P(k|k)$ and $\lambda_{max}(S)I - S \geq 0$ for any symmetric matrix S, $\overline{P}(k|k) - P(k|k) \geq 0$.

Using Lemma 9.4.2, Lemma 9.3.2, and the induction hypothesis, we obtain the following theorem. The ub-KF maintains the state error covariance which is lower-bounded by the state error covariance of the modified KF.

Theorem 9.4.2 *If the ub-KF starts with an initial covariance $\overline{P}(0|0)$, such that $\overline{P}(0|0) \geq P(0|0)$, then $\overline{P}(k|k) \geq P(k|k)$ for all $k \geq 0$.*

The convergence will be the same as followed in Section III.9.3.4 and in Theorem 9.3.4.

9.5   Fusion Algorithm

The information captured in each priori case is designed for a distributed structure. The idea is taken from [625].

Suppose there is X number of sensors. For every measurement coming from these sensors that is received in the fusion center, there is a corresponding estimation based solely on one of the sensors that is so-called virtual sensor (VS). Every estimation from the single VS then is processed through the fusion algorithm to get the optimal estimation of the state. The overall diagram of the fusion process using multiple sensors can be seen in Figure 9.1.

When estimates of the states are available, based on their prior knowledge, the problem now turns to how to combine these different estimations to get the optimal result. Fused estimation based on a series of particular sensors is computed every sampling time T_s, where the fused estimation $\hat{x}(k|k)$ is no more than an estimation coming from each sensor $\hat{x}_i(k|k)$.

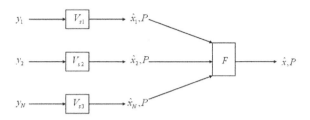

Figure 9.1: Proposed data fusion design.

Theorem 9.5.1 *For any* $k = 1, 2, ...,$ *the estimate and the estimation error covariance of* $x(k)$ *based on all the observations before time* kT *are denoted by* $\hat{x}(k|k)$ *and* $P(k|k)$; *then they can be generated by the use of the following formula:*

$$\hat{x}(k|k) = \sum_{i=1}^{N} \alpha_i(k) \hat{x}_{N|i}(k|k) \tag{9.81}$$

$$P(k|k) = (\sum_{i=1}^{N} P_{N|i}^{-1}(k|k))^{-1} \tag{9.82}$$

where,

$$\alpha_i(k) = P(k|k) P_{N|i}^{-1}(k|k) \tag{9.83}$$

where $\hat{x}_{N|i}(k|k)$ *is state estimation at the highest sample rate based on estimation from VS* i *and* $P_{N|i}(k|k)$ *is it's error covariance.*

From equation (9.83), it can be verified that:

$$P(k|k) \leq P_{N|i}(k|k) \tag{9.84}$$

which means that the fused estimation error from the estimations of different sensors will always be less or equal to the estimation error of each sensor.

9.5.1 Evaluation and Testing

The evaluation and testing has been made on an electro-hydraulic system [48]. Fault scenarios are created by using the rotational hydraulic drive in the simulation program. In these scenarios, leakage fault and controller fault are being considered.

Scenario I: Leakage Fault   In this scenario, while the system is working in real time, leakage faults are being introduced into the hydraulic fluid flow linked to the servo-valve of the system. The leakage fault is considered as $\omega_h C_{L_{leakage}} x_3(t)$ in state 3.

Scenario II: Controller Fault   In this scenario, while the system is working in real time and getting the input for driving the dynamics of the system, a fault has been introduced by increasing the torque load in the hydraulic drive, then affecting the controller;

$$-\frac{\omega_h}{\alpha} t_{L_{fault}}$$

is considered in state 2 of the system. Following [48] and the fault scenarios, the fault model of the system can be described in state space form as:

$$\dot{x}_1(t) = \omega_{max} x_2(t) \tag{9.85}$$

$$\dot{x}_2(t) = -\gamma \frac{\omega_h}{\alpha} x_2(t) + \frac{\omega_h}{\alpha} x_3(t) -$$
$$\frac{\omega_h}{\alpha} t_L - \frac{\omega_h}{\alpha} t_{L_{fault}} \tag{9.86}$$

$$\dot{x}_3(t) = -\alpha \omega_h x_2(t) - \omega_h C_L x_3(t)$$
$$+ \quad \alpha \omega_h x_4(t) \sqrt{1 - x_3(t) sigm(x_4(t))}$$
$$- \quad \omega_h C_{L_{leakage}} x_3(t) \tag{9.87}$$

$$\dot{x}_4(t) = -\frac{1}{\tau_v} x_4(t) + \frac{i(t)}{\tau_v} \tag{9.88}$$

where

$$
\begin{aligned}
x_1(t) &= \theta(t), \ x_2(t) = \frac{\dot{\theta}(t)}{\omega_{max}}, \\
x_3(t) &= \frac{P_L(t)}{P_s}, \ x_4(t) = \frac{A_v(t)}{A_v^{max}}, \\
u_1(t) &= i(t) = \frac{I(t)}{I_{max}}, \ u_2 = t_L = \frac{T_L}{P_s D_m}, \\
\gamma &= \frac{B\omega_{max}}{P_s D_m}, \\
\omega_h &= \sqrt{\frac{2\beta D_m^2}{JV}}, \\
\alpha &= \frac{(C_d A_v^{max}\sqrt{P_s \rho})J\omega_h}{P_s D_m^2}, \\
c_L &= \frac{JC_L\omega_h}{D_m^2}
\end{aligned}
$$

and $C_{L_{leakage}}$ is the leakage fault considered in state 3, $t_{L_{fault}}$ is the controller fault in the form of torque load in state 2.

Using the sign convention for $A_v(t)$ and the definition of $x_3(t)$, it follows that $0 \leq x_3(t)sigm(x_4(t)) \leq 1$. It is also noted here that $0 \leq x_3(t)sigm(x_4(t)) \leq 1$, because $P_1(t)$ and $P_2(t)$ are both positive and the condition $x_3(t)sigm(x_4(t)) = 1$ implies that $P_1(t) = P_s$ and $P_2(t) = 0$ or $P_2(t) = P_s$ and $P_1(t) = 0$, indicating zero pressure drop across the open ports of the servo-valve and thus, no flow to or from the actuator, a situation that would occur if the rotational motion of the drive are impeded.

9.5.2 Simulation results

In what follows, we present the simulation results for the proposed distributed approximate estimation with three cases of prior knowledge. The experiment has been performed on the rotational hydraulic drive system. Two sets of faults have been considered here, that is, the leakage fault in state 3 and the controller fault. First, the data collected from the plant has been initialized and the parameters have been being optimized comprising the pre-processing and normalization of the data. The comparison of results for the distributed estimation, and estimation generated from various levels of faults, and the basic profile of that particular fault has been compared. Moreover, same pattern of comparison has been followed for modified estimation filters with lower bound and upper bound. Later, computational time comparison has been shown for different results showing the effectiveness of the modified filter in all cases.

Fault 1 (Leakage): Complete prior information, with lower and upper bound filter versions

The Bayesian-Based FB Kalman filter has been simulated here for the leakage fault of the plant. Simulations have been made for the x-estimate and the covariance of each case. In the simulation, comparison of various levels of leakage, that is, no leakage, small leakage, and medium intensity of leakage faults, and the distributed estimation has been shown. It can be seen from the estimate profile in Figure 9.3 that the distributed structure is clearly performing well compared to the other profiles for complete prior information; when it comes to the covariance of modified filter implementation with the upper bound, see Figure 9.5, and for the lower bound, see Figure 9.8 for estimates of the lower bound scheme. It is performing equally well for the distributed structure. Actually, the advantage of using the modified upper and lower bound filters can be seen more clearly when we talk about the time computation as discussed in the next section.

Fault 1 (Leakage): Incomplete prior information, with lower and upper bound filter versions

In case of incomplete prior information with a leakage fault, when it comes to the covariance and estimate of the modified filter implementation with the upper bound, see in Figure 9.11 that it performs well for a distributed structure. Actually, the advantage of using the modified upper and lower bound filters can be seen more clearly when we talk about the time computation, as discussed in the next section.

Fault 1 (Leakage): No prior information, with lower and upper bound filter versions

In the case of estimation without prior information but with a leakage fault, it can be seen for the covariance profile, see Figure 9.12, that the distributed structure is clearly performing well compared to the other profiles or those without prior information, which is the worst scenario case chosen from all three as far as the prior information is concerned. This can be attributed to the impact of the H matrix in the gain K_k. Observe the covariance and estimate of modified filter implementation with upper bound, Figure 9.15 for covariance of upper bound scheme and lower bound, Figure 9.17 for covariance of lower bound scheme. It appears to be performing equally well for distributed structure.

A comparison of computation times will be reported later.

Fault 2 (Controller): Complete Prior Information, with its lower and upper bound filter versions

The Bayesian-Based FB Kalman filter has been simulated here for the controller fault of the plant, which has been introduced by increasing the torque load in the hydraulic drive, then effecting the controller. Simulations have been made for the x-estimate and the covariance of each case. In the simulation, comparison of various levels of controller faults, that is, no faults, small faults, and medium intensity of faults, and distributed estimation has been shown.

Considering Figure 9.2 through 9.9, it can be seen for the covariance profile and estimate that the distributed structure is clearly performing well compared to the other profiles like modified filter implementation with upper and lower bound schemes.

Fault 2 (Controller): Incomplete Prior Information, with its lower and upper bound filter versions

In the case of estimation without prior knowledge but with controller fault, it can be seen from Figure 9.13 through Figure 9.20 that the covariance profile and estimate, see Figure 9.14 that the distributed structure is clearly performing well compared to the other profiles without prior information. The interpretation for this is due to the full rank of the H matrix in the gain K_k. This is true for modified filter implementation with upper bound and lower bound schemes.

9.5.3 Time computation

Evaluation of the time computation of different methods is summarized below:

1. The computing device is an HP COMPAQ laptop, n × 7300 INTEL (R) core (TM) 2 CPU T 7200 @ 2 GHz with 2.5 GB ram and 500 Hard disk.

2. The number of iterations has been taken as 5 for achieving each and every of the estimate.

3. For the case of complete prior information, it can be seen from Table 9.1, that the iteration time of the basic Bayesian-Based FB Kalman filter is taking the maximum number of time for the computation.

4. Both modified filters of the upper bound and the lower bound are performing well with less computation time for leakage fault (fault 1) and controller fault (fault 2) respectively.

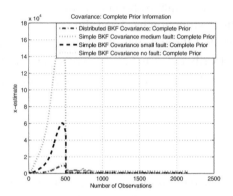

Figure 9.2: Comparison of covariance for complete prior information for a controller fault.

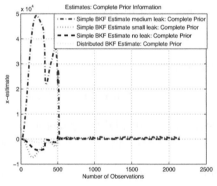

Figure 9.3: Comparison of estimates with complete prior information for a leakage fault.

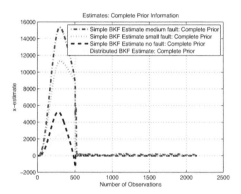

Figure 9.4: Comparison of estimates with complete prior information for a controller fault.

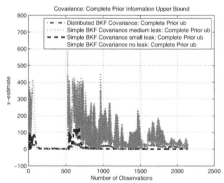

Figure 9.5: Comparison of covariance with complete prior information for a leakage fault with upper bound modified filter.

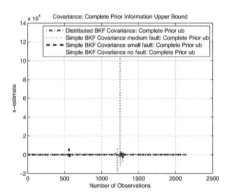

Figure 9.6: Comparison of covariance for complete prior information for a controller fault with upper bound modified filter.

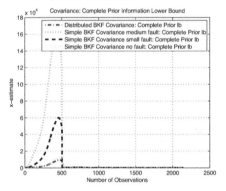

Figure 9.7: Comparison of covariance with complete prior information for a controller fault with lower bound modified filter.

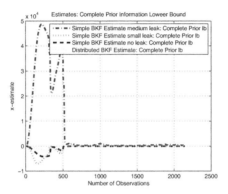

Figure 9.8: Comparison of estimates with complete prior information for a leakage fault with lower bound modified filter.

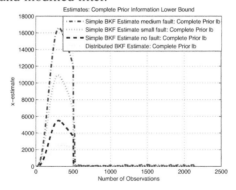

Figure 9.9: Comparison of estimates with complete prior information for a controller fault with lower bound modified filter.

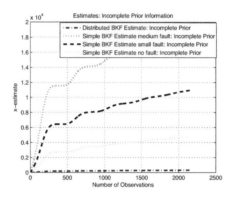

Figure 9.10: Comparison of estimates with incomplete prior information for a controller fault.

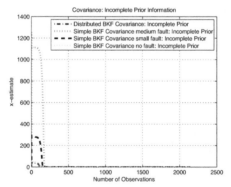

Figure 9.11: Comparison of covariance with incomplete prior information for a leakage fault with upper bound modified filter.

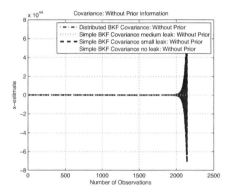

Figure 9.12: Comparison of covariance without prior information for a leakage fault.

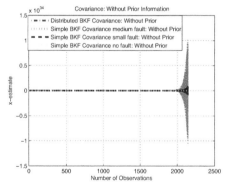

Figure 9.13: Comparison of covariance without prior information for a controller fault.

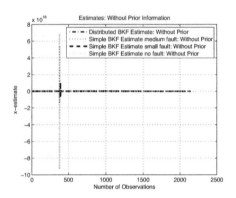

Figure 9.14: Comparison of estimates for without prior information for controller fault.

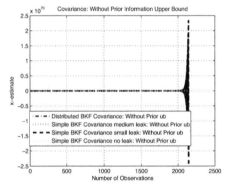

Figure 9.15: Comparison of covariance without prior information for a leakage fault with upper bound modified filter.

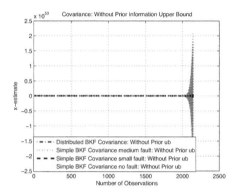

Figure 9.16: Comparison of covariance without prior information for a controller fault with upper bound modified filter.

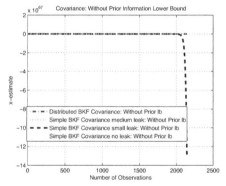

Figure 9.17: Comparison of covariance without prior information for a leakage fault with lower bound modified filter.

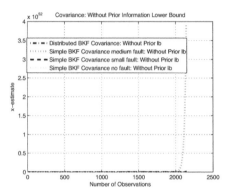

Figure 9.18: Comparison of covariance without prior information for a controller fault with lower bound modified filter.

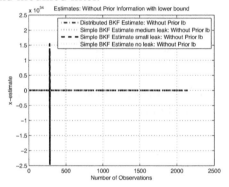

Figure 9.19: Comparison of estimates without prior information for a Leakage Fault with lower bound modified filter.

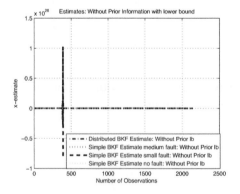

Figure 9.20: Comparison of estimates without prior information for a controller fault with lower bound modified filter.

Table 9.1: Case I: Time Computation Comparison for Complete Prior Information

FILTER	LEAKAGE FAULT	CONTROLLER FAULT
1- BAYESIAN FB KF	14.81	12.53
2- BAYESIAN FB KF+	12.23	12.22
3- BAYESIAN FB KF-	12.09	12.26

Table 9.2: Case II: Time Computation Comparison for Incomplete Prior Information

FILTER	LEAKAGE FAULT	CONTROLLER FAULT
1- BAYESIAN FB KF	13.503922	12.492827
2- BAYESIAN FB KF+	12.732579	12.191222
3- BAYESIAN FB KF-	12.939255	12.166062

5. Considering Table 9.2 and Table 9.3 for the cases of incomplete prior information and without prior information, respectively, it turns out that the results are even more crucial and critical because of their structures. Moreover, the basic Bayesian-Based FB Kalman filter is taking comparatively more time than the corresponding modified lower bound and upper bound filters.

6. The performance of the modified filters was consistent even here for both leakage fault (fault 1) and controller fault (fault 2) respectively.

In the tables, Bayesian FB KF+ means with upper bound and Bayesian FB KF- corresponds to with lower bound.

Table 9.3: Case III: Time Computation Comparison for Without Prior Information

FILTER	LEAKAGE FAULT	CONTROLLER FAULT
1- BAYESIAN FB KF	23.463690	22.445465
2- BAYESIAN FB KF+	22.926070	12.165139
3- BAYESIAN FB KF-	22.366596	21.970777

9.6 Notes

In this chapter, an approximate distributed estimation has been proposed in explicit forms using a Bayesian-Based FB Kalman filter for estimating the states of a network control system for an arbitrary number of sensors with complete, incomplete, or no prior information. The approximate estimation presents all the prior cases with an effort to minimize time complexity, and cases showing dependency on prior knowledge. Then, the algorithms were made effective by data fusion of all the knowledge in a distributed filtering architecture. The proposed scheme has been evaluated on a rotational drive-based electro-hydraulic system, using various fault scenarios, thus ensuring the effectiveness of the approach with different prior cases.

9.7 Proposed Topics

1. The use of variational approximations applied to the observation model allows to formulate the entire dynamic model as a Kalman filter. One attractive line of research is to look at adopting the idea of "approximate Kalman filtering" in implementing smoothing in the dynamic model to see if it offers improved results over filtering.

2. We have learned that the Kalman filter is a technique for estimating a time-varying state given a dynamical model for, and indirect measurements of, the state. In principle, even in the case of nonlinear state and/or measurement models, standard implementation requires only linear algebra. However, for sufficiently large-scale problems, such as arise in weather forecasting and oceanography, the matrix inversion and storage requirements of the Kalman filter are prohibitive, and hence, approximations must be made. One potential suggestion is to use the conjugate gradient iteration within the Kalman filter for quadratic minimization, as well as for obtaining low-rank, low-storage approximations of the covariance and inverse-covariance matrices required for its implementation. Carry out the analysis and indicate the merits and demerits of the approach.

3. In the literature, it was argued that the principle difficulty for applying Kalman filtering to nonlinear systems is the need to consistently predict the new state and observation of the system. One proposed research topic is to use a set of discretely sampled points to parametrize the mean and covariance to build up an approximate Kalman filter. Follow up on this idea, carry out the mathematical analysis and then evaluate the performance of the filter.

Chapter 10

Estimation via Information Matrix

In this chapter, we develop an approach for distributed estimation using an information matrix filter on a distributed tracking system in which N number of sensors are tracking the same target. The approach incorporates proposed engineered versions of information matrix filter derived from covariance intersection, weighted covariance, and Kalman-like particle filter respectively. The steady performance of these filters is evaluated with different feedback strategies. Moreover they were employed with commonly used measurement fusion methods, like measurement fusion and state-vector fusion respectively, to complete the picture. The proposed filters are then validated on an industrial utility boiler, ensuring the effectiveness and applicability of the scheme underpinning it.

10.1 Introduction

The process of state estimation yields the estimate in the form of a mean and a covariance matrix or in the form of a distribution function. In this way, it is effectively used in providing a strict surveillance system for appropriate supervision. One of the methods to achieve such estimation requires a group of distributed sensors which provides information about the local targets. The classic work of Rao and Durrant-Whyte [545] presents an approach to decentralized Kalman filtering which accomplishes globally optimal performance in the case where all sensors can communicate with all other sensors. Other estimation methods can be a sensor-less approach [546][547], or a derivative-free filtering estimation [548], a least-squares-Kalman technique [550], a robot-based autonomous estimation and detection [549], and an H_∞ filtering-based estimation made for stochastic incom-

315

plete measurements [551], to name a few. In [568], a nonlinear constrained system observed by a sensor network is analyzed and a distributed state estimation scheme based on a moving horizon estimation (MHE) is developed. The problem of second-order consensus is investigated in [569] for a class of multi-agent systems with a fixed directed topology and communication constraints, where each agent is assumed to share information only with its neighbors on some disconnected time intervals. In [570], the distributed state estimation problem is investigated for a class of sensor networks described by uncertain discrete-time dynamical systems with Markovian jumping parameters and distributed time-delays. The authors of [571] examined the filtering design problem for discrete-time Markov jump linear systems under the assumption that the transition probabilities are not completely known.

The problem of multi-target tracking utilizing information from multiple sensors has been in focus for many years [552]–[560]. While achieving this approach, many fusion algorithms and filters were derived to combine local estimates [561][562, 563, 564] to prove better efficiency and effectiveness. For example, the state vectors can be fused using weighted covariance [559, 575, 576], an information matrix [565], and covariance intersection [566, 567]. The algorithms differ with the method they used to treat the covariance. As for the performance of different algorithms, [572] shows that the performance of a weighted covariance algorithm is consistently worse compared to the measurement fusion method. Moreover, it has been pointed out in [573] that results of the weighted covariance algorithm are showing the behavior to be a maximum likelihood estimate. At the same time, Chang indicates that the information matrix approach is optimal when the tracking systems are deterministic (i.e., process noise is zero) or when full-rate communication (i.e., two sensors exchange information each time when they receive new measurements and update their respective track files) is employed [573]. Covariance intersection avoids cross-covariance computation and its fusion result will be a consistent estimate, but its conservative estimates reduce performance [567]. However, covariance intersection is also being used for simultaneous localization and mapping, to maintain the full correlation structure.

In this chapter, we have derived distributed estimation with various versions of the information matrix filter. The estimation is derived on a distributed tracking system. After achieving a distributed estimation with various versions, we have selected two methods for measurement fusion. The proposed scheme is then validated on an industrial utility boiler system, where different types of faults were introduced and were considered for the evaluation of the proposed scheme.

10.2 Problem Formulation

Consider a distributed estimation system, as in [577] in which $N(N \geq 2)$ sensors are tracking the same target. The mathematical model describing process dynamics is assumed to be linear time invariant and of the form:

$$x_{k+1} = Fx_k + Gv_k, \; k = 0, 1, 2, \dots. \tag{10.1}$$

where $x_k \in \Re^{n_1}$ is the state vector of the target at time k and F is the state transition matrix, $v_k \in \Re^{n_2}$ is zero mean white Gaussian process noise with known covariance Q, and G is the input matrix. The target is tracked by N sensors, where the measurement model of sensor $j = 1, \dots, N$ is described by:

$$z_k^j = H^j x_k + w_k^j \tag{10.2}$$

where $w_k^j \in \Re^{n_3}$ is a zero-mean white Gaussian measurement noise with covariance $\mathbf{R}_k^j$.

It is assumed that local estimates, $\hat{x}_{k|k}^j$ and $P_{k|k}^j$, where $j = 1, \dots, N$ are obtained by each sensor's information-based filter based on measurement sequence $Z_k^j = \{z_i^j, i = 1, 2, \dots, k\}$ and are optimal in the sense of minimum variance. At the end of each n sampling interval, each sensor transmits its local estimate to the fusion center where track association and fusion are performed. For the fused estimate, there are two choices: either to be sent back to the sensor to improve the local estimation performance or to store it on the fusion center. For the sake of simplicity, the dimensions of the fused track and all local tracks are assumed to be the same. The distributed track fusion problem is to generate an optimal estimate $\hat{x}_{k|k}$ from all the local track information, i.e., $\hat{x}_{k|k}^j$ and $P_{k|k}^j$, and the prior information about local and fused estimation if possible [560]. The following sections work on the derived versions of information-based filters for the distributed tracking system.

10.3 Covariance Intersection

According to the standard results of the covariance intersection in [579], the covariance intersection at the sensor is:

$$\hat{x}_{k|k} = P_{k|k}(\omega P_{k|k}^{i}{}^{-1} \hat{x}_{k|k}^i + (1 - \omega) P_{k|k}^{i}{}^{-1} \hat{x}_{k|k}^j) \tag{10.3}$$

$$K_i = \omega P_{k|k} P_{k|k}^{i}{}^{-1} \tag{10.4}$$

$$K_j = (1 - \omega) P_{k|k} P_{k|k}^{j}{}^{-1} \tag{10.5}$$

where K_i and K_j are the gains and $\omega \in [0, 1]$ and it manipulates the weights which are assigned to $\hat{x}^i_{k|k}$ and $\hat{x}^j_{k|k}$ respectively. The covariance of filtering error is given by:

$$P_{k|k} = (\omega P^i_{k|k}{}^{-1} + (1 - \omega)P^j_{k|k}{}^{-1})^{-1} \qquad (10.6)$$

or

$$P^{-1}_{k|k} = (\omega P^i_{k|k}{}^{-1} + (1 - \omega)P^j_{k|k}{}^{-1}) \qquad (10.7)$$

where $\omega = (K_i/P_{k|k}).P^i_{k|k}$ and $1 - \omega = (K_j/P_{k|k}).P^j_{k|k}$.
  Thus substituting (10.4), (10.5), and (10.7) into (10.3) yields

$$P^{-1}_{k|k}x_{k|k} = \omega P^i_{k|k}{}^{-1}\hat{x}^i_{k|k} + (1 - \omega)P^j_{k|k}{}^{-1}\hat{x}^j_{k|k} \qquad (10.8)$$

Remark 10.3.1 *Different choices of ω can be used to optimize the update with respect to different performance criteria such as minimizing the trace or determinant of $P_{k|k}$.*

10.4 Covariance Intersection Filter

In what follows, we discuss the information-based covariance intersection filter and provide the relevant features. We start with the underlying algorithm.

10.4.1 Algorithm

For the case of deriving an information-based covariance intersection filter, the target dynamic model of (10.1) and (10.2) will be of the form:

$$x_{k+1} = Fx^i_k + Fx^j_k + G\nu_k \qquad (10.9)$$
$$z^j_k = K_1 x^i_k + K_2 x^j_k + w_k \qquad (10.10)$$

The key idea of the information matrix filter is to identify the common information shared by estimates that are to be fused, and then removing the information or decorrelation is implemented. It will take into consideration the common information but not the common process noise. Under the assumption of no feedback, the estimation using an information-based filter in the case of covariance intersection is as follows:

$$
\begin{aligned}
P_{k|k}^{-1}\hat{x}_{k|k} &= P_{k|k-n}^{-1}\hat{x}_{k|k-n} + \omega P_{k|k}^{i}{}^{-1}\hat{x}_{k|k}^{i} \\
&\quad - \omega P_{k|k-n}^{i}{}^{-1}\hat{x}_{k|k-n}^{i} + (1-\omega)P_{k|k}^{j}{}^{-1}\hat{x}_{k|k}^{j} \\
&\quad - (1-\omega)P_{k|k-n}^{j}{}^{-1}\hat{x}_{k|kn}^{j} \tag{10.11}
\end{aligned}
$$

$$
\begin{aligned}
P_{k|k}^{-1} &= P_{k|k-n}^{-1} + \omega P_{k|k}^{i}{}^{-1} - \omega P_{k|k-n}^{i}{}^{-1} + (1-\omega)P_{k|k}^{j}{}^{-1} \\
&\quad - (1-\omega)P_{k|k-n}^{j}{}^{-1} \tag{10.12}
\end{aligned}
$$

where the n step fusion state prediction is:

$$
x_{k|k-n} = Fx_k^i + Fx_k^j \tag{10.13}
$$

The associated covariance is explained below. Following [580], since v_k is assumed to be the $m \times 1$ zero-mean white noise process, and x_k the $n \times 1$ so-called state vector, it can be easily seen from $x_{k+1} = Fx_k^i + Fx_k^j + Gv_k$ that the covariance matrix of x_k obeys the recursion,

$$
\Pi_{i+1} = F_k\Pi_k^i F_k^* + F_k\Pi_k^j F_k^* + G_iQ_iG_i^* \tag{10.14}
$$

where $\Pi_k^i = \mathbf{E}\, x_k^i x_k^{i*}$ and $\Pi_k^j = \mathbf{E}\, x_k^j x_k^{j*}$.

Likewise, since $\hat{x}_{k|k-n} = Fx_k^i + Fx_k^j$, it satisfies the recursion,

$$
\Sigma_{i+1} = F_k^i\Sigma_k^i F_k^{i*} + F_k^j\Sigma_k^j F_k^{j*}, \tag{10.15}
$$

where $\Sigma_k^i = \mathbf{E}\hat{x}_{k|k-1}^i\hat{x}_{k|k-1}^{i*}$ and $\Sigma_k^j = \mathbf{E}\hat{x}_{k|k-1}^j\hat{x}_{k|k-1}^{j*}$ with initial condition $\Sigma_0 = 0$. Now the orthogonal decomposition $x_i = \hat{x}_{k|k-1} +$ with $\hat{x}_{i|i-1}$, shows that $\Pi_i = \Sigma_k^i + \Sigma_k^j + P_{k|k-1}$. It is then immediate to conclude that $P_{k+1|k} = \Sigma_{k+1} - \Sigma_{k+1}^i + \Sigma_{k+1}^j$ satisfies the recursion

$$
P_{k+1|k} = F_k^i P_{k|k-1} F_k^{i*} + G_iQ_iG_i^* \tag{10.16}
$$

As for the distributed tracking system, the communication network is considered to be large; therefore, the fused state estimate and associated covariance depend upon the local estimates as:

$$
\hat{x}_{k|k-n}^i + \hat{x}_{k|k-n}^j = \hat{x}_{k|k-n} \tag{10.17}
$$

$$
P_{k|k-n}^i + P_{k|k-n}^j = P_{k|k-n} \tag{10.18}
$$

10.4.2   Complete feedback case

For the case of complete feedback, a closed form analytical solution of steady fused covariance of information-based covariance intersection filter with N sensors is derived below. From (10.9) and (10.10), it is easy to show that the following two equations hold,

$$x_k = F_k^i x_{k-n} + F_k^j x_{k-n} + \sum_{i=1}^{n} F^{n-i} G v_{k-n+i} \tag{10.19}$$

$$\begin{aligned} z_k^j &= K_1 F^i x_{k-n}^i + K_2^j F^j x_{k-n}^j + w_{k-n} + K_1 F^i G v_{k-n+i} \\ &+ K_2 F^j G v_{k-n+j} \end{aligned} \tag{10.20}$$

For the two local sensors in the covariance intersection i.e., i and j, it is possible to write

$$x_{k|k} = \omega P_{k|k} {P_{k|k}^i}^{-1} F x_{k|k}^i + (1-\omega) P_{k|k} {P_{k|k}^j}^{-1} F x_{k|k}^j \tag{10.21}$$

Using (10.21) and (10.17), we have

$$\hat{x}_{k|k} = A_n x_{k|k}^i + B_i x_{k|k}^j \tag{10.22}$$

where, $\forall\, i = 1, ..., n$, we have $A_0 = I$, $A_i = \omega A_{i-1} P_{k|k} {P_{k|k}^i}^{-1} F$, $B_i = (1 - \omega) A_{i-1} P_{k|k} {P_{k|k}^j}^{-1} F$. Under the assumption of complete feedback, (10.11) and (10.12) can be re-written as:

$$\begin{aligned} P_{k|k}^{-1} \hat{x}_{k|k} &= -(N-1) P_{k|k-n}^{-1} \hat{x}_{k|k-n} + \omega {P_{k|k}^i}^{-1} \hat{x}_{k|k}^i \\ &+ (1-\omega) {P_{k|k}^j}^{-1} \hat{x}_{k|k}^j \end{aligned} \tag{10.23}$$

$$\begin{aligned} P_{k|k}^{-1} &= -(N-1) P_{k|k-n}^{-1} + \omega {P_{k|k}^i}^{-1} \\ &+ (1-\omega) {P_{k|k}^j}^{-1} \end{aligned} \tag{10.24}$$

To compute the steady state error covariance of a fused state estimate, subtracting $P_{k|k}^{-1} x_k$, from both sides of (10.23), and substituting (10.22) yields

$$\begin{aligned} P_{k|k}^{-1} (\hat{x}_{k|k} - x_k) &= -P_{k|k}^{-1} x_k - (N-1) P_{k|k-n}^{-1} \hat{x}_{k|k-n} \\ &+ \omega {P_{k|k}^i}^{-1} \hat{x}_{k|k}^i + (1-\omega) {P_{k|k}^j}^{-1} \hat{x}_{k|k}^j \\ &= -(N-1) P_{k|k-n}^{-1} F^n (\hat{x}_{k|k-n} - x_{k-n}) \\ &- P_{k|k}^{-1} x_k - (N-1) P_{k|k-n}^{-1} F^n x_{k-n} \\ &+ P_{k|k}^{-1} [A_n x_{k|k}^i + B_i x_{k|k}^j] \end{aligned} \tag{10.25}$$

Through simple algebra manipulation and substituting (10.20) into (10.25) as:

$$
\begin{aligned}
P_{k|k}^{-1}(\hat{x}_{k|k} - x_k) &= \{-(N-1)P_{k|k-n}^{-1}F^n + P_{k|k}^{-1}A_n\} \\
&\quad \cdot (\hat{x}_{k-n|k-n} - x_{k-n}) + P_{k|k}^{-1}A_n\hat{x}_{k-n} \\
&\quad - P_{k|k}^{-1}x_k - (N-1)P_{k|k-n}^{-1}F^n x_{k-n} \\
&\quad + P_{k|k}^{-1}B_i x_{k|k}^j \\
&= \{-(N-1)P_{k|k-n}^{-1}F^n + P_{k|k}^{-1}A_n\} \\
&\quad \cdot (\hat{x}_{k-n|k-n} - x_{k-n}) + P_{k|k}^{-1}A_n\hat{x}_{k-n} \\
&\quad - (N-1)P_{k|k-n}^{-1}F^n x_{k-n} \\
&\quad + P_{k|k}^{-1}B_i w_{k-n+i} - P_{k|k}^{-1}x_k \\
&\quad + P_{k|k}^{-1}B_i(K_1 F^i x_{k-n}^i + K_2 F^j x_{k-n}^j) \\
&\quad + P_{k|k}^{-1}B_i \sum_{h=1}^{i}(K_1 + K_2) \\
&\quad \cdot F^{i-h}G v_{k-n+h},
\end{aligned}
\tag{10.26}
$$

it has been proven in [558] that A_n satisfies the following identity

$$
A_n = -\sum_{i=1}^{n} B_i K F' + F^n.
\tag{10.27}
$$

Substituting (10.27) and (10.24) into (10.26), we have

$$
\begin{aligned}
P_{k|k}^{-1}(\hat{x}_{k|k} - x_k) &= \{-(N-1)P_{k|k-n}^{-1}F^n + P_{k|k}^{-1}A_n\} \\
&\quad \cdot (\hat{x}_{k-n|k-n} - x_{k-n}) + P_{k|k}^{-1}A_n x_{k-n} \\
&\quad - (N-1)P_{k|k-n}^{-1}F^n x_{k-n} + P_{k|k}^{-1}B_i w_{k-n+i} \\
&\quad - P_{k|k}^{-1}x_k + P_{k|k}^{-1}(F^n - A_n)x_{k-n} \\
&\quad + P_{k|k}^{-1}B_i \sum_{h=1}^{i} F^{i-h}G v_{k-n+h} \\
&= \{-(N-1)P_{k|k-n}^{-1}F^n + P_{k|k}^{-1}A_n\} \\
&\quad \cdot (\hat{x}_{k-n|k-n} - x_{k-n}) + P_{k|k}^{-1}B_i w_{k-n+i} \\
&\quad + (P_{k|k}^{-1}B_i \sum_{h=i}^{n}(K_1 + K_2)F^{h-i} - P_{k|k}^{-1} \\
&\quad \cdot F^{n-i})G v_{k-n+i}
\end{aligned}
\tag{10.28}
$$

Using (10.28), showing a Lyapunov form as follows

$$\Omega_x = C_f \Omega_x C_f' + \Omega_f \tag{10.29}$$

where

$$
\begin{aligned}
C_f &= \lim_{k \to \infty} P_{k|k}(-(N-1)P_{k|k-n}^{-1}F^n + P_{k|k}^{i^{-1}}A_n^i + P_{k|k}^{j^{-1}}A_n^j), \\
\Omega_f &= W_s(k)RW_s(k)' + V_s(k)GQG'V_s(k), \\
W_s(k) &= \lim_{k \to \infty} P_{k|k}P_{k|k}^{-1}B_i, \\
V_s(k) &= \lim_{k \to \infty} P_{k|k}P_{k|k}^{-1}B_i \sum_{h=1}^{n}(K_1 + K_2)F^{h-i} \\
&\quad - P_{k|k}P_{k|k}^{-1}F^{n-i} \tag{10.30}
\end{aligned}
$$

10.4.3   Partial feedback case

In the case of partial feedback, (10.11) and (10.12) can be formulated as follows:

$$
\begin{aligned}
P_{k|k}^{-1}\hat{x}_{k|k} &= P_{k|k-n}^{-1}\hat{x}_{k|k-n} + \omega P_{k|k}^{i^{-1}}\hat{x}_{k|k}^i \\
&\quad - \omega P_{k|k-n}^{i^{-1}}\hat{x}_{k|k-n} + (1-\omega)P_{k|k}^{j^{-1}}\hat{x}_{k|k}^j \\
&\quad - (1-\omega)P_{k|k-n}^{j^{-1}}\hat{x}_{k|k-n} \tag{10.31} \\
P_{k|k}^{-1} &= P_{k|k-n}^{-1} + \omega P_{k|k}^{i^{-1}} - \omega P_{k|k-n}^{i^{-1}} + (1-\omega)P_{k|k}^{j^{-1}} \\
&\quad - (1-\omega)P_{k|k-n}^{j^{-1}} \tag{10.32}
\end{aligned}
$$

Note that changing the value of N does not alter the forms of (10.31) and (10.32) and only the length of the summation item needs to be adjusted. Like the case of complete feedback, there is also a discrete Lyapunov equation,

$$\Omega_x = C_p \Omega_x C_p' + \Omega_p \tag{10.33}$$

where

$$
\begin{aligned}
C_p &= \lim_{k \to \infty} P_{k|k}[P_{k|k}^{i^{-1}}A_n^i + P_{k|k}^{j^{-1}}A_n^j - P_{k|k-n}^{i^{-1}}F^n \\
&\quad - P_{k|k-n}^{j^{-1}}F^n + P_{k|k-n}^{-1}F^n] \tag{10.34}
\end{aligned}
$$

where Ω_p has the same definition of Ω_f as in (10.30).

10.5 Weighted Covariance

According to the standard results of covariance intersection in [579], the weighted covariance at the sensor is:

$$\hat{x}_{k|k} = A_k^i \hat{x}_{k|k}^i + A_k^j \hat{x}_{k|k}^j \tag{10.35}$$

where the weighted matrices of two local estimates are calculated as:

$$A_k^i = (P_{k|k}^j - \Sigma_{k|k}^{j,i})(P_{k|k}^i + P_{k|k}^j - \Sigma_{k|k}^{ij} - \Sigma_{k|k}^{ji})^{-1} \tag{10.36}$$

$$A_k^j = (P_{k|k}^i - \Sigma_{k|k}^{i,j})(P_{k|k}^i + P_{k|k}^j - \Sigma_{k|k}^{ij} - \Sigma_{k|k}^{ji})^{-1} \tag{10.37}$$

and the covariance of the fused estimate is computed as:

$$\begin{aligned}
P_{k|k} &= P_{k|k}^j - (P_{k|k}^j - \Sigma_{k|k}^{j,i})(P_{k|k}^i + P_{k|k}^j - \Sigma_{k|k}^{ij} - \Sigma_{k|k}^{ji})^{-1} \\
&\quad (P_{k|k}^j - \Sigma_{k|k}^{ji})^T.
\end{aligned} \tag{10.38}$$

Alternatively

$$\begin{aligned}
P_{k|k}^{-1} &= (P_{k|k}^j - (P_{k|k}^j - \Sigma_{k|k}^{j,i})(P_{k|k}^i + P_{k|k}^j - \Sigma_{k|k}^{ij} - \Sigma_{k|k}^{ji})^{-1} \\
&\quad (P_{k|k}^j - \Sigma_{k|k}^{ji})^T)^{-1}.
\end{aligned} \tag{10.39}$$

where

$$\begin{aligned}
\Sigma_{1|1}^{i,j} &= (I - K_1^i H_1^i)Q_0(I - K_1^i H_1^i)^T, \\
\Sigma_{k|k}^{i,j} &= (I - K_k^i H_k^i)F_{k-1}\Sigma_{k-1|k-1}^{i,j}F_{k-1}^T(I - K_k^i H_k^i)^T \\
&\quad + (I - K_k^i H_k^i)Q_{k-1}(I - K_k^i H_k^i)^T, \\
\Sigma_{k|k}^{j,i} &= (\Sigma_{k|k}^{i,j})^T
\end{aligned}$$

Multiplying (10.39) with (10.35) gives:

$$\begin{aligned}
P_{k|k}^{-1}\hat{x}_{k|k} &= (P_{k|k}^j - (P_{k|k}^j - \Sigma_{k|k}^{j,i})(P_{k|k}^i + P_{k|k}^j - \Sigma_{k|k}^{ij} \\
&\quad - \Sigma_{k|k}^{ji})^{-1}.(P_{k|k}^j - \Sigma_{k|k}^{ji})^T)^{-1} \\
&\quad (A_k^i \hat{x}_{k|k}^i + A_k^j \hat{x}_{k|k}^j).
\end{aligned} \tag{10.40}$$

10.5.1 Algorithm

For the case of deriving an information-based weighted covariance filter, the target dynamic model of (10.1) and (10.2) will be of the form:

$$x_{k+1} = Fx_k + Gw_k \tag{10.41}$$

$$z_k = H^i x_k + H^j x_k + v^i + v^j \tag{10.42}$$

The key idea of the information matrix filter is to identify the common information shared by estimates that are to be fused, and then removing the information or de-correlation is implemented. It will take into consideration the common information but not the common process noise. Under the assumption of no feedback, the estimation using an information-based filter in the case of a weighted covariance is as follows:

$$
\begin{aligned}
P_{k|k}^{-1}\hat{x}_{k|k} &= P_{k|k-n}^{-1}\hat{x}_{k|k-n} + (P_{k|k}^j - (P_{k|k}^j - \Sigma_{k|k}^{j,i}) \\
&\quad \cdot (P_{k|k}^i + P_{k|k}^j - \Sigma_{k|k}^{ij} - \Sigma_{k|k}^{ji})^{-1}(P_{k|k}^j \\
&\quad - \Sigma_{k|k}^{ji})^T)^{-1} \cdot (A_k^i \hat{x}_{k|k}^i + A_k^j \hat{x}_{k|k}^j) - (P_{k|k-n}^j \\
&\quad - (P_{k|k-n}^j - \Sigma_{k|k-n}^{j,i}) \cdot (P_{k|k-n}^i + P_{k|k-n}^j - \Sigma_{k|k-n}^{ij} \\
&\quad - \Sigma_{k|k-n}^{ji})^{-1} \cdot (P_{k|k-n}^j - \Sigma_{k|k-n}^{ji})^T)^{-1}(A_k^i \hat{x}_{k|k-n}^i \\
&\quad + A_k^j \hat{x}_{k|k-n}^j) \tag{10.43}
\end{aligned}
$$

$$
\begin{aligned}
P_{k|k}^{-1} &= P_{k|k-n}^{-1} + (P_{k|k}^j - (P_{k|k}^j - \Sigma_{k|k}^{j,i}) \cdot (P_{k|k}^i + P_{k|k}^j \\
&\quad - \Sigma_{k|k}^{ij} - \Sigma_{k|k}^{ji})^{-1}(P_{k|k}^j - \Sigma_{k|k}^{ji})^T)^{-1} - (P_{k|k-n}^j \\
&\quad - (P_{k|k-n}^j - \Sigma_{k|k-n}^{j,i}) \cdot (P_{k|k-n}^i + P_{k|k-n}^j - \Sigma_{k|k-n}^{ij} \\
&\quad - \Sigma_{k|k-n}^{ji})^{-1} \cdot (P_{k|k-n}^j - \Sigma_{k|k-n}^{ji})^T)^{-1} \tag{10.44}
\end{aligned}
$$

The n step fusion state prediction and associated covariance are given by:

$$
\hat{x}_{k|k-n} = F^i \hat{x}_{k-n|k-n} + F^j \hat{x}_{k-n|k-n} \tag{10.45}
$$

$$
P_{k+1|k} = F_k^i P_{k|k-1} F_k^{i*} + G_i Q_i G_i^* \tag{10.46}
$$

The fused state estimate and associated covariance depends upon the local estimates as:

$$
\hat{x}_{k|k-n}^i + \hat{x}_{k|k-n}^j = \hat{x}_{k|k-n} \tag{10.47}
$$

$$
P_{k|k-n}^i + P_{k|k-n}^j = P_{k|k-n} \tag{10.48}
$$

10.5.2 Complete feedback case

For the case of complete feedback, a closed form analytical solution of steady fused covariance of information-based covariance intersection filter with N sensors is derived below. From (10.41) and (10.42), it is easy to show that the following two equations hold,

$$x_k = F_k^i x_{k-n} + F_k^j x_{k-n} + \sum_{i=1}^{n} F^{n-i} G v_{k-n+i} \qquad (10.49)$$

$$z_k = H^i F^i x_{k-n} + H^j F^j x_{k-n} + w_{k-n+i}^i + w_{k-n+j}^j$$
$$+ \quad H^i F^i G v_{k-n+i} + H^j F^j G v_{k-n+j} \qquad (10.50)$$

For the local sensors, it is possible to write the weighted covariance as:

$$\hat{x}_{k|k} = P_{k|k}(P_{k|k}^j F \hat{x}_{k|k-n} + (P_{k|k}^j - \Sigma_{k|k}^{ji})(P_{k|k}^i + P_{k|k}^j$$
$$- \quad \Sigma_{k|k}^{ij} - \Sigma_{k|k}^{ji})^{-1}(P_{k|k}^j - \Sigma_{k|k}^{ji})^T)^{-1} P_{k|k}$$
$$\cdot \quad (A_k^i F \hat{x}_{k|k}^i + A_k^j F \hat{x}_{k|k}^j) \qquad (10.51)$$

Using (10.51) and (10.49), we have

$$\hat{x}_{k|k} = A_n P_{k|k} A_k^i F x_{k|k}^i + A_n P_{k|k} A_k^j F x_{k|k}^j \qquad (10.52)$$

where for all $i = 1, ..., n$, we have $A_0 = I$ and

$$A_i = A_{i-1} P_{k|k}(P_{k|k}^j F \hat{x}_{k|k-n}$$
$$+ \quad (P_{k|k}^j - \Sigma_{k|k}^{ji})(P_{k|k}^i + P_{k|k}^j - \Sigma_{k|k}^{ij} - \Sigma_{k|k}^{ji})^{-1}(P_{k|k}^j - \Sigma_{k|k}^{ji})^T)^{-1}$$

Under the assumption of complete feedback, (10.43) and (10.44) can be re-written as:

$$P_{k|k}^{-1} \hat{x}_{k|k} = -(N-1)P_{k|k-n}^{-1} \hat{x}_{k|k-n} + (P_{k|k}^j - (P_{k|k}^j - \Sigma_{k|k}^{j,i})$$
$$\cdot \quad (P_{k|k}^i + P_{k|k}^j - \Sigma_{k|k}^{ij} - \Sigma_{k|k}^{ji})^{-1}.(P_{k|k}^j - \Sigma_{k|k}^{ji})^T)^{-1}$$
$$\cdot \quad (A_k^i \hat{x}_{k|k}^i + A_k^j \hat{x}_{k|k}^j) \qquad (10.53)$$
$$P_{k|k}^{-1} = -(N-1)P_{k|k-n}^{-1} + (P_{k|k}^j - (P_{k|k}^j - \Sigma_{k|k}^{j,i})$$
$$\cdot \quad (P_{k|k}^i + P_{k|k}^j - \Sigma_{k|k}^{ij} - \Sigma_{k|k}^{ji})^{-1}$$
$$\cdot \quad (P_{k|k}^j - \Sigma_{k|k}^{ji})^T)^{-1} \qquad (10.54)$$

To compute the steady state error covariance of the fused state estimate, subtracting $P_{k|k}^{-1}x_k$ from both sides of (10.53) and substituting (10.52) yields

$$
\begin{aligned}
P_{k|k}^{-1}(\hat{x}_{k|k} - x_k) &= -P_{k|k}^{-1}x_k - (N-1)P_{k|k-n}^{-1}\hat{x}_{k|k-n} \\
&\quad - (N-1)P_{k|k-n}^{-1}\hat{x}_{k|k-n} + (P_{k|k}^j - (P_{k|k}^j \\
&\quad - \Sigma_{k|k}^{j,i}).(P_{k|k}^i + P_{k|k}^j - \Sigma_{k|k}^{ij} - \Sigma_{k|k}^{ji})^{-1} \\
&\quad \cdot (P_{k|k}^j - \Sigma_{k|k}^{ji})^T)^{-1}.(A_k^i\hat{x}_{k|k}^i + A_k^j\hat{x}_{k|k}^j) \\
&= -(N-1)P_{k|k-n}^{-1}F^n(\hat{x}_{k-n|k-n} - x_{k-n}) \\
&\quad - P_{k|k}^{-1}x_k - (N-1)P_{k|k-n}^{-1}F^n x_{k-n} \\
&\quad + P_{k|k}^{-1}(A_n P_{k|k}A_k^i F x_{k|k}^i \\
&\quad + A_n P_{k|k}A_k^j F\hat{x}_{k|k}^j)
\end{aligned}
\tag{10.55}
$$

Through simple algebra manipulations and substituting (10.50), we can re-write (10.55) as

$$
\begin{aligned}
P_{k|k}^{-1}(\hat{x}_{k|k} - x_k) &= (-(N-1)P_{k|k-n}^{-1}F^n + P_{k|k}^{-1}A_n P_{k|k}A_k^i F \\
&\quad + P_{k|k}^{-1}A_n P_{k|k}A_k^j F).(\hat{x}_{k-n|k-n} - \hat{x}_{k|k}^i \\
&\quad - \hat{x}_{k|k}^j) + P_{k|k}^{-1}A_n P_{k|k}A_k^i F\hat{x}_{k-n}^i + P_{k|k}^{-1} \\
&\quad \cdot A_n P_{k|k}A_k^j F\hat{x}_{k|k-n}^j - P_{k|k}^{-1}x_k \\
&\quad - (N-1)P_{k|k-n}^{-1}F^n x_{k-n}
\end{aligned}
\tag{10.56}
$$

Using (10.56), showing a Lyapunov form as follows:

$$
\Omega_x = C_f \Omega_x C_f' + \Omega_f
\tag{10.57}
$$

where

$$
\begin{aligned}
C_f &= \lim_{k \to \infty} P_{k|k}(-(N-1)P_{k|k-n}^{-1}F^n + P_{k|k}^{-1}A_n \\
&\quad \cdot P_{k|k}A_k^i F + P_{k|k}^{-1}A_n P_{k|k}A_k^j F) \\
\Omega_f &= W_s(k)RW_s(k)', \\
W_s(k) &= \lim_{k \to \infty} P_{k|k}P_{k|k}^{-1}A_n P_{k|k}(A_k^i + A_k^j)
\end{aligned}
\tag{10.58}
$$

10.5.3 Partial feedback case

In the case of partial feedback, (10.43) and (10.44) can be formulated as follows:

$$
\begin{aligned}
P_{k|k}^{-1}\hat{x}_{k|k} &= P_{k|k-n}^{-1}\hat{x}_{k|k-n} + (P_{k|k}^{j} - (P_{k|k}^{j} - \Sigma_{k|k}^{j,i}) \\
&\quad\cdot (P_{k|k}^{i} + P_{k|k}^{j} - \Sigma_{k|k}^{ij} - \Sigma_{k|k}^{ji})^{-1}(P_{k|k}^{j} \\
&\quad- \Sigma_{k|k}^{ji})^{T})^{-1}.(A_{k}^{i}\hat{x}_{k|k}^{i} + A_{k}^{j}\hat{x}_{k|k}^{j}) - (P_{k|k-n}^{j} \\
&\quad- (P_{k|k-n}^{j} - \Sigma_{k|k-n}^{j,i}).(P_{k|k-n}^{i} + P_{k|k-n}^{j} - \Sigma_{k|k-n}^{ij} \\
&\quad- \Sigma_{k|k-n}^{ji})^{-1}.(P_{k|k-n}^{j} - \Sigma_{k|k-n}^{ji})^{T})^{-1}(A_{k}^{i}\hat{x}_{k|k-n} \\
&\quad+ A_{k}^{j}\hat{x}_{k|k-n})
\end{aligned}
\tag{10.59}
$$

$$
\begin{aligned}
P_{k|k}^{-1} &= P_{k|k-n}^{-1} + (P_{k|k}^{j} - (P_{k|k}^{j} - \Sigma_{k|k}^{j,i}).(P_{k|k}^{i} + P_{k|k}^{j} \\
&\quad- \Sigma_{k|k}^{ij} - \Sigma_{k|k}^{ji})^{-1}(P_{k|k}^{j} - \Sigma_{k|k}^{ji})^{T})^{-1} \\
&\quad- (P_{k|k-n}^{j} - (P_{k|k-n}^{j} - \Sigma_{k|k-n}^{ij} \\
&\quad- \Sigma_{k|k-n}^{ji})^{-1}.(P_{k|k-n}^{j} - \Sigma_{k|k-n}^{ji})^{T})^{-1}
\end{aligned}
\tag{10.60}
$$

Note that changing the value of N does not alter the forms of (10.59) and (10.60) and only the length of the summation item needs to be adjusted. Like the case of complete feedback, there is also a discrete Lyapunov equation,

$$
\Omega_x = C_p\Omega_x C_p' + \Omega_p
\tag{10.61}
$$

where

$$
\begin{aligned}
C_p &= \lim_{k\to\infty} P_{k|k}[P_{k|k}^{i^{-1}} A_{n}^{i} P_{k|k} A_{k}^{i} F + P_{k|k}^{j^{-1}} A_{n}^{j} P_{k|k} A_{k}^{j} F \\
&\quad- P_{k|k-n}^{i^{-1}} F^{n} - P_{k|k-n}^{j^{-1}} F^{n} + P_{k|k-n}^{-1} F^{n}]
\end{aligned}
\tag{10.62}
$$

where Ω_p has the same definition of Ω_f as in (10.58).

10.6   Kalman-Like Particle Filter

In this section, we will derive an information-based Kalman-like particle filter, where the simple Kalman-like particle filter is expressed in [581]. A question arises here as to why the Kalman-like particle filter has been preferred to a basic Kalman filter. The justification for the approach w.r.t the filter is given in [581]; moreover, it is preferred here to the basic Kalman filter because of the following. (See Figure 10.2 for the comparison of estimates of a basic Kalman filter and a Kalman-like particle filter. See Figure 10.2 where it can be seen, that the mean square error is reduced in fewer iterations for a particle filter as compared to a regular Kalman filter):

According to the standards results of the Kalman-like particle filter in [581], the Kalman-like particle filter at a sensor is:

$$
\begin{aligned}
\hat{x}_{k|k} &= \hat{x}_{k|k-1} + \frac{P_k H_k^T}{H_k P_k H_k^T + \sigma_v^2}(y_k - H_k \hat{x}_{k|k-1}) \\
&= (I - \frac{P_k H_k^T}{H_k P_k H_k^T + \sigma_v^2} H_k)\hat{x}_{k|k-1} + \\
&+ \frac{P_k H_k^T}{H_k P_k H_k^T + \sigma_v^2} y_k
\end{aligned}
\tag{10.63}
$$

with covariance of the filtering error given by

$$
\begin{aligned}
P_{k|k} &= (I - \frac{P_k H_k^T}{H_k P_k H_k^T + \sigma_v^2} H_k)P_{k|k-1} \\
P_{k|k-1}^{-1} &= P_{k|k}^{-1}(I - \frac{P_k H_k^T}{H_k P_k H_k^T + \sigma_v^2} H_k)
\end{aligned}
\tag{10.64}
$$

or

$$
P_{k|k}^{-1} = P_{k|k-1}^{-1} + P_{k|k}^{-1}\frac{P_k H_k^T H_k}{H_k P_k H_k^T + \sigma_v^2}
\tag{10.65}
$$

Thus substituting (10.64) into (10.63) yields

$$
\begin{aligned}
P_{k|k}^{-1}\hat{x}_{k|k} &= P_{k|k-1}^{-1}\hat{x}_{k|k-1} + P_{k|k}^{-1} \\
&\quad (\frac{P_k H_k^T H_k}{H_k P_k H_k^T + \sigma_v^2})\hat{x}_{k|k}.
\end{aligned}
\tag{10.66}
$$

10.6.1 Algorithm

The key idea of the information matrix filter is to identify the common information shared by the estimates that are to be fused, and then removing the information or de-correlation is implemented. It will take into consideration the common information but not the common process noise. Under the assumption of no feedback, the estimation using an information-based filter in the case of the Kalman-like particle

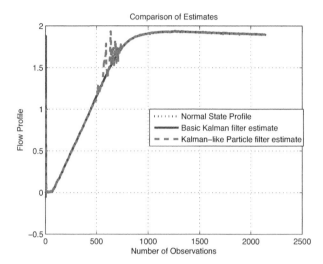

Figure 10.1: Estimates of Kalman-like particle and basic Kalman filter.

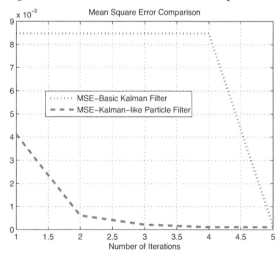

Figure 10.2: Mean square error for Kalman-like particle and basic filters.

filter is as follows:

$$
\begin{aligned}
P_{k|k}^{-1}\hat{x}_{k|k} &= P_{k|k-1}^{-1}\hat{x}_{k|k-1} + P_{k|k}^{j^{-1}} \\
&\quad (\frac{P_k^j H_k^{j^T} H_k^j}{H_k^j P_k^j H_k^{j^T} + \sigma_v^2})\hat{x}_{k|k}^j \\
&\quad - P_{k|k-n}^{j^{-1}}(\frac{P_k^j H_k^{j^T} H_k^j}{H_k^j P_k^j H_k^{j^T} + \sigma_v^2})\hat{x}_{k|k-n}^j \qquad (10.67) \\
P_{k|k}^{-1} &= P_{k|k-1}^{-1} + P_{k|k}^{j^{-1}} \\
&\quad (\frac{P_k^j H_k^{j^T} H_k^j}{H_k^j P_k^j H_k^{j^T} + \sigma_v^2}). \\
&\quad - P_{k|k-n}^{j^{-1}}(\frac{P_k^j H_k^{j^T} H_k^j}{H_k^j P_k^j H_k^{j^T} + \sigma_v^2}) \qquad (10.68)
\end{aligned}
$$

The n step fusion state prediction and associated covariance are given by:

$$
x_{k|k-n} = F^n \hat{x}_{k-n|k-n} \qquad (10.69)
$$

$$
P_{k|k-n} = F^n P_{k-n|k-n} F^{n*} + F^{n-i} G Q G^* F^{n-i^*} \qquad (10.70)
$$

where the n step fusion state prediction and associated covariance is written as:

$$
\hat{x}_{k|k-n}^j = \hat{x}_{k|k-n} \qquad (10.71)
$$

$$
P_{k|k-n}^j = P_{k|k-n} \qquad (10.72)
$$

10.6.2   Complete feedback case

For the case of complete feedback, a closed form analytical solution of steady fused covariance of an information-based Kalman-like particle filter with N sensors is derived below. From (10.1) and (10.2), it is easy to show that the following two equations hold,

$$
x_k = F_k^i x_{k-n} + F^{n-i} G v_{k-n+i} \qquad (10.73)
$$

$$
z_{k-n+i}^j = H^j F^j x_{k-n} + w_{k-n+i}^j
$$

$$
+ \sum_{h=1}^{i} H^j F^{i-h} G v_{k-n+h} \qquad (10.74)
$$

For the two local sensors in the Kalman-like particle filter, it is possible to write as follows:

$$
\begin{aligned}
\hat{x}_{k|k}^{j} &= P_{k|k}P_{k|k}^{j^{-1}}F\hat{x}_{k|k-1}^{j} + P_{k|k}^{j}P_{k|k}^{j^{-1}} \\
&\quad \frac{P_k^j H_k^{j^T} H_k^j}{H_k^j P_k^j H_k^j + \sigma_v^2}\hat{x}_{k|k}.
\end{aligned}
\tag{10.75}
$$

Utilizing (10.71) and (10.75), we have

$$
\hat{x}_{k|k}^{j} = A_n^j \hat{x}_{k-n|k-n} + \sum_{i=1}^{n} B_i^j \hat{x}_{k|k}
\tag{10.76}
$$

where, $\forall\, i = 1, ..., n$, we have $A_0^j = I$, $A_i^j = P_{k-i+1|k-i+1}P_{k-i+1|k-i+1}^{j^{-1}}F$, $B^j = A_{i-1}^j P_{k-i+1|k-i+1}^{j} P_{k-i+1|k-i+1}^{j^{-1}} (P_k^j H_k^{j^T} H_k^j/(H_k^j P_k^j H_k^j + \sigma_v^2))F$.

Under the assumption of complete feedback, (10.67) and (10.68) can be re-written as:

$$
\begin{aligned}
P_{k|k}^{-1}\hat{x}_{k|k} &= -(N-1)P_{k|k-n}^{-1}\hat{x}_{k|k-n} \\
&\quad + \sum_{j=1}^{N} P_{k|k}^{j^{-1}} \frac{P_k H_k^T H_k}{H_k P_k H_k^T + \sigma_v^2}\hat{x}_{k|k}^{j}
\end{aligned}
\tag{10.77}
$$

$$
\begin{aligned}
P_{k|k}^{-1} &= -(N-1)P_{k|k-n}^{-1} \\
&\quad + \sum_{j=1}^{N} P_{k|k}^{j^{-1}} \frac{P_k H_k^T H_k}{H_k P_k H_k^T + \sigma_v^2}
\end{aligned}
\tag{10.78}
$$

To compute the steady state error covariance of a fused state estimate, subtracting $P_{k|k}^{-1}x_k$ from both sides of (10.78) and substituting (10.76) yields

$$
\begin{aligned}
P_{k|k}^{-1}(\hat{x}_{k|k} - x_k) &= -P_{k|k}^{-1}x_k - (N-1)P_{k|k-n}^{-1}\hat{x}_{k|k-n} \\
&\quad + \sum_{j=1}^{N} P_{k|k}^{j^{-1}} \frac{P_k H_k^T H_k}{H_k P_k H_k^T + \sigma_v^2}\hat{x}_{k|k}^{j} \\
&= -(N-1)P_{k|k-n}^{-1}F^n(\hat{x}_{k|k-n} - x_{k-n}) \\
&\quad - P_{k|k}^{-1}x_k - (N-1)P_{k|k-n}^{-1}F^n x_{k-n} \\
&\quad + \sum_{j=1}^{N} P_{k|k}^{j^{-1}} \frac{P_k H_k^T H_k}{H_k P_k H_k^T + \sigma_v^2} \\
&\quad [A_n^j \hat{x}_{k-n|k-n} + \sum_{i=1}^{n} B_i^j x_{k|k}].
\end{aligned}
\tag{10.79}
$$

Through simple algebra manipulation and substituting (10.75), we can re-write (10.79) as:

$$
\begin{aligned}
P_{k|k}^{-1}(\hat{x}_{k|k} - x_k) &= (-(N-1)P_{k|k-n}^{-1}F^n \\
&+ \sum_{j=1}^{N} P_{k|k}^{j-1}\left(\frac{P_k H_k^T H_k}{H_k P_k H_k^T + \sigma_v^2}\right)A_n^j) \\
&\times (\hat{x}_{k-n|k-n} - x_{k-n}) + \sum_{j=1}^{N} P_{k|k}^{j-1} \\
&\times (\frac{P_k H_k^T H_k}{H_k P_k H_k^T + \sigma_v^2})A_n^j x_{k-n} - P_{k|k}^{-1}x_k \\
&- (N-1)P_{k|k-n}^{-1}F^n x_{k-n} \\
&+ (\sum_{j=1}^{N} P_{k|k}^{j-1} \sum_{i=1}^{n} B_i^j x_{k|k})
\end{aligned}
\tag{10.80}
$$

Using (10.80), showing a Lyapunov form as follows:

$$
\Omega_x = C_f \Omega_x C_f' + \Omega_f
\tag{10.81}
$$

where

$$
\begin{aligned}
C_f &= \lim_{k \to \infty} P_{k|k}(-(N-1)P_{k|k-n}^{-1}F^n + \sum_{j=1}^{n} P_{k|k}^{-1} \\
&\times \frac{P_k H_k^T H_k}{H_k P_k H_k^T + \sigma_v^2}A_n^j), \\
\Omega_f &= \sum_{j=1}^{N}\sum_{k=1}^{n} W_s^j(k)R^j W_s^j(k)', \\
W_s^j(k) &= \lim_{k \to \infty} P_{k|k} P_{k|k}^{j-1} B_i^j
\end{aligned}
\tag{10.82}
$$

10.6.3 Partial feedback case

In the case of partial feedback, (10.67) and (10.68) can be formulated as follows:

$$
\begin{aligned}
P_{k|k}^{-1}\hat{x}_{k|k} &= P_{k|k-n}^{-1}\hat{x}_{k|k-n} \\
&+ \sum_{j=1}^{N} P_{k|k}^{j^{-1}} \frac{P_k H_k^T H_k}{H_k P_k H_k^T + \sigma_v^2} \hat{x}_{k|k}^{j} \\
&- P_{k|k-n}^{j^{-1}} \frac{P_k^j H_k^{j^T} H_k^j}{H_k^j P_k^j H_k^{j^T} + \sigma_v^2} \hat{x}_{k|k-n}^{j} \qquad (10.83)
\end{aligned}
$$

$$
\begin{aligned}
P_{k|k}^{-1} &= P_{k|k-n}^{-1} + \sum_{j=1}^{N} P_{k|k}^{j^{-1}} \left(\frac{P_k H_k^T H_k}{H_k P_k H_k^T + \sigma_v^2} \right) \\
&- P_{k|k-n}^{j^{-1}} \left(\frac{P_k^j H_k^{j^T} H_k^j}{H_k^j P_k^j H_k^{j^T} + \sigma_v^2} \right) \qquad (10.84)
\end{aligned}
$$

Note that changing the value of N does not alter the forms of (10.83) and (10.84) and only the length of the summation item needs to be adjusted. Like the case of complete feedback, there is also a discrete Lyapunov equation,

$$
\Omega_x = C_p \Omega_x C_p' + \Omega_p \qquad (10.85)
$$

where

$$
\begin{aligned}
C_p &= \lim_{k \to \infty} P_{k|k} \Big[\sum_{j=1}^{n} \Big(P_{k|k}^{-1} \cdot \frac{P_k H_k^T H_k}{H_k P_k H_k^T + \sigma_v^2} A_n^j - P_{k|k-n}^{j^{-1}} F^n \Big) \\
&+ P_{k|k-n}^{-1} F^n \Big] \qquad (10.86)
\end{aligned}
$$

where Ω_p has the same definition of Ω_f as in (10.82).

10.7 Measurement Fusion Algorithm

The information captured in each of the information-based filter cases is designed for a distributed structure. The idea is taken from the fusion methods in [578].

Suppose there is X number of sensors. For every measurement coming from these sensors that is received in the fusion center, there is a corresponding estimation based solely on these individual sensors. The information can be structured as estimated information or prior estimated information in the following two ways, which are the measurement fusion method and the state-vector fusion method as shown in Figure 10.3 and Figure 10.4, respectively.

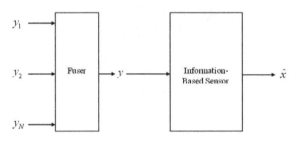

Figure 10.3: Measurement fusion for an information-based sensor.

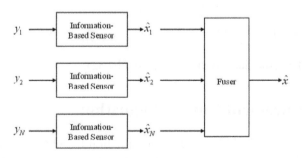

Figure 10.4: State vector fusion for an information-based sensor.

Measurement fusion method   The measurement fusion method integrates the sensor measurement information by augmenting the observation vector as follows:

$$y(k) \;=\; y^{(mf)}(k) = [y_1(k) \;\ldots\; y_N(k)]^T \tag{10.87}$$

$$C(k) \;=\; C^{(mf)}(k) = [C_1(k) \;\ldots\; C_N(k)]^T \tag{10.88}$$

$$R(k) \;=\; R^{(mf)}(k) = diag[R_1(k) \;\ldots\; R_N(k)] \tag{10.89}$$

where the superscript mf stands for the measurement fusion.

State-vector Fusion Method   The state-vector fusion method obtains the fused measurement information by weighted observation as follows:

$$y(k) \;=\; y^{(sf)}(k) = [\sum_{j=1}^{N} R_j^{-1}(k)]^{-1} \sum_{j=1}^{N} R_j^{-1}(k)y_j(k) \tag{10.90}$$

$$C(k) \;=\; C^{(sf)}(k) = [\sum_{j=1}^{N} R_j^{-1}(k)]^{-1} \sum_{j=1}^{N} R_j^{-1}(k)C_j(k) \tag{10.91}$$

$$R(k) \;=\; R^{(sf)}(k) = [\sum_{j=1}^{N} R_j^{-1}(k)]^{-1} \tag{10.92}$$

where the superscript sf stands for state-vector fusion.

10.8   Equivalence of Two Measurement Fusion Methods

Comparing (10.87)–(10.89) with (10.90)–(10.92), we note that the treatment in the measurement fusion schemes is quite different. With reference to [578], we will show here that there exists a functional equivalence between the two methods.

Theorem 10.8.1 *If the N sensors used for data fusion with different and independent noise characteristics have identical measurement matrices, i.e., $C_1(k) = C_2(k) = \ldots = C_N(k)$, then the measurement fusion method is functionally equivalent to the state-vector fusion.*

Proof 10.8.1 *The following formula in linear algebra will be used to cope with the inversion of matrices:*

$$\begin{bmatrix} A_1 & A_2 \\ A_3 & A_4 \end{bmatrix}^{-1} = \begin{bmatrix} B_1 & B_2 \\ B_3 & B_4 \end{bmatrix} \tag{10.93}$$

$$
\begin{aligned}
(A + HBH^t)^{-1} &= A^{-1} - A^{-1}H(B^{-1} \\
&+ H^T A^{-1} H)^{-1} H^T A^{-1}
\end{aligned} \tag{10.94}
$$

where $B_1 = (A_1 - A_2 A_4^{-1} A_3)^{-1}$, $B_2 = -B_1 A_2 A_4^{-1}$, $B_3 = -A_4^{-1} A_3 B_1$, and $B_4 = A_4^{-1} + A_4^{-1} A_3 B_1 A_2 A_4^{-1}$. If the information-based covariance intersection filter is used, in order to demonstrate the functional equivalence of the two measurement fusion methods, we only need to check whether the terms $(K_1 + K_2)C_k$ and $(K_1 + K_2)(k)y(k)$ in the measurement fusion method are functionally equivalent to those in the state-vector fusion method. Alternatively, if the information filter is used, then we need to check the functional equivalence between terms $C^T(k)R^{-1}(k)C(k)$ and $C^T(k)R^{-1}(k)y(k)$ in both methods.

Consider the case when the information-based covariance intersection filter is applied, and $(K_1 + K_2)^{(mf)}$ is:

$$
\begin{aligned}
(K_1 + K_2)^{(mf)}(k) = {} & \\
& \omega P^{(mf)}(k|k-1)(C^{(sf)})^T (C(k)P^i(k|k-1)C(k) \\
+ {} & R(k))^{-1} + (1-\omega)P^{(mf)}(k|k-1)(C^{(sf)})^T \\
\times {} & (C(k)P^j(k|k-1)C(k) + R(k))^{-1}
\end{aligned} \tag{10.95}
$$

where $\Xi_i^{(mf)} = (C(k)P^i(k|k-1)C(k) + R(k))^{-1}$ and $\Xi_j^{(mf)} = (C(k)P^j(k|k-1)C(k) + R(k))^{-1}$.

$$
\begin{aligned}
(K_1 + K_2)^{(mf)}(k) = {} & \\
& \omega P^{(mf)}(k|k-1)(C^{(sf)})^T \\
\times {} & \begin{bmatrix} R_1 + \Xi_i^{(mf)} & \Xi_i^{(mf)} \\ \Xi_i^{(mf)} & R_2 + \Xi_i^{(mf)} \end{bmatrix}^{-1} \\
+ {} & (1-\omega)P^{mf}(k|k-1)(C^{(sf)})^T \\
\times {} & \begin{bmatrix} R_1 + \Xi_j^{(mf)} & \Xi_j^{(mf)} \\ \Xi_j^{(mf)} & R_2 + \Xi_j^{(mf)} \end{bmatrix}^{-1}
\end{aligned} \tag{10.96}
$$

$$(K_1 + K_2)^{(mf)}(k) =$$
$$\omega P^{(mf)}(k|k-1)(C)^T[(R_2 + \Xi_i^{(mf)})^{-1} \times R_2$$

$$\times \overbrace{[R_1 + \Xi_i^{(mf)} - \Xi_i^{mf}(R_2 + \Xi_i^{(mf)})^{(mf)}]^{-1}}^{B_1},$$

$$\times \overbrace{(R_2 + \Xi_i^{(mf)})^{-1}}^{A_4} - \overbrace{(R_2 + \Xi_i^{(mf)})^{-1}}^{A_4}$$

$$\times \overbrace{R_2}^{A_3}\overbrace{[R_1 + \Xi_i^{(mf)} - \Xi_i^{(mf)}(R_2 + \Xi_i^{(mf)})^{-1}\Xi_i^{(mf)}]^{-1}}^{B_1}$$

$$\times \overbrace{\Xi_i^{(mf)}}^{A_2}\overbrace{(R_2 + \Xi_i^{(mf)})^{-1}}^{A_4} + (1 - \omega)P^{(mf)}(k|k-1)C^T$$

$$\times [(R_2 + \Xi_j^{(mf)})^{-1} \times R_2\overbrace{[R_1 + \Xi_j^{(mf)}]}^{B_1} - \Xi_j^{(mf)}(R_2$$

$$+ \Xi_j^{(mf)})^{-1}\Xi_j^{(mf)}]^{-1} \times \overbrace{(R_2 + \Xi_j^{(mf)})^{-1}}^{A_4}$$

$$- \overbrace{(R_2 + \Xi_j^{(mf)})^{-1}}^{A_4}$$

$$\times \overbrace{R_2}^{A_3}\overbrace{[R_1 + \Xi_j^{(mf)} - \Xi_i^{(mf)}(R_2 + \Xi_j^{(mf)})^{-1}\Xi_j^{(mf)}]^{-1}}^{B_1}$$

$$\times \overbrace{\Xi_j^{(mf)}}^{A_2}\overbrace{(R_2 + \Xi_j^{(mf)})^{-1}}^{A_4} \tag{10.97}$$

where as proved in [578],

$$(R_2 + \Xi_i^{(mf)})^{-1}R_2[R_1 + \Xi^{(mf)}$$
$$- \Xi_i^{(mf)}(R_2 + \Xi_i^{(mf)})^{-1}\Xi_i^{(mf)}]^{-1}$$
$$= [\Xi_i^{(mf)} + R_1(R_1 + R_2)^{-1}R_2]^{-1}R_2(R_1 + R_2)^{-1} \tag{10.98}$$

and

$$(R_2 + \Xi_i^{(mf)})^{-1} - (R_2 + \Xi_i^{(mf)})^{-1}R_2 \times [R_1 + \Xi_i^{(mf)} - \Xi^{(mf)}$$
$$\times (R_2 + \Xi_i^{(mf)})^{-1}\Xi_i^{(mf)}]^{-1}\Xi_i^{(mf)}(R_2 + \Xi_i^{(mf)})^{-1}$$
$$= [\Xi_i^{(mf)} + R_1(R_1 + R_2)^{-1}R_2]^{-1}R_1(R_1 + R_2)^{-1} \tag{10.99}$$

likewise for $\Xi_j^{(mf)}$ from (10.98) and (10.99). Based on (10.97)–(10.99), we have

$$
\begin{aligned}
(K_1+ \ K_2 \ )^{(mf)}(k) = & \\
& \omega P^{(mf)}(k|k-1)C^T \times [CP^{i^{(mf)}}(k|k-1)C^T \\
+ \ & R_1(R_1+R_2)^{-1}R_2]^{-1} \times [R_2(R_1+R_2)^{-1}, \\
& R_1(R_1+R_2)^{-1}] + (1-\omega)P^{(mf)}(k|k-1)C^T \\
\times \ & [CP^{j^{(mf)}}(k|k-1)C^T \\
+ \ & R_1(R_1+R_2)^{-1}R_2]^{-1} \times [R_2(R_1+R_2)^{-1}, \\
& R_1(R_1+R_2)^{-1}]
\end{aligned} \tag{10.100}
$$

$$
\begin{aligned}
(K_1+ \ K_2 \ )^{(mf)}(k)C^{(mf)}(k) = & \\
& \omega P^{(mf)}(k|k-1)C^T \times [CP^{i^{(mf)}}(k|k-1) \\
\times \ & C^T + R_1(R_1+R_2)^{-1}R_2]^{-1}C \\
+ \ & (1-\omega)P^{(mf)}(k|k-1)C^T \\
\times \ & [CP^{j^{(mf)}}(k|k-1)C^T \\
+ \ & R_1(R_1+R_2)^{-1}R_2]^{-1}C
\end{aligned} \tag{10.101}
$$

$$
\begin{aligned}
(K_1+ \ K_2 \ )^{(mf)}(k)y^{(mf)}(k) = & \\
& \omega P^{(mf)}(k|k-1)C^T \times [CP^{i^{(mf)}}(k|k-1) \\
\times \ & C^T + R_1(R_1+R_2)^{-1}R_2]^{-1} \\
\times \ & [R_2(R_1+R_2)^{-1} \\
\times \ & y_1(t) + R_1(R_1+R_2)^{-1}y_2(t)] + (1-\omega) \\
\times \ & P^{(mf)}(k|k-1)C^T \times [CP^{j^{(mf)}}(k|k-1)C^T \\
+ \ & R_1(R_1+R_2)^{-1}R_2]^{-1} \times [R_2(R_1+R_2)^{-1} \\
\times \ & y_1(t) + R_1(R_1+R_2)^{-1}y_2(t)]
\end{aligned} \tag{10.102}
$$

If $C_1 = C_2 = C$, then $C^{(II)} = C$, and we obtain the Kalman gain in the state-vector method as follows:

$$
\begin{aligned}
(K_1+ \ K_2 \ )^{(sf)}(k) = & \\
& \omega P^{(sf)}(k|k-1)C^T \times [CP^{i^{(sf)}}(k|k-1)C^T \\
+ \ & R_1(R_1+R_2)^{-1}R_2]^{-1} + (1-\omega)P^{(sf)}(k|k-1)C^T \\
\times \ & [CP^{j^{(sf)}}(k|k-1)C^T \\
+ \ & R_1(R_1+R_2)^{-1}R_2]^{-1}
\end{aligned} \tag{10.103}
$$

and we can derive the terms $K^{(sf)}(k)C^{(sf)}(k)$ *and* $K^{(sf)}(k)y^{(sf)}(k)$:

$$
\begin{aligned}
(K_1 + K_2)^{(sf)}(k)C^{(sf)}(k) =& \\
& \omega P^{(sf)}(k|k-1)C^T \times [CP^{i^{(sf)}}(k|k-1) \\
\times\;& C^T + R_1(R_1+R_2)^{-1}R_2]^{-1}C + (1-\omega) \\
\times\;& P^{(sf)}(k|k-1)C^T \times [CP^{j^{(sf)}}(k|k-1)C^T \\
+\;& R_1(R_1+R_2)^{-1}R_2]^{-1}C
\end{aligned}
\tag{10.104}
$$

$$
\begin{aligned}
(K_1 + K_2)^{(sf)}(k)y^{(sf)}(k) =& \\
& \omega P^{(sf)}(k|k-1)C^T \times [CP^{i^{(sf)}}(k|k-1)C^T \\
+\;& R_1(R_1+R_2)^{-1}R_2]^{-1} \times [R_2(R_1+R_2)^{-1} \\
\times\;& y_1(t) + R_1(R_1+R_2)^{-1}y_2(t)] + (1-\omega) \\
\times\;& P^{(sf)}(k|k-1)C^T \times [CP^{j^{(sf)}}(k|k-1)C^T \\
+\;& R_1(R_1+R_2)^{-1}R_2]^{-1} \times [R_2(R_1+R_2)^{-1} \\
\times\;& y_1(t) + R_1(R_1+R_2)^{-1}y_2(t)]
\end{aligned}
\tag{10.105}
$$

Note that (10.101) and (10.104) are in the same form and that (10.102) and (10.105) are also in the same form. Therefore, with the same initial conditions, i.e., $P^{(mf)}(0|0) = P^{(sf)}(0|0)$ and $\hat{x}^{(mf)}(0|0) = \hat{x}^{(sf)}(0|0)$, the Kalman filters based on the observation information generated by (10.87)–(10.89) and (10.90)–(10.92), irrespectively, will result in the same state estimate $\hat{x}(k|k)$. This means that the two measurement fusion methods are functionally equivalent in the sensor-to-sensor case.

Now, consider the case when the information filter is applied.

From (10.87)–(10.92), it is easy to prove the following equalities:

$$[C^{(mf)}(k)]^T [R^{(mf)}(k)]^{-1} C^{(mf)}(k)$$

$$= \sum_{j=1}^{N} C_j^T R_j^{-1} C_j \tag{10.106}$$

$$[C^{(mf)}(k)]^T [R^{(mf)}(k)]^{-1} y^{(mf)}(k)$$

$$= \sum_{j=1}^{N} C_j^T R_j^{-1} y_j \tag{10.107}$$

$$[C^{(sf)}(k)]^T [R^{(sf)}(k)]^{-1} C^{(sf)}(k) = [(\sum_{j=1}^{N} R_j^{-1})^{-1}$$

$$\times \sum_{j=1}^{N} R_j^{-1} C_j]^T \sum_{j=1}^{N} R_j^{-1} C_j \tag{10.108}$$

$$[C^{(sf)}(k)]^T [R^{(sf)}(k)]^{-1} y^{(sf)}(k) = [(\sum_{j=1}^{N} R_j^{-1})^{-1}$$

$$\times \sum_{j=1}^{N} R_j^{-1} C_j]^T \sum_{j=1}^{N} R_j^{-1} y_j \tag{10.109}$$

If $C_j = C, j = 1, 2, ..., N$, then we have

$$[C^{(mf)}(k)]^T [R^{(mf)}(k)]^{-1} C^{(mf)}(k)$$
$$= [C^{(sf)}(k)]^T [R^{(sf)}(k)]^{-1} C^{(sf)}(k) \tag{10.110}$$
$$[C^{(mf)}(k)]^T [R^{(mf)}(k)]^{-1} y^{(mf)}(k)$$
$$= [C^{(sf)}(k)]^T [R^{(sf)}(k)]^{-1} y^{(sf)}(k) \tag{10.111}$$

Remark 10.8.1 *The functional equivalence is proved here by considering the gain K as the center of existence for all the calculations, which can be the case for an information-based weighted covariance filter too, but not for information-based Kalman-like particle filters where the gain K is not present.*

10.9 Tracking Level Cases

In the sequel, we provide simulation studies by means of two examples:

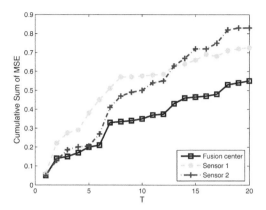

Figure 10.5: Fusion structure with feedback: covariance intersection.

10.9.1 Illustrative example 1

It becomes increasingly apparent that the sensor data fusion, in general, may be carried out at either the measurement level or the track level. For the purpose of illustration, the track level fusion is adopted and for the performance evaluation, three feedback types are considered in the sequel which are:

- Complete feedback, that is, feedback of both state vector and covariance,

- Partial feedback, meaning only the state vector and

- No feedback.

Some simulation of the system structure with feedback is illustrated in Figures 10.5–10.7.

10.9.2 Illustrative example 2

Consider the dynamical model (10.1) and (10.2) where

$$
F = \begin{bmatrix} 1 & 0 & 1 & 0 \\ 0 & 1 & 0 & 1 \\ 0 & 0 & 1 & 0 \\ 0 & 0 & 0 & 1 \end{bmatrix}, \; G = \begin{bmatrix} 1 & 0 \\ 1 & 0 \\ 0 & 1 \\ 0 & 1 \end{bmatrix}
$$

$$
H^j = \begin{bmatrix} 1 & 0 & 0 & 0 \\ 0 & 1 & 0 & 0 \end{bmatrix}
$$

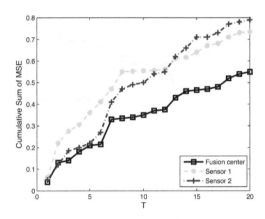

Figure 10.6: Fusion structure with feedback: information matrix.

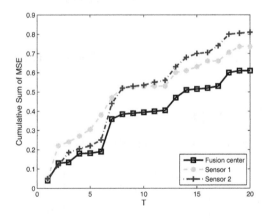

Figure 10.7: Fusion structure with feedback: weighted covariance.

Here the state vector x_k stands for $[z_k, \ y_k, \ \dot{z}_k, \ \dot{y}_k]$ with $z_k, \ y_k, \ \dot{z}_k, \ \dot{y}_k$ representing the position and velocity of the $Z-$ and $Y-$ directions respectively. Let the initial state vector x_o equal to $[10, \ 10, \ 1, \ 0]$ with v_k being a zero-mean white Gaussian noise with covariance $V = 1 * I$, and I a 2×2 identity matrix. We suppose that two local sensors observe same the flying object synchronously, measurements are in Cartesian coordinates, the processing and communication delay between local sensors and the fusion center are ignored, and w^j are zero-mean white Gaussian noise with covariance W^j, which is also independent of v_k with $W^1 = 1.1 * I$, $R^2 = 1.2 * I$. The scalar ω is set at 0.5.

In what follows, the Monte Carlo simulation results of 20 steps and 50 runs are illustrated. Simulation experiments mainly concentrate on the following comparisons:

- Fusion performance comparisons of each approach between fusion center and local sensor,

- Fusion performance comparison among the three fusion approaches on aspects of different feedback and noises,

- Fusion performance comparisons of each approach with different feedback types and fusion performance comparisons of each approach with different process noises.

Fusion performance comparisons of each approach between the fusion center and the local sensor are presented in Figures 10.8–10.10. From these figures, we find that target can be tracked more accurately and the tracking performance is improved in all kinds of the fusion methods. This proves the effectiveness of information matrix, covariance intersection and weighted covariance algorithms.

In Figures 10.8–10.10 the performance comparison of fused tracks using information matrix, weighted covariance and covariance intersection on condition of complete feedback are provided. From the ensuing results, the track performances of three algorithms show almost the same effectiveness. Look in detail, we can find that performance of information matrix is a little better than that of covariance intersection which also show better tracking effect than weighted covariance approach. We also do same simulations in tracking systems with partial feedback and with no feedback, and they show consistent results. The reason for current simulation results is likely that information matrix is an optimal algorithm when the condition of full-rate communication is satisfied, while the weighted covariance algorithm is only a maximum likelihood estimate, and not an optimal one.

Figures 10.12–10.14 illustrate the effects of different feedback types on fusion performance for each fusion approach respectively. It is quite evident that

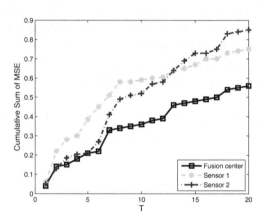

Figure 10.8: Fusion performance with complete feedback: covariance intersection.

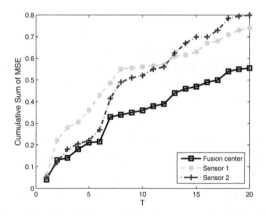

Figure 10.9: Fusion performance with complete feedback: information matrix.

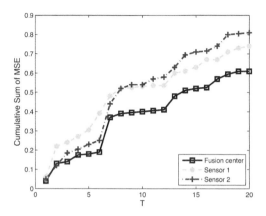

Figure 10.10: Fusion performance with complete feedback: weighted covariance.

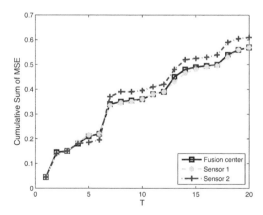

Figure 10.11: Performance comparison on complete feedback.

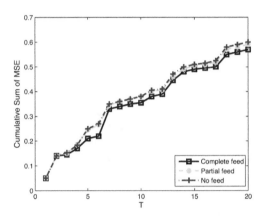

Figure 10.12: Effects of feedback type on fusion performance: covariance intersection.

all the algorithms apart from the weighted covariance approach show better performance on condition of complete feedback. For a little worse performance of
weighted covariance on condition of complete feedback, it is perhaps because an
optimal (maximum likelihood) correlation affects the quality of feedback information which misleads the local sensors.

Figures 10.15–10.17 present the effects of different process noises on fusion
performance for each fusion approach in the case of no feedback. We can easily
find that the smaller the process noise, the higher the tracking performance. Tracking systems with complete feedback and partial feedback show the same results.

10.10   Testing and Evaluation

The evaluation and testing has been made on an industrial utility boiler [472]. In
the system, the principal input variables are u_1, feedwater flow rate (kg/s); u_2, fuel
flow rate (kg/s); and u_3, attemperator spray flow rate (kg/s), the states are x_1, fluid
density, x_2, drum pressure, x_3, water flow input, x_4, fuel flow input, x_5, spray flow
input. The principal output variables are y_1, drum level (m); y_2, drum pressure
kPa; and y_3, steam temperature C^0. The schematic diagram of the utility boiler
can be seen in Figure 10.18.

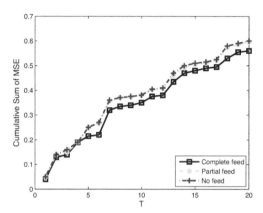

Figure 10.13: Effects of feedback type on fusion performance: information matrix.

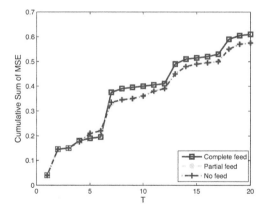

Figure 10.14: Effects of feedback type on fusion performance: weighted covariance.

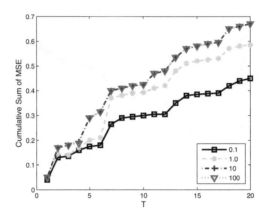

Figure 10.15: Effects of process noise on fusion performance: covariance intersection.

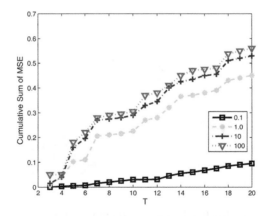

Figure 10.16: Effects of process noise on fusion performance: information matrix.

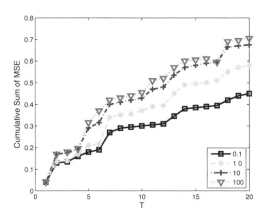

Figure 10.17: Effects of process noise on fusion performance: weighted covariance.

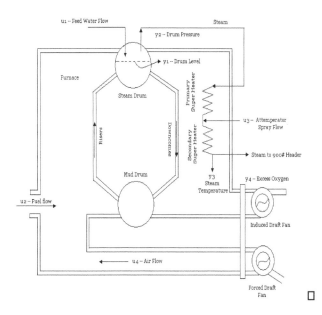

Figure 10.18: Schematic diagram of an industrial utility boiler.

10.10.1   Fault model for utility boiler

A fault model for the utility boiler is being developed. The mathematical model of the faulty utility boiler can be given as follows where faults in steam pressure are there in state 4 (fuel flow input) and 5 (spray flow input) respectively.

In the utility boiler, the steam temperature must be kept at a certain level to avoid overheating of the super-heaters. Applying a step to the water flow input (state 3), steam temperature increases and the steam temperature dynamics behave like a first order system. Applying a step to the fuel flow input (state 4), the steam temperature increases and the system behaves like a second order system. Applying a step to the spray flow input (state 5), the steam temperature decreases and the system behaves like a first order system. Then, a third order system is selected for the steam temperature model. Steam pressure is added in state 4 and 5, resulting in a more uncontrolled nonlinear system. Following [472] and the proposed fault scenarios, the fault model of the system can be described as:

$$\dot{x}_1(t) \;=\; \frac{u_1 - 0.03\sqrt{x_2^2 - (6306)^2}}{155.1411} \tag{10.112}$$

$$\dot{x}_2(t) \;=\; (-1.8506 \times 10^{-7} x_2 - 0.0024)\sqrt{x_2^2 - (6306)^2}$$
$$-0.0404 u_1 + 3.025 u_2 \tag{10.113}$$

$$\dot{x}_3(t) \;=\; -0.0211\sqrt{x_2^2 - (6306)^2} + x_4 - 0.0010967 u_1$$
$$+0.0475 u_2 + 3.1846 u_3 \tag{10.114}$$

$$\dot{x}_4(t) \;=\; 0.0015\sqrt{x_2^2 - (6306)^2} + x_5 - 0.001 u_1$$
$$+0.32 u_2 - 2.9461 u_3$$
$$+(a_{st\,pr})\sqrt{x_2^2 - (6306)^2} \tag{10.115}$$

$$\dot{x}_5(t) \;=\; -1.278 \times 10^{-3}\sqrt{x_2^2 - (6306)^2}$$
$$-0.00025831\,x_3 - 0.29747\,x_4$$
$$-0.8787621548\,x_5 - 0.00082\,u_1 - 0.2652778$$
$$u_2 + 2.491\,u_3$$
$$+(a_{st\,pr})\sqrt{x_2^2 - (6306)^2} \tag{10.116}$$

In what follows, we present simulation results for the proposed information-based versions of filters. The simulations have been performed on the utility boiler system where the faults due to steam pressure have been introduced in state, 4 and 5 respectively. First, the data generated from the simulation of the plant has been initialized and the parameters have been optimized which comprises the pre-processing and normalization of the data. The comparison of results for the distributed estimation, and normal estimation with different feedback generated from faults, and the basic profile of that particular state have been compared. Moreover, the same pattern of comparison has been followed for all the versions of information-based filters.

10.10.2 Covariance intersection filter

The information-based covariance intersection filter has been simulated here for the utility boiler steam pressure fault of state 4. Simulations have been made for the estimate of each case using the state-vector fusion method. In the simulation, comparisons of various profiles have been made i.e., a profile of normal fault-free state, an estimate of normal fault-free state, an estimate of the faulty state, and a distributed estimate based on the state-vector fusion for the different feedback strategies. The comparison of profiles mentioned above for complete feedback, partial feedback, and no feedback profile can be seen in Figure 10.19–10.21 respectively. Moreover, the one–on–one full comparison for all the feedback strategies can be seen in Figure 10.22. It can be seen that in the case of an information-based covariance intersection, the complete feedback case performs better than the partial and no feedback cases.

10.10.3 Weighted covariance filter

The information-based weighted covariance filter has been simulated here for the utility boiler steam pressure fault of state 4. Simulations have been made for the estimate of each case, using the state-vector fusion method. In the simulation, comparisons of various profiles have been made i.e., the profile of a normal fault-free state, the estimate of a normal fault-free state, the estimate of a faulty state, and a distributed estimate based on the state-vector fusion for different feedback strategies. The comparisons of profiles mentioned above for complete feedback and partial feedback profile can be seen in Figure 10.23 and 10.24 respectively. Moreover, the one–on–one full comparison of all the feedback strategies can be seen in Figure 10.25. It can be seen that in the case of an information-based weighted covariance, the no feedback case performs better than the partial feedback, and the complete feedback case has the lowest performance.

10.10.4 Kalman-like particle filter

The information-based Kalman-like particle filter has been simulated here for the utility boiler steam pressure fault of state 4. Simulations have been made for the

Table 10.1: MSE Comparison for All Information-Based Filters

FILTER	COMPLETE FB	PARTIAL FB	NO FB
CI	6.424	8.2759	8.411
WC	1.031×10^{-3}	1.0273×10^{-3}	1.0275×10^{-3}
KLPF	0.565	0.703	0.6223

estimate of each case using the state-vector fusion method. In the simulation, comparisons of various profiles have been made i.e., a profile of a normal fault-free state, the estimate of a normal fault-free state, the estimate of faulty state, and a distributed estimate based on the state-vector fusion for different feedback strategies. The comparison of profiles mentioned above for the complete feedback and partial feedback profiles can be seen in Figure 10.26 and 10.27 respectively. Moreover, the one–on–one full comparison for all the feedback strategies can be seen in Figure 10.28. It can be seen that in the case of an information-based Kalman-like particle filter, the partial feedback case performs better than the complete feedback, and the no feedback case has the lowest performance. Also, a profile comparison for the measurement fusion method can be seen in Figure 10.29 for a complete feedback case.

10.10.5   Mean square error comparison

In this section, we have made a comparison of all the versions of information-based filters with complete, partial, and no feedback respectively. It can be seen from Table 10.1 that the feedback versions are performing differently for a particular case of information-based filter. The mean square error value of complete feedback is the minimum in the case of the information-based covariance intersection filter and the Kalman-like particle filter respectively, whereas the partial feedback performs well in the case of the information-based weighted covariance filter.

Observe that the table shows the comparison of all the versions of the information-based filters, where MSE stands for mean square error, FB stands for feedback, CI stands for covariance intersection, WC stands for weighted covariance and KLPF stands for Kalman-like Particle filter.

10.11   Notes

In this chapter, distributed estimation has been proposed using various versions of information matrix filter. Different feedback strategies were evaluated and the focal point is the relation of performance and number of sensors. It is shown that

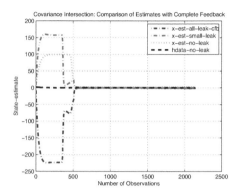

Figure 10.19: Covariance intersection: complete feedback comparison.

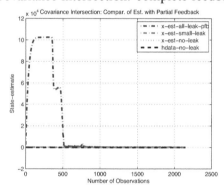

Figure 10.20: Covariance intersection: partial feedback comparison.

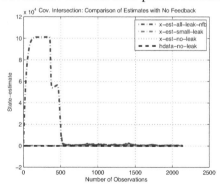

Figure 10.21: Covariance intersection: no feedback comparison.

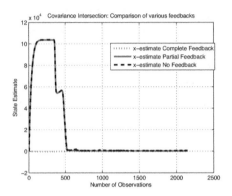

Figure 10.22: Covariance intersection: feedback comparison.

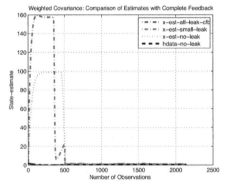

Figure 10.23: Weighted covariance: complete feedback comparison.

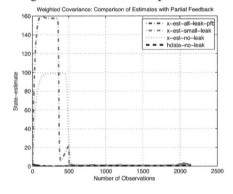

Figure 10.24: Weighted covariance: partial feedback comparison.

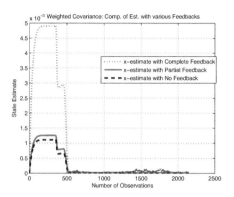

Figure 10.25: Weighted covariance: feedback comparison.

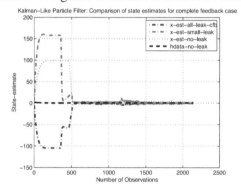

Figure 10.26: Kalman-like particle filter: complete feedback comparison.

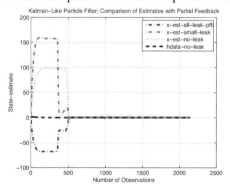

Figure 10.27: Kalman-like particle filter: partial feedback comparison.

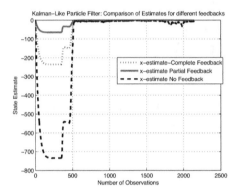

Figure 10.28: Kalman-like particle filter: feedback comparison.

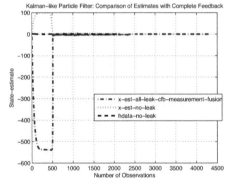

Figure 10.29: Kalman-like particle filter: complete feedback with measurement fusion method.

for algorithms, the feedback strategies perform differently i.e., the information-based covariance intersection and the Kalman-like particle filter perform better with complete feedback case, whereas an information-based weighted covariance performs better with the partial feedback case. The proposed scheme has been evaluated on an industrial boiler using fault scenarios, thus ensuring a thorough performance evaluation of the proposed filters with measurement fusion.

An evaluation is developed to compare the performance of the information matrix, weighted covariance and covariance intersection algorithms. Simulation experiments show the following results: The three fusion algorithms are proved to be effective. Using these fusion algorithms, targets can be tracked more accurately than local sensors, and tracking performance can be improved significantly. No matter which feedback types the tracking systems have, performance of the information matrix is a little better than that of the covariance intersection which also shows better tracking effect than the weighted covariance approach. Moreover, no matter what feedback types the tracking systems have, the larger the process noises the worse the tracking performance is for an indicated fusion algorithm.

10.12 Proposed Topics

1. Estimating statistical models within sensor networks requires distributed algorithms, in which both data and computation are distributed across the nodes of the network. It is known that graphical models are a natural framework for distributed inference in sensor networks. However, most learning algorithms are not distributed, requiring centralized data processing and storage. It is suggested to examine a framework for distributed parameter estimation, based on combining local and inexpensive estimators. The estimators are proposed to adopt a computationally efficient algorithm based on the maximum pseudo-likelihood estimator (MPLE). Develop the analysis and carry out simulations that justify the proposed arguments.

2. Consider the problem of decentralized state estimation for a class of linear time-invariant systems affected by stochastic disturbances and deterministic unknown input. One attractive approach is to look at the problem from a consensus estimation point of view. Therefore, it is suggested to characterize a consensus-based multi-agent Kalman estimator methodology for optimally tuning the consensus parameters while reconstructing the unknown input vector. Address all relevant issues including consensus-based communication between linear Kalman observers.

3. A major challenge in wireless sensor networks (WSNs) is the computation of parameter estimates based on distributed observations collected at individual sensors. Severe energy and bandwidth limitations call for the design and implementation of distributed algorithms that are efficient in terms of reducing communication overhead and computational cost. It is desired to examine the problem of estimating unknown deterministic parameters of linear Gaussian observation models, where the signal model is completely or partially known at individual sensors. Take into consideration the idea of recasting estimators of interest as optimizers of convex functions under a set of linear constraints and reformulate these convex optimization problems in a form that is amenable to parallel/distributed computation. Could the observation models be nonlinear?

4. Consider a wireless sensor network in which each sensor estimates the channel gains by collaborating with a few other network nodes. While performing this channel estimation, it is suggested to maintain a low average network energy consumption by employing a random sleep strategy. Then apply two estimation algorithms, one derived from the "Expectation Propagation (EP)" principle and the other the diffusion Least-Mean Squares (LMS) algorithm, in order to estimate the channel gains and compare their performance in terms of estimation error.

Chapter 11

Filtering in Sensor Networks

This chapter is concerned with the distributed filtering methods in sensor networks. It focuses on four distinct methods: H_∞ filtering, cooperative filtering, consensus filtering, and fusion filtering. In distributed H_∞ filtering, discrete-time systems with missing measurements and communication link failures are considered. In addition, the sensor measurements are unavailable randomly and the communication linked between nodes may be lost.

11.1 Distributed H_∞ Filtering

Distributed estimation or filtering in sensor networks is an important problem and has been paid much attention to research, recently. Within a network, a node can receive signals both from its sensors and the adjacent nodes, thus the strategy of fusing this information is a key point in filter design scheme. Recently, consensus protocol has been frequently employed as information fusion strategy in the research of distributed filtering problems, which can drive the estimation or measurement on a node to reach a consensus with its neighbors through local communication, [643], [644], [646], [653].

11.1.1 Introduction

In [643], several types of distributed Kalman filtering algorithms have been proposed. A distributed high-pass filter has been used to fuse the local sensor measurements such that the node can track the average measurement in the overall network. The performance analysis of the Kalman-consensus filter in [643] has been provided in [644] where a scalable suboptimal Kalman-consensus filter has been derived as well. In [646], a distributed robust filter is designed, and the H_∞

performance constraint has been considered both on the filter errors and the estimation deviations between adjacent nodes. By using the vector dissipativity theory, a sufficient condition is derived in terms of certain LMIs. The control and filtering problems of 2-D systems have been discussed widely in the past few decades and a number of results have been reported in [630], [635], [651], and [652]. In [653], a two–dimensional system-based approach for solving the distributed H_∞ filtering problem has been provided and the estimation consensus can be reached at each step by implementing the consensus updating procedure repetitiously.

In recent years, the filtering and control problems on networked control systems have been widely studied, [628], [629], [633], [634], [640], [650], [654], and the phenomenon of missing measurements is frequently discussed. A popular tool to model this phenomenon is the binary random variable sequence [647]–[649], and the error dynamics are usually in the form of a system with multiplicative noise [631], [632]. In [647], the problem of variance-constrained filtering with missing measurements has been discussed, and the derived filter can guarantee that the variance of steady-state estimation error is less than a prescribed bound. In [648], [649], the same issue has been considered in robust H_∞ filtering and control problems for stochastic systems with time delays which can be solved in terms of certain LMIs.

In [645], a H_∞-consensus performance has been established and the missing measurements phenomenon was first discussed in distributed filtering problems. The phenomenon of communication link failures has been studied in distributed consensus algorithms recently [636], [637]. In [636], the network is modeled as a Bernoulli random topology, and necessary and sufficient conditions have been established for average consensus in mean square sense and almost sure convergence when network links fail. Both the link failures and channel noise have been considered in the distributed consensus algorithm in [637], and two different compromises have been provided as a tradeoff between the average estimate bias and the variance.

11.1.2 System analysis

The topology of a sensor network is represented by a directed graph $G = (V, E, A)$ of order n with the set of nodes $V = \{1, 2, \ldots, n\}$, the set of edges $E \subseteq V \times V$, and the adjacency matrix $A = [a_{ij}]_{n \times n}$ with nonnegative adjacency element a_{ij}. An edge of G is denoted by an ordered pair (i, j). The adjacency elements associated with the edges are positive, i.e., $a_{ij} > 0 \Leftrightarrow (i, j) \in E$. The node j is called an in-neighbor of node i if $(i, j) \in E$. The set of in-neighbors of node i is denoted by N_i. Assume that G is strongly connected.

Consider the following discrete time linear time-invariant system defined on $k \in [0, N-1]$:

$$\begin{cases} x(k+1) = Ax(k) + Bw(k) \\ z(k) = Mx(k) \end{cases} \tag{11.1}$$

where $x(k) \in \Re^m$ is the system state which cannot be observed directly, $z(k) \in \Re^r$ is the output to be estimated, $w(k) \in \Re^q$ is the process noise belonging to $l_2[0, N-1]$. The initial state $x(0)$ is an unknown vector. The measurement of node $i(1 \le i \le n)$ is given as:

$$y_i(k) = \gamma_i(k)C_i x(k) + v_i(k) \tag{11.2}$$

where $y_i(k) \in \Re^{p_i}$ is the measurable output received by the node i, $v_i(k)$ is the measurement noise on sensor i belonging to $l_2[0, N-1]$, $\gamma_i(k) \in R$ is a Bernoulli distributed random variable and $\gamma_i(k) = 0$ indicates that the measurement of node i is missing at instant k. If $(i, j) \in E$, the communication link state between node i and j can be described by a Bernoulli distributed random variable $\theta_{ij}(k) \in \Re$, and $\theta_{ij}(k) = 0$ means that the communication link is lost at instant k.

Random variables $\{\gamma_i(k)\}_{k=0}^{N-1}$ and $\{\theta_{ij}(k)\}_{k=0}^{N-1}$ are both Bernoulli distributed white sequences taking values on 0 and 1 with

$$E\{\gamma_i(k)\} = Prob\{\gamma_i(k) = 1\} := \bar{\beta}_i \tag{11.3}$$

$$E\{\theta_{ij}(k)\} = Prob\{\theta_{ij}(k) = 1\} := \bar{\mu}_{ij} \tag{11.4}$$

where $\bar{\beta}_i$ and $\bar{\mu}_{ij}$ are known positive constants. $\gamma_i(k), \theta_{ij}(k), w(k)$ and $v_i(k)$ are assumed to be independent mutually to each other. Therefore, we have

$$Prob\{\gamma_i(k) = 0\} = 1 - \bar{\beta}_i \quad \sigma_{\gamma_i}^2 = E\{(\gamma_i(k) - \bar{\beta}_i)^2\} = \bar{\beta}_i(1 - \bar{\beta}_i) \tag{11.5}$$

$$Prob\{\theta_{ij}(k) = 0\} = 1 - \bar{\mu}_{ij} \quad \sigma_{\theta_{ij}}^2 = E\{(\theta_{ij}(k) - \bar{\mu}_{ij})^2\} = \bar{\mu}_{ij}(1 - \bar{\mu}_{ij}) \tag{11.6}$$

Consider the following two-step filter on node i for system (11.1) using the measurement (11.2):

$$\hat{x}_i^+(k) = \hat{x}_i(k) + G_i \sum_{j \in N_i} a_{ij}(\bar{\mu}_{ij}\hat{x}_i(k) - \chi_{ij}(k)) \tag{11.7}$$

$$\hat{x}_i(k+1) = A\hat{x}_i^+(k) + K_i(y_i(k) - \bar{\beta}_i C_i \hat{x}_i^+(k)) \tag{11.8}$$

$$\hat{z}_i(k) = M\hat{x}_i(k) \tag{11.9}$$

where $\hat{x}_i(k)$ is the estimation of system state $x(k)$, $\hat{x}_i^+(k)$ is the consensus update of $\hat{x}_i(k)$ through information exchange with the in-neighbors of node i, and

$$\chi_{ij}(k) = \theta_{ij}(k)\hat{x}_j(k)$$

is the signal received from node $j(j \in N_i)$, $\hat{z}_i(k)$ is the estimation of output $z(k)$.

In (11.7)–(11.9), that is, A, B, C_i and M are known matrices with appropriate dimensions, matrices K_i and G_i are the filter parameters to be determined on node i. Moreover, the initial state $\hat{x}_i(0)(1 \leq i \leq n)$ are assumed to be zero.

Remark 11.1.1 *The measurement model (11.2) is frequently used to represent the missing measurements and sensor faults in the research of networked control systems, which has been widely studied in the literature [645], [647]–[649]. It is introduced here to represent the randomly occurred missing measurements on sensor nodes.*

Remark 11.1.2 *Consensus protocol has been used as local information fusion strategy to fuse the signals received by node i in [653]. In Eq. (3.11), random variable θ_{ij} is introduced into the consensus updating procedure to represent the randomly occurred link failures in the local communication. The similar structures have been frequently used in the studies of distributed filtering, cooperative control and consensus problems in multi-agent systems, [723], [639].*

Denote $\tilde{z}_i(k) = z(k) - \hat{z}_i(k)$; the aim of this chapter is to design a filter for system (11.1) on each node with structure (11.7) and (11.8), such that the output estimation errors $\tilde{z}_i(k)(1 \leq i \leq n)$ satisfy the following average H_∞ performance constraint:

$$\frac{1}{n}\sum_{i=1}^{n} E_{\gamma_i,\theta_{ij}}\{\|\tilde{z}_i\|^2_{[0,N-1]}\} < \gamma^2\left(\|w\|^2_{[0,N-1]} + \frac{1}{n}\sum_{i=1}^{n}\|v_i\|^2_{[0,N-1]} + x(0)^T Rx(0)\right)$$

$$(11.10)$$

for a given disturbance attenuation level $\gamma > 0$ and positive definite weighting matrix $R = R^T > 0$.

Remark 11.1.3 *The average H_∞ performance (3.14) is a condition arising in classical H_∞ theory and has been discussed in [653]. (3.14) can be rewritten as*

$$\frac{\frac{1}{n}\sum_{i=1}^{n} E_{\gamma_i,\theta_{ij}}\{\|\tilde{z}_i\|^2_{[0,N-1]}\}}{\frac{1}{n}\sum_{i=1}^{n}\left(\|w\|^2_{[0,N-1]} + \|v_i\|^2_{[0,N-1]} + x(0)^T R_i x(0)\right)} < \gamma^2$$

which shows that the average energy gain from disturbances to estimation errors is less than a certain level γ^2. This performance constraint is reasonable for the distributed filtering problems in sensor networks because it is conservative to require all filters to meet the classical H_∞ performance constraint.

11.1.3 Simulation example 1

In what follows, an example is presented to demonstrate the effectiveness and applicability of the proposed method.

Algorithm 11.1.1 *Step 1: Given the average H_∞ performance index γ, consensus level $\varepsilon > 0$, the mathematical expectation $\bar{\beta}_i$ and $\bar{\mu}_{ij}$, and the positive definite weighting matrices R, compute the filtering gain K_i and the consensus gain G_i for all nodes through solving the LMIs . Set $k = 0$.*

Step 2: Get the consensus updating $\hat{x}_i^+(k)$ for all $1 \le i \le n$ by the formula (11.7), respectively.

Step 3: If there exists $i_0, j_0 \in \{1, \ldots, n\}$, such that

$$\|\hat{x}_i^+(k) - \hat{x}_{j0}^+(k)\| \ge \varepsilon$$

then, set $\hat{x}_i(k) \leftarrow \hat{x}_i^+(k)(1 \le i \le n)$ and go to Step 2, else go to Step 4.

Step 4: Compute the filtering updating $\hat{x}_i(k + 1)$ for all $1 \le i \le n$ by the formula (11.8), respectively. Set $k \leftarrow k + 1$.

Step 5: If $k < N$, then go to Step 2, else go to Step 6.

Step 6: Stop.

Without loss of generality, we consider a sensor network with 6 nodes as shown in Figure 11.1, whose topology is represented by a directed graph $G = (V, E, A)$ with the set of nodes $V = \{1, 2, 3, 4, 5, 6\}$, the set of edges $E = \{(1, 2), (1, 6), (2, 4), (2, 5), (3, 2), (4, 3), (4, 5), (5, 1), (5, 1), (6, 5)\}$, and the adjacency matrix

$$A = \begin{bmatrix} 0 & 1 & 0 & 0 & 0 & 1 \\ 0 & 0 & 0 & 1 & 1 & 0 \\ 0 & 1 & 0 & 0 & 0 & 0 \\ 0 & 0 & 1 & 0 & 1 & 0 \\ 1 & 1 & 0 & 0 & 0 & 0 \\ 0 & 0 & 0 & 0 & 1 & 0 \end{bmatrix}$$

Consider the following discrete-time linear system:

$$\begin{cases} x(k + 1) = \begin{bmatrix} -0.2 & -0.5 \\ 1.5 & -1 \end{bmatrix} x(k) + \begin{bmatrix} 1 \\ 1 \end{bmatrix} w(k) \\ z(k) = [0.5 \quad 0.5] x(k) \end{cases} \tag{11.11}$$

with the initial state $x(0) = [2\,2]^T$ and weighting matrix $R = I_2$. The measurement of sensor node i is

$$y_i(k) = \gamma_i(k)[-2 \quad 1]x(k) + v_i(k), \quad i = 1, 2, \ldots, 6$$

where $E\{\gamma_1\} = 0.9$ and $E\{\gamma_i\} = 1, i = 2, \ldots, 6$. Here, $w(k)$ and $v_i(k)(1 \leq i \leq n)$ are zero mean Gaussian white noise sequences with standard deviation 0.1 and 0.3, respectively. The filter initial states $\hat{x}_i(0)$ are all set to be zeros.

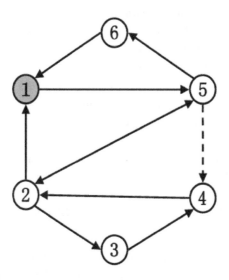

Figure 11.1: Topology structure of the sensor network.

Case 1: The sensor network has perfect communication link, i.e., $E\{\theta_{ij}\} = 1, i = 1, \ldots, n, j \in N_i$. The corresponding filter parameters are

$$K = diag\{K_1, \ K_2\}$$

$$K_1 = diag\left\{\begin{bmatrix} -0.0001 \\ -0.0374 \end{bmatrix}, \begin{bmatrix} 0.3611 \\ -0.6310 \end{bmatrix}, \begin{bmatrix} 0.3850 \\ -0.6185 \end{bmatrix}\right\},$$

$$K_2 = diag\left\{\begin{bmatrix} 0.3801 \\ -0.6209 \end{bmatrix}, \begin{bmatrix} 0.3549 \\ -0.6341 \end{bmatrix}, \begin{bmatrix} 0.3504 \\ -0.6378 \end{bmatrix}\right\}$$

$$G = diag\{G_1, \ G_2, \ G_3\}$$

$$G_1 = diag\left\{\begin{bmatrix} -0.1154 & -0.0938 \\ 0.3015 & -0.5874 \end{bmatrix}, \begin{bmatrix} -0.4852 & 0.2336 \\ -0.1189 & -0.1333 \end{bmatrix}\right\}$$

$$G_2 = diag\left\{\begin{bmatrix} -0.2238 & -0.0023 \\ -0.1356 & -0.0295 \end{bmatrix}, \begin{bmatrix} -0.2785 & -0.0408 \\ -0.1393 & -0.0443 \end{bmatrix}, \right\}$$

$$G_3 = diag\left\{\begin{bmatrix} -0.3893 & -0.0785 \\ -0.1958 & -0.1406 \end{bmatrix}, \begin{bmatrix} -0.7241 & 0.0439 \\ -0.2596 & -0.0289 \end{bmatrix}\right\}$$

In this case, N is set to 15, and the measurement of node 1 has missed when $k = 9$, i.e., $\gamma_1(9) = 0$. We have obtained that the average errors energy $\frac{1}{6}\sum_{i=1}^{6}\|\tilde{z}_i\|_{[0,14]}^2$ are 4.5633 and 4.4177 when the two-step filter (11.7)–(11.8) and Algorithm 11.1.1 are used, respectively, which indicate that repetitive consensus updating can improve the performance of filters. The results are presented in Figures 11.2–11.6, where the actual output $z(k)$ and its estimations on six nodes are presented in Figures 11.2 and 11.3, the estimation errors are plotted in Figures 11.4 and 11.5. For Algorithm 3.1.2, the repetitive consensus updating process of $\hat{x}_i^+$ when k = 10 is shown in Figure 11.6, where s is the consensus updating steps. The simulation results have confirmed that the designed filters work well.

Case 2: The sensor network has imperfect communication links and $E\{\theta_{45}\} = 0.95$, other conditions are as same as Case 1. The corresponding filter parameters

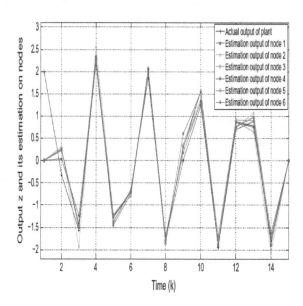

Figure 11.2: Output $z(k)$ and its estimations $\hat{z}_i(k)(1 \leq i \leq 6)$ using two-step filter (11.7)–(11.8).

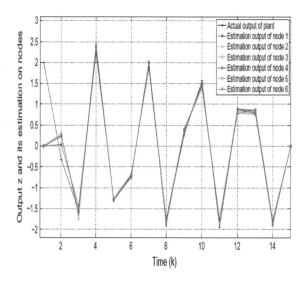

Figure 11.3: Output z(k) and its estimations $\hat{z}_i(k)(1 \leq i \leq 6)$ using Algorithm 11.1.1.

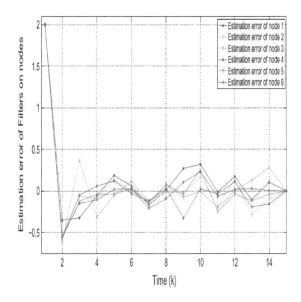

Figure 11.4: Estimation errors $\tilde{z}_i(k)(1 \leq i \leq 6)$ using two-step filter (11.7)–(11.8).

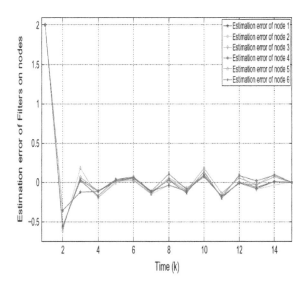

Figure 11.5: Estimation errors $\tilde{z}_i(k)(1 \leq i \leq 6)$ using Algorithm 11.1.1.

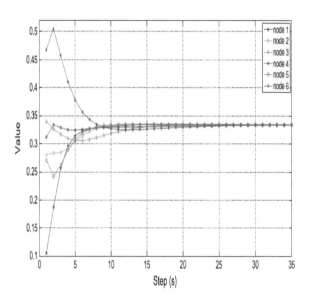

Figure 11.6: Consensus updating process of $\hat{x}_i^+(10)(1 \leq i \leq 6)$ under perfect communication using Algorithm 11.1.1.

are

$$K = diag\{K_1, \ K_2\}$$

$$K_1 = diag\left\{\begin{bmatrix} -0.0002 \\ -0.0392 \end{bmatrix}, \begin{bmatrix} 0.3585 \\ -0.6317 \end{bmatrix}, \begin{bmatrix} 0.3850 \\ -0.6163 \end{bmatrix}, \right\}$$

$$K_2 = \left\{\begin{bmatrix} 0.3728 \\ -0.6254 \end{bmatrix}, \begin{bmatrix} 0.3554 \\ -0.6333 \end{bmatrix} \begin{bmatrix} 0.3507 \\ -0.6368 \end{bmatrix}\right\}$$

$$G = diag\{G_1, \ G_2, \ G_3\}$$

$$G_1 = diag\left\{\begin{bmatrix} -0.1196 & -0.0909 \\ 0.2919 & -0.5797 \end{bmatrix}, \begin{bmatrix} -0.4712 & 0.2178 \\ -0.1255 & -0.1195 \end{bmatrix}\right\}$$

$$G_2 = diag\left\{\begin{bmatrix} -0.2026 & -0.0107 \\ -0.1106 & -0.0358 \end{bmatrix}, \begin{bmatrix} -0.1296 & -0.0010 \\ -0.0766 & -0.0147 \end{bmatrix}\right\}$$

$$G_3 = diag\left\{\begin{bmatrix} -0.3717 & -0.0679 \\ -0.1950 & -0.1186 \end{bmatrix}, \begin{bmatrix} -0.7105 & 0.3847 \\ -0.2427 & -0.0544 \end{bmatrix}\right\}$$

Based on the parameters given above, Algorithm 11.1.1 has been implemented and the average errors energy $\frac{1}{6}\sum_{i=1}^{6} \|\tilde{z}_i\|_{[0,14]}^2$ is 4.4527. Figure 11.7 plots that

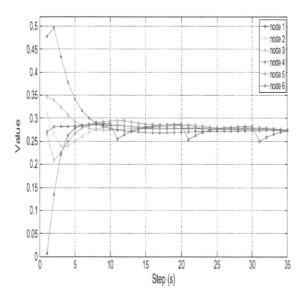

Figure 11.7: Consensus updating process of $\hat{x}_i^+(10)(1 \leq i \leq 6)$ under imperfect communication using Algorithm 11.1.1.

the consensus updating is implemented repetitiously when k = 10 under imperfect communication, and the link failures happened at step; 10, 20 and 30. This result shows that the link failures in communication decrease the performance of filters.

11.2 Distributed Cooperative Filtering

A challenge in the control of autonomous agents is to use spatially distributed information in an efficient way. It is often important to be able to track a common variable without employing a centralized strategy, since such a strategy is vulnerable to node failures. A prominent example is the distributed tracking of a moving target using a wireless sensor network (WSN). In this case, sensors have to cooperate in order to accomplish accurate information of, that is, target position, velocity, clock, etc.

11.2.1 Introduction

In recent years, several techniques for estimation using WSNs have been proposed. These techniques often rely on some physical characteristics of the wireless propagation, on the communication protocols, or any physical reading related to the signal to estimate. For example, the distance between pairs of nodes can be estimated by measuring the network connectivity [655]. A taxonomy of methods for the location estimation in WSNs from a signal processing perspective is provided in [656]. In traditional synchronization algorithms, nodes exchange data packets containing the current clock value and synchronize using received data and knowledge of the communication delay [657]. The propagation delay has been also used for the estimation of the node positions [658]. It is clear that these techniques are effective only if there are not packet losses or time-varying communication delays, if the nodes are highly connected, and if there are not clock drifts.

Collaboration can be suitable to overcome intrinsic limitations in processing only local measurements, since measurements are usually affected by noise, e.g., [659]. In fact, exploiting samples taken from different nodes and explicitly taking into account the communication, it is possible to design distributed algorithms for which nodes cooperate to achieve better estimates. A large body of literature is available on distributed sensor fusion. Here we limit the discussion to some recent and relevant contributions on consensus filters. In [660], the problem of distributed estimation of an average by a wireless sensor network is presented. It is assumed that nodes take a set of initial samples, and then iteratively exchange the averages of the samples locally collected. Each node reaches asymptotically the global average. The approach is based on a local weighted least squares method, where the weights are derived from a fast mixing Markov chain on a graph. In [415], a more general approach is investigated. The authors propose the consensus of the average of a common time-varying signal measured by each sensor, when the signal is affected by a zero-mean noise. A convex optimization problem is posed for which the authors find a set of symmetric edge weights that minimize the least mean square deviation. In [414], a related consensus filter for distributed low-pass sensor fusion has been proposed.

In the sequel, we study cooperative and distributed estimation algorithms using sensor nodes communicating through wireless transmission. Specifically, the system model takes into account time-varying signals, and we investigate how to ensure the consensus of the estimates while minimizing the variance of the error. We propose a distributed filter where the nodes compute the estimates without central coordination. The filter design includes local to guarantee the global asymptotic stability of the estimation error. Moreover, the distributed filter is scalable with respect to the network size. The algorithm can be applied e.g., for the position estimation, and temporal synchronization, as well as tracking of signals. Compared to recent relevant work [660],[415], [414] the approach hereafter is a basic improvement because we adopt a more general model of the filter structure, without resorting to the common heuristic of the Laplacian associated to the communication graph.

11.2.2 Problem formulation

Let us consider $N > 1$ nodes randomly distributed in the plane. For simplicity in exposition, we assume that each node can measure a common scalar signal $d(t)$ corrupted by additive noise:

$$u_i(k) = d(t) + \nu_i(t), \quad i = 1, ..., N \tag{11.12}$$

where $t \geq 0$ is the time where ν_i is a zero-mean Gaussian random variable. This is a common assumption to characterize the noise fluctuations and can be motivated by the central limit theorem. Defining the vectors

$$u(t) = [u_1(t), ..., u_N(t)]^T, \quad \nu(t) = [\nu_1(t), ..., \nu_N(t)]^T$$

we can rewrite (11.12) as

$$u(t) = d(t)\mathbf{1} + \nu(t), \quad \mathbf{1} := [1, ..., 1]^T \tag{11.13}$$

We assume that the covariance matrix of the random vector $\nu(t)$ is $\Omega = \omega^2 I$, so that ν_i and ν_j, for $i \neq j$, are uncorrelated.

Now since the nodes are connected through a communication network, each node has available extra data, transmitted by the neighbors, in order to reconstruct the signal $d(t)$. We thus assume that a node i builds an estimate, $x_i(t)$, of the signal $d(t)$ as

$$x_i(k) = \sum_{j=1}^{N} k_{ij}(t)x_j(t-1) + \sum_{j=1}^{N} h_{ij}(t)u_j(t) \tag{11.14}$$

Looked at in this light, each node computes an new estimate by linearly combining its previous estimate and the current measurement with previous estimates and current measurements received from neighbor's nodes. If node i is not connected with node j, then

$$k_{ij} = k_{ji} = 0, \quad h_{ij} = h_{ji} = 0, \quad \forall t \geq 0$$

Remark 11.2.1 *From the model (11.14), it is clear that one could try to design $k_{ii}(t)$ and $h_{ii}(t)$ so that a single node, without exchanging data with neighbors, is able to estimate $d(t)$. This would have the advantage of saving power for communication. However, for a single node it would require a longer time before achieving a good estimate of $d(t)$. Moreover, measurements taken too close in time, by the same node, are generally corrupted by correlated noise*

$$\boldsymbol{E}\left\{\nu_i(t)\nu_i(t\tau)\right\} = \varepsilon(\tau)$$

where $\varepsilon(\tau)$ is the autocorrelation function of the noise. Measurements taken by different nodes are instead corrupted by uncorrelated noise.

We rewrite the estimator (11.14) in a more compact way as

$$
\begin{aligned}
x(t) &= K(t)x(t-1) + H(t)u(t), \ [K(t)]_{ij} = k_{ij}(t) \\
x(t)(t) &= [x_1(t), \ldots, x_N(t)]^T, \ [H(t)]_{ij} = h_{ij}(t)
\end{aligned}
\tag{11.15}
$$

At this position, we recall the Appendix to model the communication network as an undirected graph $\mathcal{G} = (\mathcal{V}, \mathcal{E})$, where $\mathcal{V} = \{1, \ldots, N\}$ is the vertex set and $\mathcal{E} \subseteq \mathcal{V} \times \mathcal{V}$ the edge set. Note that $(i, j) \in \mathcal{E}$ implies that $(j, i) \in \mathcal{E}$ since the graph is undirected. The graph $\mathcal{G}$ is said to be connected if there is a sequence of edges in $\mathcal{E}$ that can be traversed to go from any vertex to any other vertex. We associate to each edge $(i, j) \in \mathcal{E}$ a time-varying weight $w_{ji}(t)$. In general, it may hold that the weights $w_{ij}(t)$ and $w_{ji}(t)$ are different. We introduce the adjacency matrix $W(t)$ as

$$
[W(t)]_{ij} = \begin{cases} w_{ij}(t) & if \ (i,j) \in \mathcal{E} \\ 0, & otherwise \end{cases}
$$

We say that a matrix $W(t)$ is compatible with $\mathcal{G}$, if $W(t)$ defines an adjacency matrix for $\mathcal{G}$. We denote this by $W(t) \cong \mathcal{G}$. We interpret the matrices $K(t)$ and $H(t)$ of (11.15) as the adjacency matrices of two weighted graphs, one associated to the communication of estimates $x(t)$ and the other associated to the communication of measurements $u(t)$. It is convenient to introduce the neighbors of a node i as the set $\mathbf{N}_i$ of all nodes that can communicate with i, namely

$$
\mathbf{N}_i = \{j \in \mathcal{V} : (j, i) \in \mathcal{E}\}
$$

We can now state the main problem of interest in this section. *Given a wireless sensor network modeled by a connected graph $\mathcal{G}$, find time-varying matrices $K(t)$ and $H(t)$, compatible with $\mathcal{G}$, such that the signal $d(t)$ is consistently estimated and the variance of the estimate is minimized. Moreover, the solution should be distributed in the sense that the computation of $k_{ij}(t)$ and $h_{ij}(t)$ should be performed by node i.*

11.2.3   Centralized estimation

For convenience, we assume that $x(0)$ and $u(0)$ are independent and identically distributed random variables. Letting $e(t) = x(t) - d(t)\mathbf{1}$, then we have

$$
e(t) = K(t)e(t-1) + K(t)d(t-1)\mathbf{1} + (H(t) - 1)d(t)\mathbf{1} + H(t)\nu(t) \tag{11.16}
$$

Assume that $d(t) = d(t-1) + \delta$, $|delta| < \bar{\delta} << 1$. Taking the expected value with respect to the noise $\nu(t)$, we obtain

$$\mathbf{E}[e](t) = K(t)\mathbf{E}[e](t-1)+(K(t)+H(t)-\mathbf{1})\mathbf{1}d(t-1)+(H(t)-\mathbf{1})\delta(t)\mathbf{1} \quad (11.17)$$

In the case that $\sigma_M(K(t)) < 1$, $\forall t$, the convergence of $\mathbf{E}[e(t)]$ to zero is guaranteed, whenever the conditions

$$\begin{aligned} (K(t) + H(t))\mathbf{1})\mathbf{1} &= \mathbf{1} \\ H(t)\mathbf{1} &= \mathbf{1} \end{aligned} \qquad (11.18)$$

are fulfilled, where $\sigma_M(R)$ is the largest singular value of the matrix R. Now, it is an easy task to show that the proposed filter is unbiased and the minimum variance is achieved with $K(t) = 0$ with $H(t)$ such that

$$h_{ij}(t) = h_{ji}(t) = \begin{cases} \frac{1}{\mathbf{N}_i}, & if\ j \in \mathbf{N}_i \\ 0, & otherwise \end{cases}$$

Remark 11.2.2 *In order to take advantage of the foregoing estimates we need to relax the condition on the bias. This means that, we have to tradeoff the bias for the variance reduction. By removing the condition $H(t)\mathbf{1} = \mathbf{1}$ on the matrix $H(t)$, simple analysis will show that if the signal is slowing and varying, as assumed, then the bias is negligible.*

Observe that the degree of freedom in the choice of $K(t)$ and $H(t)$, can be exploited to minimize the variance of the estimate. For this purpose we study how the covariance matrix changes with time. Let us assume that $x(t)$ and $u(t)$ are independent random vectors. Introduce the matrix

$$P(t) = \mathbf{E}\left[(e(t) - \mathbf{E}[e](t))(e(t) - \mathbf{E}[e](t))\right]^T \qquad (11.19)$$

Then

$$P(t) = K(t)P(t-1)K^T(t) + \omega^2 H(t)H^T(t) \qquad (11.20)$$

One attractive way is to choose $K(t)$ and $H(t)$ such that $P(t)$ is minimized at each time instant. Hence, we have the following minimization problem:

$$\begin{aligned} \min_{K(t),H(t)} \quad & Tr\left[K(t)P(t-1)K^T(t) + \omega^2 H(t)H^T(t)\right] \\ subject\ to \quad & (K(t) + H(t))\mathbf{1} = \mathbf{1} \\ & \sigma_M(K(t)) < 1 \\ & K(t) \cong \mathcal{G} \\ & H(t) \cong \mathcal{G} \end{aligned} \qquad (11.21)$$

This optimization problem is solved iteratively, starting with some initial guess $P(0)$. Notice that

- The objective function is quadratic in $K(t)$ and $H(t)$ for a given $P(t1)$.

- The first constraint is the linear matrix equality (11.18).

- The second constraint, which ensures that the estimation error converges to zero, can be written as a linear matrix inequality using the Schur complement.

- The last constraints impose structure on the matrices $K(t)$ and $H(t)$. They are represented by equalities to zero of some elements of these matrices.

Remark 11.2.3 *Although the optimization problem (11.21) could be solved using standard numerical optimization tools, it clearly requires a powerful central node collecting data, computing new weights, and dispatching them to the sensors. Beside the typical disadvantage of a centralized solution, which is not robust to failures, we would also have large delays due to the propagation of the data from the farthest nodes to the central node. Although this could be overcome by having directed paths from every node to the central node, this would require, in general, a total power consumption which is prohibitive for small nodes. We propose in the following a decentralized solution where each node computes its weights minimizing the variance of its estimate.*

11.2.4   Distributed estimation

In order to be able to deal with distributed estimation, we introduce some relevant notations. Let M_i denote the number of neighbors of node i, that is, M_i is the cardinality of $\mathbf{N}_i = \{i_1, ..., i_{M_i}\}$. We then collect the estimation errors available at node i in the vector $\zeta_i(t) \in \Re^{M_i}$. Next, the elements of $\zeta_i(t)$ are ordered according to the node indices:

$$\zeta_i(t) = [e_{i_1}, ..., e_{i_{M_i}}]^T, \quad i_1 < .... < i_{M_i}$$

Similarly, corresponding to the non-zero elements of row i of the matrices $K(t)$ and $H(t)$, we introduce vectors $\kappa_i(t)$, $\eta_i(t) \in \Re^{M_i}$, respectively. These vectors are then ordered according to node indices. It follows from (11.20) that the variance of $e_i(t)$ can be evaluated as

$$\mathbf{E}[e_i^2(t)] = \kappa_i^T(t)\Gamma_i(t-1)\kappa_i(t) + \omega^2\eta_i^T(t)\eta_i(t), \quad \Gamma_i = \mathbf{E}[e_i e_i^T] \qquad (11.22)$$

Now to minimize the variance of the estimation error in each node, we assume that $\kappa_i(t)$ and $\eta_i(t)$ are selected as free variables. Let us set the target as obtaining "consensus and convergence" as well. Consequently, the following optimization problem should be solved at each time t and in each node i:

$$\min_{\kappa_i(t),\eta_i(t)} \quad \kappa_i^T(t)\Gamma_i(t)\kappa_i(t) + \omega^2\eta_i^T(t)\eta_i(t)$$

$$subject\ to \quad (\kappa_i(t) + \eta_i(t))^T \mathbf{1} = 1$$

$$\sigma_M(K(t)) < 1 \quad (11.23)$$

Observe that the inequality constraint in (11.23) is global since $K(t)$ depends on all $\kappa_i(t)$, $i = 1, ..., N$ which does not allow a distributed structure. However, dropping out the time-dependency and recognizing the following mathematical arguments

$$\sigma_M(K(t)) < 1 \quad \Rightarrow \quad \lambda_M(KK^T) < 1$$

$$\lambda \in \left\{ z \in \mathbb{C} : \bigcup_{i=1}^{N} |z - \Sigma_s k_{is}^2| \leq \Sigma_{j\neq i}|\Sigma_s k_{is} k_{js}| \right\} \quad \Rightarrow$$

$$\lambda \in \left\{ z \in \mathbb{C} : \bigcup_{i=1}^{N} |z - \Sigma_s k_{is}^2| \leq \Sigma_{j\neq i}\Sigma_s |k_{is}||k_{js}| \right\} \quad \Rightarrow$$

$$\lambda \in \left\{ z \in \mathbb{C} : \bigcup_{i=1}^{N} |z - \Sigma_s k_{is}^2| < 1 - \Sigma_s k_{is}^2 \right\} \quad (11.24)$$

where we have used

$$\left(\Sigma_s |k_{is}| \right)^2 = \Sigma_s k_{is}^2 + \Sigma_s \Sigma_{s\neq\ell} |k_{is}||k_{i\ell}|$$

Taking into account the cardinality property, the foregoing argument leads to substituting the inequality $\sigma_M(K(t)) < 1$ by

$$\Sigma_s^{M_i} k_{is}^2 < 1/2, \quad \Sigma_s^{M_i}|k_{is}| < 1$$

and hence we cast (11.23) into the following problem

$$\min_{\kappa_i(t),\eta_i(t)} \quad \kappa_i^T(t)\Gamma_i(t-1)\kappa_i(t) + \omega^2\eta_i^T(t)\eta_i(t)$$

$$subject\ to \quad (\kappa_i(t) + \eta_i(t))^T \mathbf{1} = 1$$

$$\kappa_i(t) \in \Phi :=$$

$$\left\{ [\phi_1, ..., \phi_{M_i}] : \Sigma_{j=1}^{M_i}\phi_j^2 < 1/2 \wedge \Sigma_{j=1}^{M_i}|\phi_j| < 1 \right\} \quad (11.25)$$

It is easy to verify that the set Φ is convex. Observe that

- Problem (11.25) is essentially quadratic in $\kappa_i(t)$ and $\eta_i(t)$ for a given $\Gamma_i(t-1)$.

- The first constraint is a linear constraint.

- The second constraint is a nonlinear constraint

To deal with this constrained minimization, we recall arguments from mathematical optimization. Note that the cost function of the primal problem (11.25) is convex in Φ subject to a linear constraint and if we assume the solution is in the set Φ, then invoking the strong duality to transform in its dual [661]. The standard procedure is to:

1. Form the Lagrangian $\mathsf{L}(\eta, \kappa_i(t), \eta_i(t))$ using a Lagrangian multiplier ξ:

$$
\begin{aligned}
\mathsf{L}(\eta, \kappa_i(t), \eta_i(t)) &= \kappa_i^T(t)\Gamma_i(t-1)\kappa_i(t) + \omega^2\eta_i^T(t)\eta_i(t) \\
&+ \xi[(\kappa_i(t) + \eta_i(t))^T\mathbf{1} - 1], \quad \xi \in \Re \quad (11.26)
\end{aligned}
$$

2. Introduce the dual function

$$
\mathsf{G}(\eta) = \inf_{\kappa_i(t),\eta_i(t)} \mathsf{L}(\xi, \kappa_i(t), \eta_i(t)) \quad (11.27)
$$

3. The dual optimization problem becomes

$$
\max_{\xi} \mathsf{G}(\eta)
$$

4. Implementation of steps 1)–3) above yields

$$
\begin{aligned}
\kappa_i(t) &= -\frac{\xi\Gamma_i(t-1)^{-1}\mathbf{1}}{2} \\
\eta_i(t) &= -\frac{\xi\mathbf{1}}{2\omega^2} \quad (11.28)
\end{aligned}
$$

5. Minimize the dual optimization function $\mathsf{G}(\xi)$ with respect to ξ. Simple algebraic calculations yield the optimal values

$$
\begin{aligned}
\kappa_i(t) &= \frac{\Gamma_i(t-1)^{-1}\mathbf{1}}{\omega^{-2}M_i + \mathbf{1}^T\Gamma_i(t-1)\mathbf{1}} \\
\eta_i(t) &= \frac{1}{M_i + \omega^2\mathbf{1}^T\Gamma_i(t-1)\mathbf{1}} \quad (11.29)
\end{aligned}
$$

Thus (11.29) are the optimal weights for each sensor, for a given $_i(t1)$.

Remark 11.2.4 *It is left to the reader to show that the optimal weights (11.29) are feasible for the problem (11.25), which corresponds to prove that the solution belongs to the set Φ. As we have pointed out before, the optimal weights $\kappa_i(t)$ and $\eta_i(t)$ depend on the covariance matrix $\Gamma_i(t-1)$. Since each node receives measurements and estimates from the neighbors, it is possible to compute, or estimate, the covariance matrix $\Gamma_i(t-1)$ at each time step.*

11.2.5 Issues of implementation

In the implementation of the distributed cooperative algorithm, we consider that each sensor initializes the local mean estimation error $\hat{m}_\zeta$ and its local covariance matrix estimate with the noise covariance, $\hat{\Gamma_i}(0) = \omega^2 I$. The optimal weights are computed using the equations in (11.29). We then compute the optimal estimate $x_i(t)$ and proceed to implement the covariance update based on the available data.

In order to compute $\hat{m}_\zeta$ and $\hat{\Gamma_i}(t)$, an estimate of the estimation error needs to be computed. This is done by computing an average between the current available estimates and measurements. As an estimate of $d(t)$ we cannot use directly $x(t)$ since it is biased, and we can underestimate the actual error. On the other hand, using only measurements is also not a good choice, since their noise variance is very high. A reasonable compromise is to combine low variance slightly biased estimates and unbiased measurement's corrupted noises with high variance. In particular the mean and covariance matrix are estimated from the samples.

In the algorithm, the inversion of the covariance matrix should be computed. This is a routine operation in resource constrained sensor networks, since each node has generally a rather limited number of neighbors, and thus the size of the matrix $\hat{\Gamma_i}(t)$ is small. It should be emphasized that the algorithm is implemented under the assumption that each node is able to compute and communicate data within the sampling instance.

Numerical simulations have been carried out to compare the proposed algorithm with other possible approaches. Specifically, the proposed algorithm is compared with two other algorithms defined below:

Arithmetic Mean Estimator (AME)

$$k_{ij} = h_{ij} = \begin{cases} \frac{1}{2|\mathbf{N}_i|} & if \ j \in \mathbf{N}_i \\ 0, & otherwise \end{cases}$$

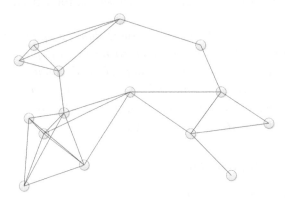

Figure 11.8: Random generated network with N = 15. Each node has 3.6 neighbors, on average.

Laplacian Based Estimator (LBE)

$$k_{ij} = h_{ij} = \begin{cases} 1 - \dfrac{|\mathbf{N}_i|}{\max_i |\mathbf{N}_i|} & if\ i = j \\ \dfrac{|\mathbf{N}_i|}{\max_i |\mathbf{N}_i|} & if\ j \in \mathbf{N}_i \\ 0, & otherwise \end{cases}$$

Two random generated networks have been considered. The first is shown in Figure 11.8, with N = 15 nodes, and the second with N = 150 nodes is reported in Figure 11.9. The signal to be tracked is $d(t) = 3sin(2t/780)cos(2t/620)$ with a noise normally distributed around $d(t)$ with variance $\omega^2 = 1.2$. The network is generated with sensors randomly distributed in a squared area of side $N/2$. Two nodes are connected if and only if their relative distance is less than $1.5\sqrt{N}$

For the two network cases, all the N realizations are shown in Figure 11.10 and Figure 11.11. The first plot of the two figures shows the signal corrupted by noise, and the second refers the realization for the arithmetic mean based estimator, the third to the Laplacian based, and the last refers to the proposed algorithm. In particular, it is possible to appreciate, visually, the improvements due to the solution proposed.

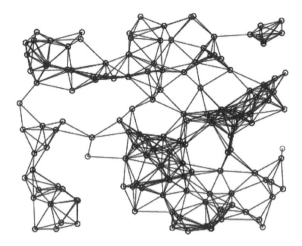

Figure 11.9: Random generated network with N = 150. Each node has an average of 8.3 neighbors.

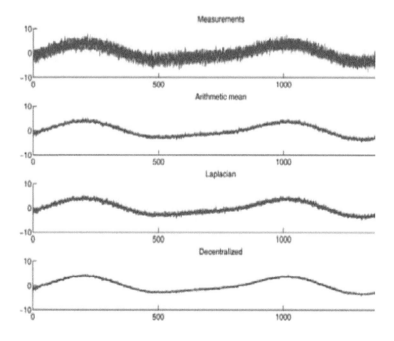

Figure 11.10: Realization of the different estimators versus time based on small network.

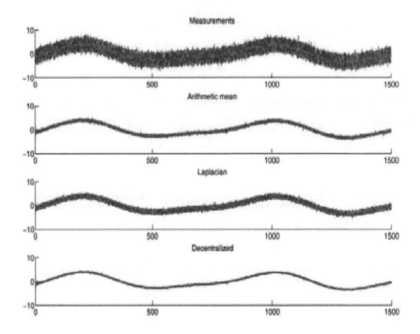

Figure 11.11: Realization of the different estimators versus time based on large network.

11.3 Distributed Consensus Filtering

11.3.1 Introduction

Recently, distributed estimation and control have been shown to be very effective for consensus [40]–[42]. In a distributed framework, each node can only communicate with its neighboring peers, and the objective of filtering or control can be achieved in a distributed way. Noticeably, the convergence analysis for distributed filtering is still lacking today, so this chapter aims to provide some basic theoretical analysis of a new class of distributed consensus filters.

A large-scale sensor network can be viewed as a complex network with each node representing a sensor and each edge performing information exchange between sensors. It would be interesting to see how synchronization of complex networks could be used in the distributed consensus filtering design. In such a network, each node communicates with its neighboring nodes to exchange information, and eventually, all the states could achieve the expected synchronization, which achieves distributed consensus filtering.

Practically, it is usually difficult to observe all the states of the target, so pinning observers may be designed in the case where the sensors can only measure partial states of the target.

In what follows, based on the theory of synchronization and consensus in complex networks and systems, some distributed consensus filters are designed. From the pinning control approach, only a small fraction of sensors are used to measure the target information, thereby achieving the intended control. Furthermore, pinning observers are designed when the sensor can observe only partial states of the target. Several theoretical results and design methods are new to the existing research literature.

11.3.2 Problem formulation

Let the target be described by the following model:

$$ds(t) = f(s(t), t)dt + v(t)d\nu \qquad (11.30)$$

where $s(t) \in \Re^m$ is the target state, $v(t) \in \Re^n$ is an external noise intensity function and $\nu(t)$ is a one-dimensional Brownian motion with

$$\begin{cases} \mathbf{E}\{d\nu(t)\} = 0 \\ \mathbf{C}\{d\nu(t)\} = 1 \end{cases}$$

where **P** stands for the variance. Model (11.30) is defined over a complete probability $\{\Omega, \mathbf{F}, \mathbf{P}\}$ with a natural filtration $\{\mathbf{F}\}$; $t \geq 0$ generated by $\{\nu(s) : 0 \leq$

$s \leq t\}$, where Ω is associated with the canonical space generated by $\nu(t)$ and $\mathbf{F}$ is the associated σ-algebra generated by $\{\nu(t)\}$ with a probability measure $\mathbf{P}$.

Next consider a sensor network of size N and assume that if sensor i can measure the signal sensor $s(t)$, then

$$y_i(t) = s(t) + \sigma_i(t)\frac{d\omega_i(t)}{dt} \qquad (11.31)$$

where $y_i(t) \in \Re^n$ is the measurable of sensor i on target $s(t)$, $\sigma_i(t)$ is an external noise intensity function of agent i, and $\omega_i(t)$ is an independent one-dimensional Brownian motion with

$$\begin{cases} \mathbf{E}\{d\omega_i(t)\} = 0 \\ \mathfrak{C}\{d\omega_i(t)\} = 1, \quad i = 1, 1, 2, ..., N \end{cases}$$

In principle, there are a large number of sensors that can measure the target $s(t)$ from observations $y_i(t), i = 1, 1, 2, ..., N$.

The objective here is to design a distributed filter to track the state $s(t)$ of the target. We assume that the sensor network allows only each sensor to communicate with the neighboring sensors. Taking the measurement y_i as the input, we consider the following filter for design

$$\dot{x}_i(t) = f(x_i(t), t) + g \sum_{j=1, j\neq i}^{N} a_{ij}(x_j(t) - x_i(t)) + u_i(t) \qquad (11.32)$$

where $x_i(t)$ is the estimation of the target $s(t)$ in sensor node i, g is the coupling strength, and $u_i(t)$ is the designed controller which is dependent on the measurement y_i. Note that if sensor i is in the sensing range of sensor j, then there is an interaction between sensor i and sensor j. This implies that $a_{ij} = a_{ji} > 0$, $(i \neq j)$; otherwise, $a_{ij} = a_{ji} = 0$. Let $a_{ii} =? \sum_{j=1, j\neq i}^{N} a_{ij}$, $for \ i = 1, 2, ..., N$. Also, let $\mathbf{N}(i) = j|aij > 0$ denote the set of neighbors of sensor i. Then, (11.32) can be written as

$$\dot{x}_i(t) = f(x_i(t), t) + g \sum_{j=1, j\neq i}^{N} a_{ij}x_j(t) + u_i(t) \qquad (11.33)$$

It follows from (11.33), it is easy to see that the sensor i can only receive estimated signals from its neighbors in $\mathbf{N}(i)$. In the sequel, we only consider the situation where the sensor network coupling matrix $A = \{aij\} \in \Re^{N \times N}$ is irreducible. For simplicity of exposition, we provide the following assumptions and definitions:

Assumption 11.3.1 *For all* $x,\ y \in \Re^n$, *there exists a constant* θ *such that*

$$(x - y)^t (f(x, t) - f(y, t)) \leq \theta\ (x - y)^t (x - y), \qquad \forall t \in \Re \qquad (11.34)$$

Assumption 11.3.2 *Both* $v(t) \in \Re^n$ *and* $\sigma_i(t) \in \Re^n$ *belongs to* $L_\infty[0, \infty)$. *This means that* $v(t)$ *and* $\sigma_i(t)$ *are bounded vector functions satisfying*

$$v^t(t)\, v(t) \ \leq \ \alpha, \qquad \forall t \in \Re, \ \ \alpha > 0 \qquad (11.35)$$

$$\sigma_i^t(t)\, \sigma_i(t) \ \leq \ \alpha, \qquad \forall t \in \Re, \ \ \sigma_i > 0, \ i = 1, 2, ..., N \qquad (11.36)$$

Assumption 11.3.3 *For all* $x \in \Re^n$, *there exists a constant* γ *such that*

$$\|(f(x, t)\| \ \leq \ \gamma, \qquad \forall t \in \Re \qquad (11.37)$$

Definition 11.3.1 *The controllers* $u_i(t)$, $i = 1, 2, ..., N,$, *are said to be* **distributed bounded consensus controllers** *if there exist constants* $\phi > 0$, $\eta_i > 0$ *and* $\mu > 0$ *such that*

$$\lim_{t \to \infty} \frac{1}{N} \boldsymbol{E}\left(\sum_{i=1}^{N} \|x_i(t) - s(t)\|^2 \right) \ \leq \ \phi\gamma + \sum_{i=1}^{N} \eta_i \beta_i + \mu\alpha \qquad (11.38)$$

In the case that $\phi = 0$, *then they are called* **distributed consensus controllers**.

Definition 11.3.2 *The filters (11.32) or (11.33) are said to be* **distributed bounded consensus filters** (**distributed consensus filters**) *if the controllers in (11.32) or (11.33) are* **distributed bounded consensus controllers** (**distributed consensus controllers**).

Of interest is to analyze the convergence of the bound for

$$\lim_{t \to \infty} \frac{1}{N} \mathbf{E}\left(\sum_{i=1}^{N} \|x_i(t) - s(t)\|^2 \right)$$

11.3.3   Filter design: fully-equipped controllers

In the sequel, we consider the case of fully-equipped controllers for all nodes. Let the linear state-feedback controllers for $i = 1, ..., N$

$$u_i(t) = -gk_i(x_i(t) - y_i(t)) \qquad (11.39)$$

where the gain $0 < k_i \in \Re$. For convenience, we introduce

$$
\begin{aligned}
A_c &= A - K, \; K = diag(k_1, \; k_2, \; ..., k_N), \; A_c \in \Re^{N \times N} = \{a_{cij}\} \\
e_i(t) &= x_i(t) - s(t), \; i = 1, ..., N
\end{aligned}
$$

Manipulating (11.30), (11.33) and (11.39), we obtain the following error dynamics:

$$
\begin{aligned}
d\dot{e}_i(t) &= \left[f(x_i(t), t) - f(s(t), t) + g \sum_{j=1}^{N} a_{ij} e_j(t) - g K_i e_i(t) \right] dt \\
&\quad - v(t) d\nu + g k_i \sigma_i(t) d\omega_i
\end{aligned}
\tag{11.40}
$$

The following theorem establishes a main result for the controller

Theorem 11.3.1 *Suppose that Assumptions 11.3.1 and 11.3.2 hold. Then the controllers (11.39) are distributed consensus controllers if*

$$
\theta + g \lambda_M(A_c) < 0
\tag{11.41}
$$

where $\lambda(A_c)$ denote the largest eigenvalue of matrix A_c. The estimated bound is given by

$$
\lim_{t \to \infty} \frac{1}{N} \mathbf{E} \left(\sum_{i=1}^{N} \|x_i(t) - s(t)\|^2 \right) \leq \frac{\alpha + (g^2/N) \sum_{i=1}^{N} k_i^2 \beta_i}{-2(\theta + g \lambda_M(A_c))}
\tag{11.42}
$$

Proof 11.3.1 *Consider the following Lyapunov function:*

$$
V(t) = \frac{1}{2} \sum_{i=1}^{N} e_i^t(t) e_i(t)
\tag{11.43}
$$

By the Ito formula [664], we obtain the following stochastic differential:

$$
dV(t) = \mathcal{L} V(t) dt + \sum_{i=1}^{N} e_i^t(t) [v(t) d\nu + g k_i \sigma_i(t) d\omega_i]
\tag{11.44}
$$

In view of Assumption 11.3.2, the weak infinitesimal operator $\mathcal{L}$ of the stochastic *process gives*

$$
\begin{aligned}
\mathcal{L}V(t) &= \sum_{i=1}^{N} \left[e_i^t(t) \left(f(x_i(t), t) - f(s(t), t) + g \sum_{j=1}^{N} a_{ij} e_j(t) - g K_i e_i(t) \right) \right. \\
&\quad \left. + \frac{1}{2} v^t(t) v(t) + \frac{g^2 k_i^2}{2} \sigma_i^t(t) \sigma_i(t) \right] \\
&\leq \theta \sum_{i=1}^{N} e_i^t(t) e_i(t) + g \sum_{i=1}^{N} \sum_{j=1}^{N} b_{ij} e_i^t(t) e_j(t) + \frac{1}{2} N\alpha + \frac{g^2}{2} \sum_{i=1}^{N} k_i^2 \beta_i \\
&\leq \left(\theta + g \lambda_M(A_c) \right) \sum_{i=1}^{N} e_i^t(t) e_i(t) + \frac{1}{2} N\alpha + \frac{g^2}{2} \sum_{i=1}^{N} k_i^2 \beta_i \qquad (11.45)
\end{aligned}
$$

By virtue of Lemma A.10.1 in the Appendix, $\lambda_M(A_c) < 0$. It follows from (11.41) that

$$
\begin{aligned}
\mathcal{L}V(t) &\leq \left(\theta + g \lambda_M(A_c) \right) \sum_{i=1}^{N} e_i^t(t) e_i(t) + \frac{1}{2} N\alpha + \frac{g^2}{2} \sum_{i=1}^{N} k_i^2 \beta_i \\
&= N \left(\theta + g \lambda_M(A_c) \right) \left(\theta + g \lambda_M(A_c) \right) \\
&\quad \times \left[\frac{1}{N} \sum_{i=1}^{N} e_i^t(t) e_i(t) - \frac{\alpha + (g^2/N) \sum_{i=1}^{N} k_i^2 \beta_i}{-2(\theta + g \lambda_M(A_c))} \right] \qquad (11.46)
\end{aligned}
$$

By the Ito formula, we have

$$
\begin{aligned}
\boldsymbol{E}V(t) - \boldsymbol{E}V(0) &= \int_0^t \frac{1}{2} \mathcal{L}V(s) ds \\
&\leq N \left(\theta + g \lambda_M(A_c) \right) \\
&\quad \times \int_0^t \left[\frac{1}{N} \sum_{i=1}^{N} e_i^t(t) e_i(t) - \frac{\alpha + (g^2/N) \sum_{i=1}^{N} k_i^2 \beta_i}{-2(\theta + g \lambda_M(A_c))} \right] \qquad (11.47)
\end{aligned}
$$

Under condition (11.42), if

$$
\frac{1}{N} \boldsymbol{E} \left[\sum_{i=1}^{N} \|e_i(t)\|^2 \right] > \frac{\alpha + (g^2/N) \sum_{i=1}^{N} k_i^2 \beta_i}{-2(\theta + g \lambda_M(A_c))}
$$

then $\boldsymbol{E}V(t) - \boldsymbol{E}V(0) < 0$, which completes the proof.

The case of identical controllers $k_i = k$, $i = 1, ..., N$ is covered by the following corollary:

Corollary 11.3.1 *Suppose that Assumptions 11.3.1 and 11.3.2 hold. Then the controllers (11.39) with $k_i = k$, $i = 1, ..., N$ are distributed consensus controllers if*

$$\theta - gk < 0 \tag{11.48}$$

The estimated bound is given by

$$\lim_{t\to\infty} \frac{1}{N} \mathbf{E}\left(\sum_{i=1}^{N} ||x_i(t) - s(t)||^2 \right) \leq \frac{\alpha + (g^2/N) \sum_{i=1}^{N} k^2 \beta_i}{2(gk - \theta)} \tag{11.49}$$

Proof 11.3.2 *Follows from Lemma A.10.2 and Theorem (11.3.1).*

Setting

$$h(k) = \frac{\alpha + (g^2/N) \sum_{i=1}^{N} k^2 \beta_i}{2(gk - \theta)}$$

Simple mathematics shows that

$$\frac{dh(k)}{dk} = \frac{((\frac{g^3}{N} \sum_{i=1}^{N} k^2 \beta_i - 2\frac{g^2}{N} \sum_{i=1}^{N} k\theta\beta_i - g\alpha)\alpha + (g^2/N)}{)} 2(gk - \theta)^2$$

The equation $dh(k)/dk = 0$ has roots

$$k_{1,2}^* = \frac{\theta \pm \sqrt{\theta^2 + \left[\frac{\alpha}{(1/N) \sum_{i=1}^{N} \beta_i}\right]}}{g}$$

Since

$$k_1^* > k > \frac{\theta}{g} > k_2^*$$

if $k_1^* > k > 0$, then $dh(k)/dk < 0$; if $k > k_1^*$, then $dh(k)/dk > 0$. This leads to a minimum value of $h(k)$ exists at $k = k_1^*$.

The important special case of a single sensor being used to track the target $N = 1$, has the following corollary

Corollary 11.3.2 *Suppose that Assumptions 11.3.1 and 11.3.2 hold. Then the controllers (11.39) with $k_i = k$, $i = 1, ..., N$ are distributed consensus controllers if*

$$\theta - gk_1 < 0 \tag{11.50}$$

The estimated bound is given by

$$\lim_{t\to\infty} \mathbf{E}\left(\sum_{i=1}^{N} ||x_i(t) - s(t)||^2 \right) \leq \frac{\alpha + g^2 k_1^2 \beta_1}{2(gk_1 - \theta)} \tag{11.51}$$

When Assumption 11.3.3 holds, we get the following result:

Corollary 11.3.3 *Suppose that Assumptions 11.3.2 and 11.3.3 hold. Then the controllers (11.39) are distributed bounded consensus controllers and the estimated bound is given by*

$$\lim_{t\to\infty} \frac{1}{N} \boldsymbol{E}\left(\sum_{i=1}^{N} ||x_i(t) - s(t)||^2\right)$$

$$\leq \left[\frac{\gamma + \sqrt{\gamma^2 - g\lambda_M(A_c)\frac{(\alpha + (g^2/N)\sum_{i=1}^{N} k_i^2\beta_i)}{2}}}{-g\lambda_M(A_c)}\right]^2 \qquad (11.52)$$

Proof 11.3.3 *Consider the following Lyapunov function (11.43). In view of Assumption 11.3.3, the weak infinitesimal operator $\mathcal{L}$ of the stochastic process gives*

$$\mathcal{L}V(t) \leq \sum_{i=1}^{N} e_i^t(t)\left[f(x_i(t), t) - f(s(t), t)\right] + g\lambda_M(A_c)\sum_{i=1}^{N} e_i^t(t)e_i(t)$$

$$+ \frac{1}{2}N\alpha + \frac{g^2}{2}\sum_{i=1}^{N} k_i^2\beta_i$$

$$\leq 2\gamma\sum_{i=1}^{N} ||e_i(t)|| + g\lambda_M(A_c)\sum_{i=1}^{N} ||e_i(t)||^2$$

$$+ \frac{1}{2}N\alpha + \frac{g^2}{2}\sum_{i=1}^{N} k_i^2\beta_i \qquad (11.53)$$

Since $\lambda_M(A_c) < 0$ by Lemma A.10.1 and noting that

$$\left[\sum_{i=1}^{N} ||e_i(t)||\right]^2 = \sum_{i=1}^{N}\sum_{j=1}^{N} ||e_i(t)||||e_j(t)||$$

$$\leq \frac{1}{2}\sum_{i=1}^{N}\sum_{j=1}^{N}\left[||e_i(t)||^2 + ||e_j(t)||^2\right] = N\sum_{i=1}^{N} ||e_i(t)||^2$$

it follows that

$$
\mathcal{L}V(t) \leq g\lambda_M(A_c)\sum_{i=1}^{N}||e_i(t)||^2 + 2\sqrt{N}\sqrt{\sum_{i=1}^{N}||e_i(t)||^2 + \frac{1}{2}N\alpha + \frac{g^2}{2}\sum_{i=1}^{N}k_i^2\beta_i}
$$

$$
= gN\lambda_M(A_c)\left[\frac{1}{N}\sum_{i=1}^{N}||e_i(t)||^2 + \frac{2\gamma}{g\lambda_M(A_c)}\sqrt{\frac{1}{N}\sum_{i=1}^{N}||e_i(t)||^2}\right.
$$

$$
\left. - \frac{(\alpha + (g^2/N)\sum_{i=1}^{N}k_i^2\beta_i)}{-2g\lambda_M(A_c)}\right] \tag{11.54}
$$

For convenience, define

$$
\eta = \sqrt{\frac{1}{N}\sum_{i=1}^{N}||e_i(t)||^2}
$$

$$
d(\eta) = \eta^2 + (\frac{2\gamma}{g\lambda_M(A_c)})\eta + \frac{(\alpha + (g^2/N)\sum_{i=1}^{N}k_i^2\beta_i)}{(-2g\lambda_M(A_c))}
$$

It is not difficult to show that the condition $d(\eta) = 0$ has two solutions

$$
\eta_{1,2} = \frac{-\gamma + \sqrt{\gamma^2 - g\lambda_M(A_c)\frac{(\alpha+(g^2/N)\sum_{i=1}^{N}k_i^2\beta_i)}{2}}}{g\lambda_M(A_c)} \tag{11.55}
$$

where $\eta_1 < 0$, $\eta_2 > 0$. If $\eta(t) \geq \eta_2$, it is easy to check that $d(\eta) \geq 0$ and therefore $EV(t) - EV(0) < 0$, which completes the proof

Remark 11.3.1 *It is interesting to observe that Corollary 11.3.2 is useful on the local level (node to node) while for networks with (N) sensors Theorem 11.3.1 and Corollary 11.3.1 achieve the desired filtering performance.*

11.3.4   Filter design: pinning controllers

To reduce the number of controlled sensors, some local feedback injections may be applied to a small fraction of sensors. This approach is known as pinning control [662]. Here, the pinning strategy is applied to a small fraction $\delta(0 < \delta < 1)$ of the sensors in network (11.32). Without loss of generality, let the first $\ell = \mathbf{I}(\delta N)$ nodes be controlled, where $\mathbf{I}(.)$ is the integer part of a real number. In this regard, the designed pinning controllers can be described by

$$u_i(t) = -gk_i(x_i(t) - y_i(t)), \quad i = 1, 2, ..., \ell,$$
$$u_i(0) = 0, \quad i = \ell + 1, \ell + 2, ..., N \qquad (11.56)$$

where the gain $0 < k_i \in \Re$. In a similar way, we introduce

$$A_a = A - K_a, \; K_a = diag(k_1, ..., k_\ell, 0, ..., 0), \; A_a \in \Re^{N \times N} = \{a_{aij}\}$$
$$e_i(t) = x_i(t) - s(t), \quad i = 1, ..., N$$

Manipulating (11.30), (11.33) and (11.56), we obtain the following error dynamics:

$$d\dot{e}_i(t) = \left[f(x_i(t), t) - f(s(t), t) + g \sum_{j=1}^{N} a_{aij} e_j(t) \right] dt$$
$$- v(t)d\nu + gk_i \sigma_i(t)d\omega_i, \quad i = 1, 2, ..., \ell$$
$$d\dot{e}_i(t) = \left[f(x_i(t), t) - f(s(t), t) + g \sum_{j=1}^{N} a_{aij} e_j(t) \right] dt$$
$$- v(t)d\nu, \quad i = \ell + 1, \ell + 2, ..., N \qquad (11.57)$$

The following theorem establishes a main result for the controller:

Theorem 11.3.2 *Suppose that Assumptions 11.3.1 and 11.3.2 hold. Then the controllers (11.56) are distributed consensus controllers if*

$$\theta + g\lambda_M(A_a) < 0 \qquad (11.58)$$

The estimated bound is given by

$$\lim_{t \to \infty} \frac{1}{N} \boldsymbol{E} \left(\sum_{i=1}^{N} ||x_i(t) - s(t)||^2 \right) \le \frac{\alpha + (g^2/N) \sum_{i=1}^{N} k_i^2 \beta_i}{-2(\theta + g\lambda_M(A_a))} \qquad (11.59)$$

Proof 11.3.4 *By Lemma A.10.1, it is known that $A_a < 0$. Hence, using Theorem 11.3.2 the proof is completed.*

Corollary 11.3.4 *Suppose that Assumptions 11.3.2 and 11.3.3 hold. Then the controllers (11.56) are distributed bounded consensus controllers and the estimated bound is given by*

$$\lim_{t \to \infty} \frac{1}{N} \boldsymbol{E} \left(\sum_{i=1}^{N} ||x_i(t) - s(t)||^2 \right) \le \left[\frac{\gamma + \sqrt{\gamma^2 - g\lambda_M(A_a) \frac{(\alpha + (g^2/N) \sum_{i=1}^{N} k_i^2 \beta_i)}{2}}}{-g\lambda_M(A_a)} \right]^2$$
$$(11.60)$$

By expressing A_a in the form

$$A_a = \begin{bmatrix} A_1 - \tilde{K} & A_2 \\ \bullet & \tilde{A} \end{bmatrix}$$

where $\tilde{K} = diag(k_1, ..., k_\ell)$, A_1, A_2 are matrices with appropriate dimensions and $\tilde{A}$ is obtained by removing the $1, 2, ..., \ell$ row-column pairs of matrix A. Following [662], one may select $\tilde{K} > \lambda_M(A_1 - A_2\tilde{A}^{-1}A_2^t)I_\ell$. Then Theorem 11.3.2 specializes to

Corollary 11.3.5 *Suppose that Assumptions 11.3.1 and 11.3.2 hold. Then the controllers (11.56) are distributed consensus controllers if*

$$\theta + g\lambda_M(\tilde{A}) < 0 \qquad (11.61)$$

The estimated bound is given by

$$\lim_{t \to \infty} \frac{1}{N}\boldsymbol{E}\left(\sum_{i=1}^{N} ||x_i(t) - s(t)||^2 \right) \leq \frac{\alpha + (g^2/N)\sum_{i=1}^{N} k_i^2 \beta_i}{-2(\theta + g\lambda_M(\tilde{A}))} \qquad (11.62)$$

Proof 11.3.5 *It is not difficult to show that if $A_a - \lambda I_N < 0$ then $\tilde{A} - \lambda I_{N-\ell} < 0$. Moreover, choosing $\tilde{K} > \lambda_M(A_1 - A_2\tilde{A}^{-1}A_2^t)I_\ell$ then by the Schur complements, we have $A_a - \lambda I_N < 0$ which completes the proof. The proof is completed.*

11.3.5   Simulation example 2

In order to illustrate the developed distributed consensus controller, we perform a simulation-based analysis on the controlled scale-free network by using the high-degree pinning scheme. In the simulated scale-free network [665], $N = 100$, and $m_0 = m = 3$, which contains about 3000 connections. In the high degree pinning scheme, one first pins the node with the highest degree and then continues to choose and pin the other nodes in monotonically decreasing order of node degrees. The target model is described by Chua's circuit [666]:

$$\begin{cases} ds_1 = [\eta(-s_1 + s_-\ell(s)_1)]dt + v_1(t)d\nu \\ ds_2 = [s_1 - s_2 + s_3]dt + v_2(t)d\nu \\ ds_3 = [\beta s_2]dt + v_3(t)d\nu \end{cases} \qquad (11.63)$$

where

$$\ell(x)_1 = bx_1 + 0.5(a - b)(|x_1 + 1| - |x_1 - 1|), \quad v_i = 0.5, i = 1, 2, 3.$$

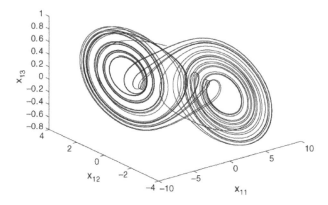

Figure 11.12: Chaotic orbits of Chua's circuit.

As shown in Figure 11.12, system (11.63) is noiseless and chaotic when $\eta = 10$, $\beta = 18$, $a = -4/3$ and $b = -3/4$. Computations show that Assumption 1 is verified with $\theta = 5.1623$. The measurement model is described by

$$y_i^+(t) = M_i s(t) + \sigma_i(t) \frac{d\omega_i(t)}{dt}, i = 1, 2, ..., \ell$$

where $\sigma_i(t) = 0.5$ and $M_i = [(1,0,0), \ (0, \ 1, \ 0), \ (0, \ 0, \ 1)$. The designed controllers have the form

$$u_i(t) = -gL_i(x_i(t) - y_i(t)), \quad i = 1, 2, ..., \ell,$$
$$u_i(0) = 0, \quad i = \ell + 1, \ell + 2, ..., N$$

with $g = 15$, $\ell = 20$ and $L_i = [(50,0,0), \ (0, 50, \ 0), \ (0, \ 0, \ 50)$. The selected M_i, L_i matrices satisfy the inequality

$$-\sum_{i=1}^{\bar{\ell}} e_i^t(t) L_l M_i e_i \leq \sum_{j=1}^{n} \sum_{k \in \mathsf{M_j}} \epsilon_{kj} e_{kj}^2$$

where $\bar{\ell} \leq \ell$, n is the number of sensors and $\mathsf{M_j}$ are the sets of all sensors that can observe the target state j, $j = 1, 2, ,,, n$. In this case, the constants are $\epsilon_{kj} = 50$, $k = 1, ..., \ell$, $j = 1, 2, 3$ Figures 11.13 and 11.14 depict the state estimate and the associated error patterns, respectively.

11.4 Distributed Fusion Filtering

This section presents a distributed fusion estimation method for estimating states of a dynamical process observed by wireless sensor networks (WSNs) with random

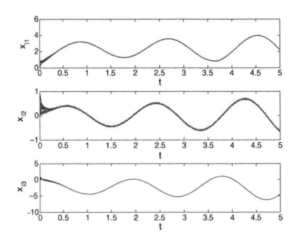

Figure 11.13: Patterns of state estimate x_i, $i = 1, ..., 100$.

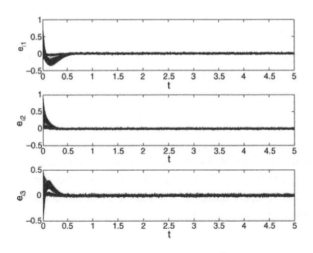

Figure 11.14: Patterns of errors e_i, $i = 1, ..., 100$.

packet losses. It is assumed that the dynamical process is not changing too rapidly, and a multi-rate scheme by which the sensors estimate states at a faster time scale and exchange information with neighbors at a slower time scale is proposed to reduce communication costs.

11.4.1 Introduction

A wireless sensor network (WSN) consists of spatially distributed autonomous sensors to cooperatively monitor physical or environmental conditions. The purpose of a WSN is to provide users access to the information of interest from data gathered by spatially distributed sensors. In most applications, users are interested in processed data that carries useful information about a physical plant rather than measured data contaminated by noises. Therefore, it is not surprising that signal estimation has been one of the most fundamental collaborative information processing problems in WSNs [670], [678].

However, it is known that the WSNs are usually severely constrained in energy, and energy-efficient methods are thus important for WSN based estimation to reduce energy consumption and to prolong network life. Consider the situation where a WSN is deployed to observe and estimate states of a dynamically changing process, but the process is not changing too rapidly. Then it is wasteful from an energy perspective for sensors to transmit every measurement to an estimator to generate estimates, and this waste is amplified by packet losses which are usually unavoidable in WSNs [673], [645], [525], [682], [647], [683], [686], [687]. Though there have been some energy-efficient estimation methods in the literature, such as the quantization method [674], [677], [685] and the data-compression method [668], [679], [688], they are not helpful in dealing with the above raised problem, because the main idea in quantization and compression is to reduce the size of a data packet and thus to reduce energy consumption in transmitting and receiving packets. Actually, a useful and straightforward approach to saving energy in the above considered estimation problem is to slow down the information transmission rate in the sensors; for example, the sensors may measure and transmit measurements with a period that is several times longer than the sampling period. This method might thus be intuitively called a transmission rate method. Few results of this method have been reported for WSN based estimation except for [675], and the main difficulty of using this method is that it may result in multi-rate estimation systems.

On the other hand, signal estimations in WSN could be done under the end-to-end information flow paradigm by communicating all the relevant data to a central collector node, e.g., a sink node. This, however, is a highly inefficient solution in WSN, because it may cause long packet delay, consume a large amount of energy and it has the potential for a critical failure point at the central collector node. An alternative solution is for the estimation to be performed in-network [669], [671], i.e., every sensor in the WSN with both sensing and computation capabilities performs not only as a sensor but also as an estimator, and it collects measurements only from its neighbors to generate estimates. It is obvious that local estimates obtained at each sensor by such a distributed in-network method are not optimal in the sense that not all the measurements in the WSN are used. Moreover, there exist disagreements among local estimates obtained at different sensors. In other words, local estimates at any two sensors may be different from each other. As pointed out in [643], such form of group disagreement regarding the signal estimates is highly undesirable for a peer-to-peer network of estimators. This gives rise to two issues that should be considered in designing a distributed estimation algorithm:

1. How could each sensor improve its local performance by making full use of limited information from its neighbors?

2. How to reduce disagreements of local estimates among different sensors?

A consensus strategy [667], [672], [643] , [302], [680], [684] and a diffusion strategy [627] have been presented in the literature to deal with the aforementioned two issues, where the consensus strategy mainly focuses on issue (1) while the diffusion strategy mainly focuses on issue (2). The main idea of the consensus strategy is that all sensors should obtain the same estimate in steady-state by using some consensus algorithms. In the diffusion strategy, both measurements and local estimates from neighboring sensors are used to generate estimates at each sensor. However, the energy-efficiency issue is not considered in both the consensus and diffusion strategies which usually require frequent information exchange among sensors to reach a common state and improve each local estimate. These motivate us to use the transmission rate method to design an energy-efficient fusion estimation method for the WSN based distributed estimation system with slowly changing dynamics and packet losses, and to provide a solution to the problems raised in issues (1) and (2).

In this section, the WSN is considered to be a peer-to-peer network without a fusion center, and every sensor in the network collects information only from its neighbors to generate estimates. A multi-rate scheme by which the sensors estimate states at a faster time scale and exchange information with neighbors at a slower time scale is proposed to reduce communication costs. Packets exchanged among the sensors may be lost during the transmission and several binary-valued white Bernoulli sequences are used to describe the random packet losses. Then, by applying a lifting technique as used in [675], [676], the multi-rate estimation system is finally modeled as a single-rate discrete-time system with multiple stochastic parameters. Based on the obtained system model, the distributed fusion estimation is carried out in two stages. At the first stage, every sensor in the WSN collects measurements from neighboring sensors to generate a local estimate, then local estimates from neighboring sensors are further collected to form a fused estimate at the second stage. By fusion of both measurements and local estimates, more information from different sensors is used to generate estimates in the two-stage method, as compared with the one-stage one where only measurements are collected to generate estimates. Therefore, the proposed two-stage estimation method helps steer each local estimate closer to the global optimal one and thus helps reduce disagreements of local estimates among different sensors. Then, by using the orthogonal projection principle and the innovation analysis approach, an estimation algorithm with a set of recursive Lyapunov and Riccati equations is presented to design the distributed estimators. The estimation performances obtained critically depend on the information transmission rate and the packet loss probabilities, and it is demonstrated by a simulation example of a maneuvering target tracking system that the time scale of information exchange among sensors can be slower, while still maintaining a satisfactory estimation performance by using the proposed estimation method.

11.4.2 Problem statement

Consider a linear discrete-time stochastic system described by the following state-space model

$$x(k_{i+1}) = A_p x(k_i) + B_p \omega_p(k_i), \quad i = 0, 1, 2, \dots \tag{11.64}$$

where $x(k_i) \in R^n$ is the system state, $\omega_p(k_i) \in R^p$ is a zero mean white noise, $h_p = k_{i+1} - k_i, \forall i = 0, 1, 2, \dots$ is the sampling period of the system (11.64). A WSN consisting of N spatially distributed sensors is deployed to collect observations of the system (11.64) according to the following observation models:

$$y_l(k_i) = C_{pl} x(k_i) + D_{pl} v_{pl}(k_i), \quad l = 1, \dots, N, \tag{11.65}$$

where $y_l(k_i) \in R^{m_l}$ is the observation collected by sensor l at time instant k_i, $v_{pl}(k_i) \in R^{q_l}$ are white measurement noises with zero means, A_p, B_p, C_{pl}, and D_{pl} are constant matrices with appropriate dimensions. $\omega_p(k_i)$ is uncorrelated with $v_{pl}(k_i)$, while $v_{pl}(k_i)$ are mutually correlated, and

$$E\{\omega_p(k_i)\omega_p^T(k_j)\} = Q_{\omega_p}\delta_{ij}, E\{v_{pl}(k_i)v_{ps}^T(k_j)\} = Q_{l,s}^{v_p}\delta_{ij}, l, s \in Z_0,$$

where $\delta_{ii} = 1$ and $\delta_{ij} = 0 (i \neq j)$.

The WSN is considered to be a peer-to-peer network; there is no fusion center in the network, and every sensor in the network acts also as an estimator. The observations are transmitted among the sensors in an ad-hoc manner via unreliable wireless communication channels and may be subject to random packet losses. We say that two sensors are connected if they can communicate directly with each other, i.e., they can communicate with each other within one hop. Notice that a sensor is always connected to itself. The set of sensors connected to a certain sensor r is called the neighborhood of sensor r and is denoted by $N_r, r \in Z_0 \triangleq \{1, \ldots, N\}$ (notice that $r \in N_r$), and the number of neighbors of sensor r is given by the number of elements of N_r, written as n_r. Denote by $L_i, j, i, j \in N_r$ the link between sensor i and sensor j in a neighborhood. Then, the random packet loss in the link $L_{i,j}$ is described by a white binary distributed random process $\alpha_{i,j}(k_i)$, where $\alpha_{i,j}(k_i) = 1$ indicates that a packet transmitted from sensor i successfully arrives at sensor j at instant k_i, while $\alpha_{i,j}(k_i) = 0$ implies that a packet is lost during the transmission from sensor i to sensor j.$\theta_{i,j}, E\{\alpha_{i,j}(k_i)\} = Prob\{\alpha_{i,j}(k_i) = 1\}$ is called the packet arriving probability (PAP), while $1 - \theta_{i,j} \triangleq 1 - E\{\alpha_{i,j}(k_i)\} = Prob\{\alpha_{i,j}(k_i) = 0\}$ is called the packet loss probability (PLP). By definition, one has $\alpha_{i,j}(k_i) = \alpha_{j,i}(k_i), \theta_{i,j} = \theta_{j,i}$, and $\theta_{i,i} = 1$. It is assumed that $\alpha_{l,r}(k_i), \forall l \in N_r, r \in Z_0$ are mutually independent and are also independent of $\omega_p(k_i), \upsilon_{pl}(k_i)$, and the initial system state. All the sensors in the network are assumed to be time-synchronized. Moreover, the sensors are time-driven, i.e., they calculate the state estimates periodically at certain time instants, and the sensors are not necessary to know the packet transmission status in the network. A structure of the distributed estimation system is shown schematically in Figure 11.15, where a distributed estimation is operated in a WSN to track a moving target.

Supposing that the dynamic of the stochastic process (11.64) is not changing too rapidly, then brutal force collection of every measurement at sampling instants k_i is a waste of energy, and this waste is amplified by packet losses. To reduce the energy waste, we suppose that every sensor r transmits measurements to its neighbors with a period h_m that is larger than the sampling period h_p. Denote $t_i, i = 0, 1, 2, \ldots$ as the measurement transmission instants, then $h_m = t_{i+1} - t_i, i = 0, 1, 2, \ldots$. Thus, every sensor in the WSN collects measurements, runs a Kalman estimator, calculates and outputs local estimates with a period h_m. In practice, one may expect to obtain estimates not only at the instances t_i but also at instances over the interval $(t_{i-1}, t_i]$, this is to say, one may expect to update the estimates at a rate that is higher than the estimate output rate. Suppose that estimates are updated at instances T_i, and $T_{i+1} - T_i = h_e, i = 0, 1, 2, \ldots$. In this generic case, the estimation system runs with three rates, namely, the measurement sampling rate (also the system state updating rate), the measurement transmitting rate (also the estimate output rate), and the estimate updating rate. In what follows, the multi-rate estimation system model will be transformed into a single-rate system model for further development by using the lifting technique.

For simplicity but without loss of generality, it is assumed that both the measurement transmitting period h_m and the estimate updating period h_e are integer multiples of the measurement sampling period h_p, and h_m is also the integer multiple of h_e. Specifically, let $h_e = ah_p$ and $h_m = bh_e$, where a and b are positive integers and chosen as small as possible in practice under the energy constraints of the sensor networks. Then, by applying the difference equation in (11.64) recursively, one obtains the following state equation with a state updating period of h_e.

$$x(T_{i+1}) = A_e x(T_i) + B_e \omega_e(T_i), i = 0, 1, 2, \ldots \tag{11.66}$$

where $A_e = A_p^a, B_e = [A_p^{a-1} B_p \ldots A_p B_p B_p]$, and

$$\omega_e(T_i) = [\omega_p^T(T_i)\omega_p^T(T_i + h_p) \ldots \omega_p^T(T_i + (a-1)h_p)]^T.$$

Similarly, applying the difference equation in (11.66) recursively leads to the following state equation with a state updating period of h_m

$$x(t_{i+1}) = A_m x(t_i) + B_m \omega_m(t_i), i = 0, 1, 2, \ldots \tag{11.67}$$

where $A_m = A_e^b, B_m = [A_e^{b-1} B_e \ldots A_e B_e B_e]$, and

$$\omega_m(T_i) = [\omega_e^T(T_i)\omega_e^T(t_i + h_e) \ldots \omega_e^T(t_i + (b-1)h_e)]^T.$$

The corresponding observation models are as follows

$$y_l(t_i) = C_{pl} x(t_i) + D_{pl} \upsilon_{pl}(t_i), \quad l = 1, \ldots, N. \tag{11.68}$$

By following the similar procedures for obtaining (11.67), one has for $j = 1, \ldots, b-1$ that

$$x(t_{i+1} - jh_e) = A_{mj} x(t_i) + B_{mj} \omega_m(t_i), \tag{11.69}$$

where $A_{mj} = A_e^{b-j}$, and

$$B_{m1} = [A_e^{b-2} B_e \ldots A_e B_e B_e 0],$$

$$\vdots$$

$$B_{m(b-1)} = [B_e 0 \ldots 0].$$

Define

$$\eta(t_i) = [x^T(t_i)x^T(t_i - h_e) \ldots x^T(t_i - (b-1)h_e)]^T,$$

then one obtains the following augmented single-rate estimation system model from (11.67)–(11.69)

$$
\begin{cases}
\eta(t_{i+1}) = A\eta(t_i) + B\omega_m(t_i), \\
y_l(t_i) = C_l\eta(t_i) + D_{pl}\upsilon_{pl}(t_i), \\
l = 1, \ldots, N, i = 0, 1, 2, \ldots
\end{cases}
\tag{11.70}
$$

where $C_l = [C_{pl}0 \ldots 0]$ and

$$
A = \begin{bmatrix}
A_m & 0 & \cdots & 0 \\
A_{m1} & 0 & \cdots & 0 \\
\vdots & 0 & \vdots & 0 \\
A_{m(b-1)} & 0 & \cdots & 0
\end{bmatrix}, \quad
B = \begin{bmatrix}
B_m \\
B_{m1} \\
\vdots \\
B_{m(b-1)}
\end{bmatrix}.
$$

The initial states $x(t_0 - jh_e), j = 0, 1, \ldots, b - 1$ are mutually uncorrelated and are also uncorrelated with $\omega_p(t_i)$ and $\upsilon_{pl}(t_i), l = 1, \ldots, N$, and satisfy $E\{x(t_0 - jh_e)\} = \bar{x}_j, E\{(x(t_0 - jh_e) - \bar{x}_j)(x(t_0 - jh_e) - \bar{x}_j)^T\} = \bar{P}_j$, where t_0 is the initial time.

At each instant t_i, every sensor collects measurements $y_l(t_i)$ from its neighbors to generate an unbiased state estimate $\hat{\eta}(t_{i+j}|t_i)$, where j is an integer, and thus the estimates $\hat{x}(t_{i+j}|t_i), \hat{x}(t_{i+j} - h_e|t_i), \ldots, \hat{x}(t_{i+j} - (b - 1)h_e|t_i)$ are obtained simultaneously in blocks. An example of the multi-rate state estimation is shown schematically in Figure 11.16, where a one-step prediction is considered, and $h_m = 2h_e, h_e = 2h_p, k_i$ are the measurement sampling instants (also the system state updating instants), T_i are the estimate updating instants, t_i are the estimate output instants (also the measurement transmitting instants). At each instant t_i, every sensor collects measurements from its neighbors, and prediction estimates $\hat{x}(t_{i+1}|t_i)$ and $\hat{x}(t_{i+1} - h_e|t_i)$ are then generated simultaneously.

Remark 11.4.1 *If $b = 1$, i.e., the estimate output rate equals the estimate updating rate ($h_m = h_e$), then $\eta(t_i)$ and $\omega_m(t_i)$ reduce to $x(T_i)$ and $\omega_e(T_i)$, respectively, while A, B and C_l reduce to A_e, B_e and C_{pl}, respectively, and thus the model (11.70) reduces to the model (11.66). Moreover, if $a = 1$ and $b = 1$, then $h_m = h_e = h_p, x(T_i)$ becomes $x(k_i), \omega_e(T_i)$ reduces to $\omega_p(k_i)$, while A_e and B_e reduce to A_p and B_p, respectively, and thus the model (11.66) reduces to the model (11.64).*

Remark 11.4.2 *An efficient method has been presented in [675] for designing multi-rate estimation systems, and the derivation of the single-rate estimation system model in (11.70) is motivated by the approach in [675]. However, the model*

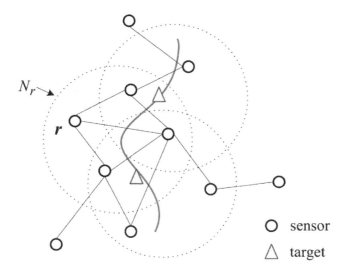

Figure 11.15: A structure of the distributed estimation system.

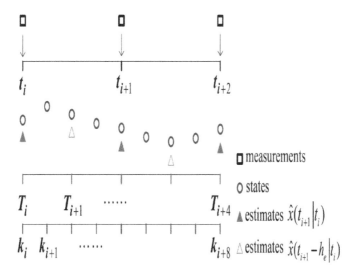

Figure 11.16: An example of the multi-rate estimation.

(11.70) is given in a different way for ease of presenting the fusion estimator de-sign procedures to be given in the following sections. Moreover, the model in (11.70) is different from that in [675] in two aspects. First, the state $x(t_i)$ is included in the augmented state $\eta(t_i)$. Notice that when filtering is considered, $\hat{x}(t_i - h_e|t_i), \ldots, \hat{x}(t_i - (b-1)h_e|t_i)$ are all delayed estimates. So, one advan-tage of the proposed model is that it can provide at least one non-delayed estimate $\hat{x}(t_i|t_i)$, and this is important in many practical applications, such as real-time moving target tracking. Second, the lifted noise $\omega_m(t_i)$ is still uncorrelated with $v_{pl}(t_i), l = 1, \ldots, N$ provided that $\omega_p(k_i)$ and $v_{pl}(k_i)$ are uncorrelated.

In what follows, an estimation system model with random packet losses will be established based on the model (11.70). Denote by $z_l(t_i)$ the measurement that sensor r receives from sensor l, then $z_l(t_i)$ might not equal $y_l(t_i)$ since $y_l(t_i)$ may be lost during the transmission. Suppose that the hold input mechanism [687] is adopted by all the sensors, i.e., sensor r will hold at its last available input when the current measurement is lost; then one has in this scenario that $z_l(t_i) = \alpha_{l,r}(t_i)y_l(t_i) + (1 - \alpha_{l,r}(t_i))z_l(t_{i-1})$. Stacking $z_l(t_i), l \in N_r$ into an augmented vector $Z_r(t_i) = col\{z_l(t_i)\}_{l \in N_r}$ which will be used to generate local estimates, one obtains

$$
\begin{aligned}
Z_r(t_i) &= col\{\alpha_{l,r}(t_i)y_l(t_i)\}_{l \in Nr} \\
&+ col\{(1 - \alpha_{l,r}(t_i))z_l(t_{i-1})\}_{l \in Nr}.
\end{aligned}
\tag{11.71}
$$

It can be seen from (11.71) that the stochastic variables $\alpha_{l,r}(t_i)$ are incorpo-rated into each element of the estimator input $Z_r(t_i)$, which makes the estima-tor design problem intractable. To remove the difficulty, some auxiliary matrices $\Pi_{l,r} = diag\{\underbrace{0, \ldots, 0}_{l-1}, I_{l,r}, \underbrace{0, \ldots, 0}_{n_r-l}\}, l \in N_r$ are introduced to rewrite $Z_r(t_i)$ in (11.71) as follows

$$
\begin{aligned}
Z_r(t_i) &= \left(\sum_{l \in N_r} \alpha_{l,r}(t_i)\Pi_{l,r} \right) Y_r(t_i) \\
&+ \left(I_r - \sum_{l \in N_r} \alpha_{l,r}(t_i)\Pi_{l,r} \right) Z_r(t_{i-1}),
\end{aligned}
\tag{11.72}
$$

where $Y_r(t_i) = col\{y_l(t_i)\}_{l \in N_r}, I_{l,r} \in R^{m_l \times m_l}$ and $I_r \in R^{\bar{m}_r \times \bar{m}_r}$ are identity matrices, and $\bar{m}_r = \sum_{l \in N_r} m_l$. Denote $G_r = col\{C_l\}_{l \in N_r}, H_r = diag\{D_{pl}\}_{l \in N_r}$, and $v_r(t_i) = col\{v_{pl}(t_i)\}_{l \in N_r}$, then $Y_r(t_i)$ is written as

$$Y_r(t_i) = G_r \eta(t_i) + H_r v_r(t_i). \tag{11.73}$$

Furthermore, denote

$$\xi_r(t_i) = [\eta^T(t_i) Z_r^T(t_{i-1})]^T, \ and \ \nu_r(t_i) = [\omega_m^T(t_i) v_r^T(t_i)]^T$$

then one obtains the following augmented system model from (11.70), (11.72) and (11.73)

$$\begin{cases} \xi_r(t_{i+1}) = \tilde{A}_r(t_i)\xi_r(t_i) + \tilde{B}_r(t_i)\nu_r(t_i), \\ Z_r(t_i) = \tilde{C}_r(t_i)\xi_r(t_i) + \tilde{v}_r(t_i), \\ r \in Z_0, i = 0, 1, 2, \dots \end{cases} \tag{11.74}$$

where $\tilde{v}_r(t_i) = \sum_{l \in N_r} \alpha_{l,r}(t_i) \Pi_{l,r} H_r v_r(t_i)$, and

$$\tilde{A}_r(t_i) = \begin{bmatrix} A & 0 \\ \sum_{l \in N_r} \alpha_{l,r}(t_i) \Pi_{l,r} G_r & l_r - \sum_{l \in N_r} \alpha_{l,r}(t_i) \Pi_{l,r} \end{bmatrix},$$

$$\tilde{B}_r(t_i) = diag \left\{ B, \sum_{l \in N_r} \alpha_{l,r}(t_i) \Pi_{l,r} H_r \right\},$$

$$\tilde{C}_r(t_i) = \begin{bmatrix} \sum_{l \in N_r} \alpha_{l,r}(t_i) \Pi_{l,r} G_r & l_r - \sum_{l \in N_r} \alpha_{l,r}(t_i) \Pi_{l,r} \end{bmatrix}.$$

Let $Q_\omega = E\{\omega_m(t_i)\omega_m^T(t_i)\}$ and $Q_{v_r} = E\{v_r(t_i)v_r^T(t_i)\}$, then

$$Q_\omega = diag\{Q_{\omega_p}\}_{a \times b}, \ Q_{v_r} = [Q_{l,s}^{v_p}], l, s \in N_r$$

To reduce energy consumption, the transmission rate method is used, and it naturally results in a multi-rate estimation system. By using the lifting technique, the multi-rate estimation system with random packet losses is finally modeled as a single-rate system with multiple stochastic parameters as in (11.74). Based on the system model (11.74), a two-stage fusion estimation method will be proposed to help to improve local estimation performance of each sensor and reduce disagreements among local estimates caused by the distributed structure of the estimation system. At each time step, every sensor collects measurements from its neighbors and runs a Kalman estimation algorithm to obtain a local estimate of the system state. At the second stage, the sensor further collects and fuses local estimates available at its neighbors to obtain a fused estimate. Thus, state estimation at each sensor based on local measurements and the further fused estimation based on the exchanged estimates among neighbors constitute the two-stage distributed fusion estimation at hand. Then, the objective of the chapter is described as follows.

The objective here is as follows: *Design distributed Kalman estimators for system (11.74) with packet losses and establish relationships between the measurement transmission rate, PLPs, and estimation performances.* The design is carried out in two stages. At the first stage, every sensor $r, r \in Z_0$ collects measurements from its neighborhood N_r and generates a local estimate $\hat{\eta}_r = g_r(y_l, \alpha_{l,r})_{l \in N_r}$, where $g_r()$ is a local Kalman estimation algorithm. At the second stage, sensor r collects local estimates from its neighborhood N_r and generates a fused estimate $\hat{\eta}_r^0 = f_r(\hat{\eta}_l, \alpha_{l,r})_{l \in N_r}$, where $f_r()$ refers to a local fusion algorithm.

11.4.3   Two-stage distributed estimation

This subsection is devoted initially to the design of the local Kalman estimation algorithm $g_r(.)$.

Taking expectations on $\tilde{A}_r(t_i)$, $\tilde{B}_r(t_i)$ and $\tilde{C}_r(t_i)$, yields, respectively,

$$\bar{A}_r \triangleq \mathbf{E}\{\tilde{A}_r(t_i)\} = \begin{bmatrix} A & 0 \\ \sum_{l \in N_r} \theta_{l,r} \Pi_{l,r} G_r & l_r - \sum_{l \in N_r} \theta_{l,r} \Pi_{l,r} \end{bmatrix},$$

$$\bar{B}_r \triangleq \mathbf{E}\{\tilde{B}_r(t_i)\} = diag\left\{ B, \sum_{l \in N_r} \theta_{l,r} \Pi_{l,r} H_r \right\},$$

$$\bar{C}_r \triangleq \mathbf{E}\{\tilde{C}_r(t_i)\} = \begin{bmatrix} \sum_{l \in N_r} \theta_{l,r} \Pi_{l,r} G_r & l_r - \sum_{l \in N_r} \theta_{l,r} \Pi_{l,r} \end{bmatrix}.$$

Denoting $A_{0l,r} = \begin{bmatrix} 0 & 0 \\ \Pi_{l,r} G_r & -\Pi_{l,r} \end{bmatrix}$ and $C_{0l,r} = [\Pi_{l,r} G_r \quad -\Pi_{l,r}]$, one obtains

$$\begin{cases} \tilde{A}_r(t_i) - \bar{A}_r = \sum_{l \in N_r} (\alpha_{l,r}(t_i) - \theta_{l,r}) A_{0l,r}, \\ \tilde{C}_r(t_i) - \bar{C}_r = \sum_{l - N_r} (\alpha_{l,r}(t_i) - \theta_{l,r}) C_{0l,r}. \end{cases} \tag{11.75}$$

Then, some lemmas which play important roles in the derivation of main results are first presented as follows.

Lemma 11.4.1 *From the distributions of $\alpha_{i,j}(t_i)$, it can be easily obtained for $\alpha_{i,j}(t_i) \neq \alpha_{r,s}(t_i), i, j, r, s \in Z_0$ that*

$$\mathbf{E}\{\alpha_{i,j}^2(t_i)\} = \theta_{i,j}, \quad \mathbf{E}\{\alpha_{i,j}(t_i)\alpha_{r,s}(t_i)\} = \theta_{i,j}\theta_{r,s},$$
$$\mathbf{E}\{(\alpha_{i,j}(t_i) - \theta_{i,j})^2\} = \theta_{i,j}(1 - \theta_{i,j}),$$
$$\mathbf{E}\{(\alpha_{i,j}(t_i) - \theta_{i,j})(\alpha_{r,s}(t_i) - \theta_{r,s})\} = 0.$$

Lemma 11.4.2 *For $r \in Z_0$, $\boldsymbol{E}\{\tilde{v}_r(t_i)\tilde{v}_r^T(t_i)\}$ satisfies*

$$\boldsymbol{E}\{\tilde{v}_r(t_i)\tilde{v}_r^T(t_i)\} \triangleq \Delta_{r,r} = \sum_{l \in N_r} \theta_{l,r} \Pi_{l,r} H_r Q_{v_r} H_r^T \Pi_{l,r}^T$$

$$+ \sum_{l \in N_r} \sum_{j \in N_r, j \neq 1} \theta_{l,r} \theta_{j,r} \Pi_{l,r} H_r Q_{v_r} H_r^T \Pi_{j,r}^T. \qquad (11.76)$$

Proof 11.4.1 **Lemma** *11.4.2 can be followed by* **Lemma** *11.4.1.*

Lemma 11.4.3 *Define the state covariance matrix as $\Xi_{r,r}(t_i)$, $\boldsymbol{E}\{\xi_r(t_i)\ \xi_r^T(t_i)\}$, then $\Xi_{r,r}(t_i)$ satisfies the following recursion:*

$$\Xi_{r,r}(t_{i+1}) = \bar{A}_r \Xi_{r,r}(t_i) \bar{A}_r^T + diag\{BQ_\omega B^T, \Delta_{r,r}\}$$

$$+ \sum_{l \in N_r} \theta_{l,r}(1 - \theta_{l,r}) A_{0l,r} \Xi_{r,r}(t_i) A_{0l,r}^T, \qquad (11.77)$$

where the initial value of $\Xi_{r,r}(t_i)$ at t_0 is given by

$$\Xi_{r,r}(t_0) = \begin{bmatrix} \Lambda_n & O_1 \\ O_1^T & O_2 \end{bmatrix},$$

$$\Lambda_\eta \triangleq \boldsymbol{E}\{\eta(t_0)\eta^T(t_0)\} = diag\{\bar{P}_0 + \bar{x}_0\bar{x}_0^T, \bar{P}_1 + \bar{x}_1\bar{x}_1^T, \dots, \bar{P}_{b-1} + \bar{x}_{b-1}\bar{x}_{b-1}^T\},$$

$$O_1 \quad in \quad R^{b_n \times \bar{m}_r}, \quad O_2 \in R^{\bar{m}_r \times \bar{m}_r}$$

are zero matrices.

Proof 11.4.2 $\xi_r(t_{i+1})$ *can be rewritten as*

$$\xi_r(t_{i+1}) = \bar{A}_r \xi_r(t_i) + (\tilde{A}_r(t_i) - \bar{A}_r)\xi_r(t_i) + \tilde{B}_r(t_i)\nu_r(t_i). \qquad (11.78)$$

Since $E\{\tilde{A}_r(t_i) - \bar{A}_r\} = 0$ and $\xi_r(t_i) \perp \nu_r(t_i)$, one has by (11.78) that

$$\Xi_{r,r}(t_{i+1}) = \bar{A}_r \Xi_{r,r}(t_i) \bar{A}_r^T$$

$$+ \boldsymbol{E}\{(\tilde{A}_r(t_i) - \bar{A}_r)\xi_r(t_i)\xi_r^T(t_i)(\tilde{A}_r(t_i) - \bar{A}_r)^T\}$$

$$+ \boldsymbol{E}\{\tilde{B}_r(t_i)\nu_r(t_i)\nu_r^T(t_i)\tilde{B}_r^T(t_i)\}. \qquad (11.79)$$

It follows from (11.75) and Lemma 11.4.1 that

$$\boldsymbol{E}\{(\tilde{A}_r(t_i) - \bar{A}_r)\xi_r(t_i)\xi_r^T(t_i)(\tilde{A}_r(t_i) - \bar{A}_r)^T\}$$

$$= \boldsymbol{E}\left\{\sum_{l \in N_r}(\alpha_{l,r}(t_i) - \theta_{l,r})^2 A_{0l,r}\xi_r(t_i)\xi_r^T(t_i)A_{0l,r}^T\right\}$$

$$+ \boldsymbol{E}\left\{\sum_{l \in N_r}\sum_{j \in N_r, j \neq 1}(\alpha_{l,r}(t_i) - \theta_{l,r})(\alpha_{j,r}(t_i) - \theta_{j,r})\right.$$

$$\left. \times A_{0l,r}\xi_r(t_i)\xi_r^T(t_i)A_{0j,r}^T\right\}$$

$$= \sum_{l \in N_r}\theta_{l,r}(1 - \theta_{l,r})A_{0l,r}\Xi_{r,r}(t_i)A_{0l,r}^T \qquad (11.80)$$

Since $\omega_m(t_i)$ and $\upsilon_r(t_i)$ are uncorrelated, one has by Lemma 11.4.2 that

$$\boldsymbol{E}\{\tilde{B}_r(t_i)\nu_r(t_i)\nu_r^T(t_i)\tilde{B}_r^T(t_i)\}$$

$$= \boldsymbol{E}\left\{diag\left\{B, \sum_{l \in N_r}\alpha_{l,r}(t_i)\Pi_{l,r}H_r\right\}\begin{bmatrix} \omega_m(t_i) \\ \upsilon_r(t_i) \end{bmatrix}\right.$$

$$\left.\begin{bmatrix} \omega_m(t_i) \\ \upsilon_r(t_i) \end{bmatrix}^T diag\left\{B, \sum_{l \in N_r}\alpha_{l,r}(t_i)\Pi_{l,r}H_r\right\}^T\right\}$$

$$= diag\{BQ_\omega B^T, \Delta_{r,r}\}. \qquad (11.81)$$

Substituting (11.80) and (11.81) into (11.79) leads to (11.77). The proof is thus completed.

With **Lemmas** 11.4.1–11.4.3 in hand, we are now ready to present design procedures for the finite horizon local Kalman estimators. Let $\hat{\xi}_r(t_i|t_i)$ and $\hat{\xi}_r(t_{i+1}|t_i)$ denote, respectively, the unbiased linear minimum MSE filtered estimate and the one-step predicted estimate of the state $\xi_r(t_i)$. Then, the recursive local Kalman filter for system (11.74) is given in the following theorem.

Theorem 11.4.1 *For system (11.74), the finite horizon local Kalman filter in the sensor* $r, r \in Z_0$ *is given by:*

$$\varepsilon_r(t_i) = Z_r(t_i) - \bar{C}_r \hat{\xi}_r(t_i|t_{i-1}), \tag{11.82}$$

$$\Omega_r(t_i) = \sum_{l \in N_r} \theta_{l,r}(1 - \theta_{l,r}) C_{0l,r} \Xi_{r,r}(t_i) C_{0l,r}^T$$

$$+ \ C_r P_{r,r}(t_i|t_{i-1}) \bar{C}_r^T + \Delta_{r,r}, \tag{11.83}$$

$$K_r(t_i) = P_{r,r}(t_i|t_{i-1}) \bar{C}_r^T \Omega_r^{-1}(t_i), \tag{11.84}$$

$$\tag{11.85}$$

$$F_r(t_i) = \left[\sum_{l \in N_r} \theta_{l,r}(1 - \theta_{l,r}) A_{0l,r} \Xi_{r,r}(t_i) C_{0l,r}^T \right.$$

$$+ \ \bar{A}_r P_{r,r}(t_i|t_{i-1}) \bar{C}_r^T + \left[\begin{array}{c} 0 \\ \Delta_{r,r} \end{array} \right] \Bigg] \Omega_r^{-1}, (t_i), \tag{11.86}$$

$$\hat{\xi}_r(t_i|t_i) = \hat{\xi}_r(t_i|t_{i-1}) + K_r(t_i)\varepsilon_r(t_i), \tag{11.87}$$

$$\hat{\xi}_r(t_{i+1}|t_i) = \bar{A}_r \hat{\xi}_r(t_i|t_{i-1}) + F_r(t_i)\varepsilon_r(t_i), \tag{11.88}$$

$$P_{r,r}(t_i|t_i) = P_{r,r}(t_i|t_{i-1}) - K_r(t_i)\Omega_r(t_i)K_r^T(t_i), \tag{11.89}$$

$$P_{r,r}(t_{i+1}|t_i) = \sum_{l \in N_r} \theta_{l,r}(1 - \theta_{l,r})(A_{0l,r} - F_r(t_i)C_{0l,r})$$

$$\times \ \Xi_{r,r}(t_i)(A_{0l,r} - F_r(t_i)C_{0l,r})^T + diag\{BQ_\omega B^T, \Delta_{r,r}\}$$

$$+ \ (\bar{A}_r - F_r(t_i)\bar{C}_r)P_{r,r}(t_i|t_{i-1})(\bar{A}_r - F_r(t_i)\bar{C}_r)^T$$

$$- \ [0\varphi_{r,r}^T(t_i)] - [0\varphi_{r,r}^T(t_i)]^T + \varrho_{r,r}(t_i), \tag{11.90}$$

where $\varepsilon_r(t_i)$ *is the innovation sequence with covariance*

$$\Omega_r(t_i) \triangleq \boldsymbol{E}\{\varepsilon_r(t_i)\varepsilon_r^T(t_i)\}, K_r(t_i)$$

and $F_r(t_i)$ *are gain matrices of the filter and the one-step predictor, respectively,* $P_{r,r}(t_i|t_i)$ *and* $P_{r,r}(t_i|t_{i-1})$ *are the covariance matrices of the filtering error and the prediction error, respectively, and the initial values of* $\hat{\xi}_r(t_i|t_{i-1})$ *and* $P_{r,r}(t_i|t_{i-1})$ *at* t_0 *are given, respectively, by*

$$
\hat{\xi}_r(t_0|t_{-1}) = \begin{bmatrix} \eta_0 \\ 0 \end{bmatrix}, \quad P_{r,r}(t_0|t_{-1}) = \begin{bmatrix} \Lambda_p & O_1 \\ O_1^T & O_2 \end{bmatrix}
$$

$$
\eta_0 := E\{\eta(t_0)\} = [\bar{x}_0^T \bar{x}_1^T \dots \bar{x}_{b-1}^T]^T, \ \Lambda_P = diag\{\bar{P}_0, \bar{P}_1, \dots, \bar{P}_{b-1}\}
$$

$$
\varphi_{r,r}(t_i) = \sum_{l \in N_r} \theta_{l,r} \Pi_{l,r} H_r Q_{vr} H_r^T \Pi_{l,r}^T F_r^T(t_i)
$$

$$
+ \sum_{l \in N_r} \sum_{j \in N_r, j \neq l} \theta_{l,r} \theta_{j,r} \Pi_{l,r} H_r Q_{vr} H_r^T \Pi_{j,r}^T F_r^T(t_i),
$$

$$
\varrho_{r,r}(t_i) = \sum_{l \in N_r} \theta_{l,r} F_r(t_i) \Pi_{l,r} H_r Q_{vr} H_r^T \Pi_{l,r}^T F_r^T(t_i)
$$

$$
+ \sum_{l \in N_r} \sum_{j \in N_r, j \neq l} \theta_{l,r} \theta_{j,r} F_r(t_i) \Pi_{l,r}
$$

$$
\times \quad H_r Q_{vr} H_r^T \Pi_{j,r}^T F_r^T(t_i).
$$

Proof 11.4.3 *The innovation $\varepsilon_r(t_i)$ is defined as*

$$
\varepsilon_r(t_i) \triangleq Z_r(t_i) - \hat{Z}_r(t_i|t_{i-1}). \tag{11.91}
$$

Taking projection of both sides of the output equation in (11.74) onto the linear space $L(Z_r(t_0), Z_r(t_1), \dots, Z_r(t_{i-1}))$ yields

$$
\hat{Z}_r(t_i|t_{i-1}) = \bar{C}_r \hat{\xi}_r(t_i|t_{i-1}) + \left(\sum_{l \in N_r} \theta_{l,r} \Pi_{l,r} H_r \right)
$$

$$
\times proj\{v_r(t_i)|Z_r(t_0), Z_r(t_1), \dots, Z_r(t_{i-1})\}. \tag{11.92}
$$

Define a set $v_r(t_i) \triangleq \bigcup_{l \in N_r} \beta_{l,r}(t_i)$, where

$$
\beta_{l,r}(t_i) = \{\alpha_{l,r}(t_0), \alpha_{l,r}(t_1), \dots, \alpha_{l,r}(t_i)\}
$$

Then, one has by (11.74) that

$$
Z_r(t_i) \in L(v_r(t_i), \nu_r(t_{i-1}), \dots, \nu_r(t_0), \xi_r(t_0))_{\vartheta r(t_i)}, \tag{11.93}
$$

where $L()_{\vartheta r(t_i)}$ denotes that the linear space $L()$ is dependent on the stochastic parameters in the set $\vartheta r(t_i)$. It follows from (11.93) that

$$
L(Z_r(t_0), Z_r(t_1), \dots, Z_r(t_{i-1})) \subset L(v_r(t_{i-1}), dots,
$$

$$
v_r(t_0), \nu_r(t_{i-2}), \dots, \nu_r(t_0), \xi_r(t_0))_{\vartheta r(t_{i-1})}. \tag{11.94}
$$

Since

$$v_r(t_i) \perp L(v_r(t_{i-1}), \ldots, v_r(t_0), \nu_r(t_{i-2}), \ldots, \nu_r(t_0), \xi_r(t_0))_{\vartheta r(t_{i-1})},$$

it follows from (11.94) that

$$v_r(t_i) \perp L(Z_r(t_0), Z_r(t_1), \ldots, Z_r(t_{i-1})). \tag{11.95}$$

Since $E\{v_r(t_i)\} = 0$, (11.95) implies that

$$proj\{v_r(t_i)|Z_r(t_0), Z_r(t_1), \ldots, Z_r(t_{i-1})\} = 0.$$

which together with (11.91) and (11.92) yields (11.82).

By the projection approach [516], one has the following equations for determining the filtered estimate $\hat{\xi}_r(t_i|t_i)$:

$$\hat{\xi}_r(t_i|t_i) = \hat{\xi}_r(t_i|t_{i-1}) + K_r(t_i)\varepsilon_r(t_i), \tag{11.96}$$

$$K_r(t_i) = E\{\xi_r(t_i)\varepsilon_r^T(t_i)\}\Omega_r^{-1}(t_i). \tag{11.97}$$

Notice that (11.96) is just the equation in (11.87). Define the prediction error as $\tilde{\xi}_r(t_i|t_{i-1}) = \xi_r(t_i) - \hat{\xi}_r(t_i|t_{i-1})$, then substituting $Z_r(t_i)$ in (11.74) into (11.82) leads to

$$\varepsilon_r(t_i) = (\tilde{C}_r(t_i) - \bar{C}_r)\xi_r(t_i) + \bar{C}_r\tilde{\xi}_r(t_i|t_{i-1}) + \tilde{v}_r(t_i). \tag{11.98}$$

Since $\xi_r(t_i) \perp v_r(t_i), \tilde{\xi}_r(t_i|t_{i-1}) \perp v_r(t_i)$ and $E\{\tilde{C}_r(t_i) - \bar{C}_r\} = 0$, one has by (11.98) that

$$
\begin{aligned}
\Omega_r(t_i) &= E\{(\tilde{C}_r(t_i) - \bar{C}_r)\xi_r(t_i)\xi_r^T(t_i)(\tilde{C}_r(t_i) - \bar{C}_r)^T\} \\
&+ E\{\bar{C}_r\tilde{\xi}_r(t_i|t_{i-1})\tilde{\xi}_r^T(t_i|t_{i-1})\bar{C}_r^T\} \\
&+ E\{\tilde{v}_r(t_i)\tilde{v}_r^T(t_i)\}.
\end{aligned} \tag{11.99}
$$

By (11.75) and **Lemma** *11.4.1, and following the similar derivation procedures as in (11.80), one obtains*

$$
\begin{aligned}
E\{(\tilde{C}_r(t_i) - \bar{C}_r)\xi_r(t_i)\xi_r^T(t_i)(\tilde{C}_r(t_i) - \bar{C}_r)^T\} \\
\sum_{l \in N_r} \theta_{l,r}(1 - \theta_{l,r})C_{0l,r}\Xi_{r,r}(t_i)C_{0l,r}^T.
\end{aligned} \tag{11.100}
$$

Then, (11.83) follows from (11.99), (11.100) and **Lemma 11.4.2.** *Substituting (11.98) into (11.97) and taking the facts* $E\{\tilde{C}_r(t_i) - \bar{C}_r\} = 0, \xi_r(t_i) \perp v_r(t_i), \hat{\xi}_r(t_i|t_{i-1}) \perp \tilde{\xi}_r(t_i|t_{i-1})$ *into account, yields*

$$
\begin{aligned}
K_r(t_i) &= E\{\xi_r(t_i)\tilde{\xi}_r(t_i|t_{i-1})\bar{C}_r^T\}\Omega_r^{-1}(t_i) \\
&= E\{(\hat{\xi}_r(t_i|t_{i-1}) + \tilde{\xi}_r(t_i|t_{i-1}))\tilde{\xi}_r(t_i|t_{i-1})\bar{C}_r^T\}\Omega_r^{-1}(t_i) \\
&= P_{r,r}(t_i|t_{i-1})\bar{C}_r^T\Omega_r^{-1}(t_i). \tag{11.101}
\end{aligned}
$$

By the projection approach [516], one has the following equations for determining the one-step predicted estimate $\hat{\xi}_r(t_{i-1|t_i})$:

$$
\hat{\xi}_r(t_{i+1}|t_i) = \hat{\xi}_r(t_{i+1}|t_{i-1}) + F_r(t_i)\varepsilon_r(t_i), \tag{11.102}
$$

$$
F_r(t_i) = E\{\xi_r(t_{i+1})\varepsilon_r^T(t_i)\}\Omega_r^{-1}(t_i). \tag{11.103}
$$

Taking both sides of the state equation in (11.74) onto the space

$$
L(Z_r(t_0), Z_r(t_1), \ldots, Z_r(t_{i-1}))
$$

yields

$$
\begin{aligned}
\hat{\xi}_r(t_{i+1}|t_{i-1}) &= \bar{A}_r\hat{\xi}_r(t_i|t_{i-1}) + \bar{B}_r \\
&\quad \times \; proj\{v_r(t_i)|Z_r(t_0), Z_r(t_1), \ldots, Z_r(t_{i-1})\}. \tag{11.104}
\end{aligned}
$$

It follows from (11.94) that $v_r(t_i) \perp L(Z_r(t_0), Z_r(t_1), \ldots, Z_r(t_{i-1}))$, *which together with the fact* $E\{v_r(t_i)\} = 0$ *leads to*

$$
proj\{v_r(t_i)|Z_r(t_0), Z_r(t_1), \ldots, Z_r(t_{i-1})\} = 0. \tag{11.105}
$$

Combining (11.102), (11.104) and (11.105) yields (11.88). Substituting the state equation in (11.74) into (11.103) yields

$$
\begin{aligned}
F_r(t_i) &= \boldsymbol{E}\{\tilde{A}_r(t_i)\xi_r(t_i)\varepsilon_r^T(t_i)\}\Omega_r^{-1}(t_i) \\
&\quad + \boldsymbol{E}\{\tilde{B}_r(t_i)v_r(t_i)\varepsilon_r^T(t_i)\}\Omega_r^{-1}(t_i). \tag{11.106}
\end{aligned}
$$

Substitute (11.98) into (11.106), then one obtains by (11.75), Lemma 11.4.1, and

$\xi_r(t_i) \perp \nu_r(t_i), \hat{\xi}_r(t_i|t_{i-1}) \perp \tilde{\xi}_r(t_i|t_{i-1})$ *and* $\boldsymbol{E}\{\alpha_{l,r}(t_i) - \theta_{l,r}\} = 0$ *that*

$$
\begin{aligned}
\boldsymbol{E}\{\tilde{A}_r(t_i)\xi_r(t_i)\varepsilon_r^T(t_i)\} &= E\{\tilde{A}_r(t_i)\xi_r(t_i)\xi_r^T(t_i)(\tilde{C}_r(t_i) - \bar{C}_r)^T \\
&+ \tilde{A}_r(t_i)\xi_r(t_i)\tilde{\xi}_r^T(t_i|t_{i-1})\bar{C}_r^T\} \\
&= \boldsymbol{E}\left\{\left[\bar{A}_r + \sum_{l\in N_r}(\alpha_{l,r}(t_i) - \theta_{l,r})A_{0l,r}\right]\xi_r(t_i)\xi_r^T(t_i) \right. \\
&\qquad \left. \times\left(\sum_{l\in N_r}(\alpha_{l,r}(t_i) - \theta_{l,r})C_{0l,r}\right)^T\right\} \\
&+ \boldsymbol{E}\{\tilde{A}_r(t_i)(\hat{\xi}_r(t_i|t_{i-1}) + \tilde{\xi}_r^T(t_i|t_{i-1}))\tilde{\xi}_r^T(t_i|t_{i-1})\bar{C}_r\} \\
&= \sum_{l\in N_r}\theta_{l,r}(1 - \theta_{l,r})A_{0l,r}\Xi_{r,r}(t_i)C_{0l,r}^T \\
&+ \bar{A}_r P_{r,r}(t_i|t_{i-1})\bar{C}_r^T. \qquad (11.107)
\end{aligned}
$$

Since $\xi_r(t_i) \perp \nu_r(t_i), \bar{\xi}_r(t_i|t_{i-1}) \perp \nu_r(t_i)$, *and* $\omega_m(t_i) \perp \upsilon_r(t_i)$, *one has by Lemma 11.4.2 that*

$$
E\{\tilde{B}_r(t_i)\nu_r(t_i)\varepsilon_r^T(t_i)\} = [0\Delta_{r,r}^T]^T. \qquad (11.108)
$$

Combining (11.106)–(11.108) leads to (11.86).

Derivation procedures for the covariance matrices $P_{r,r}(t_{i+1}|t_i)$ *and* $P_{r,r}(t_i|t_i)$ *are presented as follows. Substituting (11.88) and the state equation in (11.74) into the right–hand side of the equation* $\tilde{\xi}_r(t_{i+1}|t_i) = \xi_r(t_{i+1}) - \hat{\xi}_r(t_{i+1}|t_i)$, *one has by (11.75) and (11.98) that*

$$
\begin{aligned}
\tilde{\xi}_r(t_{i+1}|t_i) &= \left[\bar{A}_r + \sum_{l \in N_r} (\alpha_{l,r}(t_i) - \theta_{l,r}) A_{0l,r} \right] \xi_r(t_i) \\
&\quad - \bar{A}_r \tilde{\xi}_r(t_i|t_{i-1}) + \tilde{B}_r(t_i)\nu_r(t_i) - F_r(t_i)\varepsilon_r(t_i) \\
&= \bar{A}_r \tilde{\xi}_r(t_i|t_{i-1}) + \left[\sum_{l \in N_r} (\alpha_{l,r}(t_i) - \theta_{l,r}) A_{0l,r} \right. \\
&\quad \left. - F_r(t_i) \sum_{l \in N_r} (\alpha_{l,r}(t_i) - \theta_{l,r}) C_{0l,r} \right] \xi_r(t_i) \\
&\quad + \tilde{B}_r(t_i)\nu_r(t_i) - F_r(t_i)\bar{C}_r \tilde{\xi}_r(t_i|t_{i-1}) \\
&\quad - F_r(t_i) \sum_{l \in N_r} \alpha_{l,r}(t_i)\Pi_{l,r}H_r \upsilon_r(t_i) \\
&= \sum_{l \in N_r} (\alpha_{l,r}(t_i) - \theta_{l,r})(A_{0l,r} - F_r(t_i)C_{0l,r})\xi_r(t_i) \\
&\quad + (\bar{A}_r - F_r(t_i)\bar{C}_r)\tilde{\xi}_r(t_i|t_{i-1}) + \tilde{B}_r(t_i)\nu_r(t_i) \\
&\quad - \sum_{l \in N_r} \alpha_{l,r}(t_i)F_r(t_i)\Pi_{l,r}H_r \upsilon_r(t_i).
\end{aligned} \tag{11.109}
$$

*Since $\xi_r(t_i) \perp \nu_r(t_i), \xi_r(t_i) \perp \upsilon_r(t_i), \tilde{\xi}_r(t_i|t_{i-1}) \perp \nu_r(t_i), \tilde{\xi}_r(t_i|t_{i-1}) \perp \upsilon_r(t_i),$
and $E\{\alpha_{l,r}(t_i) - \theta_{l,r}\} = 0$, one has by **Lemma 11.4.1** and (11.81), and following
the similar derivation procedures as in (11.80) that*

$$
\begin{aligned}
P_{r,r}(t_{i+1}|t_i) &= \boldsymbol{E}\{\tilde{\xi}_r(t_{i+1}|t_i)\tilde{\xi}_r^T(t_{i+1}|t_i)\} \\
&= \sum_{l \in N_r} \theta_{l,r}(1 - \theta_{l,r})(A_{0l,r} - F_r(t_i)C_{0l,r})\Xi_{r,r}(t_i) \\
&\quad \times (A_{0l,r} - F_r(t_i)C_{0l,r})^T \\
&\quad + (\bar{A}_r - F_r(t_i)\bar{C}_r)P_{r,r}(t_i|t_{i-1})(\bar{A}_r - F_r(t_i)\bar{C}_r)^T \\
&\quad - \boldsymbol{E}\{\tilde{B}_r(t_i)\nu_r(t_i)\sigma_r^T(t_i)\} - \boldsymbol{E}\{\sigma_r(t_i)(\tilde{B}_r(t_i)\nu_r(t_i))^T\} \\
&\quad + diag\{BQ_\omega B^T, \Delta_{r,r}\} + \boldsymbol{E}\{\sigma_r(t_i)\sigma_r^T(t_i)\}, \tag{11.110}
\end{aligned}
$$

where $\sigma_r(t_i) = \sum_{l \in N_r} \alpha_{l,r}(t_i)F_r(t_i)\Pi_{l,r}H_r\upsilon_r(t_i)$.
    By following the similar derivation procedures as in (11.108), one obtains

$$
\boldsymbol{E}\{\tilde{B}_r(t_i)\nu_r(t_i)\sigma_r^T(t_i)\} = [0\varphi_{r,r}^T(t_i)]^T. \tag{11.111}
$$

Moreover, it follows from Lemma 11.4.1 that

$$E\{\sigma_r(t_i)\sigma_r^T(t_i)\} = \varrho_{r,r}(t_i). \tag{11.112}$$

Combining (11.110)–(11.112) leads to (11.90).

Substituting (11.87) into the right–hand side of the equation $\tilde{\xi}_r(t_i|t_i) \triangleq \xi_r(t_i) - \hat{\xi}_r(t_i|t_i)$ yields

$$\tilde{\xi}_r(t_i|t_i) = \tilde{\xi}_r(t_i|t_{i-1}) - K_r(t_i)\varepsilon_r(t_i). \tag{11.113}$$

Let $\Phi(t_i) = E\{\tilde{\xi}_r(t_i|t_{i-1})\varepsilon_r^T(t_i)\}$, then it follows from (11.113) that

$$
\begin{aligned}
P_{r,r}(t_i|t_i) &= E\{\tilde{\xi}_r(t_i|t_i)\tilde{\xi}_r^T(t_i|t_i)\} \\
&= P_{r,r}(t_i|t_{i-1}) - K_r(t_i)\Phi^T(t_i) - \Phi(t_i)K_r^T(t_i) \\
&\quad + K_r(t_i)\Omega_r(t_i)K_r^T(t_i).
\end{aligned}
\tag{11.114}
$$

Since $\hat{\xi}_r(t_i|t_{i-1}) \perp \varepsilon_r(t_i)$, one has by (11.97) that

$$
\begin{aligned}
\Phi(t_i) &= E\{(\xi_r(t_i) - \hat{\xi}_r(t_i|t_{i-1}))\varepsilon_r^T(t_i)\} \\
&= E\{\xi_r(t_i)\varepsilon_r^T(t_i)\} = K_r(t_i)\Omega_r(t_i).
\end{aligned}
\tag{11.115}
$$

Substituting (11.115) into (11.114) leads to (11.89). The proof is thus completed.

Theorem 11.4.1 provides a set of recursive equations for designing the finite horizon of local Kalman filters; as a byproduct, the local one-step predictor is also given. Denote $\hat{\eta}_r(t_{i+k}|t_i)$ and $P_{r,r}^\eta(t_{i+k}|t_i)(k = 0, 1)$, respectively, the estimate of the system state $\eta(t_i)$ and the corresponding error covariance generated at sensor r. Then, $\hat{\eta}_r(t_{i+k}|t_i)$ and $P_{r,r}^\eta(t_{i+k}|t_i)$ are given by

$$
\begin{aligned}
\hat{\eta}_r(t_{i+k}|t_i) &= [I_{bn} \quad O_1]\hat{\xi}_r(t_{i+k}|t_i) \\
P_{r,r}^\eta(t_{i+k}|t_i) &= [I_{bn} \quad O_1]P_{r,r}(t_{i+k}|t_i)[I_{bn} \quad O_1]^T
\end{aligned}
$$

respectively, where $I_{bn} \in R^{bn \times bn}$ is an identity matrix.

In **Theorem** 11.4.1, every sensor r in the WSN generates local estimates by using measurements only from its neighbors. Each local estimate thus obtained is suboptimal in the sense that not all the measurements in the WSN are used. Moreover, there may exist disagreements among local estimates at different sensors. Similar to [643], one may define some disagreement potentials as follows to characterize the disagreement of local estimates in the neighborhood $N_r(r \in Z_0)$:

$$\kappa_r(t_i) = \frac{1}{2n_r}\sum_{u,s\in N_r}\|\hat{\eta}_u(t_{i+k}|t_i) - \hat{\eta}_s(t_{i+k}|t_i)\|^2, \tag{11.116}$$

$$\psi_r(t_i) = \frac{1}{2n_r} \sum_{u,s \in N_r} [Tr(P^\eta_{u,u}(t_{i+k}|t_i)) - Tr(P^\eta_{s,s}(t_{i+k}|t_i))]^2, \qquad (11.117)$$

where $k = 0, 1$, and $\kappa_r(t_i)$ and $\psi_r(t_i)$ are the disagreement potential of estimates and the disagreement potential of estimation performances, respectively. Notice that at each time step, not only a measurement but also a local estimate is available at each sensor. Therefore, one efficient way to improve each local estimation performance and reduce the disagreement is to further collect local estimates available at neighboring sensors and then generate a fused estimate at every sensor in the WSN. This gives rise to the two-stage estimation strategy. Different from the approach in [627] where a fusion rule with scalar weights is used, a fusion criterion weighted by matrices in the linear minimum variance sense will be used in this chapter to generate fused estimates, and the main results will be presented in the following subsection.

11.4.4   Distributed fusion algorithm

In the sequel, a fusion criterion weighted by matrices in the linear minimum variance sense is applied to generate fusion estimates for every sensor $r, r \in Z_0$, and the criterion is first given in the following lemma.

Lemma 11.4.4 *([681]). Let $\hat{x}_i, i \in \bar{Z} \triangleq \{1, \ldots, m\}$ be unbiased estimates of a stochastic state vector $x \in R^n$. Let the estimation errors be $\tilde{x}_i = x_i - \hat{x}_i$. Assume that $\tilde{x}_i$ and $\tilde{x}_j, i \neq j$ are correlated, and define the covariance and cross-covariance matrices as $P_{ii} = E\{\tilde{x}_i \tilde{x}_i^T\}$ and $P_{ij} = E\{\tilde{x}_i \tilde{x}_j^T\}(i \neq j)$, respectively. Then, the optimal fusion estimate of x with matrix weights is given by*

$$\hat{x}_o = \sum_{i=1}^m A_{oi}\hat{x}_i, \qquad (11.118)$$

where the optimal matrix weights $A_{oi}, i \in \bar{Z}$ are computed by $col\{A_{oi}^T\}_{i \in \bar{Z}} = \Psi^{-1}e(e^T\Psi^{-1}e)^{-1}$, $\Psi = [P_{ij}], i, j \in \bar{Z}$ is an $nm \times nm$ symmetric positive definite matrix, and $e = \underbrace{[I_n, \ldots, I_n]}_{m}^T, I_n \in R^{n \times n}$ is an identity matrix. The corresponding covariance matrix of the fused estimation error is computed by $P_o = (e^T\Psi^{-1}e)^{-1}$, and one has that $P_o \leq P_{ii}, i \in \bar{Z}$.

When local estimates calculated by the estimators in Theorem 11.4.1 are available at the sensors in the WSN, every sensor $r, r \in Z_0$ then collects them from its neighborhood N_r to generate a fused estimate according to the fusion rule in Lemma 11.4.4. Note that the links in the WSN are subject to packet losses, local estimates $\hat{\xi}_l, l \in N_r$ may be lost during the transmission, and thus only the estimates that successfully arrive at the sensor r are used to generate the fused estimate $\hat{\eta}_{or}$ of the system state η. Let $\bar{N}_r(t_i)$ denote the index set of the estimates $\hat{\xi}_l$ that are successfully received by sensor r at instant t_i, and $\bar{n}_r(t_i)$ denote the number of elements in $\bar{N}_r(t_i)$. Then, by Lemma 11.4.4 one has the following theorem that determines the fused estimates and the corresponding covariance matrix of the estimation error at sensor $r, r \in Z_0$.

The following theorem follows directly from **Lemma** 11.4.4.

Theorem 11.4.2 *For system (11.74), the fusion estimator in the sensor $r, r \in Z_0$ is given by:*

$$\hat{\eta}_{or}(t_{i+k}|t_i) = \sum_{u \in \bar{N}_r(t_i)} \bar{A}_{ou,k}(t_i)\bar{\eta}_u(t_{i+k}|t_i), \quad k = 0, 1, \tag{11.119}$$

where

$$\hat{\eta}_u(t_{i+k}|t_i) = [I_{bn} \ O_1]\hat{\xi}_u(t_{i+k}|t_i)$$

and the optimal matrix weights

$$A_{ou,k}(t_i), u \in \bar{N}_r(t_i)$$

are computed by

$$\begin{aligned}
col\{\bar{A}_{ou,k}^T(t_i)\}_{u \in \bar{N}_r(t_i)} \\
= \Upsilon_{r,k}^{-1}(t_i)e_r(t_i)(e_r^T(t_i)\Upsilon_{r,k}^{-1}(t_i)e_r(t_i))^{-1}, \quad k = 0, 1,
\end{aligned} \tag{11.120}$$

where $\Upsilon_{r,k}(t_i) = [P_{u,s}^\eta(t_{i+k}|t_i)], u, s \in \bar{N}_r(t_i)$ is a $bn\bar{n}_r(t_i) \times bn\bar{n}_r(t_i)$ symmetric positive definite matrix, and

$$P_{u,s}^\eta(t_{i+k}|t_i) = [I_{bn} \ O_1]P_{u,s}(t_{i+k}|t_i)[I_{bn} \ O_1]^T, e_r(t_i) = \underbrace{[I_{bn}, \ldots, I_{bn}]}_{\bar{n}_r(t_i)}^T$$

The corresponding covariance matrix of the fusion estimation error is computed by $P_{or}^\eta(t_{i+k}|t_i) = (e_r^T(t_i)\Upsilon_{r,k}^{-1}(t_i)e_r(t_i))^{-1}$, and one has that

$$P_{or}^\eta(t_{i+k}|t_i) \leq P_{u,u}^\eta(t_{i+k}|t_i), u \in \bar{N}_r(t_i)$$

The estimates $\hat{\xi}_u(t_{i+k}|t_i)$ and the covariance matrices $P_{u,u}(t_{i+k}|t_i)$ are computed by the recursive equations in Theorem 11.4.1.

It can be seen from (11.120) that computation of the cross-covariance matrices $P_{u,s}(t_{i+k}|t_i), k = 0, 1, u, s \in \bar{N}_r(t_i), u \neq s$ is one of the key issues in applying the fusion estimator in Theorem 11.4.2. In what follows, computation procedures for the cross-covariances $P_{u,s}(t_{i+k}|t_i)$ will be presented, before which, some useful lemmas are first given as follows.

Lemma 11.4.5 *For any two augmented measurement noise vectors $v_u(t_i)$ and $v_s(t_i), u, s \in N_r, u \neq s$, define $Q_{v_u v_s} = E\{v_u(t_i)v_s^T(t_i)\}$. Then, one has*

$$Q_{v_u v_s} = [\zeta_{l,j}], \quad l \in N_u, j \in N_s, \tag{11.121}$$

where $\zeta_{l,j} = \begin{cases} Q_{l,l}^{v_p}, & l, j \in N_{u,s}, l = j \\ Q_{l,j}^{v_p}, & otherwise \end{cases}$ *and* $N_{u,s} = N_u \cap N_s$.

Lemma 11.4.6 *For $u, s \in N_r$ and $u \neq s$, $E\{\tilde{v}_u(t_i)\tilde{v}_s^T(t_i)\}$ satisfies*

$$\boldsymbol{E}\{\tilde{v}_u(t_i)\tilde{v}_s^T(t_i)\} \triangleq \Delta_{u,s}\theta_{s,u}(1 - \theta_{s,u})\Pi_{s,u}H_u Q_{v_u v_s}H_s^T\Pi_{u,s}^T$$
$$+ \sum_{l \in N_u}\sum_{j \in N_s}\theta_{l,u}\theta_{j,s}\Pi_{l,u}H_u Q_{v_u v_s}H_s^T\Pi_{j,s}^T. \tag{11.122}$$

Proof 11.4.4 *Since $u \in N_r$ and $s \in N_r$, i.e., sensor u and sensor s are neighbors, one has $u \in N_s$ and $s \in N_u$. Moreover, by the facts $\alpha_{s,u}(t_i) = \alpha_{u,s}(t_i)$ and $\theta_{s,u} = \theta_{u,s}$, one has by Lemma 11.4.1 that*

$$
\mathbf{E}\{\tilde{v}_u(t_i)\tilde{v}_s^T(t_i)\} = \mathbf{E}\Bigg\{\Bigg(\sum_{l\in N_u}\alpha_{l,u}(t_i)\Pi_{l,u}H_u v_u(t_i)Biggr\Bigg)
$$

$$
\times\Bigg(\sum_{j\in N_s}\alpha_{j,s}(t_i)\Pi_{j,s}H_s v_s(t_i)Biggr\Bigg)^T\Bigg\}
$$

$$
= \mathbf{E}\Bigg\{\Bigg(\sum_{l\in N_u}\alpha_{l,u}(t_i)\Pi_{l,u}H_u v_u(t_i)Biggr\Bigg)(\alpha_{u,s}(t_i)\Pi_{u,s}H_s v_s(t_i))^T\Bigg\}
$$

$$
+ \ \mathbf{E}\Bigg\{\sum_{l\in N_u}\sum_{j\in N_s,j\neq u}\alpha_{l,u}(t_i)\alpha_{j,s}(t_i)\Pi_{l,u}H_u v_u(t_i)v_s^T(t_i)H_s^T\Pi_{j,s}^T\Bigg\}
$$

$$
= \sum_{l\in N_u,l\neq s}\theta_{l,u}\theta_{u,s}\Pi_{l,u}H_u Q_{v_u v_s}H_s^T\Pi_{j,s}^T
$$

$$
+ \ \theta_{s,u}\Pi_{s,u}H_u Q_{v_u v_s}H_s^T\Pi_{u,s}^T + \sum_{l\in N_u}\sum_{j\in N_s,j\neq u}\theta_{l,u}
$$

$$
\times\theta_{j,s}\Pi_{l,u}H_u Q_{v_u v_s}H_s^T\Pi_{j,s}^T
$$

$$
= \ \theta_{s,u}(1-\theta_{s,u})\Pi_{s,u}H_u Q_{v_u v_s}H_s^T\Pi_{u,s}^T
$$

$$
+ \ \sum_{l\in N_u}\theta_{l,u}\theta_{u,s}\Pi_{l,u}H_u Q_{v_u v_s}H_s^T\Pi_{u,s}^T
$$

$$
+ \ \sum_{l\in N_u}\sum_{j\in N_s,j\neq u}\theta_{l,u}\theta_{j,s}\Pi_{l,u}H_u Q_{v_u v_s}H_s^T\Pi_{j,s}^T
$$

$$
= \ \Delta_{u,s}. \tag{11.123}
$$

which completes the proof.

Lemma 11.4.7 *Define the state cross-covariance matrix as*

$$
\Xi_{u,s}(t_i) \triangleq \mathbf{E}\{\xi_u(t_i)\xi_s^T(t_i)\}
$$

where $u, s \in N_r$ and $u \neq s$. Then $\Xi_{u,s}(t_i)$ satisfies the following recursion:

$$
\begin{aligned}
\Xi_{u,s}(t_{i+1}) &= \theta_{s,u}(1-\theta_{s,u})A_{0s,u}\Xi_{u,s}(t_i)A_{0u,s}^T \\
&\quad + \ \bar{A}_u\Xi_{u,s}(t_i)\bar{A}_s^T + diag\{BQ_\omega B^T,\Delta_{u,s}\}, \tag{11.124}
\end{aligned}
$$

where the initial value of $\Xi_{u,s}(t_i)$ at t_0 is given by $\Xi_{u,s}(t_0) = \Xi_{u,u}(t_0)$.

Proof 11.4.5 *It follows from (11.78) that*

$$
\begin{aligned}
\Xi_{u,s}(t_{i+1}) &= E\{\Xi_u(t_{i+1})\Xi_s^T(t_{i+1})\} \\
&= \bar{A}_u \Xi_{u,s}(t_i)\bar{A}_s^T + \chi_1 + \chi_2,
\end{aligned}
\tag{11.125}
$$

where

$$
\begin{aligned}
\chi_1 &= \boldsymbol{E}\{(\tilde{A}_u(t_i) - \tilde{A}_u)\xi_u(t_i)\xi_s^T(t_i)(\tilde{A}_s(t_i) - \tilde{A}_s)^T\}, \\
\chi_2 &= \boldsymbol{E}\{\tilde{B}_u(t_i)\nu_u(t_i)\nu_s^T(t_i)\tilde{B}_s^T(t_i)\}
\end{aligned}
$$

*Noting $u \in N_s, s \in N_u$ and $u \neq s$, and by (11.75), **Lemma 11.4.1** and the facts $\alpha_{s,u}(t_i) = \alpha_{u,s}(t_i)$ and $\theta_{s,u} = \theta_{u,s}$, one obtains that*

$$
\begin{aligned}
\chi_1 &= \boldsymbol{E}\left\{\left[\sum_{l \in N_u}(\alpha_{l,u}(t_i) - \theta_{l,u})A_{0l,u}\right]\xi_u(t_i)\xi_s^T(t_i)\right. \\
&\quad \left. \times \left[\sum_{j \in N_s}(\alpha_{j,s}(t_i) - \theta_{j,s})A_{0j,s}\right]^T\right\} \\
&= \theta_{s,u}(1 - \theta_{s,u})A_{0s,u}\Xi_{u,s}(t_i)A_{0u,s}^T.
\end{aligned}
\tag{11.126}
$$

*Since $\omega_m(t_i)$ and $\upsilon_l(t_i), l \in N_r$ are uncorrelated, one has by **Lemma 11.4.6** that*

$$
\begin{aligned}
\chi_2 &= \boldsymbol{E}\left\{diag\left\{B, \sum_{l \in N_u}\alpha_{l,u}(t_i)\Pi_{l,u}H_u\right\}\begin{bmatrix} \omega_m(t_i) \\ \upsilon_u(t_i) \end{bmatrix}\right. \\
&\quad \left. \times \begin{bmatrix} \omega_m(t_i) \\ \upsilon_s(t_i) \end{bmatrix}^T diag\left\{B, \sum_{j \in N_s}\alpha_{j,s}(t_i)\Pi_{j,s}H_s\right\}^T\right\} \\
&= diag\{BQ_\omega B^T, \Delta_{u,s}\}.
\end{aligned}
\tag{11.127}
$$

Substituting (11.126) and (11.127) into (11.125) leads to (11.124). The proof is thus completed.

A set of recursive equations for calculating the cross-covariances

$$
P_{u,s}(t_{i+k}|t_i), k = 0, 1, u, s \in N_r, u \neq s
$$

is now presented in the following theorem based on **Lemmas** 11.4.5–11.4.7.

Theorem 11.4.3 *For system (11.74), the cross-covariance of local Kalman estimation errors between the sensors* u *and* s *in the neighborhood* $N_r, r \in Z_0$ *satisfies the following recursive equations:*

$$P_{u,s}(t_i|t_i) = \Gamma_1 + \Gamma_2 + \Gamma_3, \tag{11.128}$$

$$
\begin{aligned}
P_{u,s}(t_{i+1}|t_i) &= \Gamma_4 + \Gamma_5 + \Gamma_6 + diag\{BQ_\omega B^T, \Delta_{u,s}\} \\
&\quad - [0 \quad \varphi_{u,s}^T(t_i)]^T.[0 \quad \pi_{u,s}(t_i)],
\end{aligned} \tag{11.129}
$$

where

$$
\begin{aligned}
\Gamma_1 &= (\bar{I}_u - K_u(t_i)\bar{C}_u)P_{u,s}(t_i|t_{i-1})(\bar{I}_s - K_s(t_i)\bar{C}_s)^T, \\
\Gamma_2 &= \theta_{s,u}(1 - \theta_{s,u})K_u(t_i)C_{0s,u}\Xi_{u,s}(t_i)C_{0u,s}^T K_s^T(t_i), \\
\Gamma_3 &= \theta_{s,u}(1 - \theta_{s,u})K_u(t_i)\Pi_{s,u}H_u Q_{v_u v_s}H_s^T\Pi_{u,s}^T K_s^T(t_i) \\
&\quad + \sum_{l \in N_u}\sum_{j \in N_s}\theta_{l,u}\theta_{j,s}K_u(t_i)\Pi_{l,u}H_u Q_{v_u v_s} \\
&\quad \times H_s^T\Pi_{j,s}^T K_s^T(t_i), \\
\Gamma_4 &= \theta_{s,u}(1 - \theta_{s,u})(A_{0s,u} - F_u(t_i)C_{0s,u})\Xi_{u,s}(t_i) \\
&\quad \times (A_{0u,s} - F_s(t_i)C_{0u,s})^T, \\
\Gamma_5 &= (\bar{A}_u - F_u(t_i)\bar{C}_u)P_{u,s}(t_i|t_{i-1})(\bar{A}_s - F_s(t_i)\bar{C}_s)^T, \\
\Gamma_6 &= \theta_{s,u}(1 - \theta_{s,u})F_u(t_i)\Pi_{s,u}H_u Q_{v_u v_s}H_s^T\Pi_{u,s}^T F_s^T(t_i) \\
&\quad + \sum_{l \in N_u}\sum_{j \in N_s}\theta_{l,u}\theta_{j,s}F_u(t_i)\Pi_{l,u} \\
&\quad \times H_u Q_{v_u v_s}H_s^T\Pi_{j,s}^T F_s^T(t_i), \\
\varphi_{u,s}(t_i) &= \theta_{s,u}(1 - \theta_{s,u})\Pi_{s,u}H_u Q_{v_u v_s}H_s^T\Pi_{u,s}^T F_s^T(t_i) \\
&\quad + \sum_{l \in N_u}\sum_{j \in N_s}\theta_{l,u}\theta_{j,s}\Pi_{l,u}H_u Q_{v_u v_s}H_s^T\Pi_{j,s}^T F_s^T(t_i), \\
\pi_{u,s}(t_i) &= \theta_{s,u}(1 - \theta_{s,u})F_u(t_i)\Pi_{s,u}H_u Q_{v_u v_s}H_s^T\Pi_{u,s}^T \\
&\quad + \sum_{l \in N_u}\sum_{j \in N_s}\theta_{l,u}\theta_{j,s}F_u(t_i)\Pi_{l,u}H_u Q_{v_u v_s}H_s^T\Pi_{j,s}^T
\end{aligned}
$$

and $\bar{I}_u \in R^{bn+\bar{m}_u}$ *and* $\bar{I}_s \in R^{bn+\bar{m}_s}$ *are identity matrices,* $Q_{v_u v_s}$ *and* $\Delta_{u,s}$ *are given by (11.121) and (11.122), respectively, and* $\Xi_{u,s}(t_i)$ *is computed by (11.124), the initial value of* $P_{u,s}(t_i|t_{i-1})$ *at* t_0 *is given by*

$$P_{u,s}(t_0|t_{-1}) = P_{u,u}(t_0|t_{-1})$$

Proof 11.4.6 *Substituting (11.82) into (11.87) yields*

$$\hat{\xi}_u(t_i|t_i) = (\bar{I}_u - K_u(t_i)\bar{C}_u)\hat{\xi}_u(t_i|t_{i-1}) + K_u(t_i)Z_u(t_i), \quad u \in N_r. \quad (11.130)$$

Substituting the output equation in (11.74) into (11.130) leads to

$$
\begin{aligned}
\hat{\xi}_u(t_i|t_i) &= (\bar{I}_u - K_u(t_i)\bar{C}_u)\hat{\xi}_u(t_i|t_{i-1}) \\
&+ K_u(t_i)\tilde{C}_u(t_i)\xi_u(t_i) \\
&+ \sum_{l \in N_u} \alpha_{l,u}(t_i)K_u(t_i)\Pi_{l,u}H_u v_u(t_i) \\
&= (\bar{I}_u - K_u(t_i)\bar{C}_u)\hat{\xi}_u(t_i|t_{i-1}) + K_u(t_i)\bar{C}_u)\hat{\xi}_u(t_i) \\
&+ K_u(t_i)(\tilde{C}_u(t_i) - \bar{c}_u)\xi_u(t_i) \\
&+ \sum_{l \in N_u} \alpha_{l,u}(t_i)K_u(t_i)\Pi_{l,u}H_u v_u(t_i) \\
&= \hat{\xi}_u(t_i|t_i) + K_u(t_i)\bar{C}_u\tilde{\xi}_u(t_i|t_{i-1}) \\
&+ K_u(t_i)(\tilde{C}_u(t_i) - \bar{C}_u)\xi_u(t_i) \\
&+ \sum_{l \in N_u} \alpha_{l,u}(t_i)K_u(t_i)\Pi_{l,u}H_u v_u(t_i). \quad (11.131)
\end{aligned}
$$

Subtracting $\xi_u(t_i)$ from both sides of (11.131) and taking (11.75) into account, yields

$$
\begin{aligned}
\tilde{\xi}_u(t_i|t_i) &= (\bar{I}_u - K_u(t_i)\bar{C}_u)\tilde{\xi}_u(t_i|t_{i-1}) \\
&- \sum_{l \in N_u} (\alpha_{l,u}(t_i) - \theta_{l,u})K_u(t_i)C_{0l,u}\xi_u(t_i) \\
&- \sum_{l \in N_u} \alpha_{l,u}(t_i)K_u(t_i)\Pi_{l,u}H_u v_u(t_i). \quad (11.132)
\end{aligned}
$$

Since $\tilde{\xi}_u(t_i|t_{i-1})$ consists of the linear combination of

$$\{\omega_m(t_{i-2}), \ldots, \omega_m(t_0), v_u(t_{i-1}), \ldots, v_u(t_1), \xi_u(t_0)\}$$

applying the projection property [516] and following the similar derivation procedures as in (11.93)–(11.95) in the Appendix, one has $\tilde{\xi}_u(t_i|t_{i-1}) \perp v_s(t_i)$. Moreover, since $\xi_u(t_i) \perp v_s(t_i)$, $E\{\alpha_{l,u}(t_i) - \theta_{l,u}\} = 0$, and $E\{\alpha_{j,s}(t_i) - \theta_{j,s}\} = 0, l \in N_u, j \in N_s$, one has by (11.132) that

$$
\begin{aligned}
P_{u,s}(t_i|t_i) &= E\{\tilde{\xi}_u(t_i|t_i)\tilde{\xi}_s^T(t_i|t_i)\} \\
&= (\bar{I}_u - K_u(t_i)\bar{C}_u)P_{u,s}(t_i|t_{i-1})(\bar{I}_s - K_s(t_i)\bar{C}_s)^T \\
&+ \chi_3 + \chi_4, \quad (11.133)
\end{aligned}
$$

where

$$\chi_3 = E\left\{\left[\sum_{l\in N_u}(\alpha_{l,u}(t_i) - \theta_{l,u})K_u(t_i)C_{0l,u}\right]\xi_u(t_i)\right.$$

$$\left.\times\xi_s^T(t_i)\left[\sum_{l\in N_s}(\alpha_{j,s}(t_i) - \theta_{j,s})K_s(t_i)C_{0j,s}\right]^T\right\}$$

$$\chi_4 = E\left\{\left(\sum_{l\in N_u}\alpha_{l,u}(t_i)K_u(t_i)\Pi_{l,u}H_u\right)v_u(t_i)v_s^T(t_i)\right.$$

$$\left.\times\left(\sum_{l\in N_s}\alpha_{j,s}(t_i)K_s(t_i)\Pi_{j,s}H_s\right)^T\right\}$$

Since $u \in N_s, s \in N_u$ and $u \neq s$, one obtains by Lemma 11.4.1 and $\alpha_{s,u}(t_i) = \alpha_{u,s}(t_i)$ and $\theta_{s,u} = \theta_{u,s}$ that

$$\chi_3 = \Gamma_2. \tag{11.134}$$

By following the similar derivation procedures as in the proof of Lemma 11.4.6, one has that

$$\chi_4 = \Gamma_3. \tag{11.135}$$

Combining (11.133)–(11.135) leads to (11.128).

Notice that one has to calculate $P_{u,s}(t_i|t_{i-1})$ in computing $P_{u,s}(t_i|t_i)$. Since $\xi_u(t_i) \perp \nu_s(t_i), \xi_u(t_i) \perp v_s(t_i), \tilde{\xi}_u(t_i|t_{i-1}) \perp \nu_s(t_i), \tilde{\xi}_u(t_i|t_{i-1}) \perp v_s(t_i), E\{\alpha_{l,u}(t_i) - \theta_{l,u}\} = 0$ and $E\{\alpha_{j,s}(t_i) - \theta_{j,s}\} = 0, l \in N_u, j \in N_s$, one has by (11.109) in the Appendix, (11.127), Lemma 11.4.1 and following the similar derivation procedures as in (11.126) that

$$\begin{aligned}P_{u,s}(t_{i+1}|t_i) &= \Gamma_4 + \Gamma_5 + diag\{BQ_\omega B^T, \Delta_{u,s}\}\\ &+ E\{\rho u(t_i)\rho_s^T(t_i)\} - E\{\tilde{B}_u(t_i)\nu_u(t_i)\rho_s^T(t_i)\}\\ &- E\{\rho u(t_i)(\tilde{B}_s(t_i)\nu_s(t_i))^T\},\end{aligned} \tag{11.136}$$

where

$$\rho_u(t_i) = \sum_{l\in N_u}\alpha_{l,u}(t_i)F_u(t_i)\Pi_{l,u}H_u v_u(t_i),$$

$$\rho_s(t_i) = \sum_{l\in N_s}\alpha_{j,s}(t_i)F_s(t_i)\Pi_{j,s}H_s v_s(t_i).$$

By following the similar derivation procedures as in the proof of Lemma 11.4.6, one obtains

$$E\{\rho u(t_i)\rho_s^T(t_i)\} = \Gamma_6.$$ (11.137)

By following the similar derivation procedures as in (11.127) and Lemma 11.4.6, one has

$$E\{\tilde{B}_u(t_i)\nu_u(t_i)\rho_s^T(t_i)\} = [0\varphi_{u,s}^T(t_i)]^T,$$ (11.138)

$$E\{\rho u(t_i)(\tilde{B}_s(t_i)\nu_s(t_i))^T\} = [0\pi_{u,s}(t_i)].$$ (11.139)

Combining (11.136)–(11.139) yields (11.129). The proof is thus completed.

Remark 11.4.3 *By fusing local estimates, more measurements from different sensors are used to generate fused estimates at every sensor, which helps improve the local estimation performance and reduce the disagreement of local estimates. Similar to (11.116) and (11.117), one may define some disagreement potentials as follows to characterize the performance of the distributed estimation algorithm in Theorems 11.4.2 and 11.4.3:*

$$\kappa_r^o(t_i) = \frac{1}{2nr} \sum_{u,s \in N_r} \|\hat{\eta}_{ou}(t_{i+k}|t_i) - \hat{\eta}_{os}(t_{i+k}|t_i)\|^2,$$ (11.140)

$$\psi_r^o(t_i) = \frac{1}{2nr} \sum_{u,s \in N_r} [Tr(P_{ou}^\eta(t_{i+k}|t_i)) - Tr(P_{os}^\eta(t_{i+k}|t_i))]^2,$$ (11.141)

where $k = 0, 1$, $\kappa_r^o(t_i)$ and $\psi_r^o(t_i)$ are, respectively, the disagreement potential of the fused estimates and the disagreement potential of the fused estimation performances in the neighborhood N_r, and some smaller κ_r^o and ψ_r^o imply a better performance of the estimation algorithm in Theorems 11.4.2 and 11.4.3.

Remark 11.4.4 *It can be seen from Theorems 11.4.1–11.4.3 that the estimation performance assessed by the error covariances critically depends on the PLPs and the parameter b that determines the measurement transmission rate, and thus one may see how the packet loss and the measurement transmission rate can affect the estimation performance by applying the algorithms in Theorems 11.4.1–11.4.3.*

Remark 11.4.5 *The proposed two-stage fusion estimation needs more computation and communication costs as compared with the one-stage one. Nevertheless, the multi-rate scheme helps reduce communication costs significantly, since the* *transmission rate of the measurements and local estimates is slowed down, and it is well known that computation consumes much less energy than communication in WSNs. Energy saved from the multi-rate scheme can be used to implement the second stage fusion estimation which helps improve estimation performance. Thus, the two-stage estimation may achieve a better performance without consuming more energy than the one-stage estimation.*

11.4.5 Simulation example 3

To demonstrate the effectiveness of the proposed estimator design method, simulations of a maneuvering target tracking system [675] are presented in the sequel, where the target's position and the velocity evolve according to the state space model in (11.64) with

$$A_p = \begin{bmatrix} 1 & h_p \\ 0 & 1 \end{bmatrix}, \quad B_p = \sqrt{10} \begin{bmatrix} h_p^2/2 \\ h_p \end{bmatrix} \tag{11.142}$$

where h_p is the sampling period. The state is $x(k_i) = [x_p^T(k_i) \ x_v^T(k_i)]^T$, where $x_p(k_i)$ and $x_v(k_i)$ are the position and the velocity of the maneuvering target at time k_i, respectively. Suppose that the target is not moving too fast, and we take $h_p = 0.5s$ in the simulation.

A wireless sensor network with 12 sensor nodes is deployed to monitor the target, and the topology of the WSN is shown in Figure 11.17. The wireless links in the WSN may be subject to random packet losses. Suppose that only the position of the target is measurable, and the observation equations of the sensors are given by (11.65), where $v_{pl}(k_i) = c_l\omega_0(k_i) + v_{0l}(k_i), \omega_0(k_i)$ is a zero mean white noise with variance $Q_{\omega_0}, v_{0l}(k_i)$ are zero mean white noises with variances $Q_{v_{0l}}, v_{0l}(k_i)$ are mutually uncorrelated and are independent of $\omega_0(k_i), \omega_0(k_i)$ and $v_{0l}(k_i)$ are uncorrelated with $\omega_p(k_i), C_{p1} = [1 \ 0], C_{p2} = [0.8 \ 0], C_{p3} = [0.7 \ 0], C_{p4} = [0.6 \ 0], C_{p5} = [0.5 \ 0], C_{p6} = [0.4 \ 0], C_{p7} = [0.3 \ 0], C_{p8} = [0.2 \ 0], C_{p9} = [1 \ 0], C_{p10} = [0.8 \ 0], C_{p11} = [0.6 \ 0], C_{p12} = [0.7 \ 0]$, and $D_{pl} = 1, l = 1, \ldots, 12$. It can be easily calculated that $Q_{l,l}^{vp} = c_l^2 Q_{\omega_0} + Q_{v_{0l}}$ and $Q_{l,s}^{vp} = c_l c_s Q_{\omega_0}, l \neq s, l, s = 1, \ldots, 12$. In the simulation, we take $c_l = 0.1l, Q_{\omega_p} = 0.1, Q_{\omega_0} = 1, Q_{v_{01}} = 0.4, Q_{v_{02}} = 0.7, Q_{v_{03}} = 0.4, Q_{v_{04}} = 0.4, Q_{v_{05}} = 0.3, Q_{v_{06}} = 0.2, Q_{v_{07}} = 0.3, Q_{v_{08}} = 0.3, Q_{v_{09}} = 0.5, Q_{v_{10}} = 0.4, Q_{v_{11}} = 0.3, Q_{v_{12}} = 0.1$.

It can be seen from the topology of the WSN that sensors 2 and 5 are directly connected to sensor 1, and thus they are neighbors of the sensor 1, and the neighborhood N_1 consists of three sensors, and they are sensors 1, 2, 5. In what follows, estimation at the sensors in neighborhood N_1 will be considered to show the effectiveness of the proposed estimator design. At each instant t_i, sensor 1 collects measurements from itself and sensors 2 and 5 to generate local estimates; then at the second stage, sensor 1 collects local estimates from itself and sensors 2 and 5 to form fused estimates.

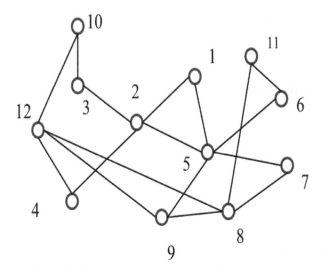

Figure 11.17: The network topology with $N = 12$ sensor nodes.

We first consider the situation where $a = b = 2$, i.e., the sensors in N_1 collect measurements from their neighborhoods and generate estimates with period $h_m = 2s$ which is four times the sampling period, and the estimates are updated with period $h_e = 1s$ which is two times the sampling period. By slowing down the measurement transmission rate and the estimate updating rate, one may expect to save energies consumed in communications and computations. The PLPs in the links $L_{2,1}$ and $L_{5,1}$ are supposed to be $1 - \theta_{2,1} = 1 - \theta_{5,1} = 0.2$. The initial time is $t_0 = 0$, and the initial state is given by $x(0) = x(-1) = [1 \ 0.5]^T$, and $\bar{x}_0 = \bar{x}_1 = [1.5 \ 1.0]^T$, $\bar{P}_0 = \bar{P}_1 = diag\{0.25, 0.25\}$.

By applying Theorems 11.4.1–11.4.3, the true values and the filtered fusion estimates of the target positions obtained at sensor 1 are depicted in Figure 11.18(a), while Figure 11.18(b) depicts the true values and the filtered fusion estimates of the target velocities. It can be seen that the sensor 1 is able to track the maneuvering target well in the presence of random packet losses and with a slow measurement transmission rate. Figure 11.19(a) shows the individual estimation performance (assessed by the trace of estimation error covariance) of every sensor in the neighborhood N_1. It can be seen from Figure 11.19(a) that the estimation performance at sensor 1 is improved by using the two-stage fusion strategy, and the fusion estimator outperforms each of its local estimators.

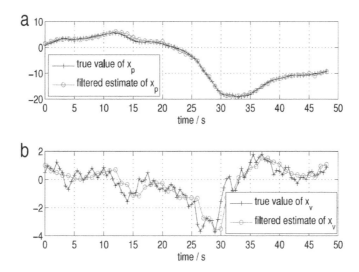

Figure 11.18: True values and fused estimates (obtained at sensor 1) of the target positions and velocities with $a = 2, b = 2, \theta_{2,1} = \theta_{5,1} = 0.5$.

The advantage of the two-stage fusion estimation strategy is further shown in Figures 11.19(b) and 11.20. In Figure 11.19(b), the filtering performances obtained by using measurements only from sensor 1 and by using measurements from neighbors of sensor 1 are depicted. It can be seen that the estimation performance may be improved by using more measurements from different sensors. The red curve shows the fusion estimation performance obtained by using the proposed two-stage estimation strategy. It can be seen that the estimation performance can be further improved by fusing local estimates from its neighborhood. The disagreement of estimates and disagreement of estimation performances obtained by two estimation strategies (one-stage estimation and two-stage estimation) are shown, respectively, in Figure 11.20(a) and (b). It is clearly shown by Figure 11.20 that both the disagreement of estimate and the disagreement of estimation performance are significantly reduced by using the two-stage estimation strategy, confirming that the proposed two-stage strategy is an efficient way to address the issues (1) and (2) discussed in the Introduction.

Notice that the two-stage estimation usually causes more communication costs as compared with the normal one-stage estimation, because, besides the measurements, local estimates in the neighborhood should also be transmitted among sensors in the group to generate a fused estimate. Fortunately, by slowing down the measurement transmission and estimate updating rates, energy can be saved to implement the two-stage estimation. In this way, the two-stage strategy may improve each local estimation performance and reduce the disagreement of estimates among different sensors, without consuming more energy than the normal one-stage strategy. An example is shown in Figure 11.21 which depicts filtering performances obtained at sensor 1 with $\theta_{2,1} = \theta_{5,1} = 0.9$. The blue curve in Figure 11.21 shows the filtering performance obtained by using the one-stage estimation strategy with $a = 2$ and $b = 1$, i.e., sensor 1 collects measurements from sensors 2 and 5 and generates estimates with a period of 1 s, and thus totally 4 times of measurement transmissions and 2 times of estimate computations are involved over every 2 s by using the one-stage estimation. The red curve in Figure 11.21 shows the filtering performance obtained by using the two-stage estimation strategy with a = 4 and b = 1, i.e., sensor 1 collects not only measurements but also the local estimates from sensors 2 and 5 and generates fused estimates with a period of 2 s, and therefore totally 4 times of measurement transmissions and 2 times of estimate computations are involved over every 2 s by using the two-stage estimation. It thus can be observed from Figure 11.21 that, though the two strategies consume the same communication and computation costs, the two-stage estimation is able to provide better performance than the one-stage estimation, confirming that the two-stage strategy may outperform the one-stage one without increasing energy consumption due to the benefits from slowing down the measurement transmission rate.

In what follows, we will show how the packet loss and the measurement transmission period may affect the estimation performances. Figure 11.22 shows the filtering performances of the sensors in N_1 with different PLPs, and Figure 11.23 shows filtering performances of the sensors in N_1 with different measurement transmission periods. It can be seen from Figures 11.22 and 11.23 that packet loss degrades estimation performance and a smaller measurement transmission period leads to a better estimation performance, which are as expected, and demonstrate the effectiveness of the proposed estimator design method.

11.5   Distributed Filtering over Finite Horizon

In this section, the distributed H_∞-consensus filtering problem is investigated for a class of discrete time-varying nonlinear systems on a finite horizon. The topology of the sensor networks is assumed to be Markovian switching, and the missing measurements (packet dropouts) problem is also considered. Based on the recursive linear matrix inequalities (RLMI), an effective distributed H_∞-consensus filter is designed, which is suitable for online computation.

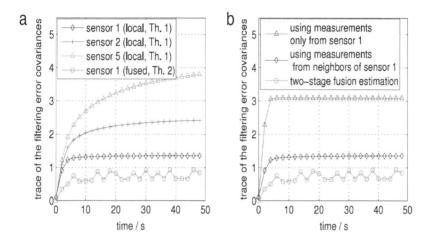

Figure 11.19: (a) Estimation performances obtained at the sensors in the neighborhood N_1 with $a = 2, b = 2, \theta_{2,1} = \theta_{5,1} = 0.8$; (b) estimation performances of sensors 1 with different estimation strategies, $a = 2, b = 2, \theta_{2,1} = \theta_{5,1} = 0.8$.

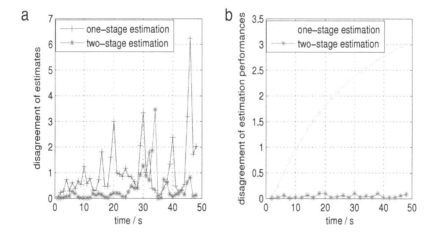

Figure 11.20: Comparison of disagreement potentials in two estimation strategies, $a = 2, b = 2, \theta_{2,1} = \theta_{5,1} = 0.8$.

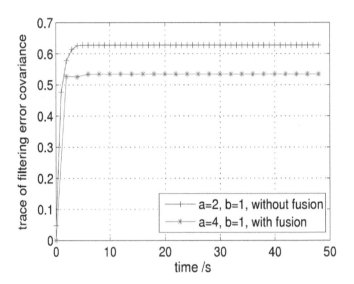

Figure 11.21: Comparison of estimation performance and energy consumption in two estimation strategies, $\theta_{2,1} = \theta_{5,1} = 0.9$.

11.5.1   Introduction

In recent years, sensor networks have been more and more attractive due to their wide applications in information acquisition and processing, signal detection, intelligent robotics, environment monitoring and so on. Sensor networks are a class of special multi-agent systems, where each sensor node exchanges information with its neighbors to perform the information acquisition or processing tasks collaboratively. The distributed filters in the sensor networks context have their own advantages in robustness and fault tolerance. According to different environments and application occasions, the topologies of sensor networks take different forms, for example, fixed, switching, and self-organized. Correspondingly, the complex coupling in information exchange between the neighbors brings great challenge in distributed filters analysis and design [294], [404], [638].

On the other hand, a great number of researchers have devoted their attention to the theory and applications of multi-agent systems, especially the synchronization and consensus problems. Various types of multi-agent systems, continuous and discrete-time, have been investigated and conditions or protocols for the consensus are given [691, 692], [693, 694], [696] and [695].

For multi-agent systems with switching topologies, the aforementioned work concentrates mostly on the arbitrary switching topologies. However, for wireless sensor networks, in many cases one can model the topologies to be switching in a

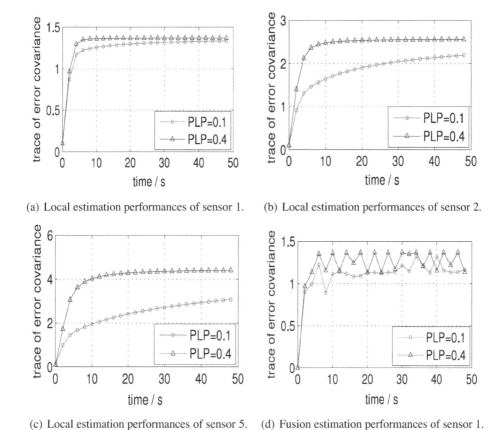

(a) Local estimation performances of sensor 1. (b) Local estimation performances of sensor 2.

(c) Local estimation performances of sensor 5. (d) Fusion estimation performances of sensor 1.

Figure 11.22: Estimation performances of sensors in N_1 with different PLPs, $a = 2, b = 2$.

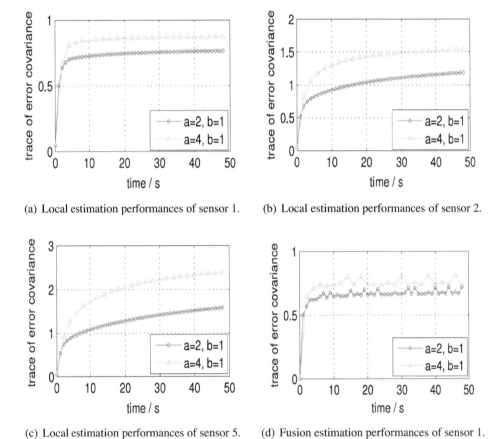

(a) Local estimation performances of sensor 1.   (b) Local estimation performances of sensor 2.

(c) Local estimation performances of sensor 5.   (d) Fusion estimation performances of sensor 1.

Figure 11.23: Estimation performances of sensors in N_1 with different measurement transmission periods, $\theta_{2,1} = \theta_{5,1} = 0.8$.

stochastic way, and as is known to us, the Markov process has its own advantage in describing many stochastic switching phenomena [697] and [698].

The filter's design problem has been an ongoing topic for many decades. Conventional Kalman filters assume that the disturbance is known Gaussian which is not always practical. Therefore the robust and/OR H_∞ filtering approaches have been developed to deal with unknown noise with bounded energy in the past decades, see [701], [702], [703], [708], [709] [704], and [707]. Time-varying systems are often a suitable model for multi-agent systems due to the introduction of the communication networks. Also, the presence of the communication networks makes finite horizon filter a more desirable solution due to its less requirement in its online implementation, [710], [711], and [706].

There is little literature about the problems of filtering in sensor networks with switching topologies, which is more practical in engineering with the development of wireless senor networks. It is also noted that the topology switching influences the information exchange between the sensor nodes involved, and correspondingly may have an effect on the measurements of them. Hence in this chapter, to make our results be more practical, it is reasonable to assume that the switching signal and the measured output of the sensors are not independent. Moreover, the burden of the networks must be taken into consideration in the filter's design, because a heavy burden may result in serious delays, frequent data missing, and even the network paralysis. Hence, how to design consensus filters with less data transmitted between the sensors is an important and interesting problem. Hereafter, we aim at dealing with the distributed H_∞ consensus filtering for a class of discrete time-varying nonlinear systems with sensor networks under Markov switching topologies. The missing measurements (packets dropouts) are also considered.

11.5.2 Problem description

In this section, we shall first provide some preliminaries of graph theory which will be used to describe the topologies of the sensor networks, then we shall present the initial discrete time-varying nonlinear systems to be investigated. Some useful lemmas are also to be given in this section.

A sensor networks with fixed topology could be denoted by a directed graph $\mathcal{G} = \{\mathcal{V}, \xi, \mathcal{A}\}$, with $\mathcal{V} = \{v_1, v_2, ..., v_n\}$ denoting all the sensor nodes involved, and v_i being the ith one. ξ is the set of all the directed edges of the topology graph, which obviously satisfies $\xi \subset \mathcal{V} \times \mathcal{V} = \{(v_i, v_j) : \forall v_i, v_j \in \mathcal{V}\}$. $\mathcal{A} = [a_{ij}]_{n \times n}$ is the adjacency matrix, reflecting the connection between the sensors. An edge $e_{ij} = (v_i, v_j)$ denotes a directed edge from j to i, and $e_{ij} \in \xi$ holds if and only if $a_{ij} > 0$. It is seen that a sensor node is always its own neighbor under this assumption. Moreover, the notation $\mathcal{N}_i$ is used in the chapter to denote all the neighbors of the

ith sensor, i.e., $\mathcal{N}_i = \{v_j : e_{ij} \in \xi\}$. In practical systems, due to the fact that the wireless sensors are playing a more and more important role, the topologies for the sensor networks are usually stochastic switching while the wireless sensors networks are reorganized. To be more reasonable, we model the topologies to be a set of graphs governed by a Markov chain, which are denoted by $\mathcal{G}(r_k)$, with r_k being a discrete-time, finite modes, ergodic Markov chain, taking values in the set $S = \{1, 2.., M\}$, and the one–step transition probability is denoted by $Prob = [\lambda_{st}]_{M \times M}$, $s, t \in S$, where λ_{st} implies the transition probability from mode s to mode t. Hence, the sensor networks with switching topologies could be denoted by $\mathcal{G}(r_k) = \{\mathcal{V}, \xi(r_k), \mathcal{A}(r_k)\}$, $\xi(r_k) = \{e_{ij}(r_k)\}$, and $\mathcal{A}(r_k) = [a_{ij}(r_k)]_{n \times n}$, $r_k \in S$.

In the sequel, let us consider the following discrete time-varying nonlinear systems defined on the finite interval $[0, N - 1]$.

$$(\Sigma :) \quad \begin{aligned} x(k+1) &= A(k)x(k) + B(k)f(x(k)) + C(k)w(k) \\ z(k) &= M(k)x(k) \end{aligned} \qquad (11.143)$$

and the measured signal output of each sensor node is given by

$$y_i(k) = \gamma_i(k)\Big[D_i(k)x(k) + E_i(k)g(x(k))\Big] + F_i(k)w(k) \qquad (11.144)$$

where $x(k) \in \Re^{n_x}$ is the state vector which is immeasurable, $z(k) \in \Re^{n_z}$ is the output to be estimated, $w(k) \in \Re^{n_w}$ is the exogenous disturbance input which is assumed to belong $l_2[0 \ N - 1]$, and it is assumed to be independent of the Markov process r_k. The nonlinear function $f(x(k)), g(x(k)) : \Re^{n_x} \to \Re^{n_x}$ are assumed to be satisfying the following sector nonlinearity described as

$$\big[f(x) - f(y) - K_1(x - y)\big]^t \ \times \ \big[f(x) - f(y) - K_2(x - y)\big]$$
$$\leq \ 0 \ \forall x, y \in \Re^{n_x} \qquad (11.145)$$

$$\big[g(x) - g(y) - L_1(x - y)\big]^t \ \times \ \big[g(x) - g(y) - L_2(x - y)\big]$$
$$\leq \ 0 \ \forall x, y \in \Re^{n_x} \qquad (11.146)$$

both of which satisfy the zero initial condition, i.e., $f(0) = 0$, $g(0) = 0$, and K_1, K_2, L_1, L_2 are all known matrices satisfying $K_1 - K_2 < 0$, $L_1 - L_2 < 0$. $A(k), B(k), C(k), M(k)$ are all known matrices with compatible dimensions. $\gamma_i(k)$ is assumed to be a random variable satisfying any discrete probabilistic distributions on the interval $[0, 1]$, mutually independent with each other with mathematical expectation α_i and variance σ_i^2, and it is also assumed to be independent of the exogenous disturbance $w(k)$.

Remark 11.5.1 *The missing measurements phenomenon has received much atten-*
tion in the filtering problems. In the earlier literature, it is usually assumed to
be described by a random variable with Bernoulli distribution; one can refer to
[503, 700], and the reference therein. However, some scholars have found that
it is not so reasonable, because in practice the transmitted information could be
neither completely missing nor completely received , but only a part of the initial
information could be transmitted successfully. To overcome the shortcomings, a
more general model was proposed in [705], in which the missing phenomenon is
described by random variables satisfying any probability distribution on interval
$[0, 1]$. So in this chapter, we adopt the general model which is of more practical sig-
nificance. And it is seen that the model could cover the Bernoulli distributed model
as a special case. Moreover, it is noted that we did not assume the independence
between the Markov process r_k and $\gamma_i(k)$, due to the fact that the mutual infor-
mation transmission under different topologies could influence the measurements,
hence the obtained results should be more realistic.

Due to the mutual influence on information transmission between the sensors,
for each sensor node, in this chapter, the filter to be designed is assumed to be of
the following form

$$
\begin{aligned}
\hat{x}_i\,(k\,+\,1) &= H_{ii}\,(k, r_k)\hat{x}_i(k)\,+\,N_i\,(k, r_k)y_i\,(k)\,+\,u_i(k, r_k) \\
\hat{z}_i(k) &= \hat{M}_i(k, r_k)\hat{x}_i(k)
\end{aligned}
\tag{11.147}
$$

It is noted that compared with the forms usually chosen for filter design, the
additional mode-dependent term $u(k, r_k)$ reflects the mutual influence of the sensor
nodes, which is time-switching with the topologies. We shall use the following
linear consensus-like protocol, the similar form of which is widely used in the
investigation of multi-agent systems, see [691, 692, 693, 694] and the reference
therein.

$$
u_i(k, r_k) = \sum_{j \in \mathcal{N}_i(r_k)\backslash\{i\}} a_{ij}(r_k)H_{ij}(k, r_k)(\hat{x}_i(k) - \hat{x}_j(k))
\tag{11.148}
$$

Remark 11.5.2 *It is worth mentioning that in (11.147) a set of mode-dependent*
distributed linear filters are to be designed for the initial discrete time-varying
nonlinear system, which renders much practical significance for application. It is
also seen that only the estimation $\hat{x}_i$ of the sensors is transmitted in the sensor
networks in (11.147), which implies that less data is needed and the burden of the
networks will be reduced. As is formulated above, the similar form of u is very

popular in the investigation of consensus problems for multi-agent systems. On the other hand, without the connection influence between the sensor nodes of the networks (that is to say, $a_{ij}(r_k) = 0$), the form of the filers is transformed to be (the mode-independent case)

$$\hat{x}_i(k+1) = H_{ii}(k)\hat{x}_i(k) + N_i(k)y_i(k)$$

which is quite usually chosen for H_∞ filtering in the existed literature, see [705], [706] for discrete-time case; and [708] for the corresponding continuous-time counterpart. Moreover, we have that

$$
\begin{aligned}
u_i(k, r_k) &= \sum_{j \in \mathcal{N}_i(r_k) \backslash \{i\}} a_{ij}(r_k) H_{ij}(k, r_k)(\hat{x}_i - \hat{x}_j) \\
&= \sum_{j \in \mathcal{N}_i(r_k)} a_{ij}(r_k) H_{ij}(k, r_k)(\hat{x}_i - \hat{x}_j)
\end{aligned}
\tag{11.149}
$$

The latter one is seen to be more simple for analysis and expression, which will be used in the subsequent analysis.

By now, for the formulation convenience, we shall firstly define the new introduced variables

$$\xi(k) = vec^T\left\{x^T(k), \hat{x}_1^T(k), ..., \hat{x}_n^T(k)\right\}, \quad \eta(k) = vec^T\left\{z_{1err}^T, ..., z_{ierr}^T, ..., z_{nerr}^T\right\}$$

where z_{ierr} is defined as

$$z_{ierr} = z_i - \hat{z}_i, \ i \in [1 \ n]$$

Then we shall denote

$$
\begin{aligned}
\widetilde{H(k, r_k)} &= diag\left\{\sum_{j \in \mathcal{N}_i(r_k)} a_{ij}(r_k) H_{ij}(k, r_k) + H_{ii}(k, r_k)\right\} \\
\overline{H}(k, r_k) &= \left[a_{ij}(r_k) H_{ij}(k, r_k)\right]_{n \times n} \\
\overline{D}_1(k, r_k) &= vec^T\left\{\alpha_i(N_i(k, r_k)D_i(k))^T\right\} \\
\overline{D}_2(k, r_k) &= vec^T\left\{\left(\gamma_i(k) - \alpha_i\right)\left(N_i(k, r_k)D_i(k)\right)^T\right\} \\
\overline{A}_1(k, r_k) &= \begin{bmatrix} A(k) & \mathbf{0} \\ \overline{D}_1(k, r_k) & \widetilde{H(k, r_k)} - \overline{H}(k, r_k) \end{bmatrix} \\
\overline{A}_2(k, r_k) &= \begin{bmatrix} \mathbf{0} & \mathbf{0} \\ \overline{D}_2(k, r_k) & \mathbf{0} \end{bmatrix} \\
h(x) &= \begin{bmatrix} f(x(k)) \\ g(x(k)) \end{bmatrix}
\end{aligned}
$$

$$
\overline{E}_1(k, r_k) = vec^T \left\{ \alpha_i (N_i(k, r_k) E_i(k))^T \right\}
$$

$$
\overline{E}_2(k, r_k) = vec^T \left\{ \left(\gamma_i(k) - \alpha_i \right) \left(N_i(k, r_k) E_i(k) \right)^T \right\}
$$

$$
\overline{B}_1(k, r_k) = \begin{bmatrix} B(k) & \mathbf{0} \\ \mathbf{0} & \overline{E}_1(k, r_k) \end{bmatrix}
$$

$$
\overline{B}_2(k, r_k) = \begin{bmatrix} \mathbf{0} & \mathbf{0} \\ \mathbf{0} & \overline{E}_2(k, r_k) \end{bmatrix}
$$

$$
\overline{F}(k, r_k) = vec^T \left\{ \left(N_i(k, r_k) F_i(k) \right)^T \right\}
$$

$$
\overline{C}(k, r_k) = \begin{bmatrix} C(k) \\ \overline{F}(k, r_k) \end{bmatrix}
$$

$$
\overline{M}_1(k) = vec_n^T \{ M^T(k) \}
$$

$$
\overline{M}_2(k, r_k) = diag_n \{ \hat{M}_i(k, r_k) \}
$$

$$
\overline{M}(k, r_k) = \begin{bmatrix} \overline{M}_1(k) & -\overline{M}_2(k, r_k) \end{bmatrix}
$$

$$
\overline{d}_{2i}(k, r_k) = vec^T \left\{ \underbrace{\mathbf{0} \ldots \mathbf{0}}_{i-1}, \left(N_i(k, r_k) D_i(k) \right)^T, \underbrace{\mathbf{0} \ldots \mathbf{0}}_{n-i} \right\}
$$

$$
\overline{a}_{2i}(k, r_k) = \begin{bmatrix} \mathbf{0} & \mathbf{0} \\ \overline{d}_{2i}(k, r_k) & \mathbf{0} \end{bmatrix}
$$

$$
\overline{D}_2(k, r_k) = \sum_{i=1}^{n} \left(\gamma_i(k) - \alpha_i \right) \overline{d}_{2i}(k, r_k)
$$

$$
\overline{A}_2(k, r) = \sum_{i=1}^{n} \left(\gamma_i(k) - \alpha_i \right) \overline{a}_{2i}(k, r_k)
$$

$$
\overline{e}_{2i}(k, r_k) = vec^T \left\{ \underbrace{\mathbf{0} \ldots \mathbf{0}}_{i-1}, \left(N_i(k, r_k) E_i(k) \right)^T, \underbrace{\mathbf{0} \ldots \mathbf{0}}_{n-i} \right\}
$$

$$\bar{b}_{2i}(k, r_k) \;=\; \begin{bmatrix} \mathbf{0} & \mathbf{0} \\ \mathbf{0} & \bar{e}_{2i}(k, r_k) \end{bmatrix}$$

$$\overline{E}_2(k, r_k) \;=\; \sum_{i=1}^{n} \Big(\gamma_i(k) - \alpha_i\Big)\bar{e}_{2i}(k, r_k)$$

$$\overline{B}_2(k, r_k) \;=\; \sum_{i=1}^{n} \Big(\gamma_i(k) - \alpha_i\Big)\bar{b}_{2i}(k, r_k)$$

$$\mathcal{H} \;=\; [\mathbf{I}_{n_x \times n_x}, \underbrace{\mathbf{0}, \ldots, \mathbf{0}}_{n}]^T$$

$$\mathcal{G} \;=\; \begin{bmatrix} \mathbf{0}_{n_x \times n_x} \\ I_{n_x \times n_x} \end{bmatrix} \quad \mathcal{J} = \begin{bmatrix} I_{n_x \times n_x} \\ \mathbf{0}_{n_x \times n_x} \end{bmatrix} \quad \mathcal{I} = \begin{bmatrix} vec_n\{I\} \\ -diag_n\{I\} \end{bmatrix}$$

Then the augmented system could be expressed as

$$\xi(k+1) \;=\; \Big(\overline{A}_1(k, r_k) + \overline{A}_2(k, r_k)\Big)\xi(k)$$

$$+\; \Big(\overline{B}_1(k, r_k) + \overline{B}_2(k, r_k)\Big)h(x(k)) + \overline{C}(k, r_k)w(k)$$

$$\eta(k) \;=\; \overline{M}(k, r_k)\xi(k) \tag{11.150}$$

In the sequel, we are focusing on designing the distributed H_∞-consensus filters for each sensor node of the sensor networks, such that the filtering errors could satisfy the H_∞ performance on an average consensus. That is to say, it is not required that the filtering error of each sensor node satisfies the H_∞ performance, but the average error is the main target. Obviously, the filtering systems with sensor networks have the advantage of better robustness and fault tolerance compared with the traditional single sensor case.

Definition 11.5.1 *[711] The filtering errors z_{ierr} (filtering error for the ith sensor node, $i \in [1\ n]$) are said to satisfy the H_∞ -consensus performance constraints if the following inequalities hold*

$$\frac{1}{n}\sum_{i=1}^{n}\|z_{ierr}\|_{\mathbb{E}_2}^2 \leq \gamma^2\Big\{\|w\|_2^2 + \frac{1}{n}\sum_{i=1}^{n}e_i^T(0)S_ie_i(0)\Big\} \tag{11.151}$$

where $\|z_{ierr}\|_{\mathbb{E}_2}^2 = \mathbb{E}\Big(\sum_{k=0}^{N-1}\|z_{ierr}(k)\|^2\Big)$, and $e_i(0) = \hat{x}_i(0) - x(0)$, for some given disturbance attenuation level $\gamma > 0$ and some given positive definite matrices $S_i = S_i^T > 0$ ($i \in [1\ n]$). Moreover, it is noted that the H_∞ consensus performance could be rewritten as $\|\eta\|_{\mathbb{E}_2}^2 \leq \gamma^2\Big\{n\|w\|_2^2 + e^T(0)Re(0)\Big\}$, where $e(0) = vec^T\{e_i^T(0)\}$, and $R = diag\{S_i\}$.

Lemma 11.5.1 *[713] Let X and Y be any n–dimensional real vectors, and let P be an $n \times n$ positive semi-definite matrix. Then, the following matrix inequality holds:*

$$2X^T PY \leq X^T PX + Y^T PY$$

Lemma 11.5.2 *[715] Let $(\Omega, \mathbf{F}, P)$ be the complete probability space, X, Y are both random variables satisfying $X \in \mathcal{F}$ and $Y \in \mathbf{F}$ (are integrable with respect to $\mathbf{F}$), $\mathbf{G} \subset \mathcal{F}$ denotes a sub-field, if $X \leq Y$ a.e. holds, then $\mathbb{E}\{X|\mathbf{G}\} \leq \mathbb{E}\{Y|\mathbf{G}\}$, where $\mathbb{E}\{\cdot|\mathbf{G}\}$ denotes the conditional expectation with respect to $\mathbf{G}$.*

Lemma 11.5.3 *Let $(\Omega, \mathbf{F}, P)$ be the complete probability space, X be a integrable random variable satisfying the arbitrary probability distribution on interval $[0, 1]$, with mathematical expectation α. $\mathbf{G} \subset \mathbf{F}$ is a sub-field. We conclude that $\mathbb{E}\{(X - \alpha)^2|\mathbf{G}\} \leq 1$*

Proof 11.5.1 *Noting that for a integrable random variable on interval $[0, 1]$, it always holds $(X - \alpha)^2 \leq 1$, the proof follows directly from **Lemma** 11.5.2.*

11.5.3 Performance analysis

In this section, the analysis of the H_∞ consensus performance is presented, and then sufficient conditions are obtained under which the H_∞ consensus performance could be guaranteed for the given consensus filters.

For clarity, in the rest of the chapter, we shall use s, t to denote the modes for the Markov process, i, j for the individual sensor nodes, and k for the time–steps, if they are not explicitly specified .

For convenience, first we shall denote

$$\overline{P}(k,s) = \sum_{t \in S} \lambda_{st} P(k,t)$$

$$\overline{\xi}(k) = vec^T \left\{ \xi^T(k), h^T(x(k)), w^T(k) \right\}$$

$$\mathcal{D}(k,s) = vec^T \left\{ \mathbf{0}, vec\left\{ \sqrt{5}\left(N_i(k,s)D_i(k)\right)^T \right\} \right\}$$

$$\mathcal{E}(k,s) = vec^T \left\{ \mathbf{0}, vec\left\{ \sqrt{5}\left(N_i(k,s)E_i(k)\right)^T \right\} \right\}$$

$$\Lambda_{11}(k,s) = A_1^T(k,s)\overline{P}(k+1,s)\overline{A}_1(k,s) + \overline{M}^T(k,s)\overline{M}(k,s) - P(k,s)$$

$$\Lambda_{21}(k,s) = \overline{B}_1^T(k,s)\overline{P}(k+1,s)\overline{A}_1(k,s)$$

$$\Lambda_{22}(k,s) = \overline{B}_1^T(k,s)\overline{P}(k+1,s)\overline{B}_1(k,s)$$

$$\Lambda_{33}(k,s) = \overline{C}^T(k,s)\overline{P}(k+1,s)\overline{C}(k,s) - n\gamma^2 \times I$$

$$\Omega_{11}(k,s) = 3\overline{A}_1^T(k,s)\overline{P}(k+1,s)\overline{A}_1(k,s) + \overline{M}^T(k,s)\overline{M}(k,s)$$
$$- \quad P(k,s) + \mathcal{H}\Big(\rho(k,s)I\Big)\mathcal{H}^T$$
$$- \quad \lambda(k,s)\mathcal{H}\frac{K_1^T K_2 + K_2^T K_1}{2}\mathcal{H}^T - \mu(k,s)\mathcal{H}\frac{L_1^T L_2 + L_2^T L_1}{2}\mathcal{H}^T$$

$$\Omega_{21}(k,s) = \overline{B}_1^T(k,s)\overline{P}(k+1,s)\overline{A}_1(k,s) + \lambda(k,s)\mathcal{J}\frac{K_1 + K_2}{2}\mathcal{H}^T$$
$$+ \quad \mu(k,s)\mathcal{G}\frac{L_1 + L_2}{2}\mathcal{H}^T$$

$$\Omega_{22}(k,s) = 3\overline{B}_1^T(k,s)\overline{P}(k+1,s)\overline{B}_1(k,s) + \mathcal{G}\Big(\sigma(k,s) \times I\Big)\mathcal{G}^T$$
$$- \quad \lambda(k,s)\mathcal{J} \times I \times \mathcal{J}^T - \mu(k,s)\mathcal{G} \times I \times \mathcal{G}^T$$

$$\Omega_{31}(k,s) = \overline{C}^T(k,s)\overline{P}(k+1,s)\overline{A}_1(k,s)$$

$$\Omega_{23}(k,s) = \overline{C}^T(k,s)\overline{P}(k+1,s)\overline{B}_1(k,s)$$

$$\Omega_{33}(k,s) = 3\overline{C}^T(k,s)\overline{P}(k+1,s)\overline{C}(k,s) - n\gamma^2 \times I$$

$$\Lambda(k,s) = \begin{bmatrix} \Lambda_{11}(k,s) & * & * \\ \Lambda_{21}(k,s) & \Lambda_{22}(k,s) & * \\ \Omega_{31}(k,s) & \Omega_{32}(k,s) & \Lambda_{33}(k,s) \end{bmatrix}$$

Theorem 11.5.1 *Consider the discrete time-varying nonlinear systems to be filtered given by (Σ), the filters of form (11.147) with the filtering parameters*

$H_{ij}(k, s)$, $N_i(k, s)$, $\hat{M}_i(k, s)$ *for each sensor node known as a prior, and the consensus protocol (11.148) being given, then we could conclude that the H_∞ consensus performance given in Definition 11.5.1 could be satisfied if there exist sets of positive definite constants $\rho(k, s)$, $\sigma(k, s)$, $\lambda(k, s)$, $\mu(k, s)$, positive definite matrices $P_i(k, s)$, ($i \in [0\ n]$, $s \in S$, $k \in [0, N - 1]$) such that the following recursive linear matrix inequalities (RLMI) are satisfied, with the initial condition $\xi^T(0)P(0, r_0)\xi^T(0) \leq \gamma^2 \times \xi^T(0)\mathcal{I}R\mathcal{I}^T\xi(0)$, where R is the block diagonal matrix determined by a set of given matrix S_i as in Definition 11.5.1, and $P(k, s) = diag\{P_0(k, s), P_1(k, s), ..., P_n(k, s)\}$.*

$$\begin{bmatrix} -\rho(k, s) \times I & * \\ \mathcal{D}(k) & -\overline{P}^{-1}(k + 1, s) \end{bmatrix} \leq 0 \qquad (11.152)$$

$$\begin{bmatrix} -\sigma(k, s) \times I & * \\ \mathcal{E}(k) & -\overline{P}^{-1}(k + 1, s) \end{bmatrix} \leq 0 \qquad (11.153)$$

$$\Omega(k, s) = \begin{bmatrix} \Omega_{11}(k, s) & * & * \\ \Omega_{21}(k, s) & \Omega_{22}(k, s) & * \\ \Omega_{31}(k, s) & \Omega_{32}(k, s) & \Omega_{33}(k, s) \end{bmatrix} \leq 0 \quad (11.154)$$

Proof 11.5.2 *For $r_k = s$, let us choose the following Lyapunov function defined as*

$$V(\xi(k), r_k) = \xi^T(k)P(k, s)\xi(k) \qquad (11.155)$$

where $P(k, s) = diag\{P_0(k, s), ..., P_n(k, s)\}$. Noting that r_k, $\xi(k)$ are $\sigma(r_k = s, \xi(k))$-measurable, where $\sigma(r_k = s, \xi(k)$ denotes the created σ-field, we shall *have*

$$E\Big\{\Delta V(\xi(k), r_k)\Big|\xi(k), r_k = s\Big\} + \mathbb{E}\Big\{\|\eta(k)\|^2\Big\} - n\gamma^2\|w\|^2$$

$$= \quad E\Big\{\overline{\xi}^T(k)\Lambda(k, s)\overline{\xi}(k)$$

$$+ \quad 2\xi^T(k)\overline{A}_1^T(k, s)\overline{P}(k+1, s)\overline{A}_2(k, s)\xi(k)$$
$$+ \quad \xi^T(k)\overline{A}_2^T(k, s)\overline{P}(k+1, s)\overline{A}_2(k, s)\xi(k)$$
$$+ \quad 2\xi^T(k)\overline{A}_2^T(k, s)\overline{P}(k+1, s)\overline{B}_1(k, s)h(x(k))$$
$$+ \quad 2\xi^T(k)\overline{A}_2^T(k, s)\overline{P}(k+1, s)\overline{B}_2(k, s)h(x(k))$$
$$+ \quad 2\xi^T(k)\overline{A}_2^T(k, s)\overline{P}(k+1, s)\overline{C}(k, s)w(k)$$
$$+ \quad 2\xi^T(k)\overline{A}_1^T(k, s)\overline{P}(k+1, s)\overline{B}_2(k, s)h(x(k))$$
$$+ \quad 2h^T(x(k))\overline{B}_1^T(k, s)\overline{P}(k+1, s)\overline{B}_2(k, s)h(x(k))$$
$$+ \quad h^T(x(k))\overline{B}_2^T(k, s)\overline{P}(k+1, s)\overline{B}_2(k, s)h(x(k))$$
$$+ \quad 2h^T(x(k))\overline{B}_2^T(k, s)\overline{P}(k+1, s)\overline{C}(k, s)w(k)\Big|\xi(k), r_k = s\Big\}$$

It is noted that under the assumption, $\overline{A}_2(k, s)$ and $\overline{B}_2(k, s)$ are neither unrelated with r_k, nor measurable with respect to $\sigma(r_k = s, \xi(k))$.

In terms of Lemma 11.5.1, we have

$$2\xi^T(k)\overline{A}_1^T(k, s)$$
$$\times \quad \overline{P}(k+1, s)\overline{A}_2(k, s)\xi(k)$$
$$\leq \quad \xi^T(k)\overline{A}_1^T(k, s)\overline{P}(k+1, s)\overline{A}_1(k, s)\xi(k)$$
$$+ \quad \xi^T(k)\overline{A}_2^T(k, s)\overline{P}(k+1, s)\overline{A}_2(k, s)\xi(k)$$

Noting the form of $P(k, s)$ we choose, and from lemma 11.5.3, we could obtain that

$$E\Big\{\xi^T(k)\overline{A}_2^T(k, s)\overline{P}(k+1, s)\overline{A}_2(k, s)\xi(k)\Big|\xi(k), r_k = s\Big\}$$

$$\leq \quad E\Big\{\sum_{i=1}^{n}\big(\gamma_i(k) - \alpha_i\big)^2\xi^T(k)\overline{a}_{2i}^T(k, s)\overline{P}(k+1, s)\overline{a}_{2i}(k, s)\xi(k)\Big|\xi(k), r_k = s\Big\}$$

$$\leq \quad \sum_{i=1}^{n}\xi^T(k)\overline{a}_{2i}^T(k, s)\overline{P}(k+1, s)\overline{a}_{2i}(k, s)\xi(k)$$

Following the similar steps to deal with the other terms, and after tedious computation, we have

$$
\boldsymbol{E}\Big\{\Delta V(\xi(k), r_k)\Big|\xi(k), r_k = s\Big\} + \mathbb{E}\Big\{\|\eta(k)\|^2\Big\} - n\gamma^2\|w\|^2
$$

$$
\leq \quad \boldsymbol{E}\Big\{\overline{\xi}^T(k)\Lambda(k, s)\overline{\xi}(k)
$$

$$
+ \quad 2\xi^T(k)\overline{A}_1^T(k, s)\overline{P}(k + 1, s)\overline{A}_1(k, s)\xi(k)
$$

$$
+ \quad 2h^T(x(k))\overline{B}_1^T(k, s)\overline{P}(k+1, s)\overline{B}_1(k, s)h(x(k))
$$

$$
+ \quad 2w^T(k)\overline{C}^T(k, s)\overline{P}(k+1, s)\overline{C}(k, s)w(k)
$$

$$
+ \quad 5\xi^T(k)\overline{A}_2^T(k, s)\overline{P}(k+1, s)\overline{A}_2(k, s)\xi(k)
$$

$$
+ \quad 5h^T(x(k))\overline{B}_2^T(k, s)\overline{P}(k+1, s)\overline{B}_2(k, s)h(x(k))\Big|\xi(k), r_k = s\Big\}
$$

Proceeding further,

$$
\boldsymbol{E}\Big\{\Delta V(\xi(k), r_k)\Big|\xi(k), r_k = s\Big\} + \mathbb{E}\Big\{\|\eta(k)\|^2\Big\} - n\gamma^2\|w\|^2
$$

$$
\leq \quad \overline{\xi}^T(k)\Lambda(k, s)\overline{\xi}(k) + 2\xi^T(k)\overline{A}_1^T(k, s)\overline{P}(k+1, s)\overline{A}_1(k, s)\xi(k)
$$

$$
+ \quad 2h^T(x(k))\overline{B}_1^T(k, s)\overline{P}(k+1, s)\overline{B}_1(k, s)h(x(k))
$$

$$
+ \quad 2w^T(k)\overline{C}^T(k, s)\overline{P}(k+1, s)\overline{C}(k, s)w(k)
$$

$$
+ \quad 5\sum_{i=1}^{n}\xi^T(k)\overline{a}_{2i}^T(k, s)\overline{P}(k+1, s)\overline{a}_{2i}(k, s)\xi(k)
$$

$$
+ \quad 5\sum_{i=1}^{n}h^T(x(k))\overline{b}_{2i}^T(k, s)\overline{P}(k+1, s)\overline{b}_{2i}(k, s)h(x(k)) \qquad (11.156)
$$

We see that (11.152) and (11.153) respectively imply that

$$
\boldsymbol{E}\Big\{5\sum_{i=1}^{n}\xi^T(k)\overline{a}_{2i}^T(k)\overline{P}(k+1, s)\overline{a}_{2i}(k)\xi(k)\Big\}
$$

$$
\leq \quad \boldsymbol{E}\Big\{\xi^T(k)\mathcal{H}\Big(\rho(k, s) \times I\Big)\mathcal{H}^T\xi(k)\Big\} \qquad (11.157)
$$

$$
\boldsymbol{E}\Big\{5\sum_{i=1}^{n}h^T(x(k))\overline{b}_{2i}^T(k)\overline{P}(k+1, s)\overline{b}_{2i}(k)h(x(k))\Big\}
$$

$$
\leq \quad E\Big\{h^T(x(k))\mathcal{G}\Big(\sigma(k, s) \times I\Big)\mathcal{G}^T h(x(k))\Big\} \qquad (11.158)
$$

Moreover, if we denote

$$\zeta(k) = vec^T\left\{\xi^T(k), h^T\left(x(k)\right)\right\}$$

then from (11.145) and (11.146), we have

$$\zeta^T(k)\left[\begin{array}{cc} \mathcal{H}\frac{K_1^T K_2 + K_2^T K_1}{2}\mathcal{H}^T & * \\ -\mathcal{J}\frac{K_1 + K_2}{2}\mathcal{H}^T & \mathcal{J} \times I \times \mathcal{J}^T \end{array}\right]\zeta(k) \leq 0 \quad (11.159)$$

and

$$\zeta^T(k)\left[\begin{array}{cc} \mathcal{H}\frac{L_1^T L_2 + L_2^T L_1}{2}\mathcal{H}^T & * \\ -\mathcal{G}\frac{L_1 + L_2}{2}\mathcal{H}^T & \mathcal{G} \times I \times \mathcal{G}^T \end{array}\right]\zeta(k) \leq 0 \quad (11.160)$$

From(11.156)–(11.160), one has that if (11.152), (11.153), and (11.154) hold, it follows

$$\boldsymbol{E}\left\{\Delta V(\xi(k), r_k)|\xi(k), r_k = s\right\} + \mathbb{E}\left\{\|\eta(k)\|^2\right\} - n\gamma^2\|w\|^2$$
$$\leq \boldsymbol{E}\left\{\overline{\xi}^T(k)\Omega(k,s)\overline{\xi}(k)\right\} \leq 0$$

*Hence, the H_∞ performance defined in **Definition** 11.5.1 could be given by*

$$\|\eta\|_{\mathbb{E}_2}^2 - \gamma^2\left\{n\|w\|_2^2 + e^T(0)Re(0)\right\}$$
$$\leq \boldsymbol{E}\left\{\sum_{k=0}^{N-1}\overline{\xi}^T(k)\Omega(k,s)\overline{\xi}(k)\right\} - \mathbb{E}\left\{\xi^T(N)P(N,s)\xi(N)\right\}$$
$$+ \boldsymbol{E}\left\{\xi^T(0)\left[P(0,s) - \gamma^2\mathcal{I}R\mathcal{I}^T\right]\xi(0)\right\}$$

Noting that $P(N,s) > 0$, and from the initial conditions, we could see that the H_∞ consensus performance is guaranteed for the initial system (Σ) with given consensus filters for the sensor networks, that completes the proof.

Remark 11.5.3 *It is worth pointing out that in most of the existing literature, it is often supposed that the stochastic processes are mutually independent if there exist more than one process to govern the investigated systems. However, in this chapter, to be more realistic, we have rejected the independence between the Markov process r_k which governs the topologies of the sensor networks and $\gamma_i(k)$ which describes the missing measurements phenomenon, because the switching of the topologies could change the mutual information exchange between the sensor nodes, hence has an impact on the measurements. If independence is assumed, less conservative numerical results should be achieved, but less practical significance will be rendered correspondingly.*

11.5.4 Distributed H_∞ consensus filters design

In the previous subsection, we have solved the analysis of the H_∞ consensus filters. In the following, we shall go on to investigate the H_∞ consensus filters design problems. That is to say, given the initial system (Σ) to be estimated, and the measured outputs (11.144), then we shall accomplish the design of the filter parameters $H_{ij}(k,s)$, $N_i(k,s)$, $\hat{M}_i(k,s)$ for each sensor node of the sensor networks, such that the distributed H_∞ consensus performance is satisfied.

Also, for presentation convenience, we shall denote

$$\widetilde{\mathcal{W}(k,r_k)} = diag\{\sum_{j\in\mathcal{N}_i} a_{ij}(r_k)\mathcal{W}_{ij}(k,r_k) + \mathcal{W}_{ii}(k,r_k)\}$$

$$\overline{\mathcal{W}}(k,r_k) = \Big[a_{ij}(r_k)\mathcal{W}_{ij}(k,r_k)\Big]_{n\times n}$$

$$\overline{\mathcal{D}}_1(k,r_k) = vec^T\Big\{\alpha_i\Big(\mathcal{V}_i(k,r_k)D_i(k)\Big)^T\Big\}$$

$$\overline{\mathcal{A}}_1(k,r_k) = \left[\begin{array}{cc} \overline{P}_0(k+1,r_k)A(k) & \mathbf{0} \\ \overline{\mathcal{D}}_1(k,r_k) & \widetilde{\mathcal{W}(k,r_k)} - \overline{\mathcal{W}}(k,r_k) \end{array}\right]$$

$$\overline{\mathcal{E}}_1(k) = vec^T\Big\{\alpha_i\Big(\mathcal{V}_i(k,r_k)E_i(k)\Big)^T\Big\}$$

$$\overline{\mathcal{B}}_1(k,r_k) = \left[\begin{array}{cc} \overline{P}_0(k+1,r_k)B(k) & \mathbf{0} \\ \mathbf{0} & \overline{\mathcal{E}}_1(k,r_k) \end{array}\right]$$

$$\widetilde{\mathcal{D}}(k,r_k) = vec^T\Big\{\mathbf{0}, vec\Big\{\sqrt{5}\Big(\mathcal{V}_i(k,r_k)D_i(k)\Big)^T\Big\}\Big\}$$

$$\widetilde{\mathcal{E}}(k,r_k) = vec^T\Big\{\mathbf{0}, vec\Big\{\sqrt{5}\Big(\mathcal{V}_i(k,r_k)E_i(k)\Big)^T\Big\}\Big\}$$

$$\overline{\mathcal{F}}(k,r_k) = vec^T\Big\{\Big(\mathcal{V}_i(k,r_k)F_i(k)\Big)^T\Big\}$$

$$\overline{\mathcal{C}}(k,r_k) = \left[\begin{array}{c} \overline{P}_0(k+1,r_k)C(k) \\ \overline{\mathcal{F}}(k,r_k) \end{array}\right]$$

$$\Xi_{11}(k,s) = \left[\begin{array}{ccc} \Upsilon_{11}(k,s) & * & * \\ \Upsilon_{21}(k,s) & \Upsilon_{22}(k,s) & * \\ \Upsilon_{31}(k,s) & \Upsilon_{32}(k,s) & \Upsilon_{33}(k,s) \end{array}\right]$$

$$
\begin{aligned}
\Upsilon_{11}(k,s) &= -P(k,s) + \mathcal{H}\Big(\rho(k,s) \times I\Big)\mathcal{H}^T \\
&\quad - \lambda(k,s) \times \mathcal{H}\frac{K_1^T K_2 + K_2^T K_1}{2}\mathcal{H}^T \\
&\quad - \mu(k,s)\mathcal{H}\frac{L_1^T L_2 + L_2^T L_1}{2}\mathcal{H}^T \\
\Upsilon_{21}(k,s) &= \lambda(k,s)\mathcal{J}\frac{K_1 + K_2}{2}\mathcal{H}^T + \mu(k,s)\mathcal{G}\frac{L_1 + L_2}{2}\mathcal{H}^T \\
\Upsilon_{22}(k,s) &= \mathcal{G}\Big(\sigma(k,s) \times I\Big)\mathcal{G}^T - \lambda(k,s)\mathcal{J} \times I \times \mathcal{J}^T \\
&\quad - \mu(k,s)\mathcal{G} \times I \times \mathcal{G}^T \\
\Upsilon_{31}(k,s) &= 0 \quad \Upsilon_{32}(k,s) = 0 \quad \Upsilon_{33}(k,s) = -n \times \gamma^2 \times I \\
\Xi_{21}(k,s) &= \begin{bmatrix} \overline{\mathcal{A}}_1(k,s) & \overline{\mathcal{B}}_1(k,s) & \overline{\mathcal{C}}(k,s) \\ \overline{M}(k,s) & 0 & 0 \end{bmatrix} \\
\Xi_{22}(k,s) &= \begin{bmatrix} -\overline{P}(k+1,s) & * \\ 0 & -I \end{bmatrix} \\
\Xi_{31}(k,s) &= \begin{bmatrix} \overline{\mathcal{A}}_1(k,s) & 0 & 0 \\ 0 & \overline{\mathcal{B}}_1(k,s) & 0 \\ 0 & 0 & \overline{\mathcal{C}}(k,s) \end{bmatrix} \\
\Xi_{33}(k,s) &= diag_3\Big\{ -\frac{1}{2}\overline{P}(k+1,s)\Big\}
\end{aligned}
$$

Theorem 11.5.2 *Given the initial discrete time-varying nonlinear system* (Σ) *to be filtered, the outputs for every sensor node of the sensor networks (11.144), and also the switching topologies of the sensor networks driven by a ergodic Markov chain r_k, which is described by the transmission probability $[\lambda_{st}]_{s,t \in S}$ with the initial mode r_0, if there exist families of positive constants $\rho(k,s)$, $\sigma(k,s)$, $\lambda(k,s)$, $\mu(k,s)$, real-valued matrices $\mathcal{W}_{ij}(k,s)$, $\mathcal{V}_i(k,s)$, $\hat{M}(k,s)$, ($s \in S$, $i,j \in [1 \ \ n]$, $k \in [0 \ \ N-1]$), and positive definite matrices $P_i(k,s)$, ($s \in S$, $i \in [0 \ \ n]$, $k \in [0 \ \ N]$), such that families of the following recursive linear matrix inequalities (RLMI) hold*

$$
\begin{bmatrix} -\rho(k,s) \times I & * \\ \widetilde{\mathcal{D}}(k,s) & -\overline{P}(k+1,s) \end{bmatrix} \leq 0 \tag{11.161}
$$

$$
\begin{bmatrix} -\sigma(k,s) \times I & * \\ \widetilde{\mathcal{E}}(k,s) & -\overline{P}(k+1,s) \end{bmatrix} \leq 0 \tag{11.162}
$$

$$
\Xi(k,s) = \begin{bmatrix} \Xi_{11}(k,s) & * & * \\ \Xi_{21}(k,s) & \Xi_{22}(k,s) & * \\ \Xi_{31}(k,s) & 0 & \Xi_{33}(k,s) \end{bmatrix} \leq 0 \tag{11.163}
$$

under the initial condition

$$\xi^T(0)\left[P(0,s) - \gamma^2 \mathcal{I}R\mathcal{I}^T\right]\xi(0) \leq 0 \tag{11.164}$$

Then the designed distributed H_∞ consensus filters are solved which ensure the H_∞ consensus performance, the corresponding recursive parameters could be given by

$$H_{ij}(k,s) = \overline{P}_i^{-1}(k+1,s)\mathcal{W}_{ij}(k,s)$$

$$N_i(k,s) = \overline{P}_i^{-1}(k+1,s)\mathcal{V}_i(k,s)$$

and $\hat{M}_i(k,s)$ is solved as an entry of (11.163), $i,j \in [1\ n]$, $k \in [0\ N-1]$, $s \in S$.

Proof 11.5.3 *It is noted that*

$$P(k,s) = diag\{P_0(k,s), P_1(k,s), ..., P_n(k,s)\}$$

hence, one has

$$\overline{P}(k+1,s) = diag\{\overline{P}_0(k+1,s), \overline{P}_1(k+1,s), ..., \overline{P}_n(k+1,s)\}$$

Then, based on theorem 11.5.1, due to the Schur complement lemma and some computation, the proof could be completed.

Up to now, we have solved both the analysis of the H_∞ consensus filter and the filter design problems. Finally, we shall given the following distributed H_∞ consensus filters design procedure upon the obtained results.

| *The distributed H_∞ consensus filters design algorithm:* |

Step 1: Given the disturbance attenuation index γ, performance index matrices S_i, $i \in [1\ n]$, the initial mode s and the networks topologies, the initial state $x(0)$, and $\hat{x}_i(0)$, $i \in [1\ n]$, choose the set of $P_i(0,s)$, $i \in [0\ n]$, such that (11.164) is satisfied; then set k=0.

Step 2: If $k < N$, go to *Step 3*; else go to *Step 4*.

Step 3: For the sampling instant k, $k = 1,2...,N-1$, solve (11.161) $\sim$ (11.163), obtain $P_i(k+1,s)$, $\mathcal{W}_{ij}(k,s)$, $\mathcal{V}_i(k,s)$, then update the filter parameters by $H_{ij}(k,s) = \overline{P}_i^{-1}(k+1,s)\mathcal{W}_{ij}(k,s)$, $N_i(k,s) = \overline{P}_i^{-1}(k+1,s)\mathcal{V}_i(k,s)$, with $\hat{M}(k,s)$ being solved directly from (11.163). Set $k = k+1$, and go to step 2.

Step 4: Exit.

11.5.5 Simulation example 4

In this section, we will present an example to show the effectiveness of the proposed distributed consensus filters approaches.

Consider the following discrete time-varying nonlinear systems with the following parameters in form of (11.143).

$$A(k) = \begin{bmatrix} 0.3sink & 0.2 \\ 0.8sin2k & 0.7sin3k \end{bmatrix} B(k) = \begin{bmatrix} 0.2sink & 0 \\ 0.4sin5k & 0.3 \end{bmatrix}$$

$$C(k) = \begin{bmatrix} sink \\ 0.5 \end{bmatrix} M(k) = [0.5\ 0.4sin3k]$$

The exogenous disturbance input $w(k)$ is chosen as $w(k) = 0.2cosk$, which is obviously square summable on a finite horizon.

Consider a sensor network which contains 3 sensor nodes. The topologies of the networks are assumed to have 2 modes, and be governed by the following Markov chain, with transition probability $Prob = \begin{bmatrix} 0.6 & 0.4 \\ 0.3 & 0.7 \end{bmatrix}$. The information exchange between the sensors in each mode could be described by figure 11.24. It

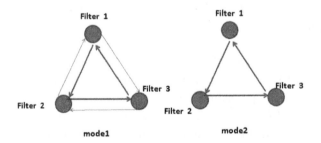

Figure 11.24: The topologies of the sensor networks.

is assumed that the adjacency matrix reflected in the figure could be given as

$$\mathcal{A}_{mode\ 1} = \begin{bmatrix} 1 & 1 & 1 \\ 1 & 1 & 1 \\ 1 & 1 & 1 \end{bmatrix} \mathcal{A}_{mode\ 2} = \begin{bmatrix} 1 & 0 & 1 \\ 1 & 1 & 0 \\ 0 & 1 & 1 \end{bmatrix}$$

The measured output of the sensors are assumed to be

$$y_i(k) = \gamma_i(k)\Big[[0.2\ 0.2sin3k]x(k) + [0.1\ 0.2sin2k]g(x(k))\Big] + 0.1sinkw(k)$$

The discrete random variable $\gamma_i(k)$ is assumed to be described by

$$Prob(\gamma_1(k)) = \begin{cases} 0.1 & \gamma_1(k) = 0 \\ 0.3 & \gamma_1(k) = 0.5 \\ 0.6 & \gamma_1(k) = 1 \end{cases}$$

$$Prob(\gamma_2(k)) = \begin{cases} 0.2 & \gamma_2(k) = 0 \\ 0.4 & \gamma_2(k) = 0.6 \\ 0.4 & \gamma_2(k) = 1 \end{cases}$$

$$Prob(\gamma_3(k)) = \begin{cases} 0.1 & \gamma_3(k) = 0 \\ 0.1 & \gamma_3(k) = 0.1 \\ 0.8 & \gamma_3(k) = 1 \end{cases}$$

The nonlinear function $f(x(k)) = g(x(k)) = [0.03x_1 + 0.01sinx_1 \quad 0.04x_2 + 0.01sin2x_2]^T$. Then the sector nonlinearity parameters could be chosen as

$$K_1 = L_1 = \begin{bmatrix} 0.02 & 0 \\ 0 & 0.02 \end{bmatrix} \quad K_2 = L_2 = \begin{bmatrix} 0.04 & 0 \\ 0 & 0.06 \end{bmatrix}$$

And the H_∞ disturbance attenuation level γ is given $\gamma = 1$, the initial state of the investigated system $x(0) = [-0.2 \quad 0.3]^T$, and for the each filter $\hat{x}_1(0) = [0.1 \quad 0.2]^T$, $\hat{x}_2(0) = [0.2 \quad 0.3]^T$, $\hat{x}_3(0) = [0.3 \quad -0.3]^T$. Let $S_1 = S_2 = S_3 = diag\{2, 2\}$, choose

$$P_0(0, 1) = P_1(0, 1) = P_2(0, 1) = P_3(0, 1) = P_0(0, 2)$$

$$= P_1(0, 2) = P_2(0, 2) = P_3(0, 2) = \begin{bmatrix} 1.1497 & 0.0501 \\ 0.0501 & 1.1079 \end{bmatrix}$$

to satisfy the initial condition (11.164). Then the distributed consensus filters parameters could be solved recursively by (11.161) $\sim$ (11.163).

The numerical results are shown in Figure 11.25–Figure 11.27. Figure 11.25 shows one possible sample path of the Markov process which governs the topology switching of the sensor networks. Figure 11.26 shows the system output and the estimations from the filters, and Figure 11.27 shows the estimation errors for each filter respectively, from which one can see that the designed distributed filters could perform effectively. Hence the numerical example has shown the validity of our approach.

11.6 Notes

An energy-efficient distributed fusion estimation algorithm has been developed in this chapter for estimating states of discrete–time linear stochastic systems with

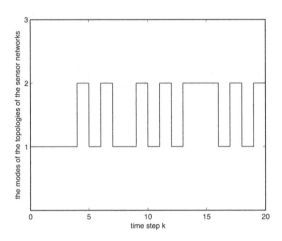

Figure 11.25: The modes of the topologies of the sensor networks.

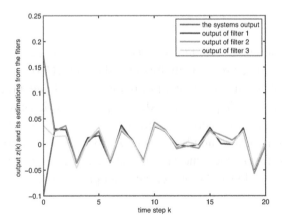

Figure 11.26: The system output and the estimations of the filters.

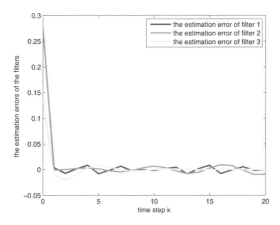

Figure 11.27: The estimation errors for the filters.

slowly changing dynamics and random packet losses in a WSN environment. A transmission rate method was proposed to reduce energy consumption in exchanging information among sensors, and to support a two-stage fusion estimate and reduce disagreements of the local estimates. It is shown that the obtained estimation performance critically depends on the measurement transmission rate and the packet loss probabilities and that the time scale of information exchange among sensors can be slower, while still maintaining a satisfactory estimation performance. A simulation example of a maneuvering target tracking system has demonstrated the effectiveness of the proposed estimation method.

In this chapter, it is assumed that each sensor in the network uses the measurements at previous time–step from its neighbors if the ones at the current time–step are not available. However, the effect of previous measurements might be negative for the local estimation at the current time–step. In this case, an alternative for the sensors is to give up nodes with unavailable measurements and just select those with available measurements at the local estimation stage. Nevertheless, the dimension of each local estimation system will be time-varying by using such a strategy, which adds much more difficulty to the modeling and design of the estimation system. This remains to be one of our future research directions.

The distributed H_∞ consensus filtering problem is investigated for a class of discrete time-varying nonlinear systems with missing measurements. The topology of the sensor networks is assumed to be stochastic switching governed by an ergodic Markov chain, and the missing measurements are assumed to be governed by an individual Bernoulli random variable for each sensor node, which is more realistic in practice. Sufficient conditions are given under which the H_∞ consensus filtering performance is satisfied with given filter parameters for every sensor node; the H_∞ consensus filters are solved by a group of recursive linear matrix inequalities (RLMI), which is suitable for online computation.

11.7 Proposed Topics

1. From the literature, it is well-known that improvement of the quality of tracking by mobile sensors (or agents) leads to the emergence of flocking behavior. It is desired to study the problem of distributed estimation for mobile ad hoc networks (MANETs) and to address distributed target tracking for mobile sensor networks with a dynamic topology.

2. Autonomous mobile sensor networks are employed to measure large-scale environmental fields. Yet an optimal strategy for mission design addressing both the cooperative motion control and the cooperative sensing is still an open problem. An important research topic is to develop a family of cooperative filters that combines measurements from a small number of mobile sensor platforms to cooperatively explore a static planar scalar field.

3. Of a particular interest is to consider the distributed tracking problem of a nonlinear dynamical system via networked sensors. The sensors communicate with each other by means of a multi-hop protocol over a communication network. Under standard assumptions, in-network processing algorithms deal with arbitrary network topology and then extend these results to account for communication delays and packet losses. Show the conditions under which these algorithms are optimal in the linear setting and achieve centralized performance. Identify the merits and demerits of the proposed techniques.

4. An appealing research subject involves investigating a moving-target tracking problem with sensor networks where each sensor node has a sensor to observe the target and a processor to estimate the target position. To enable efficient operation, there is a wireless communication capability, but with limited range and able only to communicate with neighbors. Consider that the moving target is assumed to be an intelligent agent, which is "smart" enough to escape from the detection by maximizing the estimation error. This "adversary behavior" makes the target tracking problem more difficult. It is therefore suggested to formulate this target estimation problem as a zero-sum game and use a distributed version of the minimax filter for multiple sensor nodes to estimate the target, position. Derive the features of the proposed technique.

Appendix

In this appendix, we collect some useful mathematical inequalities and lemmas which have been extensively used throughout the book.

A.1   A Glossary of Terminology and Notations

In this section, we assemble the terminologies and notations used throughout the book, with the objective of paving the way to the technical development of subsequent chapters. These terminologies and notations are quite standard in the scientific media and only vary in form or character.

A.1.1   General terms

As a start, matrices as $n \times m$ dimensional arrays of elements with n-rows and m-columns are represented by capital letters while vectors as n-tuples or columns (unless otherwise specified) and scalars (single elements) are represented by lower case letters. We use $\mathbf{R}, \mathbf{R}_+$, $\mathbf{R}^n$ and $\mathbf{R}^{n \times m}$ to denote the set of reals, positive real numbers, real n-tuples (vectors) and real $n \times m$ matrices, respectively. Alternatively, $\mathbf{R}^n$ is called the Euclidean space and is equipped with the vector-norm as $||x|| \overset{\Delta}{=} \sqrt{[x_1^2 + \cdots x_n^2]}$. The terms $f(t), g(s)$ denote, respectively, scalar-valued functions of the real variables t and s. The quantities $\dot{x}, \ddot{x}$ are the first and the second derivative of x with respect to time, respectively. The symbols $[.,.], (.,.], (.,.)$ denote, respectively, closed, semiclosed, and open intervals; that is $t \in (a, b] \Rightarrow a < t \le b$. The open left-half $(\equiv \{s : Re(s) \le 0\}$, the open proper left-half $(\equiv \{s : Re(s) < 0\}$ and the open proper right-half $(\equiv \{s : Re(s) > 0\}$ of the complex plane are represented by $\mathbb{C}^\dagger$, $\mathbb{C}^-$ and $\mathbb{C}^+$, respectively. We use $\mathcal{U}_p \in \Re^{n \times n}$ and $\mathcal{U}_k \in \Re^{m \times \ell}$ to denote, respectively, the set of uncertain plant perturbations ΔA of the nominal dynamical system A and the set of uncertain controller perturbations ΔK of the nominal controller gain K. The

Lebesgue space $\mathcal{L}_2[0, \infty)$ consists of square integrable functions on the interval $[0, \infty)$ and equipped with the norm

$$||p||_2 \triangleq \left[\int_0^\infty p^t(\tau)\, p(\tau)\, d\tau \right]^{1/2} \tag{A.1}$$

For any square matrix W of arbitrary dimension $n \times n$, let W^t, W^{-1}, $\lambda(W)$, $r(W)$, $tr(W)$, $det(W)$, $\rho(W)$ and $||W||$ denote, respectively, the transpose, the inverse, the spectrum, the rank, the trace, the determinant, the spectral radius, and the induced norm defined by

$$||W|| \triangleq \lambda(W\, W^t)^{1/2} \tag{A.2}$$

We use $W > 0$ $(\geq, <, \leq 0)$ to denote a symmetric positive definite (positive semidefinite, negative, negative semidefinite matrix W with $\lambda_m(W)$ and $\lambda_M(W)$ being the minimum and maximum eigenvalues of W. Frequently, I stands for the identity matrix with appropriate dimension, $W^\dagger$ denotes the pseudo-inverse of W and $diag(W_1, \cdots, W_p)$ stands for the block-diagonal matrix

$$\begin{bmatrix} W_1 & 0 & \cdots & 0 \\ 0 & W_2 & \cdots & 0 \\ 0 & \cdots & \cdots & 0 \\ 0 & \cdots & \cdots & W_p \end{bmatrix} \tag{A.3}$$

A.1.2 Functional differential equations

Let $\mathfrak{C}_{n,\tau} = \mathfrak{C}([-\tau, 0], \mathbf{R}^n)$ denote the Banach space of continuous vector functions mapping the interval $[-\tau, 0]$ into $\mathbf{R}^n$ with the topology of uniform convergence and designate the norm of an element $\phi \in \mathfrak{C}_{n,\tau}$ by

$$||\phi||_* \triangleq \sup_{\theta \in [-\tau, 0]} ||\phi(\theta)||_2 \tag{A.4}$$

If $\alpha \in \mathbf{R}, d \geq 0$ and $x \in \mathfrak{C}([\alpha - \tau, \alpha + d], \mathbf{R}^n)$ then for any $t \in [\alpha, \alpha + d]$, we let $x_t \in \mathfrak{C}$ be defined by $x_t(\theta) := x(t + \theta)$, $-\tau \leq \theta \leq 0$. If $\mathcal{D} \subset \Re \times \mathfrak{C}, f : \mathbf{D} \to \mathbf{R}^n$ is a given function, the relation $\dot{x}(t) = f(t, x_t)$ is a retarded functional differential equation (RFDE) on $\mathbf{D}$ where $x_t(t), t \geq t_o$ denotes the restriction of $x(\cdot)$ to the interval $[t - \tau, t]$ translated to $[-\tau, 0]$. Here, $\tau > 0$ is termed the *delay factor*. In the sequel, if $\alpha \in \mathbf{R}$, $d \geq 0$ and $x \in \mathfrak{C}([\alpha - \tau, \alpha + d], \Re^n)$ then for any

$t \in [\alpha, \alpha + d]$, we let $x_t \in \mathfrak{C}$ be defined by $x_t(\theta) \stackrel{\Delta}{=} x(t + \theta)$, $-\tau \leq \theta \leq 0$. In addition, if $\mathbf{D} \subset \mathbf{R} \times \mathfrak{C}, f : \mathbf{D} \rightarrow \mathbf{R}^n$ is given function, then the relation

$$\dot{x}(t) = f(t, x_t) \tag{A.5}$$

is a retarded functional differential equation (RFDE) on $\mathbf{D}$ where $x_t, t \geq t_0$ denotes the restriction of $x(.)$ on the interval $[t - \tau, t]$ translated to $[-\tau, 0]$. Here $\tau > 0$ is termed the **state-delay factor**. A function x is said to be a **solution** of (A.5) on $[\alpha - \tau, \alpha + d]$ if there $\alpha \in \mathbf{R}$ and $d > 0$ such that

$$x \in \mathfrak{C}([\alpha - \tau, \alpha + d], \mathbf{R}^n), \quad (t, x_t) \in \mathbf{D}, \quad t \in [\alpha, \alpha + d] \tag{A.6}$$

and $x(t)$ satisfies (A.5) for $t \in [\alpha, \alpha + d]$. For a given $\alpha \in \mathbf{R}$, $\phi \in \mathfrak{C}$, $x(\alpha, \phi, f)$ is said to be a solution of (A.5) with **initial value** ϕ at α.

A.2 Stability Notions

In this section, we present some definitions and results pertaining to the stability of dynamical systems.

Definition A.2.1 *A function of x and t is a* **carathedory function** *if, for all $t \in \Re$, it is continuous in x and for all $x \in \Re^n$, it is Lebesgue measurable in t.*

A.2.1 Practical stabilizability

Given the uncertain dynamical system

$$\begin{aligned}
\dot{x}(t) &= [A + \Delta A(r) + M]x(t) + [B + \Delta B(s)]u(t) \\
&+ Cv(t) + H(t, x, r), \quad x(0) = x_o \tag{A.7} \\
y(t) &= x(t) + w(t) \tag{A.8}
\end{aligned}$$

where $x \in \Re^n$, $u \in \Re^m$, $y = \in \Re^n$, $v \in \Re^s$, $w \in \Re^n$ are the state, control, measured state, disturbance, and measurement error of the system, respectively, and $r \in \Re^p$, $s \in \Re^q$ are the uncertainty vectors. System (A.7)–(A.8) is said to be *practically stabilizable* if, given $\mathbf{d} > 0$, there is a control law $g(.,.) : \Re^m \times \Re \rightarrow \Re^m$, for which, given any admissible uncertainties r, s, disturbances $w \in \Re^n$, $v \in \Re^s$, any initial time $t_o \in \Re$ and any initial state $x_o \in \Re^n$, the following conditions hold

1. *The closed-loop system*

$$\begin{aligned}
\dot{x}(t) &= [A + \Delta A(r) + M]x(t) + [B + \Delta B(s)]g(y, t) \\
&\quad + Cv(t) + H(t, x, r)
\end{aligned} \tag{A.9}$$

   possesses a solution $x(.) : [t_o, t_1] \to \Re^n$, $x(t_o) = x_o$

2. *Given any $\nu > 0$ and any solution $x(.) : [t_o, t_1] \to \Re^n$, $x(t_o) = x_o$ of system (A.9) with $\|x_o\| \le \nu$, there is a constant $d(\nu) > 0$ such that $\|x(t)\| \le d(\nu)$, $\forall t \in [t_o, t_1]$*

3. *Every solution $x(.) : [t_o, t_1] \to \Re^n$ can be continued over $[t_o, \infty)$*

4. *Given any $\bar{d} \ge \mathbf{d}$, any $\nu > 0$ and solution $x(.) : [t_o, t_1] \to \Re^n$, $x(t_o) = x_o$ of system (A.9) with $\|x_o\| \le \nu$, there exists a finite time $T(\bar{d}, \nu) < \infty$, possibly dependent on ν but not on t_o, such that $\|x(t)\| \le \bar{d}$, $\forall t \ge t_o + T(\bar{d}, \nu)$.*

5. *Given any $d \ge \mathbf{d}$ and any solution $x(.) : [t_o, t_1] \to \Re^n$, $x(t_o) = x_o$ of system (A.9) there is a constant $\delta(d) > 0$ such that $\|x(t_o)\| \le \delta d$ implies $\|x(t)\| \le \bar{d}$, $\forall t \ge t_o$.*

A.2.2   Razumikhin stability

A continuous function $\alpha : [0, a) \longmapsto [0, \infty)$ is said to belong to class $\mathcal{K}$ if it is strictly increasing and $\alpha(0) = 0$. Further, it is said to belong to class $\mathcal{K}_\infty$ if $a = \infty$ and $\lim_{r \to \infty} \alpha(r) = \infty$.

Consider a time-delay system

$$\dot{x}(t) = f(t, x(t - d(t))) \tag{A.10}$$

with an initial condition

$$x(t) = (t), \quad t \in [-\bar{d}, 0]$$

where the function vector $f : \Re^+ \times \mathcal{C}_{[-\bar{d}, 0]} \mapsto \Re^n$ takes $\mathcal{R} \times$ (bounded sets of $\mathcal{C}_{[-\bar{d}, 0]}$) into bounded sets in $\Re^n$; $d(t)$ is the time-varying delay and $d := \sup_{t \in \Re^+} \{d(t)\} < \infty$. The symbol $\mathcal{C}_{[a,b]}$ represents the set of $\Re^n$-valued continuous function on $[a, b]$.

Lemma A.2.1 *If there exist class $\mathcal{K}_\infty$ functions $\zeta_1(.)$ and $\zeta_2(.)$, a class $\mathcal{K}$ function $\zeta_3(.)$ and a function $V_1(.) : [-\bar{d}, \infty] \times \Re^n \mapsto \Re^+$ satisfying*

$$\zeta_1(\|x\|) \le V_1(t, x) \le \zeta_2(\|x\|), \quad t \in \Re^+, \quad x \in \Re^n$$

such that the time derivative of V_1 along the solution of the system (A.10) satisfies

$$\dot{V}_1(t, x) \leq -\zeta_3(\|x\|) \quad \text{if} \quad V_1(t + d, x(t + d)) \leq V_1(t, x(t)) \qquad \text{(A.11)}$$

for any $d \in [-\bar{d}, 0]$, then system (A.10) is uniformly stable. If in addition,

$$\zeta_3(\tau) > 0, \ \tau > 0$$

and there exists a continuous non-decreasing function $\xi(\tau) > 0$, $\tau > 0$ such that (A.11) is strengthened to

$$\dot{V}_1(t, x) \leq -\zeta_3(\|x\|) \quad \text{if} \quad V_1(t + d, x(t + d)) \leq \xi(V_1(t, x(t))) \qquad \text{(A.12)}$$

for any $d \in [-\bar{d}, 0]$, then system (A.10) is uniformly asymptotically stable. Further, if in addition, $\lim_{\tau \to \infty} \zeta_1(\tau) = \infty$, then system (A.10) is globally uniformly asymptotically stable.

The proof of this lemma can be found in [717].

Lemma A.2.2 *Consider system (A.10). If there exists a function*

$$V_o(x) = x^t P x, \ P > 0$$

such that for $d \in [-\bar{d}, 0]$ the time derivative of V_o along the solution of system (A.10) satisfies

$$\dot{V}_o(t, x) \leq -q_1 \|x\|^2 \text{if} \quad V_o(x(t + d)) \leq q_2 V_o(x(t)) \qquad \text{(A.13)}$$

for some constants $q_1 > 0$ and $q_2 > 1$, then the system (A.10) is globally uniformly asymptotically stable.

Proof A.2.1 *Since $P > 0$, it is clear that*

$$\lambda_{\min}(P)\|x\|^2 \leq V_o(x) \leq \lambda_{\max}(P)\|x\|^2$$

Let $\zeta_1(\tau) = \lambda_{\min}(P)\tau^2$ and $\zeta_2(\tau) = \lambda_{\max}(P)\tau^2$. It is easy to see that both $\zeta_1(.)$ and $zeta_2(.)$ are class $\mathcal{K}_\infty$ functions and

$$\zeta_1(\|x\|) \leq V_0(x) \leq \zeta_2(\|x\|), \quad x\Re^n$$

Further, let $zeta_3() = -q_1 \tau^2$ and $\xi(\tau) = q_2 \tau$. It is evident from $q_1 > 0$ and $q_2 > 1$ that for $\tau > 0$.

$$\xi(\tau) > \text{and}\zeta_3(\tau) > 0$$

Hence, the conclusion follows from (A.13).

A.3   Delay Patterns

Systems with time delay have attracted the interest of many researchers since the early 1900s. In the 1940s, some theorems were developed to check the stability of time delay systems in the frequency domain. The corresponding theorems in the time domain appeared in the 1950s and 1960s. In the last 20 years, the improvement in the computation tools gave an opportunity to develop new methods to check the stability of time delay systems.

The available tools to check the stability of time delay systems can be classified in two categories: delay-independent methods or delay-dependent methods. Delay-independent stability methods check whether the stability of a time delay system is preserved for a delay of any size or not. The methods in this category try to check if the magnitude of the delayed states does not affect the stability of the system, no matter what the value of that delay is. These methods are easier to derive, but they suffer some conservatism because: not all the systems have insignificant delayed states; in many cases the delay is fixed, and so applying these methods imposes unnecessary conditions and introduces additional complications; and lastly, delay-independent stability methods can be used only when the delay has a destabilizing effect. For these very reasons, many researchers have shifted their interests to the investigation of delay-dependent stability methods.

In contrast to delay-independent stability methods, delay-dependent stability methods require some information about the delay. This information serves one of the following two purposes:

- to check whether a given system, with some dynamics and delay information, is stable or not; or

- to check the maximum duration of delays in the presence of which a given system, with some dynamics, can preserve its stability.

Generally, the second purpose is used to qualify any developed method. For implementation purposes, the conditions for time delay systems can only be sufficient. Different methods give different sets of conditions. In research, the commonly used delay types are:

1. Fixed Delay

   $\tau = \rho,\ \rho = \text{constant}.$

2. Unknown Time-varying delay with an upper-bound

   $0 \leq \tau(t) \leq \rho,\ \rho = \text{constant}.$

3. Unknown time-varying delay with an upper-bound on its value and an upper-bound on its rate of change

   $0 \leq \tau(t) \leq \rho, \rho = $ constant,

   $\dot{\tau}(t) \leq \mu, \mu = $ constant.

4. Delay that varies within some interval

   $h_1 \leq \tau(t) \leq h_2, h_1, \ h_2 = $ constant.

5. Delay that varies within some interval with an upper-bound on its rate of change

   $h_1 \leq \tau(t) \leq h_2, h_1, \ h_2 = $ constant,

   $\dot{\tau}(t) \leq \mu, \mu = $ constant.

A.4 Lyapunov Stability Theorems

Based on Lyapunov's stability theory, there are two main theorems to check the stability of time delay systems: the Lyapunov-Razumikhin theorem and the Lyapunov-Krasovskii theorem.

A.4.1 Lyapunov–Razumikhin theorem

Because the evolution of the states in time delay systems depends on the current and previous states' values, their Lyapunov functions should become functionals (more details in Lyapunov-Krasovskii method discussed in the next section). The functional may complicate the formulation of the conditions and their analysis. To avoid such complications, Razumikhin developed a theorem which will construct Lyapunov functions but not as functionals. To apply the Razumikhin theorem, one should build a Lyapunov function $V(x(t))$. This $V(x(t))$ is equal to zero when $x(t) = 0$ and positive otherwise. The theorem does not require $\dot{V}$ to be less than zero always, but only when $V(x(t))$ becomes greater than or equal to a threshold $\bar{V}$. $\bar{V}$ is given by:

$$\bar{V} = \max_{\theta \in [-\tau, 0]} V(x(t + \theta))$$

Based on this condition, one provides the following theorem statement:

Theorem A.4.1 *Suppose f is a functional that takes time t and initial values x_t and gives a vector of n states $\dot{x}$, u, v and w are class $\mathcal{K}$ functions $u(s)$ and $v(s)$*

are positive for $s > 0$ and $u(0) = v(0) = 0$, v is strictly increasing. If there exists a continuously differentiable function $V : \mathcal{R} \times \mathcal{R}^n \rightarrow \mathcal{R}$ such that:

$$u(\|x\|) \leq V(t, x) \leq v(\|x\|) \tag{A.14}$$

and the time derivative of $V(x(t))$ along the solution $x(t)$ satisfies $\dot{V}(t, x) \leq -w(\|x\|)$ whenever $\bar{V} = V(t + \theta, x(t + \theta)) \leq V(t, x(t)), \theta \in [-\tau, 0]$; then the system is uniformly stable. If in addition $w(s) > 0$ for $s > 0$ and there exists a continuous non-decreasing function $p(s) > s$ for $s > 0$ such that $\dot{V}(t, x) \leq w(\|x\|)$ whenever $V(t + \theta, x(t + \theta)) \leq p(V(t, x(t)))$ for $\theta \in [-\tau, 0]$ then the system is uniformly asymptotically stable.

Here $\bar{V}$ serves as a measure for $V(x(t))$ in the interval from $t - \tau$ to t. If $V(x(t))$ is less than $\bar{V}$, $\dot{V}$ could be greater than zero. On the other hand, if $V(x(t))$ becomes greater than or equal to $\bar{V}$, then $\dot{V}$ must be less than zero, such that V will not grow beyond limits. In other words, according to the Razumikhin theorem, $\dot{V}$ need not be always less than zero, but the following conditions should be satisfied:

$$\dot{V} + a(V(x) - \bar{V}) \leq 0 \tag{A.15}$$

for $a > 0$. Therefore, there are three cases for the system to be stable:

1. $\dot{V} < 0$ and $V(x(t)) \geq V$. Here the states do not grow in magnitude;

2. $\dot{V} > 0$ but $V(x(t)) < V$. In this case, although $\dot{V}$ is positive (the values of the states increase), the Lyapunov function is limited by an upper bound; and

3. a case where both terms are negative.

The states may not reach the origin, but they are contained in some domain. To ensure the asymptotic stability, the condition should be:

$$\dot{V} + a(p(V(x(t))) - \bar{V}) < 0, \quad a > 0 \tag{A.16}$$

where $p(.)$ is a function with the property: $p(s) > s$.

This condition implies that when the system reaches some value which makes $p(V(x(t))) = \bar{V}$, then $\dot{V}$ should be negative and $V(x(t))$ will not reach $\bar{V}$. In the coming interval τ, $V(x)$ will never reach the old $\bar{V}(\bar{V}_{old})$. The maximum value of V in this interval is the new $\bar{V}(\bar{V}_{new})$ which is less $\bar{V}_{old}$. With the passage of time, V keeps decreasing until the states reach the origin.

A.4.2 Lyapunov–Krasovskii theorem

While Razumikhin's theorem is based on constructing Lyapunov functions, the Lyapunov-Krasovskii theorem constructs functionals instead. Based on the Lyapunov theorem's concept, the function V is a measure of the system's internal energy. In time delay systems, the internal energy depends on the value of x_t, and it is reasonable to construct V which is a function of x_t (which is also a function). Because V is a function of another function, it becomes a functional. To ensure asymptotic stability, $\dot{V}$ should always be less than zero. The Lyapunov-Krasovskii theorem is discussed in more detail in the following section.

The remaining advantage of Razumikhin-based methods over Krasovskii is their relative simplicity, but Lyapunov-Krasovskii gives less conservative results. Before discussing the theorem, we have to define the following notations:

$$
\begin{aligned}
\phi &= x_t \\
\|\phi\|_c &= \max_{\theta \in [-\tau, 0]} x(t + \theta)
\end{aligned}
\tag{A.17}
$$

The statement of the Lyapunov–Krasovskii theorem is:

Theorem A.4.2 *Suppose f is a functional that takes time t and initial values x_t and gives a vector of n states $\dot{x}$, u, v and w are class $\mathcal{K}$ functions $u(s)$ and $v(s)$ are positive for $s > 0$ and $u(0) = v(0) = 0$, v is strictly increasing. If there exists a continuously differentiable function V such that:*

$$
u(\|\phi\|) \leq V(t, x_t) \leq v(\|\phi\|_c)
\tag{A.18}
$$

and the time derivative of V along the solution $x(t)$ satisfies $\dot{V}(t, x_t) \leq -w(\|\phi\|)$ for $\theta \in [-\tau, 0]$; then the system is uniformly stable. If in addition $w(s) > 0$ for $s > 0$ then the system is uniformly asymptotically stable.

It is clear that V is a functional and that $\dot{V}$ must always be negative.

As a conclusion of the section, this book will make use of the Lyapunov-Krasovskii theorem to check the delay-dependent stability of uncertain continuous and discrete-time Networked systems. Since the stability of an NCS depends on the occurrence of delays, the occurrence of delays throughout this book is assumed to be governed by Bernoulli's Binomial distribution with varying probabilities.

A.4.3 Some Lyapunov–Krasovskii functionals

In this section, we provide some Lyapunov-Krasovskii functionals and their time-derivatives (or time-differences) which are of common use in stability studies throughout the text. This includes both continuous and discrete cases. First, we consider the continuous case:

$$V_1(x) = x^t P x + \int_{-\tau}^{0} x^t(t+\theta) Q x(t+\theta) \, d\theta \tag{A.19}$$

$$V_2(x) = \int_{-\tau}^{0} \left[\int_{t+\theta}^{t} x^t(\alpha) R x(\alpha) \, d\alpha \right] d\theta \tag{A.20}$$

$$V_3(x) = \int_{-\tau}^{0} \left[\int_{t+\theta}^{t} \dot{x}^t(\alpha) W \dot{x}(\alpha) \, d\alpha \right] d\theta \tag{A.21}$$

where x is the state vector, τ is a constant delay factor and the matrices $0 < P^t = P$, $0 < Q^t = Q$, $0 < R^t = R$, $0 < W^t = W$ are appropriate weighting factors.

Standard matrix manipulations lead to

$$\dot{V}_1(x) = \dot{x}^t P x + x^t P \dot{x} + x^t(t) Q x(t) - x^t(t-\tau) Q x(t-\tau) \tag{A.22}$$

$$\dot{V}_2(x) = \int_{-\tau}^{0} \left[x^t(t) R x(t) - x^t(t+\alpha) R x(t+\alpha) \right] d\theta$$

$$= \tau \, x^t(t) R x(t) - \int_{-\tau}^{0} x^t(t+\theta) R x(t+\theta) \right] d\theta \tag{A.23}$$

$$\dot{V}_3(x) = \tau \, \dot{x}^t(t) W x(t) - \int_{t-\tau}^{t} \dot{x}^t(\alpha) W \dot{x}(\alpha) \, d\alpha \tag{A.24}$$

Next, we provide some general-form of discrete Lyapunov-Krasovskii functionals and their first-differences which can be used in stability studies of discrete-time throughout the text.

$$V(k) = V_o(k) + V_a(k) + V_c(k) + V_m(k) + V_n(k)$$

$$V_o(k) = x^t(k) \mathcal{P}_\sigma x(k), \quad V_a(k) = \sum_{j=k-d(k)}^{k-1} x^t(j) \mathcal{Q}_\sigma x(j),$$

$$V_c(k) = \sum_{j=k-d_m}^{k-1} x^t(j) \mathcal{Z}_\sigma x(j) + \sum_{j=k-d_M}^{k-1} x^t(j) \mathcal{S}_\sigma x(j),$$

$$V_m(k) = \sum_{j=-d_M+1}^{-d_m} \sum_{m=k+j}^{k-1} x^t(m) \mathcal{Q}_\sigma x(m)$$

$$V_n(k) = \sum_{j=-d_M}^{-d_m-1} \sum_{m=k+j}^{k-1} \delta x^t(m) \mathcal{R}_{a\sigma} \delta x(m)$$

$$+ \sum_{j=-d_M}^{-1} \sum_{m=k+j}^{k-1} \delta x^t(m) \mathcal{R}_{c\sigma} \delta x(m) \tag{A.25}$$

where

$$
0 < \mathcal{P}_\sigma = \sum_{j=1}^{N} \lambda_j \mathcal{P}_j, \ \ 0 < \mathcal{Q}_\sigma = \sum_{j=1}^{N} \lambda_j \mathcal{Q}_j, \ \ 0 < \mathcal{S}_\sigma = \sum_{j=1}^{N} \lambda_j \mathcal{S}_j,
$$

$$
0 < \mathcal{Z}_\sigma = \sum_{j=1}^{N} \lambda_j \mathcal{Z}_j, \ \ 0 < \mathcal{R}_{a\sigma} = \sum_{j=1}^{N} \lambda_j \mathcal{R}_{aj},
$$

$$
0 < \mathcal{R}_{c\sigma} = \sum_{j=1}^{N} \lambda_j \mathcal{R}_{cj} \tag{A.26}
$$

are weighting matrices of appropriate dimensions. Now consider a class of discrete-time systems with interval-like time-delays that can be described by:

$$
\begin{aligned}
x(k+1) &= A_\sigma x(k) + D_\sigma x(k - d_k) + \Gamma_\sigma \omega(k) \\
z(k) &= C_\sigma x(k) + G_\sigma x(k - d_k) + \Sigma_\sigma \omega(k)
\end{aligned} \tag{A.27}
$$

where $x(k) \in \Re^n$ is the state, $z(k) \in \Re^q$ is the controlled output and $\omega(k) \in \Re^p$ is the external disturbance, which is assumed to belong to $\ell_2[0, \infty)$. In the sequel, it is assumed that d_k is time-varying and satisfying

$$
d_m \leq d_k \leq d_M \tag{A.28}
$$

where the bounds $d_m > 0$ and $d_M > 0$ are constant scalars. The system matrices containing uncertainties which belong to a real convex bounded polytopic model of the type

$$
[A_\sigma, \ D_\sigma, ..., \Sigma_\sigma] \in \widehat{\Xi}_\lambda := \left\{ [A_\lambda, \ D_\lambda, ..., \Sigma_\lambda] \right.
$$

$$
\left. = \sum_{j=1}^{N} \lambda_j [A_j, \ D_j, ..., \Sigma_j], \lambda \in \Lambda \right\} \tag{A.29}
$$

where Λ is the unit simplex

$$
\Lambda := \left\{ (\lambda_1, \cdots, \lambda_N) : \sum_{j=1}^{N} \lambda_j = 1 , \ \lambda_j \geq 0 \right\} \tag{A.30}
$$

Define the vertex set $\mathcal{N} = \{1, ..., N\}$. We use $\{A, ..., \Sigma\}$ to imply generic system matrices and $\{A_j, ..., \Sigma_j, \ j \in \mathcal{N}\}$ to represent the respective values at the vertices. In what follows, we provide a definition of exponential stability of system (A.27):

A straightforward computation gives the first-difference of $\Delta V(k) = V(k + 1) - V(k)$ along the solutions of (A.27) with $\omega(k) \equiv 0$ as:

$$
\begin{aligned}
\Delta V_o(k) &= x^t(k+1)\mathcal{P}_\sigma x(k+1) - x^t(k)\mathcal{P}_\sigma x(k) \\
&= [A_\sigma x(k) + D_\sigma x(k - d_k)]^t \mathcal{P}_\sigma [A_\sigma x(k) + D_\sigma x(k - d_k)] \\
&\quad - x^t(k)\mathcal{P}_\sigma x(k) \\
\Delta V_a(k) &\leq x^t(k)\mathcal{Q}x(k) - x^t(k - d(k))\mathcal{Q}x(k - d(k)) \\
&\quad + \sum_{j=k-d_M+1}^{k-d_m} x^t(j)\mathcal{Q}x(j) \\
\Delta V_c(k) &= x^t(k)\mathcal{Z}x(k) - x^t(k - d_m)\mathcal{Z}x(k - d_m) + x^t(k)\mathcal{S}x(k) \\
&\quad - x^t(k - d_M)\mathcal{S}x(k - d_M) \\
\Delta V_m(k) &= (d_M - d_m)x^t(k)\mathcal{Q}x(k) - \sum_{j=k-d_M+1}^{k-d_m} x^t(k)\mathcal{Q}x(k) \\
\Delta V_n(k) &= (d_M - d_m)\delta x^t(k)\mathcal{R}_a\delta x(k) + d_M \delta x^t(k)\mathcal{R}_c\delta x(k) \\
&\quad - \sum_{j=k-d_M}^{k-d_m-1} \delta x^t(j)\mathcal{R}_a\delta x(j) - \sum_{j=k-d_M}^{k-1} \delta x^t(j)\mathcal{R}_c\delta x(j) \quad \text{(A.31)}
\end{aligned}
$$

A.5   Algebraic Graph Theory

In this section, some preliminary knowledge of graph theory [411] is introduced for the analysis throughout the book. Briefly stated, "graph theory" is a very useful mathematical tool in the research of multi-agent systems. The topology of a communication network can be expressed by a graph, either directed or undirected, according to whether the information flow is unidirectional or bidirectional.

Let $G(V, E, A)$ be a weighted directed graph (digraph) of order n, where $V = \{s_1, \ldots, s_n\}$ is the set of nodes, $E \subseteq V \times V$ is the set of edges, and $A = [a_{ij}] \in \Re^{n \times n}$ is a weighted adjacency matrix. The node indexes belong to a finite index set $I = \{1, 2, \ldots, n\}$. An edge of G is denoted by $e_{ij} = (s_i, s_j)$, where the first element s_i of the e_{ij} is said to be the tail of the edge and the other s_j to be the head. The adjacency elements associated with the edges are positive, that is $e_{ij} \in E \Leftrightarrow a_{ij} > 0$. If a directed graph has the property that $a_{ij} = a_{ji}$ for any $i, j \in I$, the directed graph is called undirected. The Laplacian with the directed graph is defined as $L = \Delta - A \in \Re^{n \times n}$, where $\Delta = [\Delta_{ij}]$ is a diagonal matrix with $\Delta_{ii} = \sum_{j=1}^{n} a_{ij}$. An important fact of L is that all the row sums of L are zero and thus 1 is an eigenvector of L associated with the zero eigenvalue. The set of neighbors of node s_i is denoted by $N_i = \{s_j \in V : (s_i, s_j) \in E\}$. A directed path is a sequence of ordered edges of the form $(s_{i1}, s_{i2}), (s_{i2}, s_{i3}), \ldots$, where $s_{ij} \in V$ in a directed graph. A directed graph is said to be strongly connected, if there is a directed path from every node to every other node. Moreover, a directed graph is said to have spanning trees, if there exists a node such that there is a directed path from every other node to this node.

A.5.1 Basic results

Lemma A.5.1 *[722] If the graph G has a spanning tree, then its Laplacian L has the following properties:*

1. *Zero is a simple eigenvalue of L, and $\mathbf{1_n}$ is the corresponding eigenvector, that is $L\mathbf{1_n} = \mathbf{0}$.*

2. *The rest $n - 1$ eigenvalues all have positive real parts. In particular, if the graph G is undirected, then all these eigenvalues are positive and real.*

Lemma A.5.2 *[723] Consider a directed graph G. Let $D \in \Re^{n \times |E|}$ be the 01-matrix with rows and columns indexed by the nodes and edges of G, and $E \in \Re^{|E| \times n}$ be the 01-matrix with rows and columns indexed by the edges and nodes of G, such that*

$$D_{uf} = \begin{cases} 1 & \text{if the node } u \text{ is the tail of the edge } f \\ 0 & \text{otherwise} \end{cases}$$

$$E_{fu} = \begin{cases} 1 & \text{if the node } u \text{ is the head of the edge } f \\ 0 & \text{otherwise} \end{cases}$$

where $|E|$ is the number of the edges. Let $Q = diag\{q_1, q_2, \ldots, q_{|E|}\}$, where $q_p(p = 1, \ldots, |E|)$ is the weight of the pth edge of G (i.e., the value of the adjacency matrix on the pth edge). Then the Laplacian of G can be transformed into $L = DQ(D^T - E)$.

A.5.2 Laplacian spectrum of graphs

This section is a concise review of the relationship between the eigenvalues of a Laplacian matrix and the topology of the associated graph. We refer the reader to [25], [26] for a comprehensive treatment of the topic. We list a collection of properties associated with undirected graph Laplacians and adjacency matrices, which will be used in subsequent sections of the paper.

A graph $\mathcal{G}$ is defined as

$$\mathcal{G} = (\mathcal{V}, \mathcal{A}) \tag{A.32}$$

where $\mathcal{V}$ is the set of nodes (or vertices) $\mathcal{V} = \{1, \ldots, N\}$ and $\mathcal{A} \subseteq \mathcal{V} \times \mathcal{V}$ the set of edges (i, j) with $i \in \mathcal{V}, j \in \mathcal{V}$. The degree d_j of a graph vertex j is the number of edges which start from j. Let $d_{\max}(\mathcal{G})$ denote the maximum vertex degree of the graph $\mathcal{G}$.

A.5.3 Properties of adjacency matrix

We denote $\mathbf{A}(\mathcal{G})$ by the (0, 1) adjacency matrix of the graph $\mathcal{G}$. Let $\mathbf{A}_{ij} \in \mathbb{R}$ be its i, j element, then $\mathbf{A}_{i,i} = 0, \forall i = 1, \ldots, N$, $\mathbf{A}_{i,j} = 0$ if $(i, j) \notin \mathcal{A}$ and $\mathbf{A}_{i,j} = 1$ if $(i, j) \in \mathcal{A}, \forall i, j = 1, \ldots, N, i \neq j$. We will focus on *undirected* graphs, for which the adjacency matrix is symmetric.

Let $\mathcal{S}(\mathbf{A}(\mathcal{G})) = \{\lambda_1(\mathbf{A}(\mathcal{G})), \ldots, \lambda_N(\mathbf{A}(\mathcal{G}))\}$ be the spectrum of the adjacency matrix associated with an undirected graph $\mathcal{G}$ arranged in non-decreasing semi-order.

1) *Property 1:* $\lambda_N(\mathbf{A}(\mathcal{G})) \leq d_{\max}(\mathcal{G})$.

This property together with Proposition 1 implies

2) *Property 2:* $\gamma_i \geq 0, \forall \gamma_i \in \mathcal{S}(d_{\max} I_N - \mathbf{A})$.

We define the Laplacian matrix of a graph $\mathcal{G}$ in the following way:

$$L(\mathcal{G}) = \mathbf{D}(\mathcal{G}) - \mathbf{A}(\mathcal{G}) \tag{A.33}$$

where $\mathbf{D}(\mathcal{G})$ is the diagonal matrix of vertex degrees d_i (also called the valence matrix). Eigenvalues of Laplacian matrices have been widely studied by graph theorists. Their properties are strongly related to the structural properties of their associated graphs. Every Laplacian matrix is a singular matrix. By Gershgorin's theorem [724], the real part of each nonzero eigenvalue of $L(\mathcal{G})$ is strictly positive.

For undirected graphs, $L(\mathcal{G})$ is a symmetric, positive semidefinite matrix, which has only real eigenvalues. Let $\mathcal{S}(L(\mathcal{G})) = \{\lambda_1(L(\mathcal{G})), \ldots, \lambda_N(L(\mathcal{G}))\}$ be the spectrum of the Laplacian matrix L associated with an undirected graph $\mathcal{G}$ arranged in non-decreasing semi-order. Then,

3) *Property 3:*

1. $\lambda_1(L(\mathcal{G})) = 0$ with corresponding eigenvector of all ones, and $\lambda_2(L(\mathcal{G}))$ iff $\mathcal{G}$ is connected. In fact, the multiplicity of 0 as an eigenvalue of $L(\mathcal{G})$ is equal to the number of connected components of $\mathcal{G}$.

2. The modulus of $\lambda_i(L(\mathcal{G}))$, $i = 1, \ldots, N$ is less then N.

The second smallest Laplacian eigenvalue $\lambda_2(L(\mathcal{G}))$ of graphs is probably the most important information contained in the spectrum of a graph. This eigenvalue, called the algebraic connectivity of the graph, is related to several important graph invariants, and it has been extensively investigated.

Let $L(\mathcal{G})$ be the Laplacian of a graph $\mathcal{G}$ with N vertices and with maximal vertex degree $d_{\max}(\mathcal{G})$. Then properties of $\lambda_2(L(\mathcal{G}))$ include
4) *Property 4:*

1. $\lambda_2(L(\mathcal{G})) \leq (N/(N-1)) \min\{d(v), v \in \mathcal{V}\}$;

2. $\lambda_2(L(\mathcal{G})) \leq v(\mathcal{G}) \leq \eta(\mathcal{G})$;

3. $\lambda_2(L(\mathcal{G})) \geq 2\eta(\mathcal{G})(1 - \cos(\pi/N))$;

4. $\lambda_2(L(\mathcal{G})) \geq 2(\cos\frac{\pi}{N} - \cos 2\frac{\pi}{N})\eta(\mathcal{G}) - 2\cos\frac{\pi}{N}(1 - \cos\frac{\pi}{N})d_{\max}(\mathcal{G})$

where $v(\mathcal{G})$ is the vertex connectivity of the graph $\mathcal{G}$ (the size of a smallest set of vertices whose removal renders $\mathcal{G}$ disconnected) and $\eta(\mathcal{G})$ is the edge connectivity of the graph $\mathcal{G}$ (the size of a smallest set of edges whose removal renders $\mathcal{G}$ disconnected) [725].

Further relationships between the graph topology and Laplacian eigenvalue locations are discussed in [726] for undirected graphs. Spectral characterization of Laplacian matrices for directed graphs can be found in [724].

A lemma about Laplacian L associated with a balanced digraph G is given hereafter:

Lemma A.5.3 *If G is balanced, then there exists a unitary matrix*

$$V = \begin{pmatrix} \frac{1}{\sqrt{n}} & * & \cdots & * \\ \frac{1}{\sqrt{n}} & * & \cdots & * \\ \vdots & \vdots & & \vdots \\ \frac{1}{\sqrt{n}} & * & \cdots & * \end{pmatrix} \in C^{m \times n} \tag{A.34}$$

such that

$$V^*LV = \begin{pmatrix} 0 & \\ & H \end{pmatrix} = \Lambda \in C^{n \times n},$$

$$H \in C^{(n-1) \times (n-1)} \tag{A.35}$$

Moreover, if G has a globally reachable node, $H + H^*$ is positive definite.

Proof A.5.1 *Let $V = [\zeta_1, \zeta_2, \ldots, \zeta_n]$ be a unitary matrix where $\zeta_i \in C^n (i = 1, \ldots, n)$ are the column vectors of V and*

$$\zeta_1 = (1/\sqrt{n})1 = (1/\sqrt{n}, 1/\sqrt{n}, \ldots, 1/\sqrt{n})^T$$

Notice that if G is balanced, it implies that $\zeta_1^* L = 0$. Then we have

$$
\begin{aligned}
V^* L V &= V^* L[\zeta_1, \zeta_2, \ldots, \zeta_n] \\
&= \begin{pmatrix} \zeta_1^* \\ \zeta_2^* \\ \vdots \\ \zeta_n^* \end{pmatrix} [0_n, L\zeta_2, \ldots, L\zeta_n] \\
&= \begin{pmatrix} 0 & 0_{n-2}^T \\ \bullet & H \end{pmatrix}
\end{aligned}
$$

Furthermore, if G has a globally reachable node, then $L + L^T$ is positive semi-definite, see **Theorem** *7 in [41]. Hence, $V^*(L+L^T)V$ is also positive semidefinite. From Lemma [727], zero is a simple eigenvalue of L and, therefore, $H + H^*$ is positive definite.*

A.6 Minimum Mean Square Estimate

Given a random variable Y that depends on another random variable X. Of interest is the *minimum mean square error estimate* (MMSEE) which, simply stated, is $\hat{X}$ the estimate of X such that the mean square error given by

$$
\mathbf{E}\left[X - \hat{X}\right]^2
$$

is minimized where the expectation is taken over the random variables X and Y. One of the standard results is given below:

Proposition A.6.1 *The minimum mean square error estimate is given by the conditional expectation $\mathbf{E}[X|Y = y]$.*

Proof A.6.1 *Consider the functional form of the estimator as $g(Y)$. Let $f_{X,Y}(x,y)$ denote the joint probability density function of X and Y. Then the cost function is given by*

$$
\begin{aligned}
C := \mathbf{E}\left[X - \hat{X}\right]^2 &= \int_x \int_y (x - g(y))^2 f_{X,Y}(x,y)\, dx\, dy \\
&= \int_y dy\, f_Y(y) \int_x (x - g(y))^2 f_{X|Y}(x|y)\, dx
\end{aligned}
$$

Taking the derivative of the cost function with respect to the function $g(y)$:

$$\frac{\partial C}{\partial g(y)} = \int_y dy \, f_Y(y) \int_x 2(x - g(y)) f_{X|Y}(x|y) \, dx$$

$$= 2 \int_y dy \, f_Y(y) \left(g(y) - \int_x x f_{X|Y}(x|y) \, dx \right)$$

$$= 2 \int_y dy \, f_Y(y) \left(g(y) - \boldsymbol{E}[X|Y = y] \right)$$

Therefore the only stationary point is $g(y) = \boldsymbol{E}[X|Y = y]$ and can be easily verified that it is a minimum.

Remark A.6.1 *It is noted that the result established in* **Proposition** *A.6.1 holds for vector random variables as well. Observe that MMSE estimates are important because for Gaussian variables, they coincide with the Maximum likelihood (ML) estimates. It is a standard result that for Gaussian variables, the MMSE estimate is linear in the state value.*

In what follows, we will assume zero mean values for all the random variables with R_X being the covariance of X and R_{XY} being the cross-covariance between X and Y.

Proposition A.6.2 *The best linear MMSE estimate of X given $Y = y$ is*

$$\hat{x} = R_{XY} R_Y^{-1} y$$

with the error covariance

$$P = R_X - R_{XY} R_Y^{-1} R_{YX}$$

Proof A.6.2 *Let the estimate be $\hat{x} = Ky$. Then the error covariance is*

$$P := \boldsymbol{E}\left[(x - Ky)(x - Ky)^T\right]$$

$$= R_X - K R_{YX} - R_{XY} K^t + K R_Y K^T$$

Differentiating P with respect to K and setting it equal to zero yields

$$-2 R_{XY} + 2K R_Y^{-1}$$

The result follows immediately.

Extending **Proposition** A.6.1 to the case of linear measurements $y = Hx + v$, we have the following standard result.

Proposition A.6.3 *Let* $y = Hx + v$, *where* H *is a constant matrix and* v *is a zero mean Gaussian noise with covariance* R_V *independent of* X. *Then the MMSE estimate of* X *given* $Y = y$ *is*

$$\hat{x} = R_X H^T (H R_X H^T + R_V)^{-1} y$$

with the corresponding error covariance

$$P = R_X - R_X H^T (H R_X H^T + R_V)^{-1} H R_X$$

A.7   Gronwall–Bellman Inequalities

In this section we provide the continuous version of the well-known Gronwall-Bellman inequalities, which play a useful role in the study of qualitative as well as quantitative properties of solutions of differential equations, such as boundedness, stability, existence, uniqueness, continuous dependence and so on.

Lemma A.7.1 *Let* $\lambda : [a, b] \to \Re$ *be continuous and* $\mu : [a, b] \to \Re$ *be continuous and nonnegative. If a continuous function* $y : [a, b] \to \Re$ *satisfies*

$$y(t) \leq \lambda(t) + \int_a^t \mu(s) y(s) \, ds$$

for $a \leq t \leq b$ *then on the same interval*

$$y(t) \leq \lambda(t) + \int_a^t \mu(s) y(s) \, exp \Big[\int_s^t \mu(\tau) d\tau \Big] ds$$

In particular if $\lambda(t) \equiv \lambda$ *is a constant then*

$$y(t) \leq \lambda exp \Big[\int_a^t \mu(\tau) d\tau \Big]$$

If, in addition, $\mu(t) \equiv \mu \geq 0$ *is a constant then*

$$y(t) \leq \lambda exp[\mu(t - a)]$$

Proof A.7.1 *Let*

$$
\begin{aligned}
z(t) &= \int_a^t \mu(s) y(s) \, ds \\
v(t) &= z(t) + \lambda(t) - y(t) \geq 0
\end{aligned}
$$

Then z is differentiable and

$$\dot{z} = \mu(t)y(t) = \mu(t)z(t) + \mu(t)\lambda(t) - \mu(t)v(t)$$

This is a scalar linear state equation with a state transition function

$$\phi(t,s) = exp\Big[\int_s^t \mu(\tau)d\tau\Big]$$

Since $z(a) = 0$, we have

$$z(t) = \int_a^t \phi(t,s)[\mu(s)\lambda(s) - \mu(s)v(s)]\,ds$$

The term

$$\int_a^t \phi(t,s)\mu(s)v(s)\,ds$$

is nonnegative. Therefore

$$z(t) \le \int_a^t exp\Big[\int_s^t \mu(\tau)d\tau\Big]\mu(s)\lambda(s)\,ds$$

Since $y(t) \le \lambda + z(t)$, this completes the proof in the general case.
  In the special case when $\lambda(t) \equiv \lambda$, we have

$$
\begin{aligned}
\int_a^t \mu(s)\,exp\Big[\int_s^t \mu(\tau)d\tau\Big]\,ds &= -\int_a^t \frac{d}{ds}\Big\{exp\Big[\int_s^t \mu(\tau)d\tau\Big]\Big\}\,ds \\
&= -\Big\{exp\Big[\int_s^t \mu(\tau)d\tau\Big]\Big\}\big|_{|s=a}^{|s=t} \\
&= -1 + exp\Big[\int_s^t \mu(\tau)d\tau\Big]
\end{aligned}
$$

which proves the lemma when λ is a constant. The proof when both λ and μ are constants follows by integration.

A.8 Basic Inequalities

All mathematical inequalities are proved for completeness. They are termed facts due to their high frequency of usage in the analytical developments.

A.8.1   Inequality 1

For any real matrices Σ_1 , Σ_2 and Σ_3 with appropriate dimensions and $\Sigma_3^t \Sigma_3 \leq I$, it follows that

$$\Sigma_1 \Sigma_3 \Sigma_2 + \Sigma_2^t \Sigma_3^t \Sigma_1^t \leq \alpha \, \Sigma_1 \Sigma_1^t + \alpha^{-1} \, \Sigma_2^t \Sigma_2, \quad \forall \alpha > 0$$

Proof A.8.1 *This inequality can be proved as follows. Since $\Phi^t \Phi \geq 0$ holds for any matrix Φ, then take Φ as*

$$\Phi = [\alpha^{1/2} \, \Sigma_1 - \alpha^{-1/2} \, \Sigma_2]$$

Expansion of $\Phi^t \Phi \geq 0$ gives $\forall \alpha > 0$

$$\alpha \, \Sigma_1 \Sigma_1^t + \alpha^{-1} \, \Sigma_2^t \Sigma_2 - \Sigma_1^t \Sigma_2 - \Sigma_2^t \Sigma_1 \geq 0$$

which by simple arrangement yields the desired result.

A.8.2   Inequality 2

Let Σ_1, Σ_2, Σ_3 and $0 < R = R^t$ be real constant matrices of compatible dimensions and $H(t)$ be a real matrix function satisfying $H^t(t)H(t) \leq I$. Then for any $\rho > 0$ satisfying $\rho \Sigma_2^t \Sigma_2 < R$, the following matrix inequality holds:

$$(\Sigma_3 + \Sigma_1 H(t)\Sigma_2)R^{-1}(\Sigma_3^t + \Sigma_2^t H^t(t)\Sigma_1^t) \leq \rho^{-1}\Sigma_1\Sigma_1^t + \Sigma_3\left(R - \rho\Sigma_2^t\Sigma_2\right)^{-1}\Sigma_3^t$$

Proof A.8.2 *The proof of this inequality proceeds like the previous one by considering that*

$$\Phi = [(\rho^{-1} \, \Sigma_2\Sigma_2^t)^{-1/2}\Sigma_2 R^{-1}\Sigma_3^t - (\rho^{-1} \, \Sigma_2\Sigma_2^t)^{-1/2}H^t(t)\Sigma_1^t]$$

Recall the following results
$$\rho\Sigma_2^t\Sigma_2 < R,$$

$$[R - \rho\Sigma_2^t\Sigma_2]^{-1} = [R^{-1} + R^{-1}\Sigma_2^t[\rho^{-1}I - \Sigma_2 R^{-1}\Sigma_2^t]^{-1}\Sigma_2 R^{-1}\Sigma_2$$

and

$$H^t(t)H(t) \leq I \Longrightarrow H(t)H^t(t) \leq I$$

Expansion of $\Phi^t\Phi \geq 0$ under the condition $\rho\Sigma_2^t\Sigma_2 < R$ with standard matrix manipulations gives

$$
\begin{aligned}
\Sigma_3 R^{-1}\Sigma_2^t H^t(t)\Sigma_1^t + \Sigma_1 H(t)\Sigma_2 R^{-1}\Sigma_3^t + \Sigma_1 H(t)\Sigma_2\Sigma_2^t H^t(t)\Sigma_1^t &\leq \\
\rho^{-1}\Sigma_1 H(t)H^t(t)\Sigma_1^t + \Sigma_3^t R^{-1}\Sigma_2[\rho^{-1}I\ \Sigma_2\Sigma_2^t]^{-1}\Sigma_2 R^{-1}\Sigma_3^t &\Longrightarrow \\
(\Sigma_3 + \Sigma_1 H(t)\Sigma_2)R^{-1}(\Sigma_3^t + \Sigma_2^t H^t(t)\Sigma_1^t) - \Sigma_3 R^{-1}\Sigma_3^t &\leq \\
\rho^{-1}\Sigma_1 H(t)H^t(t)\Sigma_1^t + \Sigma_3^t R^{-1}\Sigma_2[\rho^{-1}I\ -\ \Sigma_2\Sigma_2^t]^{-1}\Sigma_2 R^{-1}\Sigma_3^t &\Longrightarrow \\
(\Sigma_3 + \Sigma_1 H(t)\Sigma_2)R^{-1}(\Sigma_3^t + \Sigma_2^t H^t(t)\Sigma_1^t) &\leq \\
\Sigma_3[R^{-1} + \Sigma_2[\rho^{-1}I\ -\ \Sigma_2\Sigma_2^t]^{-1}\Sigma_2 R^{-1}]\Sigma_3^t\ + & \\
\rho^{-1}\Sigma_1 H(t)H^t(t)\Sigma_1^t &= \\
\rho^{-1}\Sigma_1 H(t)H^t(t)\Sigma_1^t + \Sigma_3\left(R - \rho\Sigma_2^t\Sigma_2\right)^{-1}\Sigma_3^t &
\end{aligned}
$$

which completes the proof.

A.8.3 Inequality 3

For any real vectors β, ρ and any matrix $Q^t = Q > 0$ with appropriate dimensions, it follows that

$$
-2\rho^t\beta \ \leq\ \rho^t\,Q\,\rho + \beta^t\,Q^{-1}\,\beta
$$

Proof A.8.3 *Starting from the fact that*

$$
[\rho + Q^{-1}\beta]^t\,Q\,[\rho + Q^{-1}\beta] \geq 0\ ,\ \ Q > 0
$$

which when expanded and arranged yields the desired result.

A.8.4 Inequality 4 (Schur complements)

Given a matrix Ω composed of constant matrices Ω_1, Ω_2, Ω_3 where $\Omega_1 = \Omega_1^t$ and $0 < \Omega_2 = \Omega_2^t$ as follows

$$
\Omega \ =\ \begin{bmatrix} \Omega_1 & \Omega_3 \\ \Omega_3^t & \Omega_2 \end{bmatrix}
$$

We have the following results

    (A) $\Omega \geq 0$ if and only if either

$$
\begin{cases}
\Omega_2 \geq 0 \\
\Pi = \Upsilon\Omega_2 \\
\Omega_1 - \Upsilon\,\Omega_2\,\Upsilon^t \geq 0
\end{cases}
\tag{A.36}
$$

or

$$\left\{ \begin{array}{c} \Omega_1 \geq 0 \\ \Pi = \Omega_1 \Lambda \\ \Omega_2 - \Lambda^t \Omega_1 \Lambda \geq 0 \end{array} \right. \tag{A.37}$$

hold where Λ, Υ are some matrices of compatible dimensions.

(B) $\Omega > 0$ if and only if either

$$\left\{ \begin{array}{c} \Omega_2 > 0 \\ \Omega_1 - \Omega_3 \, \Omega_2^{-1} \, \Omega_3^t > 0 \end{array} \right.$$

or

$$\left\{ \begin{array}{c} \Omega_1 \geq 0 \\ \Omega_2 - \Omega_3^t \, \Omega_1^{-1} \, \Omega_3 > 0 \end{array} \right.$$

hold where Λ, Υ are some matrices of compatible dimensions.

In this regard, matrix $\Omega_3 \, \Omega_2^{-1} \, \Omega_3^t$ is often called the Schur complement $\Omega_1(\Omega_2)$ in Ω.

Proof A.8.4 *(A) To prove (A.36), we first note that $\Omega_2 \geq 0$ is necessary. Let $z^t = [z_1^t \;\; z_2^t]$ be a vector partitioned in accordance with Ω. Thus we have*

$$z^t \, \Omega \, z = z_1^t \Omega_1 z_1 + 2z_1^t \Omega_3 z_2 + z_2^t \Omega_2 z_2 \tag{A.38}$$

Select z_2 such that $\Omega_2 z_2 = 0$. If $\Omega_3 z_2 \neq 0$, let $z_1 = -\pi \Omega_3 z_2$, $\pi > 0$. Then it follows that

$$z^t \, \Omega \, z = \pi^2 \, z_2^t \Omega_3^t \Omega_1 \Omega_3 z_2 - 2\pi \, z_2^t \Omega_3^t \Omega_3 z_2$$

which is negative for a sufficiently small $\pi > 0$. We thus conclude $\Omega_1 z_2 = 0$ which then leads to $\Omega_3 z_2 = 0$, $\forall z_2$ and consequently

$$\Omega_3 = \Upsilon \, \Omega_2 \tag{A.39}$$

for some Υ.

Since $\Omega \geq 0$, the quadratic term $z^t \, \Omega \, z$ possesses a minimum over z_2 for any z_1. By differentiating $z^t \, \Omega \, z$ from (A.38) wrt z_2^t, we get

$$\frac{\partial(z^t \, \Omega \, z)}{\partial z_2^t} = 2\Omega_3^t \, z_1 + 2\Omega_2 \, z_2 = 2\Omega_2 \, \Upsilon^t \, z_1 + 2\Omega_2 \, z_2$$

Setting the derivative to zero yields

$$\Omega_2 \, \Upsilon \, z_1 = -\Omega_2 \, z_2 \tag{A.40}$$

Using (A.39) and (A.40) in (A.38), it follows that the minimum of $z^t \, \Omega \, z$ over z_2 for any z_1 is given by

$$\min_{z_2} \; z^t \, \Omega \, z = z_1^t [\Omega_1 \; - \; \Upsilon \, \Omega_2 \, \Upsilon^t] z_1$$

which prove the necessity of $\Omega_1 \; - \; \Upsilon \, \Omega_2 \, \Upsilon^t \geq 0$.

On the other hand, we note that the conditions (A.36) are necessary for $\Omega \geq 0$ and since together they imply that the minimum of $z^t \, \Omega \, z$ over z_2 for any z_1 is nonnegative, they are also sufficient.

Using similar argument, conditions (A.37) can be derived as those of (A.36) by starting with Ω_1.

The proof of **(B)** *follows as direct corollary of* **(A)**.

A.8.5 Inequality 5

For any quantities u and v of equal dimensions and for all $\eta_t = i \in \mathcal{S}$, it follows that the following inequality holds

$$||u \, + \, v||^2 \; \leq \; [1 + \beta^{-1}] \, ||u||^2 \; + \; [1 + \beta] ||v||^2 \qquad \text{(A.41)}$$

for any scalar $\beta > 0, \quad i \in \mathcal{S}$

Proof A.8.5 *Since*

$$\begin{aligned} [u \, + \, v]^t \, [u \, + \, v] = \\ u^t \, u \; + \; v^t \, v \; + \; 2 \, u^t \, v \end{aligned} \qquad \text{(A.42)}$$

It follows by taking norm of both sides of (A.42) for all $i \in \mathcal{S}$ that

$$||u \, + \, v||^2 \; \leq \; ||u||^2 \; + \; ||v||^2 \; + \; 2 \, ||u^t \, v|| \qquad \text{(A.43)}$$

We know from the triangle inequality that

$$2 \, ||u^t \, v|| \; \leq \; \beta^{-1} \, ||u||^2 \; + \; \beta \, ||v||^2 \qquad \text{(A.44)}$$

On substituting (A.44) into (A.43), it yields (A.41).

A.8.6 Inequality 6

Given matrices $0 < \mathcal{Q}^t = \mathcal{Q}, \; \mathcal{P} = \mathcal{P}^t$, then it follows that

$$- \mathcal{P} \mathcal{Q}^{-1} \mathcal{P} \; \leq \; - 2 \, \mathcal{P} \; + \; \mathcal{Q} \qquad \text{(A.45)}$$

This can be easily established by considering the algebraic inequality

$$(\mathcal{P} - \mathcal{Q})^t \mathcal{Q}^{-1} (\mathcal{P} - \mathcal{Q}) \geq 0$$

and expanding to get

$$\mathcal{P}\mathcal{Q}^{-1}\mathcal{P} - 2\mathcal{P} + \mathcal{Q} \geq 0 \qquad (A.46)$$

which when manipulating, yields (A.45). An important special case is obtained when $\mathcal{P} \equiv I$, that is

$$- \mathcal{Q}^{-1} \leq - 2 I + \mathcal{Q} \qquad (A.47)$$

This inequality proves useful when using Schur complements to eliminate the quantity $\mathcal{Q}^{-1}$ from the diagonal of an LMI without alleviating additional math operations.

A.8.7 Bounding lemmas

A basic inequality that has been frequently used in the stability analysis of time-delay systems is called *Jensen's Inequality* or *the Integral Inequality*, a detailed account of which is available in [718]:

Lemma A.8.1 : *For any constant matrix $0 < \Sigma \in \Re^{n \times n}$, scalar $\tau_* < \tau(t) < \tau^+$ and vector function $\dot{x} : [-\tau^+, -\tau_*] \to \Re^n$ such that the following integration is well-defined, then it holds that*

$$-(\tau^+ - \tau_*) \int_{t-\tau^+}^{t-\tau_*} \dot{x}^t(s) \Sigma \dot{x}(s) ds \leq \begin{bmatrix} x(t - \tau_*) \\ x(t - \tau^+) \end{bmatrix}^t \begin{bmatrix} -\Sigma & \Sigma \\ \bullet & -\Sigma \end{bmatrix} \begin{bmatrix} x(t - \tau_*) \\ x(t - \tau^+) \end{bmatrix}$$

Building on Lemma A.8.1, the following lemma specifies a particular inequality for quadratic terms

Lemma A.8.2 : *For any constant matrix $0 < \Sigma \in \Re^{n \times n}$, scalar $\tau_* < \tau(t) < \tau^+$ and vector function $\dot{x} : [-\tau^+, -\tau_*] \to \Re^n$ such that the following integration is well-defined, then it holds that*

$$- (\tau^+ - \tau_*) \int_{t-\tau^+}^{t-\tau_*} \dot{x}^t(s) \Sigma \dot{x}(s) ds \leq \xi^t(t) \Upsilon \xi(t)$$

$$\xi(t) = \begin{bmatrix} x(t - \tau_*) \\ x(t - \tau(t)) \\ x(t - \tau^+) \end{bmatrix}^t, \quad \Upsilon = \begin{bmatrix} -\Sigma & \Sigma & 0 \\ \bullet & -2\Sigma & \Sigma \\ \bullet & \bullet & -\Sigma \end{bmatrix}$$

Proof A.8.6 *Considering the case* $\tau_* < \tau(t) < \tau^+$ *and applying the Leibniz-Newton formula, it follows that*

$$
\begin{aligned}
&- (\tau^+ - \tau_*) \int_{t-\tau^+}^{t-\tau_*} \dot{x}^t(s) \, \Sigma \, \dot{x}(s) \, ds \; - \; (\tau^+ - \tau_*) \int_{t-\tau(t)}^{t-\tau_*} \Big[\dot{x}^t(s) \, \Sigma \, \dot{x}(s) \, ds \\
&+ \int_{t-\tau^+}^{t-\tau(t)} \dot{x}^t(s) \, \Sigma \, \dot{x}(s) \, ds \Big] \\
\leq \; &- (\tau(t) - \tau_*) \int_{t-\tau(t)}^{t-\tau_*} \Big[\dot{x}^t(s) \, \Sigma \, \dot{x}(s) \, ds \\
&- (\tau^+ - \tau(t)) \int_{t-\tau^+}^{t-\tau(t)} \dot{x}^t(s) \, \Sigma \, \dot{x}(s) \, ds \Big] \\
\leq \; &- \int_{t-\tau(t)}^{t-\tau_*} \dot{x}^t(s) \, ds \, \Sigma \, \int_{t-\tau(t)}^{t-\tau_*} \dot{x}^t(s) \, ds \\
&- \int_{t-\tau_+}^{t-\tau(t)} \dot{x}^t(s) \, ds \, \Sigma \, \int_{t-\tau_+}^{t-\tau(t)} \dot{x}^t(s) \, ds \\
= \; &[x(t-\tau_*) - x(t-\tau(t))]^t \, \Sigma \, [x(t-\tau_*) - x(t-\tau(t))] \\
&- [x(t-\tau(t)) - x(t-\tau^+)]^t \, \Sigma \, [x(t-\tau(t)) - x(t-\tau^+)]
\end{aligned}
$$

which completes the proof.

Lemma A.8.3 : *For any constant matrix* $0 < \Sigma \in \Re^{n \times n}$, *scalar* η, *any* $t \in [0, \infty)$, *and vector function* $g : [t - \eta, t] \to \Re^n$ *such that the following integration is well-defined, then it holds that*

$$
\left(\int_{t-\eta}^t g(s) \, ds \right)^t \Sigma \int_{t-\eta}^t g(s) \, ds \leq \eta \int_{t-\eta}^t g^t(s) \, \Sigma \, g(s) \, ds \qquad \text{(A.48)}
$$

Proof A.8.7 *It is simple to show that for any* $s \in [t - \eta, t]$, $t \in [0, \infty)$, *and Schur complements*

$$
\begin{bmatrix} g^t(s) \Sigma g(s) & g^t(s) \\ \bullet & \Sigma^{-1} \end{bmatrix} \geq 0,
$$

Upon integration, we have

$$
\begin{bmatrix} \int_{t-\eta}^t g^t(s) \Sigma g(s) ds & \int_{t-\eta}^t g^t(s) ds \\ \bullet & \eta \Sigma \end{bmatrix} \geq 0,
$$

By Schur complements, we obtain inequality (A.48).

The following lemmas show how to produce equivalent LMIs by an elimination procedure.

Lemma A.8.4 : *There exists $\mathcal{X}$ such that*

$$\begin{bmatrix} \mathcal{P} & \mathcal{Q} & \mathcal{X} \\ \bullet & \mathcal{R} & \mathcal{Z} \\ \bullet & \bullet & \mathcal{S} \end{bmatrix} > 0 \tag{A.49}$$

if and only if

$$\begin{bmatrix} \mathcal{P} & \mathcal{Q} \\ \bullet & \mathcal{R} \end{bmatrix} > 0, \quad \begin{bmatrix} \mathcal{R} & \mathcal{Z} \\ \bullet & \mathcal{S} \end{bmatrix} > 0 \tag{A.50}$$

Proof A.8.8 *Since LMIs (A.50) form sub-blocks on the principal diagonal of LMI (A.49), necessity is established. To show sufficiency, apply the congruence transformation*

$$\begin{bmatrix} I & 0 & 0 \\ \bullet & I & 0 \\ 0 & -V^t R^{-1} & I \end{bmatrix}$$

to LMI (A.49), it is evident that (A.49) is equivalent to

$$\begin{bmatrix} \mathcal{P} & \mathcal{Q} & \mathcal{X} - \mathcal{Q}\mathcal{R}^{-1}\mathcal{Z} \\ \bullet & \mathcal{R} & 0 \\ \bullet & \bullet & \mathcal{S} - \mathcal{Z}^t \mathcal{R}^{-1} \mathcal{Z} \end{bmatrix} > 0 \tag{A.51}$$

Clearly (A.50) is satisfied for $\mathcal{X} = \mathcal{Q}\mathcal{R}^{-1}\mathcal{Z}$ if (A.50) is satisfied in view of Schur complements.

Lemma A.8.5 : *There exists $\mathcal{X}$ such that*

$$\begin{bmatrix} \mathcal{P} & \mathcal{Q} + \mathcal{X}\mathcal{G} & \mathcal{X} \\ \bullet & \mathcal{R} & \mathcal{Z} \\ \bullet & \bullet & \mathcal{S} \end{bmatrix} > 0 \tag{A.52}$$

if and only if

$$\begin{bmatrix} \mathcal{P} & \mathcal{Q} \\ \bullet & \mathcal{R} - \mathcal{V}\mathcal{G} - \mathcal{G}^t\mathcal{V}^t + \mathcal{G}^t\mathcal{Z}\mathcal{G} \end{bmatrix} > 0,$$

$$\begin{bmatrix} \mathcal{R} - \mathcal{V}\mathcal{G} - \mathcal{G}^t\mathcal{V}^t + \mathcal{G}^t\mathcal{Z}\mathcal{G} & \mathcal{V} - \mathcal{G}^t\mathcal{Z} \\ \bullet & \mathcal{Z} \end{bmatrix} > 0 \tag{A.53}$$

Proof A.8.9 *Applying the congruence transformation*

$$
\begin{bmatrix}
I & 0 & 0 \\
0 & I & 0 \\
0 & -\mathcal{G} & I
\end{bmatrix}
$$

to LMI (A.52) and using **Lemma** *A.8.4, we readily obtain the results.*

Lemma A.8.6 *: There exists* $0 < \mathcal{X}^t = \mathcal{X}$ *such that*

$$
\begin{bmatrix}
\mathcal{P}_a + \mathcal{X} & \mathcal{Q}_a \\
\bullet & \mathcal{R}_a
\end{bmatrix} > 0,
$$

$$
\begin{bmatrix}
\mathcal{P}_c - \mathcal{X} & \mathcal{Q}_c \\
\bullet & \mathcal{R}_c
\end{bmatrix} > 0 \tag{A.54}
$$

if and only if

$$
\begin{bmatrix}
\mathcal{P}_a + \mathcal{P}_c & \mathcal{Q}_a & \mathcal{Q}_c \\
\bullet & \mathcal{R}_a & 0 \\
\bullet & \bullet & \mathcal{R}_c
\end{bmatrix} > 0 \tag{A.55}
$$

Proof A.8.10 *It is obvious from Schur complements that LMI (A.55) is equivalent to*

$$
\mathcal{R}_a > 0, \quad \mathcal{R}_c > 0
$$
$$
\Xi = \mathcal{P}_a + \mathcal{P}_c - \mathcal{Q}_a \mathcal{R}_a^{-1} \mathcal{Q}_a^t - \mathcal{Q}_c \mathcal{R}_c^{-1} \mathcal{Q}_c^t > 0 \tag{A.56}
$$

On the other hand, LMI (A.54) is equivalent to

$$
\mathcal{R}_a > 0, \quad \mathcal{R}_c > 0
$$
$$
\Xi_a = \mathcal{P}_a + \mathcal{X} - \mathcal{Q}_a \mathcal{R}_a^{-1} \mathcal{Q}_a^t > 0,
$$
$$
\Xi_c = \mathcal{P}_c - \mathcal{X} - \mathcal{Q}_c \mathcal{R}_c^{-1} \mathcal{Q}_c^t > 0 \tag{A.57}
$$

It is readily evident from (A.56) and (A.57) that $\Xi = \Xi_a + \Xi_c$ *and hence the existence of* $\mathcal{X}$ *satisfying (A.57) implies (A.56). By the same token, if (A.56) is satisfied,* $\mathcal{X} = \mathcal{Q}_a \mathcal{R}_a^{-1} \mathcal{Q}_a^t - \mathcal{P}_a - \frac{1}{2}\Xi$ *yields* $\Xi_a = \Xi_c = \Xi_a = \frac{1}{2}\Xi$ *and (A.57) is satisfied.*

Lemma A.8.7 *(The* **S** *Procedure) [720] :*
Denote the set $\mathsf{Z} = \{z\}$ *and let* $\mathcal{F}(z)$, $\mathcal{Y}_1(z)$, $\mathcal{Y}_2(z)$, $\ldots$, $\mathcal{Y}_k(z)$ *be some functionals or functions. Define domain* D *as*

$$
\mathsf{D} = \{z \in \mathsf{Z} : \mathcal{Y}_1(z) \geq 0, \ \mathcal{Y}_2(z) \geq 0, ..., \mathcal{Y}_k(z) \geq 0\}
$$

and the two following conditions:

(I) $\mathcal{F}(z) > 0,\ \forall\, z \in \mathsf{D}$,

(II) $\exists\, \varepsilon_1 \geq 0,\ \varepsilon_2 \geq 0, ..., \varepsilon_k \geq 0$ *such that* $\mathcal{S}(\varepsilon, z) = \mathcal{F}(z) - \sum_{j=1}^{k} \varepsilon_j\, \mathcal{Y}_j(z) > 0\ \forall\, z \in \mathsf{Z}$

Then (II) *implies* (I).

A.9   Linear Matrix Inequalities

It has been shown that a wide variety of problems arising in system and control theory can conveniently reduced to a few standard convex or quasi convex optimization problems involving linear matrix inequalities (LMIs). The resulting optimization problems can then be solved numerically very efficiently using commercially available interior-point methods.

A.9.1   Basics

One of the earliest LMIs arises in Lyapunov theory. It is well-known that the differential equation

$$\dot{x}(t) = A\, x(t) \tag{A.58}$$

has all of its trajectories converge to zero (stable) id and only if there exists a matrix $P > 0$ such that

$$A^t P + A P < 0 \tag{A.59}$$

This leads to the LMI formulation of stability, that is , *a linear time-invariant system is asymptotically stable if and only if there exists a matrix* $0 < P = P^t$ *satisfying the LMIs*

$$A^t P + A P < 0\ ,\quad P > 0$$

Given a vector variable $x \in \Re^n$ and a set of matrices $0 < G_j = G_j^t \in \Re^{n \times n}$, $j = 0, ..., p$, then a basic compact formulation of a linear matrix inequality is

$$G(x) := G_0 + \sum_{j=1}^{p} x_j\, G_j > 0 \tag{A.60}$$

Notice that (A.60) implies that $v^t G(x) v > 0\ \forall 0 \neq v \in \Re^n$. More importantly, the set $\{x\, | G(x) > 0$ is convex. Nonlinear (convex) inequalities are converted to LMI form using Schur complements in the sense that

$$\begin{bmatrix} Q(x) & S(x) \\ \bullet & R(x) \end{bmatrix} > 0 \tag{A.61}$$

where $Q(x) = Q^t(x)$, $R(x) = R^t(x)$, $S(x)$ depend affinely on x, is equivalent to

$$R(x) > 0 \quad , \quad Q(x) - S(x)R^{-1}(x)S^t(x) > 0 \tag{A.62}$$

More generally, the constraint

$$Tr[S^t(x) \, P^{-1}(x) \, S(x)] < 1 \quad , \quad P(x) > 0$$

where $P(x) = P^t(x) \in \Re^{n \times n}$, $S(x) \in \Re^{n \times p}$ depend affinely on x, is handled by introducing a new (slack) matrix variable $Y(x) = Y^t(x) \in\in \Re^{p \times p}$ and the LMI (in x and Y):

$$TrY < 1 \quad , \quad \begin{bmatrix} Y & S(x) \\ \bullet & P(x) \end{bmatrix} > 0 \tag{A.63}$$

Most of the time, our LMI variables are matrices. It should be clear from the foregoing discussions that a quadratic matrix inequality (QMI) in the variable P can be readily expressed as linear matrix inequality (LMI) in the same variable.

A.9.2 Some standard problems

Here we provide some common convex problems that we encountered throughout the monograph. Given an LMI $G(x) > 0$, the corresponding LMI problem (LMIP) is to
  find a feasible $x \equiv x^f$ such that $G(x^f) > 0$,
  or *determine that the LMI is infeasible.*
  It is obvious that this is a convex feasibility problem.
  The generalized eigenvalue problem (GEVP) is to minimize the maximum generalized eigenvalue of a pair of matrices that depend affinely on a variable, subject to an LMI constraint. GEVP has the general form

$$\begin{aligned} minimize \ & \lambda \\ subject \ to \ \ & \lambda B(x) - A(x) > 0 \quad , \quad B(x) > 0, \\ & C(x) > 0 \end{aligned} \tag{A.64}$$

where A, B, C are symmetric matrices that are affine functions of x. Equivalently stated

$$\begin{aligned} minimize \ & \lambda_M[A(x), B(x)] \\ subject \ to \ \ & B(x) > 0 \ , \ C(x) > 0 \end{aligned} \tag{A.65}$$

where $\lambda_M[X, Y]$ denotes the largest generalized eigenvalue of the pencil $\lambda Y - X$ with $Y > 0$. This is a quasi convex optimization problem since the constraint is convex and the objective, $\lambda_M[A(x), B(x)]$, is quasi convex.

The eigenvalue problem (EVP) is to minimize the maximum eigenvalue of a matrix that depends affinely on a variable, subject to an LMI constraint. EVP has the general form

$$minimize \ \lambda$$
$$subject \ to \ \ \lambda I - A(x) > 0 \ \ , \ \ B(x) > 0 \ \ \text{(A.66)}$$

where A, B are symmetric matrices that are affine functions of the optimization variable x. This is a convex optimization problem.

EVPs can appear in the equivalent form of minimizing a linear function subject to an LMI, that is

$$minimize \ \ c^t x$$
$$subject \ to \ \ G(x) > 0 \qquad \qquad \text{(A.67)}$$

where $G(x)$ is an affine function of x. Examples of $G(x)$ include

$$PA + A^t P + C^t C + \gamma^{-1} PBB^t P < 0 \ \ , \ \ P > 0$$

It should be stressed that the standard problems (LMIPs, GEVPs, EVPs) are tractable, from both theoretical and practical viewpoints:

They can be solved in polynomial-time.

They can be solved in practice very efficiently, using commercial software.

A.9.3   S-procedure

In some design applications, we faced the constraint that some quadratic function be negative whenever some other quadratic function is negative. In such cases, this constraint can be expressed as an LMI in the data variables defining the quadratic functions.

Let $G_o, ..., G_p$ be quadratic functions of the variable $\xi \in \Re^n$:

$$G_j(\xi) := \xi^t R_j \xi + 2u_j^t \xi + v_j \ , \ \ j = 0, ..., p, \ \ R_j = R_j^t$$

We consider the following condition on $G_o, ..., G_p$:

$$G_o(\xi) \leq 0 \ \forall \xi \ \ such \ that \ \ G_j(\xi) \geq 0, \ \ j = 0, ..., p \qquad \text{(A.68)}$$

It is readily evident that if there exist scalars $\omega_1 \geq 0, ..., \omega_p \geq 0$ such that

$$\forall \xi, \quad G_o(\xi) - \sum_{j=1}^{p} \omega_j \, G_j(\xi) \geq 0 \tag{A.69}$$

then inequality (A.68) holds. Observe that if the functions $G_o, ..., G_p$ are affine, then Farkas lemma states that (A.68) and (A.69) are equivalent. Interestingly enough, inequality (A.69) can written as

$$\begin{bmatrix} R_o & u_o \\ \bullet & v_o \end{bmatrix} - \sum_{j=1}^{p} \omega_j \begin{bmatrix} R_j & u_j \\ \bullet & v_j \end{bmatrix} \geq 0 \tag{A.70}$$

The foregoing discussions were stated for non strict inequalities. In case of strict inequality, we let $R_o, ..., R_p \in \Re^{n \times n}$ be symmetric matrices with the following qualifications

$$\xi^t R_o \xi > 0 \ \forall \xi \quad such \ that \quad \xi^t G_j \xi \geq 0, \quad j = 0, ..., p \tag{A.71}$$

Once again, it is obvious that there exist scalars $\omega_1 \geq 0, ..., \omega_p \geq 0$ such that

$$\forall \xi, \quad G_o(\xi) - \sum_{j=1}^{p} \omega_j \, G_j(\xi) > 0 \tag{A.72}$$

then inequality (A.71) holds. Observe that (A.72) is an LMI in the variables

$$R_o, \omega_1, ..., \omega_p$$

It should be remarked that the **S**-procedure which deals with non strict inequalities allows the inclusion of constant and linear terms. In the strict version, only quadratic functions can be used.

A.10 Some Formulas on Matrix Inverses

This concerns some useful formulas for inverting of matrix expressions in terms of the inverses of its constituents.

A.10.1 Inverse of block matrices

Let A be a square matrix of appropriate dimension and partitioned in the form

$$A = \begin{bmatrix} A_1 & A_2 \\ A_3 & A_4 \end{bmatrix} \tag{A.73}$$

where both A_1 and A_4 are square matrices. If A_1 is invertible, then

$$\Delta_1 = A_4 - A_3 \, A_1^{-1} \, A_2$$

is called the Schur complement of A_1. Alternatively, if A_4 is invertible, then

$$\Delta_4 = A_1 - A_2 \, A_4^{-1} \, A_3$$

is called the Schur complement of A_4.

It is well-known that matrix A is invertible if and only if either

$$A_1 \quad and \quad \Delta_1 \quad are \ invertible,$$

or

$$A_4 \quad and \quad \Delta_4 \quad are \ invertible.$$

Specifically, we have the following equivalent expressions

$$\left[\begin{array}{cc} A_1 & A_2 \\ A_3 & A_4 \end{array} \right]^{-1} = \left[\begin{array}{cc} \Upsilon_1 & -A_1^{-1} A_2 \Delta_1^{-1} \\ -\Delta_1^{-1} A_3 A_1^{-1} & \Delta_1^{-1} \end{array} \right] \tag{A.74}$$

or

$$\left[\begin{array}{cc} A_1 & A_2 \\ A_3 & A_4 \end{array} \right]^{-1} = \left[\begin{array}{cc} \Delta_4^{-1} & -\Delta_4^{-1} A_2 A_4^{-1} \\ -A_4^{-1} A_3 \Delta_4^{-1} & \Upsilon_4 \end{array} \right] \tag{A.75}$$

where

$$\begin{aligned} \Upsilon_1 &= A_1^{-1} + A_1^{-1} A_2 \Delta_1^{-1} A_3 A_1^{-1} \\ \Upsilon_4 &= A_4^{-1} + A_4^{-1} A_3 \Delta_4^{-1} A_2 A_4^{-1} \end{aligned} \tag{A.76}$$

Important special cases are

$$\left[\begin{array}{cc} A_1 & 0 \\ A_3 & A_4 \end{array} \right]^{-1} = \left[\begin{array}{cc} A_1^{-1} & 0 \\ -A_4^{-1} A_3 A_1^{-1} & A_4^{-1} \end{array} \right] \tag{A.77}$$

and

$$\left[\begin{array}{cc} A_1 & A_2 \\ 0 & A_4 \end{array} \right]^{-1} = \left[\begin{array}{cc} A_1^{-1} & -A_1^{-1} A_2 A_4^{-1} \\ 0 & A_4^{-1} \end{array} \right] \tag{A.78}$$

A.10.2   Matrix inversion lemma

Let $A \in \Re^{n \times n}$ and $C \in \Re^{m \times m}$ be nonsingular matrices. By using the definition of matrix inverse, it can be easily verified that

$$[A + B \, C \, D]^{-1} = A^{-1} - A^{-1} B \, [D \, A^{-1} B + C^{-1}]^{-1} \, D A^{-1} \tag{A.79}$$

A.10.3 Irreducible matrices

We call matrix $A = \{a_{ij}\} \in \Re^{N \times N}$ irreducible if

$$a_{ij} = a_{ij} \geq 0, \; for i \neq j, \; and \sum_{j=1}^{N} a_{ij} = 0, \; \forall \, i = 1, 2, ..., N$$

and $Rank(A) = N - 1$.

Lemma A.10.1 : *If matrix A is irreducible, then all eigenvalues of the matrix*

$$\tilde{A} = \begin{bmatrix} a_{11} - \varepsilon & a_{12} & \cdots & a_{1N} \\ \bullet & a_{22} & \cdots & a_{2N} \\ \vdots & \vdots & \ddots & \vdots \\ a_{N1} & a_{N2} & \cdots & a_{NN} \end{bmatrix}$$

are negative for any positive constant ε.

Proof A.10.1 *Since A is irreducible, there exists at least a positive element in the first column. Without loss of generality, we can assume $a_{21} > 0$. Let be the matrix $\tilde{A}_1$ obtained by excluding the first row and first column of $\tilde{A}$. Then, $\tilde{A}_1$ has the same structure as $\tilde{A}$ [663].*

Suppose that λ is an eigenvalue of $\tilde{A}$, $v = [v_1, ..., v_N]^t$ is the corresponding eigenvector, and $|v_m| = \max_{j=1,...,N} |v_j|$. It is clear that if v is an eigenvector, then $-v$ is also an eigenvector. Thus, without loss of generality, we can assume that $v_m > 0$ and $|v_m| = \max_{j=1,...,N} |v_j|$. Now if $m = 1$, then

$$\begin{aligned} \sum_{j=1}^{N} \tilde{a}_{1j} v_j &= -\varepsilon v_1 + \sum_{j=1}^{N} a_{1j} v_j \\ &\leq -\varepsilon v_1 + \sum_{j=1}^{N} a_{1j} |v_j| \\ &< -\varepsilon v_1 < 0 \end{aligned}$$

which means that $\lambda < 0$. Alternatively, if $m > 1$, then

$$\lambda v_m = \sum_{j=1}^{N} \tilde{a}_{mj} v_j \leq \tilde{a}_{mm} v_m + \sum_{j=1}^{N} \tilde{a}_{mj} |v_j| \leq 0$$

which means that $\lambda \leq 0$. *By the assumption that* $A = \{a_{ij}\} \in \Re^{N \times N}$ *is an irreducible matrix with* $Rank(A) = N - 1$ *we are led to* $v = [v_m, ..., v_m]^t$. *This is impossible since*

$$\sum_{j=1}^{N} \tilde{a}_{1j} v_j = -\varepsilon v_m < 0 \Rightarrow \lambda < 0$$

which completes the proof.

Lemma A.10.2 : *Assume that an undirected network is irreducible. Then matrix A has an eigenvalue of zero with an algebraic multiplicity of one, and all the other eigenvalues are negative:*

$$0 = \lambda_1(A) > \lambda_2(A) \leq \cdots \lambda_N(A)$$

Bibliography

[1] T. Bass, "Intrusion detection systems and multi–sensor data fusion," *Comm. ACM*, vol. 43, no. 4, pp. 99–105, April 2000.

[2] C. Staterlis and B. Maglaris, "Towards multi–sensor data fusion for DoS detection," *Proc. the 2004 ACM Symposium on Applied Computing*, ACM Press, Nicosia, Cyprus, pp. 439–446, 2004.

[3] D. L. Hall and J. Llinas, "An introduction to multi–sensor data fusion," *Proc. IEEE*, vol. 85, no, 1, pp. 6–23, 1997.

[4] C. Intanagonwiwat, R. Govindan, D. Estrin, J. Heidemann and F. Silva, "Directed diffusion for wireless sensor networking," *IEEE/ACM Trans. Netw.*, vol. 11, no. 1, pp. 2–16, 2003.

[5] L. A. Klein, *Sensor and Data Fusion Concepts and Applications*, vol. TT14. SPIE Optical Engineering Press, 1993.

[6] B. Krishnamachari and S. Iyengar, "Distributed Bayesian algorithms for fault-tolerant event region detection in wireless sensor networks," *IEEE Trans. Comput.*, vol. 53, no. 3, pp. 241–250, 2004.

[7] R. C. Luo and M. G. Kay (Eds.), *Multi–sensor Integration and Fusion for Intelligent Machines and Systems*, Reissue edition Computer Engineering and Computer Science, Ablex Publishing, New Jersey, USA, 1995.

[8] R. C. Luo, C. -C. Yih and K. L. Su, "Multi–sensor fusion and integration: approaches, applications, and future research directions," *IEEE Sensors J.*, vol. 2, no. 2, pp. 107–119, 2002.

[9] G. Brokmann, B. March, D. Romhild and A. Steinke, "Integrated multi–sensors for industrial humidity measurement," *Proc. the IEEE Int. Conference on multi–sensor Fusion and Integration for Intelligent Systems*, Baden-Baden, Germany, pp. 201–203, 2001.

[10] S. R. Madden, M. J. Franklin, J. M. Hellerstein and W. Hong, "TAG: a Tiny Aggregation service for ad-hoc sensor networks," *ACM SIGOPS Oper. Syst. Rev.*, vol. 36, pp. 131–146, 2002.

[11] S. R. Madden, M. J. Franklin, J. M. Hellerstein and W. Hong, "TinyDB: An acqusitional query processing system for sensor networks," *ACM Trans. Database Syst.*, vol. 30, no. 1, pp. 122–173, 2005.

[12] E. F. Nakamura, F. G. Nakamura, C. M. Figueiredo and A. A. Loureiro, "Using information fusion to assist data dissemination in wireless sensor networks," *Telecomm. Syst.*, vol. 30,, pp. 237–254, November 2005.

[13] A. Savvides, C. Han and M. B. Strivastava, "The n-hop multilateration primitive for node localization," *Mobile Netw. Appl.*, vol. 8, no. 4, pp. 443-451, August 2003.

[14] U. S. Department of Defence, *Data Fusion Lexicon*, Data Fusion Subpanel of the Joint Directors of Laboratories, Technical Panel for C3 (F. E. White, Code 4202, NOSC, San Diego, CA), 1991.

[15] A. Woo, T. Tong and D. Culler, "Taming the underlying challenges of reliable multihop routing in sensor networks," *Proc. the 1st Int. Conference on Embedded Network Sensor Systems (SenSys'03)*, pp. 14–27, 2003.

[16] I. Akyildiz, W. Su, Y. Sankarasubramniam, and E. Cayirci. "A survey on sensor networks," *IEEE Communications Magazine*, pp. 102–114, August 2002.

[17] J. Zhao, R. Govindan and D. Estrin, "Residual energy scans for monitoring wireless sensor networks," *Proc. the IEEE Wireless Communications and Networking Conference (WCNC02)*, vol. 1, Orlando, FL, pp. 356–362, 2002.

[18] R. Willett, A. Martin and R. Nowak, "Backcasting: Adaptive sampling for sensor networks," *Proc. the 3rd Int. Symposium on Information Processing in Sensor Networks (IPSN04)*, pp. 124–133, 2004.

[19] R. Nowak, U. Mitra and R. Willett, "Estimating inhomogeneous fields using wireless sensor networks," *IEEE J. Select. Areas Comm.*, vol. 22, no. 6, pp. 999–1006, 2004.

[20] A. Singh, R. Nowak and P. Ramanathan, "Active learning for adaptive mobile sensing networks," *Proc. the 5th Int. Conference on Information Processing in Sensor Networks (IPSN06)*, pp. 60–68, 2006.

[21] H. F. Duttant-Whyte, "Sensor models and multi–sensor integration," *Inter. J. Robotics Res.*, vol. 7, no. 6, pp. 97–113, 1988.

[22] F. Nakamura, A.F. Loureiro, and C. Frery, "Data fusion for wireless sensor networks: methods, models, and classifications," *ACM Computing Surveys*, vol. 39, No. 3, Article 9, August 2007.

[23] R. R. Brooks and S. Iyengar, *Multi–sensor Fusion: Fundamentals and Applications with Software*, Prentice Hall, Upper Saddle River, NJ, 2003.

[24] C. Intanagonwiwat, R. Govindan, and D. Estrin, Directed diffusion: A scalable and robust communication paradigm for sensor networks, *Proc. the 6th ACM Annual International Conference on Mobile Computing and Networking*, Boston, MA, USA, pp. 56–67, August 2000.

[25] L. Krishnamachari, D. Estrin, and S. Wicker, "The impact of data aggregation in wireless sensor networks," *Proc. the 22nd International Conference on Distributed Computing Systems Workshops*, Vienna, Austria, pp. 575–578, July 2002.

[26] A. Woo, T. Tong, and D. Culler, "Taming the underlying challenges of reliable multihop routing in sensor networks," *Proc. the 1st Int. Conference on Embedded Network Sensor Systems*, Los Angeles, pp. 14–27, November 2003.

[27] A. Abdelgawad and M. Bayoumi, "Data fusion in WSN," *Resource-Aware Data Fusion Algorithms for Wireless Sensor Networks*, Ch. 2, Springer, NY, 2012.

[28] B. V. Dasarathy, "Sensor fusion potential exploitation-innovative architectures and illustrative applications," *Proc. the IEEE*, vol. 85, pp. 24–38, January 1997.

[29] J. Polastre, J. Hill, and D. Culler, "Versatile low power media access for wireless sensor networks," *Proc. the 2nd ACM Int. Conference on Embedded Networked Sensor Systems*, Baltimore, USA, pp. 95–107, November 2004.

[30] J. Llinas and D. L. Hall, "An introduction to multi–sensor data fusion," *Proc. the IEEE Int. Symposium on Circuits and Systems*, Monterrey, CA, USA, pp. 537–540, 1998.

[31] L. Wald, "Some terms of reference in data fusion," *IEEE Trans. Geoscience and Remote Sensing,* vol. 37, pp. 1190–1193, May 1999.

[32] R. C. Luo and M. G. Kay, "multi–sensor integration and fusion in intelligent systems," *IEEE Trans. Systems, Man and Cybernetics*, vol. 19, pp. 901–931, October 1989.

[33] K. Kalpakis, K. Dasgupta, and P. Namjoshi, "Efficient algorithms for maximum lifetime data gathering and aggregation," *The Int. J. Computer and Telecommunications Networking*, vol. 42, pp. 697–716, August 2003.

[34] A. Boulis, S. Ganeriwal, and M. B. Srivastava, "Aggregation in sensor networks: an energy accuracy trade-off," *Proc. the 1st IEEE Int. Workshop on Sensor Network Protocols and Applications*, Anchorage, AK, USA, pp. 128–138, May 2003.

[35] A. Hoover and B. D. Olsen, "A real-time occupancy map from multiple video streams," *Proc. the IEEE Int. Conference on Robotics and Automation*, Detroit, MI, USA, pp. 2261–2266, May 1999.

[36] Y. J. Zhao, R. Govindan, and D. Estrin, "Residual energy scan for monitoring sensor networks," *Proc. the IEEE Wireless Communications and Networking Conference*, Orlando, Florida, USA, pp. 356–362, March 2002.

[37] X. Luo, M. Dong, and Y. Huang, "On distributed fault-tolerant detection in wireless sensor networks," *IEEE Trans. Computers*, vol. 55, pp. 58–70, January 2006.

[38] J. Polastre, J. Hill, and D. Culler, "Versatile low power media access for wireless sensor networks," *Proc. the 2nd ACM International Conference on Embedded Networked Sensor Systems*, Baltimore, USA, pp. 95–107, November 2004.

[39] B. Krishnamachari and S. Iyengar, "Distributed Bayesian algorithms for fault-tolerant event region detection in wireless sensor networks," *IEEE Trans. Computers*, vol. 53, pp. 241–250, March 2004.

[40] D. P. Spanos, R. O. Saber, and R. M. Murray, "Approximate distributed Kalman filtering in sensor networks with quantifiable performance," *Proc. the Fourth Int. Symposium on Information Processing in Sensor Networks*, pp. 133–139, April 2005.

[41] R. O. Saber and R. M. Murray, "Consensus problems in networks of agents with switching topology and time-delays," *IEEE Trans. Automatic Control*, vol. 49(9), pp. 1520–1533, September 2004.

[42] R. O. Saber and R. M. Murray, "Consensus protocols for networks of dynamic agents," *Proc. the American Control Conference*, vol. 2, pp. 951–956, June 2003.

[43] T. Y. Al-Naffouri, "An EM-based forward-backward Kalman filter for the estimation of time-variant channels in OFDM," *IEEE Trans. Signal Processing*, vol. 55, no. 7, pp. 3924–3930, 2007.

[44] P. M. Frank, "Fault diagnosis in dynamic systems using analytical and knowledge-based redundancy: survey and some new results" , *Automatica*, vol. 26, pp. 459–474, 1990.

[45] J. Gertler, L. Weihua, Y. Huang and T. McAvoy,"Isolation enhanced principal component analysis," *A. I. Ch. E. Journal*, vol. 45, pp. 323–334, 1999.

[46] H. Hammouri, P. Kabore, S. Othman and J. Biston, "Failure diagnosis and non-linear observer application to a hydraulic process," *J. the Franklin Institute*, vol. 339, pp. 455–478, 2002.

[47] N. Hassan, D. Theilliol, J. -C. Ponsart and A. Chamseddine, *Fault-tolerant control systems: design and practical applications*, Springer, London, 2009.

[48] C. Kaddissi, J.-P. Kenne and M. Saad, "Identification and real time control of an electro-hydraulic servo system based on nonlinear backstepping," *IEEE/ASME Trans. Mechatronics*, vol. 12, pp. 12–22, 2007.

[49] P. Kabore and H. Wang, "Design of fault diagnosis filters and fault-tolerant control for a class of nonlinear systems," *IEEE Trans. Automatic Control*, vol. 46, pp. 1805–1810, 2001.

[50] S. LeQuoc, R. M. H. Cheng, and K. H. Leung, "Tuning an electro hydraulic servo valve to obtain a high amplitude ratio and a low resonance peak," *J. Fluid Control*, vol. 20, pp. 30–39, 1990.

[51] J. MacGregor and T. Kourti, "Statistical process control of multi-variate processes," *J. Quality Technology*, vol. 28, 1996, pp. 409–428.

[52] P. Mhaskar, C. McFall, A. Gani, P. Christofides and J. Davis, "Isolation and handling of actuator faults in nonlinear systems," *Automatica*, vol. 44, pp. 53–62, 2008.

[53] B. Mogens, M. Kinnaert, J. Lunze and M. Staroswiecki, *Diagnosis and fault tolerant control*, 2nd Edition. Springer, 2006.

[54] A. Negiz and A. Cinar, "Statistical monitoring of multi-variable dynamic processes with state-space models," *A. I. Ch. E. Journal*, vol. 43, pp. 2002–2020, 1997.

[55] A. Raich and A. Cinar, "Statistical process monitoring and disturbance diagnosis in multi-variable continuous processes," *A. I. Ch. E. Journal*, vol. 42, pp. 995–1009, 1996.

[56] J. Romagnoli and A. Palazoglu, *Introduction to Process Control*, CRC Press, Boca Raton, 2006.

[57] V. Venkatasubramanian, R. Rengaswamy, K. Yin and S. Kavuri, "A review of process fault detection and diagnosis part I: quantitative model-based methods," *Computers and Chemical Engineering*, vol. 27, pp. 293–311, 2003.

[58] V. Venkatasubramanian, R. Rengaswamy, S. Kavuri and K. Yin, "A review of process fault detection and diagnosis part III: process history based-methods," *Computers and Chemical Engineering*, vol. 27, pp. 327–346, 2003.

[59] G. T. Huang, "Casting the wireless sensor net," *Technology Review*, pp. 51-56, July 2003. *www.technologyreview.com*.

[60] A. Mainwaring, J. Polastre, R. Szewczyk, D. Culler, and J. Anderson, "Wireless sensor networks for habitat monitoring," *Proc. the 1st ACM Workshop on Wireless Sensor Networks and Applications*, Atlanta, GA, September 2002.

[61] "A wireless sensor networks bibliography," *http : //ceng.usc.edu/ anrg/SensorNetBib.html*.

[62] F. Xia, W. H. Zhao, Y. X. Sun and Y. C. Tian, "Fuzzy logic control based QoS management in wireless sensor/actuator networks," *Sensors*, vol. 7(12), pp. 3179–3191, 2007.

[63] F. Xia, Y. C. Tian, Y. C. Li and Y. X. Sun, "Wireless sensor/actuator network design for mobile control applications," *Sensors*, vol. 7(10), pp. 2157–2173, 2007.

[64] A. Rezgui and M. Eltoweissy, "Service-oriented sensor-actuator networks," *IEEE Communications Magazine*, vol. 45(12), pp. 92–100, 2007.

[65] I. F. Akyildiz and I. H. Kasimoglu, "Wireless sensor and actuator networks: research challenges," *Ad Hoc Networks*, vol. 2(4), pp. 351–367, 2004.

[66] "NSF workshop on cyber-physical systems," *http : //varma.ece.cmu.edu/cps/*, Oct. 2006.

[67] M. A. El-Gendy, A. Bose and K. G. Shin, "Evolution of the Internet QoS and support for soft realtime applications," *Proc. the IEEE*, vol. 91(7), pp. 1086–1104, 2003.

[68] W. Heinzelman, A. Chandrakasan and H. Balakrishnan, "An application-specific protocol architecture for wireless micro sensor networks," *IEEE Trans. Wireless Communications,* vol. 1, no. 4, pp. 660–670, 2002.

[69] J. Yick, B. Mukherjee and D. Ghosal, "Wireless sensor network survey," *Computer Networks* vol. 52, no.12, pp. 2292–2330, 2008.

[70] I. D. Schizas, A. Ribeiro, and G. B. Giannakis, "Consensus in Ad Hoc WSNs with noisy links-Part I: Distributed estimation of deterministic signals," *IEEE Trans. Signal Process.*, vol. 56, no. 1, pp. 342–356, 2008.

[71] A. Ribeiro and G. Giannakis, "Bandwidth-constrained distributed estimation for wireless sensor networks, Part I Gaussian PDF," *IEEE Trans. Signal Processing*, vol. 54, no. 3, pp. 1131–1143, 2006.

[72] T.-Y Wang. and L.-Y. Chang, "Cooperative information aggregation for distributed estimation in wireless sensor networks," *IEEE Trans. Signal Processing*, vol. 59, no. 8, pp. 3876–3888, 2011.

[73] Y. Wang, X. Wang, B. Xie, D. Wang, and D. P. Agrawal, "Intrusion detection in homogeneous and heterogeneous wireless sensor networks," *IEEE Trans. Mobile Computing,* vol. 7, no. 6, pp. 698–711, 2008.

[74] I. F. Akyildiz, W. Su, Y. Sankarasubramaniam, and E. Cayirci, "A survey on sensor networks," *IEEE Comm. Magazine,* vol. 40, no. 8, pp. 102–114, 2002.

[75] T. M. Mubarak, S. A. Sattar, G. A. Rao and M. Sajitha, "Intrusion detection: an energy efficient approach in heterogeneous WSN," *Emerging Trends in Electrical and Computer Technology (ICETECT)*, pp. 1092–1096, 2011.

[76] M. Yarvis, N. Kushalnagar, H. Singh, A. Rangarajan, "Exploiting heterogeneity in sensor networks," *Proc. IEEE Conf. Computer Communications (INFOCOM)*, Miami, USA, pp. 878–890, 2005.

[77] H. Dong, Z. Wang, and Huijun Gao, "Distributed filtering for a class of time-varying systems over sensor networks with quantization errors and successive packet dropouts," *IEEE Trans. Signal Processing,* vol. 60, no. 6, pp. 3164–3173, 2002.

[78] M. Farina, G. Ferrari-Trecate, and R. Scattolini, "Distributed moving horizon estimation for linear constrained systems," *IEEE Trans. Automatic Control,* vol. 55, no. 11, pp. 2462–2475, 2010.

[79] U. Munz, A. Papachristodoulou, and F. Allgower, "Robust consensus controller design for nonlinear relative degree two multi-agent systems with communication constraints," *IEEE Trans. Automatic Control,* vol. 56, no. 1, pp. 145–151, 2011.

[80] S. Ramanan and J. M. Walsh, "Distributed estimation of channel gains in wireless sensor networks," *IEEE Trans. Signal Processing,* vol. 58, no. 6, pp. 3097–3107, 2010.

[81] G. Scutari, S. Barbarossa, and L. Pescosolido, "Distributed decision through self-synchronizing sensor networks in the presence of propagation delays and asymmetric channels," *IEEE Trans. Signal Processing,* vol. 56, no. 4, pp. 1667–1684, 2008.

[82] B. Shen, Z. Wang, Y. S. Hung, and G. Chesi, "Distributed H_∞ filtering for polynomial nonlinear stochastic systems in sensor networks," *IEEE Trans. Industrial Electronics,* vol. 58, no. 5, pp. 1971–1979, 2011.

[83] W. Xiao, S. Zhang, 1. Lin, and C. K. Tham, "Energy-efficient adaptive sensor scheduling for target tracking in wireless sensor networks," *J. Control Theory and Applications,* vol. 8, pp. 86–92, 2010.

[84] Y. H. Kim and A. Ortega, "Quantizer design for source localization in sensor networks," *IEEE Trans. Signal Processing,* vol. 59, pp. 5577–5588, 2011.

[85] D. W. Casbeer and R. Beard, "Distributed information filtering using consensus filters," *Proc. American Control Conference,* Hyatt Regency Riverfront, St. Louis, MO, USA, pp. 1882–1887, 2009.

[86] U. A. Khan and J. M. F. Moura, "Distributing the Kalman filter for large-scale systems," *IEEE Trans. Signal Processing,* vol. 56, no. 10, pp. 4919–4935, 2008.

[87] T. Fortmann, Y. Bar-Shalom, M. Scheffe and S. Gelfand, "Detection thresholds for tracking in clutter–A connection between estimation and signal processing," *IEEE Trans. Automatic Control*, vol. 30, no. 3, pp. 221–229, 1985.

[88] I. B. Rhodes, "A tutorial introduction to estimation and filtering," *IEEE Trans. Automatic Control*, vol. 16, no. 6, pp. 688–706, 1971.

[89] M. B. Prokhorov, and V. K. Saul'ev, "The Kalman-Bucy method of optimal filtering and its generalizations." *J. Soviet Mathematics*, vol. 12, no. 3 pp. 354–380, 1979.

[90] I. D. Schizas, G. Mateos, and G. B. Giannakis, "Distributed LMS for consensus-based in-network adaptive processing," *IEEE Trans. Signal Processing*, vol. 8, no. 6, pp. 2365–2381, 2009.

[91] J. -J. Xiao, S. Cui, Z.-Q. Luo, and A. J. Goldsmith, "Linear coherent decentralized estimation," *IEEE Trans. Signal Processing.*, vol. 56, no. 2, pp. 757–770, 2008.

[92] D. E. Quevedo, A. Ahlen, and l. Ostergaard, "Energy efficient state estimation with wireless sensors through the use of predictive power control and coding," *IEEE Trans. Signal Processing*, vol. 58, no. 9, pp. 4811–4823, 2010.

[93] L. M. Kaplan, "Global node selection for localization in a distributed sensor network," *IEEE Trans. Aerospace Elect. Systems*, vol. 42, no. 1, pp. 113–135, 2006.

[94] A. G. O. Mutambara, *Decentralized Estimation and Control for multi–sensor Systems,* CRC, Boca Raton, FL, 1998.

[95] J. Manyika and H. Durrant-Whyte, *Data Fusion and Sensor Management: A Decentralized Information-Theoretic Approach,* Prentice-Hall, NJ, 1994.

[96] Y. Bar-Shalom and X. Li, *Estimation and Tracking: Principles, Techniques and Software,* Artech House, Boston, MA, 1993.

[97] L. Moreau, "Stability of multiagent systems with time-dependent communication links," *IEEE Trans. Automatic Control,* vol. 50, no. 2, pp. 169–182, 2005.

[98] C. Berge and A. Ghouila-Houri, *Programming, Games and Transportation Networks.* John Wiley and Sons, 1965.

[99] Z. Lin, *Coupled dynamic systems: from structure towards stability and sta-bilizability,* Ph. D. dissertation, University of Toronto, Toronto, Canada, 2005.

[100] J. Liu, J. Reich, and F. Zhao, "Collaborative in-network processing for target tracking," *EURASIP J. Appl. Signal Process.*, vol. 2003, pp. 378–391, 2003.

[101] H. Wang, K. Yao, G. Pottie, and D. Estrin (2004). "Entropy-based sensor selection heuristic for target localization," *Proc. 3rd Int. Symp. Inf. Process. Sensor Netw. IPSN,* pp. 36–45, 2004.

[102] T. Vercauteren and X. Wang, "Decentralized sigma-point information filters for target tracking in collaborative sensor networks," *IEEE Trans. Signal Processing,* vol. 53, no. 8, pp. 2997–3009, 2005.

[103] T. Rappaport, *Wireless Communications: Principles and Practice,* Second ed., Prentice-Hall, NJ, 2002.

[104] Y. Bar-Shalom, X. R. Li, and T. Kirubarajan, *Estimation With Applications to Tracking and Navigation.* Wiley, New York, 2001.

[105] D. Tse, *Fundamentals of Wireless Communication*, Pramod Viswanath (Editor), Cambridge University Press, 2005.

[106] M. R. Zoghi and M. H. Kahaei, "Target tracking in collaborative sensor networks by using a decentralized leader-based scheme," *Information Sciences, Signal Processing and its Applications*, 2007.

[107] R. O. Saber and J. S. Shamma, "Consensus filters for sensor networks and distributed sensor fusion," *Proc. the 44th IEEE Conference on Decision and Control and European Control Conference (CDC-ECC '05)*, Dec. 12–15 2005, pp. 6698–6703, 2005.

[108] Y. and K. A. Hoo, "Stability analysis for closed-loop management of a reservoir based on identification of reduced-order nonlinear model," *Systems Science and Control Engineering,* vol. 1, no. 1, pp. 12–19, 2013.

[109] A. Mehrasi, H. R. Karimi and K. D. Thoben, "Integration of supply networks for customization with modularity in cloud and make-to-upgrade strategy," *Systems Science and Control Engineering,* vol. 1, no. 1, pp. 28–42, 2013.

[110] D. Dong, Z. Wang, H. Dong and H. Shu, "Distributed H_∞ state estimation with stochastic parameters and nonlinearities through sensor networks: the finite-horizon case," *Automatica*, vol. 48, no. 8, pp. 1575–1585, 2012.

[111] H. Dong, Z. Wang and H. Gao, "Distributed H_∞ filtering for a class of Markovian jump nonlinear time-delay systems over lossy sensor networks," *IEEE Trans. Industrial Electronics*, vol. 60, no. 10, pp. 4665–4672, 2013.

[112] J. Liang, Z. Wang, B. Chen and X. Liu, "Distributed state estimation in sensor networks with randomly occurring nonlinearities subject to time-delays," *ACM Trans. Sensor Networks*, vol. 9, no. 1, art no. 4, 2012.

[113] B. Shen, Z. Wang, and X. Liu, "A stochastic sampled-data approach to distributed H_∞ filtering in sensor networks," *IEEE Trans. Circuits and Systems- Part I*, vol. 58, no. 9, pp. 2237–2246, 2011.

[114] R. A. Adrian, "Sensor management," *Proc. the AIAA/IEEE Digital Avionics System Conference*, Ft. Worth, TX, pp. 32–37, 1993.

[115] S. Blackman, R. Popoli, "Sensor management," *Design and Analysis of Modern Tracking Systems*, Artech House, Norwood, MA, pp. 967–1068, 1999.

[116] C. Chong, S. Mori and K. Chang, "Distributed multitarget multi–sensor tracking," in *Multitarget Multi–sensor Tracking: Applications and Advances*, B. Shalom (Ed.), Chapter 8, Artech House, Boston, 1990.

[117] B. V. Dasarathy, *Decision Fusion*, IEEE Computer Society Press, Silver Spring, MD, 1994.

[118] H. Durrant-Whyte and M. Stevens, "Data fusion in decentralized sensing networks," *Proc. the International Conference on Information Fusion*, pp. WeA3-1924, 2001.

[119] C. Giraud and B. Jouvencel, "Sensor selection in a fusion process: a fuzzy approach," *Proc. the IEEE Conference on multi–sensor Fusion and Integration for Intelligent Systems*, Las Vegas, NV, pp. 599-606, 1994.

[120] P. Greenway and R. Deaves, "Sensor management using the decentralized Kalman filter," *Proc. the SPIE, Sensor Fusion VII*, Boston, MA, pp. 216-225, 1994.

[121] S. Julier and J. K. Uhlmann, "General Decentralized Data Fusion with Covariance Intersection (CI)," *Handbook of Data Fusion*, D. Hall, J. Llinas (Eds.), CRC Press, Boca Raton, Chapter 12, pp. 12–1-25, 2001.

[122] M. Kalandros and L. Y. Pao, "Controlling target estimate covariance in centralized multi–sensor systems," *Proc. American Control Conference*, Philadelphia, PA, pp. 2749-2753, 1998.

[123] M. Kalandros and L. Y. Pao, "Randomization and super-heuristics in choosing sensor sets for target tracking applications," *Proc. the IEEE Conference on Decision and Control*, Phoenix, AZ, pp. 1803-1808, 1999.

[124] S. Kristensen, *Sensor Planning with Bayesian Decision Analysis*, Ph. D. Dissertation, Laboratory of Image Analysis, Aalborg University, Aalborg, Denmark, 1996.

[125] R. Malhotra, "Temporal considerations in sensor management," *Proc. the IEEE National Aerospace and Electronics Conference*, pp. 86-93, 1995.

[126] J. M. Molina Lopez, F. J. Jimenez Rodriguez and J. R. Casar Corredera, "Fuzzy reasoning for multi–sensor management," *Proc. the IEEE Int. Conference on Systems, Man and Cybernetics*, pp. 1398-1403, 1995.

[127] J. M. Molina Lopez, F. J. Jimenez Rodriguez and J. R. Casar Corredera, "Symbolic processing for coordinated task management in multi-radar surveillance networks," *Proc. the Int. Conference on Information Fusion*, pp. 725–732, 1998.

[128] S. Musick and R. Malhotra, "Chasing the elusive sensor manager," *Proc. the IEEE National Aerospace and Electronics Conference*, pp. 606-613, 1994.

[129] G. W. Ng and K. H. Ng, "Sensor management what, why and how," *Information Fusion*, vol. 1(2), pp. 67-75, 2000.

[130] C. L. Pittard, D. E. Brown and D. E. Sappington, "SPA: Sensor placement analyzer, An approach to the sensor placement problem," *Technical Report, Department of Systems Engineering*, University of Virginia, 1994.

[131] C. G. Schaefer and K. J. Hintz, "Sensor management in a sensor rich environment," *Proc. the SPIE International Symposium on Aerospace/Defense Sensing and Control*, Orlando, FL, vol. 4052, pp. 48–57, 2000.

[132] A. Steinberg and C. Bowman, "Revisions to the JDL Data Fusion Model," *Handbook of Data Fusion*, D. Hall, J. Llinas (Eds.), CRC Press, Boca Raton, Chapter 2, pp. 2–1-19, 2001.

[133] P. Svensson, "Information fusion in the future swedish RMA defense," *Swedish Journal of Military Technology*, vol. 69(3), pp. 2-8, 2000.

[134] K. A. Tarabanis, P. K. Allen and R. Y. Tsai, "A survey of sensor planning in computer vision," *IEEE Trans. Robotics and Automation*, vol. 11(1), pp. 86-104, 1995.

[135] R. Wesson et al., *Network Structures for Distributed Situation Assessment*, Readings in Distributed Artificial Intelligence, A. Bond, L. Gasser (Eds.), Morgan Kaufman, Los Allos, CA, pp. 71-89, 1988.

[136] F. E. White, "Managing data fusion system in joint and coalition warfare," *Proc. the European Conference on Data Fusion*, Malvern, UK, pp. 49–52, 1998.

[137] I. F. Akyildiz, W. Su, Y. Sankarasubramaniam, and E. Cayirci, "A survey on sensor networks," *IEEE Communications Magazine*, vol. 40(8), pp. 102–114, August n b2002.

[138] A. Perrig, R. Szewczyk, J. D. Tygar, V. Wen, and D. E. Culler, "Spins: security protocols for sensor networks," *Wireless Networking*, vol. 8(5), pp. 521–534, 2002.

[139] S. Capkun and J.-P. Hubaux, "Secure positioning in wireless networks," *IEEE J. Selected Areas in Communications*, vol. 24(2), pp. 221–232, 2006.

[140] D. Liu and P. Ning, "Efficient distribution of key chain commitments for broadcast authentication in distributed sensor networks," *Proc. the 10th Annual Network and Distributed System Security Symposium*, pp. 263–276, 2003.

[141] D. Liu and P. Ning, "Multilevel μTESLA: broadcast authentication for distributed sensor networks," *Trans. Embedded Computing Sys.*, vol. 3(4), pp. 800–836, 2004.

[142] C. Intanagonwiwat, R. Govindan, D. Estrin, "Directed diffusion: a scalable and robust communication paradigm for sensor networks," *Proc. the ACM MobiCom'00*, Boston, MA, pp. 56–67, 2000.

[143] D. Estrin, R. Govindan, J. Heidemann, S. Kumar, "Next century challenges: scalable coordination in sensor networks," *Proc. ACM MobiCom99*, Washingtion, USA, pp. 263–270, 1999.

[144] J. Agre, L. Clare, "An integrated architecture for cooperative sensing networks," *IEEE Computer Magazine*, pp. 106–108, 2000.

[145] A. Cerpa, J. Elson, M. Hamilton, J. Zhao, "Habitat monitoring: application driver for wireless communications technology," *ACM SIGCOMM2000*, Costa Rica, April 2001.

[146] W. R. Heinzelman, J. Kulik, H. Balakrishnan, "Adaptive protocols for information dissemination in wireless sensor networks," *Proc. the ACM MobiCom99*, Seattle, Washington, pp. 174–185, 1999.

[147] T. H. Keitt, D. L. Urban, B. T. Milne, "Detecting critical scales in fragmented landscapes," *Conservation Ecology*, vol. 1(1), 1997.

[148] S. Slijepcevic, M. Potkonjak, "Power efficient organization of wireless sensor networks," *IEEE Int. Conference on Communications–ICC01*, Helsinki, Finland, June 2001.

[149] A. Chandrakasan, R. Amirtharajah, S. Cho, J. Goodman, G. Konduri, J. Kulik, W. Rabiner, A. Wang, "Design considerations for distributed microsensor systems," *Proc. the IEEE 1999 Custom Integrated Circuits Conference*, San Diego, CA, pp. 279–286, May 1999.

[150] R. Colwell, *Testimony of Dr. Rita Colwell, Director, National Science Foundation, Before the Basic Research Subcommitte, House Science Committe, Hearing on Remote Sensing as a Research and Management Tool*, September 1998.

[151] G. Coyle et al., "Home telecare for the elderly," *Journal Telemedicine and Telecare*, vol. 1, pp. 183–184, 1995.

[152] P. Bonnet, J. Gehrke, P. Seshadri, "Querying the physical world," *IEEE Personal Communications*, pp. 10–15, 2000.

[153] N. Bulusu, D. Estrin, L. Girod, J. Heidemann, "Scalable coordination for wireless sensor networks: self-configuring localization systems," *Proc. Int. Symposium on Communication Theory and Applications (ISCTA 2001)*, Ambleside, UK, July 2001.

[154] M. Gell-Mann, "What is complexity?," *Complexity*, vol. 1 (1), 1995.

[155] M. P. Hamilton, "Hummercams, robots, and the virtual reserve," *Directors Notebook*, February 6, 2000, available from $http : //www.jamesreserve.edu/news.html$.

[156] T. Imielinski, S. Goel, "DataSpace: querying and monitoring deeply networked collections in physical space," *ACM Int. Workshop on Data Engineering for Wireless and Mobile Access MobiDE 1999*, Seattle, Washington, pp. 44–51, 1999.

[157] Y. H. Nam et al., "Development of remote diagnosis system integrating digital telemetry for medicine," *Proc. Int. Conference IEEE-EMBS*, Hong Kong, pp. 1170–1173, 1998.

[158] N. Noury, T. Herve, V. Rialle, G. Virone, E. Mercier, G. Morey, A. Moro, T. Porcheron, "Monitoring behavior in home using a smart fall sensor," *Proc. the IEEE-EMBS Special Topic Conference on Micro Technologies in Medicine and Biology*, pp. 607–610, 2000.

[159] B. Sibbald, "Use computerized systems to cut adverse drug events," *CMAJ: Canadian Medical Association Journal*, vol. 164(13), p. 1878, 2001.

[160] B. Warneke, B. Liebowitz, K.S.J. Pister, "Smart dust: communicating with a cubic-millimeter computer," *IEEE Computer*, pp. 2–9, 2001.

[161] *http : //www.alertsystems.org.*

[162] M. Ogawa et al., "Fully automated biosignal acquisition in daily routine through 1 month," *Proc. Int. Conference on IEEE-EMBS*, Hong Kong, pp. 1947–1950, 1998.

[163] P. Bauer, M. Sichitiu, R. Istepanian, K. Premaratne, "The mobile patient: wireless distributed sensor networks for patient monitoring and care," *Proc. IEEE EMBS Int. Conference on Information Technology Applications in Biomedicine*, pp. 17–21, 2000.

[164] P. Johnson et al., "Remote continuous physiological monitoring in the home," *Journal Telemedicin Telecare*, vol. 2(2), pp. 107–113, 1996.

[165] J. Yick, B. Mukherjee and D. Ghosal, "Wireless sensor network survey," *Computer Networks*, vol. 52, pp. 2292–2330, 2008.

[166] G. Simon, M. Maroti, A. Ledeczi, G. Balogh, B. Kusy, A. Nadas, G. Pap, J. Sallai and K. Frampton, "Sensor network-based countersniper system," *Proc. the Second International Conference on Embedded Networked Sensor Systems (Sensys)*, Baltimore, MD, 2004.

[167] J. Yick, B. Mukherjee, D. Ghosal, "Analysis of a prediction-based mobility adaptive tracking algorithm," *Proc. the IEEE Second Int. Conference on Broadband Networks (BROADNETS)*, Boston, 2005.

[168] M. Castillo-Effen, D. H. Quintela, R. Jordan, W. Westhoff and W. Moreno, "Wireless sensor networks for flash-flood alerting," *Proc. the Fifth IEEE Int. Caracas Conference on Devices, Circuits, and Systems*, Dominican Republic, 2004.

[169] T. Gao, D. Greenspan, M. Welsh, R. R. Juang and A. Alm, "Vital signs monitoring and patient tracking over a wireless network," *Proc. the 27th IEEE EMBS Annual International Conference*, 2005.

[170] K. Lorincz, D. Malan, T. R. F. Fulford-Jones, A. Nawoj, A. Clavel, V. Shnayder, G. Mainland, M. Welsh and S. Moulton, "Sensor networks for emergency response: challenges and opportunities," *IEEE Pervasive Computing: Pervasive Computing for First Response (Special Issue)*, October–December 2004.

[171] G. Wener-Allen, K. Lorincz, M. Ruiz, O. Marcillo, J. Johnson, J. Lees, M. Walsh, "Deploying a wireless sensor network on an active volcano," *IEEE Internet Computing: Data-Driven Applications in Sensor Networks (Special Issue)*, March/April 2006.

[172] V. Raghunathan, A. Kansai, J. Hse, J. Friedman and M. Srivastava, "Design considerations for solar energy harvesting wireless embedded systems," *Proc. the IPSN*, pp. 457–462, 2005.

[173] P. Zhang, C. M. Sadler, S. A. Lyon and M. Martonosi, "Hardware design experiences in ZebraNet," *Proc. the SenSys04*, Baltimore, MD, 2004.

[174] S. Roundy, J. M. Rabaey and P. K. Wright, *Energy Scavenging for Wireless Sensor Networks*, Springer-Verlag, New York, 2004.

[175] M. Rahimi, H. Shah, G. S. Sukhatme, J. Heideman and D. Estrin, "Studying the feasibility of energy harvesting in mobile sensor network," *Proc. the IEEE ICRA*, pp. 19–24, 2003.

[176] A. Kansai and M. B. Srivastava, "An environmental energy harvesting framework for sensor networks," *Proc. the Int. Symposiumon Low Power Electronics and Design*, pp. 481–486, 2003.

[177] S. Toumpis and T. Tassiulas, "Optimal deployment of large wireless sensor networks," *IEEE Trans. Information Theory*, vol. 52, pp. 2935–2953, 2006.

[178] J. Yick, G. Pasternack, B. Mukherjee and D. Ghosal, "Placement of network services in sensor networks," *Int. J. Wireless and Mobile Computing (IJWMC), Self-Organization Routing and Information, Integration in Wireless Sensor Networks (Special Issue)*, vol. 1, pp. 101–112, 2006.

[179] D. Pompili, T. Melodia and I.F. Akyildiz, "Deployment analysis in underwater acoustic wireless sensor networks," *WUWNet*, Los Angeles, CA, 2006.

[180] I. F. Akyildiz and E. P. Stuntebeck, "Wireless underground sensor networks: research challenges," *Ad-Hoc Networks*, vol. 4, pp. 669–686, 2006.

[181] M. Li, Y. Liu, "Underground structure monitoring with wireless sensor networks," *Proc. the IPSN*, Cambridge, MA, 2007.

[182] I. F. Akyildiz, D. Pompili and T. Melodia, "Challenges for efficient communication in underwater acoustic sensor networks," *ACM Sigbed Review*, vol. 1(2), pp. 3–8, 2004.

[183] J. Heidemann, Y. Li, A. Syed, J. Wills and W. Ye, "Underwater sensor networking: research challenges and potential applications," *Proc. the Technical Report ISI-TR-2005-603, USC/Information Sciences Institute*, 2005.

[184] I. F. Akyildiz, T. Melodia, K.R. Chowdhury, "A survey on wireless multimedia sensor networks," *Computer Networks*, vol. 51, pp. 921–960, 2007.

[185] J. M. Kahn, R. H. Katz, K. S. J. Pister, "Next century challenges: mobile networking for smart dust," *Proc. of the ACM MobiCom99*, Washington, USA, pp. 271–278, 1999.

[186] G. J. Pottie, W. J. Kaiser, "Wireless integrated network sensors," *Communications of the ACM*, vol. 43 (5), pp. 551–558, 2000.

[187] W. R. Heinzelman et al., "Energy-scalable algorithms and protocols for wireless sensor networks," *Proc. the International Conference on Acoustics, Speech, and Signal Processing (ICASSP'00)*, Istanbul, Turkey, June 2000.

[188] R. Min et al., "An architecture for a power aware distributed microsensor node," *Proc. the IEEE Workshop on signal processing systems (SIPS'00)*, October 2000.

[189] A. Woo, D. Culler, "A transmission control scheme for media access in sensor networks," *Proc. the 7th Annual ACM/IEEE International Conference on MobilenComputing and Networking (Mobicom'01)*, Rome, Italy, July 2001.

[190] W. Ye, J. Heidemann, D. Estrin, "An energy-efficient MAC protocol for wireless sensor networks," *Proc. IEEE Info-com 2002*, New York, June 2002.

[191] E. Shih et al., "Physical layer driven protocol and algorithm design for energy-efficient wireless sensor networks," *Proc. the 7th Annual ACM/IEEE Int. Conference on Mobile Computing and Networking (Mobicom'01)*, Rome, Italy, July 2001.

[192] K. Akkaya and M. Younis, "A survey on routing protocols for wireless sensor networks," *Ad Hoc Networks*, vol. 3, pp. 325–349, 2005.

[193] L. Subramanian, R. H. Katz, "An architecture for building self configurable systems," *Proc. IEEE/ACM Workshop on Mobile Ad Hoc Networking and Computing*, Boston, MA, August 2000.

[194] F. Ye et al., "A two-tier data dissemination model for large–scale wireless sensor networks," *Proc. Mobicom'02*, Atlanta, GA, September, 2002.

[195] K. Sohrabi et al., "Protocols for self-organization of a wireless sensor network," *IEEE Personal Communications*, vol. 7 (5), pp. 16–27, 2000.

[196] W. Heinzelman, A. Chandrakasan, H. Balakrishnan, "Energy-efficient communication protocol for wireless sensor networks," *Proc. the Hawaii Int. Conference System Sciences*, Hawaii, January 2000.

[197] M. Younis, M. Youssef, K. Arisha, "Energy-aware routing in cluster-based sensor networks," *Proc. the 10th IEEE/ACM International Symposium on Modeling, Analysis and Simulation of Computer and Telecommunication Systems (MASCOTS2002)*, Fort Worth, TX, October 2002.

[198] A. Manjeshwar, D. P. Agrawal, "TEEN: a protocol for enhanced efficiency in wireless sensor networks," *Proc. the 1st International Workshop on Parallel and Distributed Computing Issues in Wireless Networks and Mobile Computing*, San Francisco, CA, April 2001.

[199] R. H. Katz, J. M. Kahn, K. S. J. Pister, "Mobile networking for smart dust," *Proc. the 5th Annual ACM/IEEE International Conference on Mobile Computing and Networking (MobiCom'99)*, Seattle, WA, August 1999.

[200] C. Intanagonwiwat, R. Govindan, D. Estrin, "Directed diffusion: a scalable and robust communication paradigm for sensor networks," *Proc. the 6th Annual ACM/IEEE Int. Conference on Mobile Computing and Networking (MobiCom'00)*, Boston, MA, August 2000.

[201] D. Estrin et al., "Next century challenges: scalable coordination in sensor networks," *Proc. the 5th annual ACM/IEEE International Conference on Mobile Computing and Networking (MobiCom'99)*, Seattle, WA, August 1999.

[202] S. Lindsey, C. S. Raghavendra, "PEGASIS: power efficient gathering in sensor information systems," *Proc. the IEEE Aerospace Conference*, Big Sky, Montana, March 2002.

[203] S. Lindsey, C. S. Raghavendra, K. Sivalingam, "Data gathering in sensor networks using the energy delay metric," *Proc. the IPDPS Workshop on Issues in Wireless Networks and Mobile Computing*, San Francisco, CA, April 2001.

[204] K. Akkaya, M. Younis, "An energy-aware QoS routing protocol for wireless sensor networks," *Proc. the IEEE Workshop on Mobile and Wireless Networks (MWN 2003)*, Providence, RI, May 2003.

[205] B. Krishnamachari, D. Estrin, S. Wicker, "Modeling data centric routing in wireless sensor networks," *Proc. the IEEE INFOCOM*, New York, June 2002.

[206] Y. Yao, J. Gehrke, "The cougar approach to in-network query processing in sensor networks," *SIGMOD Record*, September 2002.

[207] W. Heinzelman, J. Kulik, H. Balakrishnan, "Adaptive protocols for information dissemination in wireless sensor networks," *Proc. the 5th Annual ACM/IEEE Int. Conference on Mobile Computing and Networking (Mobi-Com'99)*, Seattle, WA, August 1999.

[208] D. Braginsky, D. Estrin, "Rumor routing algorithm for sensor networks," *Pro. the First Workshop on Sensor Networks and Applications (WSNA)*, Atlanta, GA, October 2002.

[209] C. Schurgers, M. B. Srivastava, "Energy efficient routing in wireless sensor networks," *The MILCOM Proc. Communications for Network-Centric Operations: Creating the Information Force*, McLean, VA, 2001.

[210] M. Chu, H. Haussecker, F. Zhao, "Scalable information driven sensor querying and routing for ad hoc heterogeneous sensor networks," *The Int. J. High Performance Computing Applications*, vol. 16(3), pp. 293–313, 2002.

[211] R. Shah, J. Rabaey, "Energy aware routing for low energy ad hoc sensor networks," *Proc. the IEEE Wireless Communications and Networking Conference (WCNC)*, Orlando, FL, March 2002.

[212] N. Sadagopan et al., "The ACQUIRE mechanism for efficient querying in sensor networks," *Proc. the First International Workshop on Sensor Network Protocol and Applications*, Anchorage, AK, May 2003.

[213] S. Hedetniemi, A. Liestman, "A survey of gossiping and broadcasting in communication networks," *Networks*, vol. 18(4), pp. 319–349, 1988.

[214] J. P. Hespanha, P. Naghshtabrizi and X. Yonggang, "A survey of recent results in networked control systems," *Proc. the IEEE*, vol. 95, pp. 138–162, 2007.

[215] A. Ulusoy, O. Gurbuz, and A. Onat, "Wireless model-based predictive networked control system over cooperative wireless network," *IEEE Trans. Industrial Informatics*, vol. 7, no. 1, pp. 41–51, February 2011.

[216] M. Bjorkbom, S. Nethi, L. M. Eriksson and R. Jantti, "Wireless control system design and co-simulation," *Control Engineering Practice*, vol. 19, pp. 1075–1086, September 2011.

[217] E. O. Elliot, "Estimates of error rates for codes on burst-noise channels," *Bell Syst. Tech. J.*, vol. 42, pp. 1977–1997, September 1963.

[218] K. Pahlavan and P. Krishnamurthy, *Principles of Wireless Networks: A Unified Approach*, Prentice-Hall, NJ, 2002.

[219] D. Yahong, R. Zheng and C. Xiao, "Improved models for the generation of multiple uncorrelated Rayleigh fading waveforms," *IEEE Communications Letters*, vol. 6, no. 6, Jun. 2002.

[220] G. P. Liu, Y. Xia, J. Chen, D. Rees, and W. Hu, "Networked predictive control of systems with random network delays in both forward and feedback channels," *IEEE Trans. Industrial Electronics*, vol. 54, no. 3, pp. 1282–1296, Jun. 2007.

[221] A. Chamaken, and L. Litz, "Joint design of control and communication in wireless networked control systems: A case study," *Proc. American Control Conference*, pp. 1835–1840, July 2010.

[222] K. Akkaya and M. Younis, "A survey on routing protocols for wireless sensor networks," *Ad Hoc Networks*, vol. 3, pp. 325–349, 2005.

[223] Y. Xu, J. Heidemann, D. Estrin, "Geography-informed energy conservation for ad hoc routing," *Proc. the 7th Annual ACM/IEEE Int. Conference on Mobile Computing and Networking (MobiCom'01)*, Rome, Italy, July 2001.

[224] V. Rodoplu, T. H. Ming, "Minimum energy mobile wireless networks," *IEEE Journal of Selected Areas in Communications*, vol. 17 (8), pp. 1333–1344, 1999.

[225] A. Manjeshwar, D.P. Agrawal, "APTEEN: a hybrid protocol for efficient routing and comprehensive information retrieval in wireless sensor networks," *Proc. the 2nd Int. Workshop on Parallel and Distributed Computing Issues in Wireless Networks and Mobile computing*, Ft. Lauderdale, FL, April 2002.

[226] , L. Moreau, "Stability of multiagent systems with time-dependent communication links," *IEEE Trans. Automatic Control*, vol. 50 (2), pp. 169–182, 2005.

[227] J. -Y. Chen, G. Pandurangan, and D. Xu, "Robust computation of aggregates in wireless sensor networks: distributed randomized algorithms and analysis," *Proc. Fourth Int. Symposium on Information Processing in Sensor Networks*, pp. 348–355, April 2005.

[228] D. D. Luo and Y. M. Zhu, "Applications of random parameter matrices Kalman filtering in Uncertain observation and multi-model systems," *Proc. the Int. Federation of the Automatic Control*, Seoul, Korea, July 6–11, 2008.

[229] Y. M. Zhu, Z. S. You, J. Zhao, K. -S. Zhang and R. X. Li, "The optimality for the distributed Kalman filtering fusion with feedback," *Automatica*, vol. 37, pp. 1489–1493, 2001.

[230] T. Jiang, I. Matei and J. S. Baras, "A trust-based distributed Kalman filtering approach for mode estimation in power systems," *Proc. the First Workshop on Secure Control Systems (SCS)*, pp. 1–6, Stockholm, Sweden, April 12, 2010.

[231] O. R. Saber, "Distributed Kalman filtering for sensor networks," *Proc. the 46th IEEE Conference on Decision and Control*, pp. 5492–5498, 2007.

[232] E. B. Song, Y. M. Zhu, J. Zhou and Z. S. You, "Optimality Kalman filtering fusion with cross-correlated sensor noises," *Automatica*, vol. 43, pp. 1450–1456, 2007.

[233] B. S. Rao and D- H. F. Whyte, "Fully decentralized algorithm for multi–sensor Kalman filtering," *IEE Proc. Control Theory and Applications, Part D*, vol. 138, pp. 413–420, September 1991.

[234] P. Jiang, J. Zhou and Y. Zhu, "Globally optimal Kalman filtering with finite-time correlated noises," *49th IEEE Conference on Decision and Control*, Hilton Atlanta Hotel, Atlanta, GA, USA, pp. 5007–5012, December 15–17, 2010.

[235] F. Govaers and W. Koch, "Distributed Kalman filter fusion at arbitrary in-stants of time," *2010 13th Conference on Information Fusion (FUSION)*, Sensor Data and Inf. Fusion, Univ. of Bonn, Bonn, Germany, pp. 1–8, 2010.

[236] Z. L. Deng, Y. Gao, L. Mao, Y. Li and G. Hao, "New approach to infor-mation fusion steady-state Kalman filtering," *Automatica*, vol. 41(10), pp. 1695–1707, 2005.

[237] L. Yingting and Z. Yunmin, "Distributed Kalman filtering fusion with packet loss or intermittent communications from local estimators to fusion Center," *Proc. the 29th Chinese Control Conference*, July 29-31, Beijing, China, pp. 4768–4775, 2010.

[238] O. R. Saber, "Distributed Kalman filter with consensus filters," *Proc. the 44th Embedded IEEE Conference on Decision and Control*, Sevilla, Spain, pp. 8179–8184, December 2005.

[239] S. Kirti and A. Scaglione, "Scalable distributed Kalman filtering through consensus," *Proc. the IEEE International Conference on Acoustics, Speech and Signal Processing, ICASSP*, April 2008.

[240] L. Schenato, R. Carli, A. Chiuso and S. Zampieri, "Distributed Kalman fil-tering based on consensus strategies," *IEEE J. Selected Areas in Communi-cations*, vol. 26(4), pp. 622–633, May 2008.

[241] S. XiaoJing, L. YingTing, Z. YunMin and S. EnBin, "Globally optimal dis-tributed Kalman filtering fusion," *Science China Information Sciences*, vol. 55(3): 512–529, DOI: 10.1007/s11432-011-4538-7, March 2012.

[242] R. B. Quirino, C. P. Bottura and F. J. T. Costa, "A computational structure for parallel and distributed Kalman filtering," *Proc. 12th Brazilian Automatic Control Conference*, Uberlandia-MG, Brazil, 14–18 September 1998.

[243] N. D. Assimakis, G. Tziallas and A. Koutsonikolas, "Optimal distributed Kalman filter," *Technical Report*, pp. 1–19, 2004.

[244] A. K. Usman and M. F. M. Jose, "Distributing the Kalman filter for large-scale systems," *IEEE Trans. Signal Processing*, vol. 56(10), pp. 4919–4935, October 2008.

[245] V. Saligrama and D. Castanon, "Reliable distributed estimation with in-termittent communications," *Proc. 45th IEEE Conf. Decision Control*, San Diego, CA, pp. 6763–6768, December 2006.

[246] O. R. Saber, "Distributed Kalman filters with embedded consensus filters," *Proc. 44th IEEE Conf. Decision Control*, Seville, Spain, pp. 8179–8184, December 2005.

[247] O. R. Saber and J. Shamma, "Consensus filters for sensor networks and distributed sensor fusion," *Proc. 44th IEEE Conf. Decision Control*, Seville, Spain, pp. 6698–6703, December 2005.

[248] U. A. Kahn and J. M. F. Moura, "Distributed Kalman filters in sensor networks: bipartite fusion graphs," *Proc. 15th IEEE Workshop Statistical Signal Processing*, Madison, WI, pp. 700–704, August 26–29, 2007.

[249] L. Xiao and S. Boyd, "Fast linear iterations for distributed averaging," *Syst. Controls Lett.*, vol. 53(1), pp. 65–78, April 2004.

[250] R. Carli, A. Chiuso, L. Schenato and S. Zampieri, "Distributed Kalman filtering based on consensus strategies," *Tech. Report – Information Engineering Dept., Univ. of Padova*, Padova, Italy, 2007.

[251] L. Xiao and S. Boyd, "Designing fast distributed iterations via semi-definite programming," *Proc. the Workshop on Large Scale Nonlinear and Semidefinite Programming*, Waterloo, Canada, May 2004.

[252] S. Kar and J. M. F. Moura, "Sensor networks with random links: Topology design for distributed consensus," *IEEE Trans. Signal Process.*, vol. 56(7), pp. 3315–3326, July 2008.

[253] W. Koch, "Exact update formulae for distributed Kalman filtering and retrodiction at arbitrary communication rates," *Proc. of the 12th ISIF Int. Conference on Information Fusion*, Seattle, USA, July 6–9, 2009.

[254] L. W. Fong, "Multi–sensor track-to-track fusion Using simplified maximum likelihood estimator for maneuvering target tracking," *Proc. IEEE*, pp. 36–41, 2007.

[255] K. C. Chang, R. K. Saha and Y. B. Shalom, "On optimal track-to-track fusion," *IEEE Trans. Aerospace and Electronic Systems*, vol. 33(4), pp. 1271–1276, May 1997.

[256] L. W. Fong, "Integrated track-to-track fusion with modified probabilistic neural network," *Proc. the CIE Int. Conference on Radar*, Shanghai, China, pp. 1869–1873, October 2006.

[257] R. B. Quirino and C. P. Bottura, "An approach for distributed Kalman filtering," *SBA Controle and Automacoa*, vol. 12(1), pp. 19–28, 2001.

[258] H. R. Hashmipour, S. Roy and A. J. Laub, "Decentralized structures for parallel Kalman filtering," *IEEE Trans. Automat. Control*, vol. 33(1), pp. 88–93, 1998.

[259] Y. B. Shalom, "On the track-to-track correlation problem," *IEEE Trans. Automatic Control*, vol. 26(2), pp. 571–572, 1981.

[260] N. A. Carlson, "Federated square root filter for decentralized parallel processes," *IEEE Trans. Aerospace Electron. Systems*, pp. 26(3), pp. 517–525, 1990.

[261] S. L. Sun, "multi–sensor optimal information fusion Kalman filter with application," *Aerospace Sci. Technol.*, vol. 8(1), pp. 57–62, 2004.

[262] S. L. Sun and Z. L. Deng, "multi–sensor optimal information fusion Kalman filter," *Automatica*, vol. 40(6), pp. 1017–1023, 2004.

[263] S. L. Sun, "multi–sensor information fusion white noise filter weighted by scalars based on Kalman predictor," *Automatica*, vol. 40(8), pp. 1447–1453, 2004.

[264] S. L. Sun, "multi–sensor optimal information fusion input white noise deconvolution estimators," *IEEE Trans. Systems Man, Cybernet.*, vol. 34(4), pp. 1886–1893, 2004.

[265] P. Barooah1, W. J. Russell and J. P. Hespanha, "Approximate distributed Kalman filtering for cooperative multi-agent localization," *Proc. the 6th IEEE international conference on Distributed Computing in Sensor Systems, DCOSS'10*, Santa Barbara, CA, pp. 102–115, 2010.

[266] P. Alriksson and A. Rantzer, "Distributed Kalman filtering using weighted averaging," *Proc. the 17th Int. Symposium on Mathematical Theory of Networks and Systems (MTNS)*, 2006.

[267] P. Alriksson and A. Rantzer, "Experimental evaluation of a distributed Kalman filter algorithm," *Proc. the 46th IEEE Conference on Decision and Control*, pp. 5499–5504, December 2007.

[268] P. Ogren, E. Fiorelli and N. E. Leonard, "Cooperative control of mobile sensor networks: adaptive gradient climbing in a distributed GMM approximation for multiple targets localization and tracking in wireless sensor networks," *Proc. the Fourth Int. Symposium on Information Processing in Sensor Networks*, April 2005.

[269] H. Medeiros, J., Park and A. C. Kak, "Distributed object tracking Using a cluster-based Kalman filter in wireless camera networks," *IEEE J. Selected Topics in Signal Processing*, vol. 2(4), pp. 448–463, August 2008.

[270] I. D. Schizas, G. B. Giannakis, S. I., Roumeliotis and A. Ribeiro, "Anytime optimal distributed Kalman filtering and smoothing," *SSP'07*, pp. 368–372, 2007.

[271] E. Franco, O. R. Saber, T. Parisini and M. M. Polycarpou, "Distributed fault diagnosis using sensor networks and consensus-based filters," *Proc. the 45th IEEE Conference on Decision and Control*, Manchester Grand Hyatt Hotel, San Diego, CA, USA, pp. 386–391, December 13–15, 2006.

[272] J. Ma amd S. Sun, "Self-Tuning information fusion reduced-Order Kalman predictors for Stochastic Singular Systems," *Proc. the 6th World Congress on Intelligent Control and Automation*, Dalian, China, pp. 1524–1528, June 21–23, 2006.

[273] E. Song, Y. Zhu, J. Zhou and C. Sichuan, "The optimality of Kalman filtering fusion with cross-correlated sensor Noises," *Proc. the 43rd IEEE Conference on Decision and Control*, Atlantis, Paradise Island, Bahamas, pp. 4637–4642, December 14–17, 2004.

[274] Y. M. Zhu, Z. S. You, J. Zhao, K. S. Zhang and R. X. Li, "The optimality of the distributed Kalman filter with feedback," *Automatica*, vol. 37, pp. 1489–1493, 2001.

[275] A. Abdelgawad and M. Bayoumi, "Low-power distributed Kalman filter for wireless sensor networks," *EURASIP Journal on Embedded Systems*, vol. 2011, Article ID 693150, doi:10.1155/2011/693150, pp. 1–11.

[276] S. Kirti and A. Scaglione, "Scalable distributed Kalman filtering through consensus," *Proc. the IEEE Inte. Conference on Acoustics, Speech and Signal Processing (ICASSP'08)*, March-April 2008, pp. 2725–2728.

[277] U. A. Khan and J. M. F. Moura, "Model distribution for distributed Kalman filters: a graph theoretic approach," *Proceedings of the 41st Asilomar Con-*

ference on Signals, Systems and Computers (ACSSC'07), pp. 611–615, November 2007.

[278] S. M. Azizi and K. Khorasani, "A distributed Kalman filter for actuator fault estimation of deep space formation flying satellites," *Proc. the 3rd IEEE International Systems Conference*, pp. 354–359, March 2009.

[279] M. Bai, R. A. Freeman and K. M. Lync, "Distributed Kalman filtering using the internal model average consensus estimator', *Proc. the American Control Conference*, O'Farrell Street, San Francisco, CA, USA, pp. 1500–1505, June 29–July 01, 2011.

[280] J. Cortes, "Distributed kriged Kalman filter for spatial estimation," *IEEE Trans. Automatic Control*, vol. 54(12), pp. 281–2827, December 2009.

[281] Y. Zhu, Z. You, J. Zhao, K. Zhang, K., and R. X. Li, "The optimality for the distributed Kalman filtering fusion," *Automatica*, vol. 37(9), pp. 1489–1493, 2001.

[282] Y. Gao, G. L. Tao and Z. L. Deng, "Decoupled distributed Kalman fuser for descriptor systems," *Signal Processing*, vol. 88, pp. 1261–1270, 2008.

[283] Z. Lendeka, B. Babuska and B. Schuttera, "Distributed Kalman filtering for cascaded systems," *Engineering Applications of Artificial Intelligence*, vol. 21, pp. 457–469, 2008.

[284] Z. L. Deng, J. W. Ma and M. Gao, "Two-sensor self-tuning information fusion Kalman filter," *Science Technology and Engineering*, vol. 3, pp. 321–324, 2003.

[285] S. L. Sun, "multi–sensor information fusion white noise filter weighted by scalars based on Kalman predictor," *Automatica*, vol. 40(8), pp. 1447–1453, 2004.

[286] M. J. Coates and B. N. Oreshkin, "Asynchronous distributed particle filter via decentralized evaluation of gaussian products," *Proc. ISIF Int. Conf. Inf. Fusion*, 2010.

[287] D. Wang, K. Niu, Z. He and B. Tian, "A novel OFDM channel estimation method based On Kalman filtering and distributed compressed sensing," *Proc. the IEEE 21st Int. Symposium on Personal Indoor and Mobile Radio Communications*, pp. 1080–1090, 2010.

[288] J. Cortes, "Distributed kriged Kalman filter for spatial estimation," *IEEE Trans. Automatic Control*, vol. 54(12), pp. 2816–2827, December 2009.

[289] D. Bickson, O. Shental and D. Dolev, "Distributed Kalman filter via gaussian belief propagation," *Proc. the 46th Annual Allerton Conference*, Allerton House, UIUC, Illinois, USA, pp. 628–635, September 23–26, 2008.

[290] G. F. Schwarzenberg, U. Mayer, N. V. Ruiter and U. D. Hanebeck, "3D reflectivity reconstruction by Means of spatially distributed Kalman Filters', *Proc. IEEE Int. Conference on multi–sensor Fusion and Integration for Intelligent Systems*, Seoul, Korea, pp. 384–391, August 20–22, 2008.

[291] E. J. Msechu, A. Ribeiro, S. I. Roumeliotis and G. B. Giannakis, "Distributed Kalman filtering based on quantized innovations," *ICASSP'08*, pp. 3293–3296, 2008.

[292] E. J. Msechu, S. I. Roumeliotis, A. Ribeiro and G. B. Giannakis, "Distributed quantized Kalman filtering with scalable communication cost," *IEEE Trans. on Sig. Proc.*, August 2007, available at http://spincom.ece.umn.edu/journal.html.

[293] S. Kirti and A. Scaglione, "Scalable distributed Kalman filtering through consensus," *ICASSP'8*, pp. 2725–2728, 2008.

[294] O. R. Saber, "Distributed Kalman filter with embedded consensus filters," *Proc. 44th IEEE Conference on Decision and Control*, pp. 8179–8184, Dec. 2005,

[295] A. Ribeiro, G. B. Giannakis and S. I. Roumeliotis, "SOI-KF: distributed Kalman filtering with low-cost communications using the sign of innovations," *Proc. Int. Conf. Acoustics, Speech, Signal Processing*, Toulouse, France, pp. IV-153–IV-156, May 14–19, 2006.

[296] W. Yang, X. Wendong and L. H. Xie, "Diffusion-based EM algorithm for distributed estimation of Gaussian mixtures in wireless sensor networks," *Sensors*, vol. 11, pp. 6297–6316, 2011.

[297] F. Cattivelli and A. H. Sayed, "Diffusion strategies for distributed Kalman filtering and smoothing," *IEEE Trans. Autom. Control*, vol. 55, pp. 2069–2084, 2010.

[298] F. Cattivelli, C. G. Lopes and A. H. Sayed, "Diffusion strategies for distributed Kalman filtering: formulation and performance analysis," *Proc. Cognitive Information Processing*, Santorini, Greece, June 2008.

[299] I. Schizas, S. I. Roumeliotis, G. B. Giannakis and A. Ribeiro, "Anytime optimal distributed Kalman filtering and smoothing," *Proc. IEEE Workshop on Statistical Signal Process.*, Madison, WI, pp. 368–372, August 2007.

[300] U. A. Khan and J. M. F. Moura, "Distributing the Kalman filter for large scale systems," *IEEE Trans. on Signal Processing*, vol. 56(10), Part 1, pp. 4919–4935, October 2008.

[301] D. Gu, "Distributed EM algorithm for gaussian mixtures in sensor networks," *IEEE Trans. Neural Networks*, vol. 19(7), pp. 1154–1166, July 2008.

[302] R. O. Saber, "Distributed Kalman filter with embedded consensus filters," *Proc. 44th IEEE Conf. Decision Control*, Dec. 12–15, pp. 8179–8184, 2005.

[303] A. Ribeiro, G. B. Giannakis and S. I. Roumeliotis, "SOI-KF: distributed Kalman filtering with low-cost communications Using the sign of innovations," *IEEE Trans. Signal Processing*, vol. 54(12), pp. 4782–4795, December 2006.

[304] S. S. Li, "Distributed optimal component fusion deconvolution filtering, *Signal Processing*, vol. 87, pp. 202–209, 2007.

[305] Z. L. Deng and R. B. Qi, "Multi–sensor information fusion suboptimal steady-state Kalman filter," *Chinese Science Abstracts*, vol. 6(2), pp. 183–184, 2000.

[306] G. Qiang and C. J. Harris, "Comparison of two measurement fusion methods for Kalman filter-based multi–sensor data fusion," *IEEE Trans. Aerospace and Electronic Systems*, vol. 37(1), pp. 273–280, 2001.

[307] D. Willner, C. B. Chang and K. P. Dunn, "Kalman filter algorithm for a multi–sensor system," *Proc IEEE Conf. Decision and Control*, 1976.

[308] R. X. Li, "Optimal linear estimation fusion– part VII: dynamic systems," *Int. Conf. Information Fusion*, Cairns, Australia,, pp. 455–462, July 2003.

[309] K. -S. Zhang, R. X. Li, P. Zhang and H. -F. Li, "Optimal linear estimation fusion–part VI: sensor data compression," *Proc. the Int. Conf. Information Fusion*, Cairns, Australia, July 2003.

[310] R. X. Li, K. -S. Zhang, J. Zhao and Y. M. Zhu, "Optimal linear estimation fusion–part V: relationships," *Proc. the Int. Conf. on Information Fusion*, pp. 497–504, Annapolis, MD, USA, July 2002.

[311] R. X. Li, and K. -S. Zhang, "Optimal linear estimation fusion–Part IV: optimality and efficiency of distributed fusion," *Proc. the Int. Conf. Information Fusion*, Montreal, QC, Canada, pp. WeB1.19–WeB1.26, Aug. 2001.

[312] R. X. Li, and p. Zhang, "Optimal linear estimation fusion–Part III: cross-correlation of local estimation errors," *Proc. the Int. Conf. Information Fusion*, Montreal, QC, Canada, pp. WeB1.11–WeB1.18, Aug. 2001.

[313] R. X. Li and J, Wang, "Unified optimal linear estimation fusion–Part II: discussions and examples," *Proc. the Int. Conf. Information Fusion*, Paris, France, pp. MoC2.18–MoC2.25, July 2000.

[314] R. X. Li, Y. M. Zhu and C. Z. Han, "Unified optimal linear estimation fusion–part I: unified models and fusion rules," *Proc. the Int. Conf. on Information Fusion*, pp. MoC2.–MoC2.17, Paris, France, July 2000.

[315] R. X. Li, Y. M. Zhu, J. Wang, and C. Z. Han, "Optimal linear estimation fusion–part I: unified fusion rules," *IEEE. Trans. Information Theory*, vol. 49(9), pp. 2192–2208, 2003.

[316] C. Z. Han, J. Wang and R. X. Li, "Fused state estimation of linear continuous-time system with multi–sensor asynchronous measurements," *Proc. 2001 Int. Conf. on Information Fusion*, pp. FrB1.3–FrB1.9, Montreal, QC, Canada, August 2001.

[317] D. Aleksandar and Q. Kun, "Decentralized random-field estimation for sensor networks using quantized spatially correlated data and fusion-center feedback," *IEEE Trans. Signal Processing*, vol. 56(12), pp. 6069–6085, December 2008.

[318] A. Dogandzic and K. Qiu, "Estimating a random field in sensor networks using quantized spatially correlated data and fusion-center feedback," *Proc. 42nd Asilomar Conf. Signals, Syst. Comput.*, Pacific Grove, CA, pp. 1943–1947, October 2008.

[319] Z. Q. Hong, Y. Z. Hong and J. Hong, "Fusion algorithm of correlated local estimates," *Aerospace Science and Technology*, vol. 8, pp. 619–626, 2004.

[320] S. L. Sun, "Multi–sensor optimal information fusion Kalman filters with applications," *Aerospace Science and Technology*, vol. 8, pp. 57–62, 2004.

[321] Z. Yunmin, L. Xianrong, R. X. Li, and Juan, Z., "Linear minimum variance estimation fusion," *Science in China Ser. F Information Sciences*, vol. 47(6), pp. 728–740, 2004.

[322] J. H. Yoon and V. Shin, "Comparison analysis of distributed receding horizon filters for linear discrete-time systems with uncertainties', *International Journal of Systems Control*, vol. 1(2), pp. 48–56, 2010.

[323] S. L. Sun and Z. L. Deng, "multi–sensor information fusion Kalman filter weighted by scalars for systems with colored measurement noises," *Journ. Dynam. Syst, Measurem., Contr.*, vol. 127(12), pp. 663–667, 2005.

[324] Q. Xiangdong and C. Kuochu, "Information matrix fusion with feedback versus number of sensors," *Proc. the 29th Chinese Control Conference*, Beijing, China, pp. 1836–1843, 2010.

[325] R. X. Li, "Optimal linear estimation fusion for multi–sensor dynamic Systems," *Proc. Workshop on Multiple Hypothesis Tracking - A Tribute to Sam Blackman*, San Diego, CA, USA, May 2003.

[326] K. -S. Zhang, R. X. Li and Y. -M. Zhu, "Optimal update with out-of-sequence observations for distributed filtering," *IEEE Trans. Signal Processing*, vol. 53(6), pp. 1992–2004, June 2005.

[327] Y. Zhu, G. Yu and R. X. Li, "multi–sensor statistical interval estimation fusion," *Proc. 2002 SPIE Conf. Sensor Fusion: Architectures, Algorithms, and Applications VI*, vol. 47, pp. 31–36, Orlando, FL, April 2002.

[328] K. Chang, T. Zhi and R. Saha, "Performance evaluation of track fusion with information matrix filter," *IEEE Trans. Aerospace and Electronic Systems*, vol. 38(2), pp. 455–466, 2002.

[329] Z. Duan and R. X. Li, "Optimal distributed estimation fusion with transformed Data," *Proc. 11th International Conference on Information Fusion*, pp. 1291–1297, June 30–July 3, 2008.

[330] K. C. Chang, Z. Tian and S. Mori, "Performance evaluation for MAP state estimate fusion," *IEEE Trans. Aerospace and Electronic Systems*, vol. 40(2), pp. 706–714, April 2004.

[331] Z. Duan and R. X. Li, "The optimality of a class of distributed estimation fusion algorithm," *Proc. the 11th Int. Conference on Information Fusion*, pp. 1285–1290, June 30-July 3, 2008.

[332] B. Noack, D. Lyons, M. Nagel and U. D. Hanebeck, "Nonlinear information filtering for distributed multi–sensor data fusion," *Proc. American Control Conference*, San Francisco, CA, USA, pp. 4846–4852, June 29 - July 01, 2011.

[333] Teng, Z., Jing, M., and Shuli, S., "Distributed fusion filter for discrete-time stochastic systems with Uncertain observation and correlated noises," *Proc. 8th IEEE International Conference on Control and Automation*, Xiamen, China, pp. 704–708, June 9-11, 2010.

[334] S. L. Sun, "multi–sensor optimal information fusion Kalman filter with application," *Aerospace Science and Technology*, vol. 8(1), pp. 57–62, January 2004.

[335] S. L. Sun and Z. L. Deng, "multi–sensor optimal information fusion Kalman filter," *Automatica*, vol. 40(6), pp. 1017–1023, June 2004.

[336] S. L. Sun, "multi–sensor optimal information fusion input white noise deconvolution estimators," *IEEE Transactions on Systems, Man, And Cybernetics*, vol. 34(4), pp. 1886–1893, August 2004.

[337] Y. Gao, C. J. Ran and Z. L. Deng, "Weighted measurement fusion Kalman filter with correlated measurement noise and its global optimality," *Proc. the Int. Colloquium information Fusion 2007*, Xi'an, China, pp. 195–200, August 22–25, 2007.

[338] B. Zhu and S. Sastry, "Data fusion assurance for the Kalman filter in Uncertain networks," *Proc. the Fourth Int. Conference on Information Assurance and Security*, pp. 115–119, 2004.

[339] S. L. Sun, "multi–sensor information fusion white noise filter weighted by scalars based on Kalman predictor," *Automatica*, vol. 40(8), pp. 1447–1453, 2004.

[340] S. L. Sun, "multi–sensor optimal information fusion input white noise deconvolution estimators," *IEEE Trans. Systems, Man, and Cybernetics*, vol. 34(4), pp. 1886–1893, 2004.

[341] S. Sun and C. Zhang, "Optimal information fusion distributed smoother for discrete multichannel ARMA signals," *IEE Proc. Vis. Image Signal Process.*, vol. 152(5), pp. 583–589, October 2005.

[342] Y. Zhu, E. Song, J. Zhou and Z. You, "Optimal dimensionality reduction of sensor data in multi–sensor estimation fusion," *IEEE Trans. Signal Processing*, vol. 53(5), pp. 1631–1639, May 2005.

[343] Z. Deng and R. Qi, "Multi–sensor information fusion sub-optimal steady-state Kalman filter," *Chinese Sciences Abstracts*, vol. 6(2), pp. 183–184, 2000.

[344] K. C. Chang, Z. Tian and R. K. Saha, "Performance evaluation of track fusion with information filter," *Proc. Int. Conference on Multisource Multi-sensor Information Fusion*, pp. 648–655, July 1998.

[345] R. K. Saha and K. C. Chang, "An efficient algorithm for multi–sensor track fusion," *IEEE Trans. Aerospace and Electronic Systems*, vol. 34(1), pp. 200–210, 1998.

[346] R. K. Saha, "Track-to-track fusion with dissimilar sensors," *IEEE Trans. Aerospace and Electronic Systems*, vol. 32(3), pp. 1021–1029, 1996.

[347] S. Li, and M. Jing, "Distributed reduced-order optimal fusion Kalman filters for stochastic singular systems," *Acta Automatica Sinica*, vol. 32(2), pp. 286–290, March 2006.

[348] G. Qiang and Y. Song-nian, "Distributed multi–sensor data fusion based on Kalman filtering and the parallel implementation," *J. Shanghai University (English Edition)*, vol. 10(2), pp. 118–122, Article ID: 1007-6417(2006)02-0118-05, 2006.

[349] Y. M. Zhu, Z. S. You and J. Zhao, "The optimality for the distributed Kalman filter with feedback," *Automatica*, vol. 37, pp. 1489–1493, 2001.

[350] M. H. Zhao,, Z. M. Zhu and S. Meng, "Suboptimal distributed Kalman filtering fusion with feedback," *J Syst Engin Electron*, vol. 4, pp. 746–749, 2005.

[351] C. Y. Chong, S. Mori and K. C. Chang, "Information fusion in distributed sensor networks," *Proc. the American Control Conference*, pp. 830–835, 1985.

[352] K. C. Chou, A. S., Willsky and A. Benveniste, "Multiscale recursive estimation, data fusion, and regularization," *IEEE Trans. Automatic Control*, vol. 39(3), pp. 464–474, March 1994.

[353] J. Xu, E. Song, Y. Luo and Y. Zhu, "Optimal distributed Kalman filtering fusion algorithm Without Invertibility of estimation error and sensor noise covariances," *IEEE Signal Processing Letters*, vol. 19(1), pp. 55–58, January 2012.

[354] X. Shen, Y. Zhu, E. Song and Y. Luo, "Minimizing Euclidean state estimation error for linear Uncertain dynamic systems based on multi–sensor and multi-algorithm fusion," *IEEE Trans. Information Theory*, vol. 57(10), pp. 7131–7146, October 2011.

[355] C. L. Wen, Q. B. Ge and X. L. Feng, "Optimal recursive fusion estimator for asynchronous system," *Proc. the 7th Asian Control Conference*, Hong Kong, China, pp. 148–153, August 27-29, 2009.

[356] L. -L. Ong, B. Upcroft, M. Ridley, T., Bailey, S. Sukkarieh and D. H. Whyte, "Consistent methods for decentralized data fusion using particle filters," *Proc. the IEEE Int. Conference on multi–sensor Fusion and Integration for Intelligent Systems (MFI'06)*, pp. 85–91, 2006.

[357] M. J. Coates and B. N. Oreshkin, "Asynchronous distributed particle filter via decentralized evaluation of Gaussian products', In: *Proc. ISIF Int. Conf. Inf. Fusion*, 2010.

[358] D. Gu, S. Junxi, H., Zhen and L. Hongzuo, "Consensus based distributed particle filter in sensor networks," *IEEE Int. Conference on Information and Automation*, pp. 302–307, 2008.

[359] B. N. Oreshkin, X. Liu and M. J. Coates, "Efficient delay-tolerant particle filtering," *IEEE Trans. Signal Processing*, vol. 59(7), pp. 3369–3381, July 2011.

[360] M. Orton and A. Marrs, "Storage efficient particle filters for the out-of-sequence measurement problem," *Proc. IEE Colloq. Target Tracking: Algorithms and Appl.*, Enschede, The Netherlands, October 2001.

[361] M. Orton and A. Marrs, "Particle filters for tracking with out-of-sequence measurements," *IEEE Trans. Aerosp. Electron. Syst.*, vol. 41(2), pp. 693–702, February 2005.

[362] W. Zhang, X. Huang and M. Wang, "Out-of-sequence measurement algorithm based on Gaussian particle filter," *Inf. Technol. J.*, vol. 9(5), pp. 942–948, May 2010.

[363] X. Liu, B. N. Oreshkin and M. J. Coates, "Efficient delay-tolerant particle filtering through selective processing of out-of-sequence measurements," *Proc. ISIF Int. Conf. Inf. Fusion*, Edinburgh, U.K., July 2010.

[364] U. Orguner and F. Gustafsson, "Storage efficient particle filters for the out-of-sequence measurement problem," *Proc. ISIF Int. Conf. Inf. Fusion*, Cologne, Germany, Jul. 2008.

[365] D. K. Tasoulis, N. M. Adam and D. J. Hand, "Selective fusion of out-of-sequence measurements," *Info. Fusion*, vol. 11(2), pp. 183–191, Apr. 2010.

[366] R. T. Sukhavasi and B. Hassibi, "The Kalman like particle filter : optimal estimation With quantized innovations/measurements," *Proc. Joint 48th IEEE Conference on Decision and Control and 28th Chinese Control Conference*, Shanghai, P.R. China, pp. 4446–4451, December 16-18, 2009.

[367] F. Aounallah, R. Amara and M. T. H. Alouane, "Particle filtering based on sign of innovation for distributed estimation in binary wireless sensor networks," *SPAWC'08*, pp. 629–633, 2008.

[368] L. Zuo, K. Mehrotra, P. K. Varshney and C. K. Mohan, "Bandwidth-efficient target tracking In distributed sensor networks using particle filters," *Proc. 9th International Conference on Information Fusion*, pp. 1–4, 10-13 July 2006.

[369] X. Sheng and Y. H. Hu, "Distributed particle filters for wireless sensor network target tracking," *ICASSP'05*, pp. IV-845–IV-848, 2005.

[370] M. Coates, "Distributed particle filters for sensor networks," *Information Processing in Sensor Networks, IPSN*, Springer, pp. 99–107, 2004.

[371] A. Mohammadi and A. Asif, "Consensus-based disributed unscented particle filter," *Proc. the IEEE Statistical Signal Processing Workshop (SSP)*, pp. 28–30, pp. 237–240, June 2011.

[372] M. Mallick, T. Kirubarajan and S. Arulampalam, "Out-of-sequence measurement processing for tracking ground target using particle flters," *Proc. the IEEE Aerospace Conference*, Big Sky, MT, USA, vol. 4, pp. 1809–1818, March 2002.

[373] M. Mallick and A. Marrs, "Comparison of the KF and particle filter based out-of-sequence measurement filtering algorithms," *Proc. the Sixth Int. Conference on Information Fusion*, Cairns, Qld., Australia, vol. 1, pp. 422–429, July 2003.

[374] T. Gui-Li, H. G. Xue and D. Zi-Li, "Self-tuning distributed fusion Kalman filter With asymptotic global optimality," *Proc. the 29th Chinese Control Conference*, Beijing, China, pp. 1268–1272, July 29-31, 2010.

[375] Z. L. Deng and C. B. Li, "Self-tuning information fusion Kalman predictor weighted by diagonal matrices and its convergence analysis," *Acta Automatic Sinica*, vol. 33(2), pp. 156–163, 2007.

[376] C. J. Ran, G. L. Tao, J. F. Liu and Z. L. Deng, "Self-tuning decoupled fusion Kalman predictor and its convergence analysis," *IEEE Sensors Journal*, vol. 9(12), pp. 2024–2032, 2009.

[377] Y. Gao, W. J. Jia, X. J. Sun and Z. L. Deng, "Self-tuning multi–sensor weighted measurement fusion Kalman filter," *IEEE Tran. Aerospace and Electronic System*, vol. 45(1), 179–191, 2009.

[378] C. Ran and Z. Deng, "Self-tuning Weighted measurement fusion Kalman filter and its convergence analysis," *Proc. Joint 48th IEEE Conference on Decision and Control and 28th Chinese Control Conference*, Shanghai, P.R. China, pp. 1830–1835, December 16–18, 2009.

[379] Z. L. Deng, Y. Gao, C. B. Li and G. Hao, "Self-tuning decoupled information fusion Wiener state component filters and their convergence," *Automatica*, vol. 44, pp. 685–695, 2008.

[380] X. J. Sun, P. Zhang and Z. L. Deng,"Self-tuning decoupled fusion Kalman filter based on Riccati equation," *Frontiers of Electrical and Electronic Engineering in China*, vol. 3(4), pp. 459–464, 2008.

[381] F. Aounallah, R. Amara and M. T. -H. Alouane, "Particle filtering based on sign of innovation for tracking a jump karkovian motion in a binary WSN," *Proc. 3rd Int. Conference on Sensor Technologies and Applications*, pp. 252–255, 2009.

[382] F. Aounallah, R. Amara and M. T. -H. Alouane, "Particle filtering based on sign of innovation for distributed estimation in binary wireless sensor networks," *IEEE Workshop SPAWC'08*, Recife, Brasil, July 2008.

[383] T. J. Moir and M. J. Grimble, "Optimal self-tuning filtering, prediction and smoothing for discrete multivarianble processes," *IEEE Trans. Automatic Control*, vol. 29(2), pp. 128–137, 1984.

[384] Z. Li, D. Bo and C. Li., "Self-tuning information fusion Kalman predictor weighted by diagonal matrices and its convergence analysis," *Acta Automatica Sinica*, vol. 33(2), pp. 156–163, February 2007.

[385] Y. Gao, C. J. Ran, X. J. Sun and Z. L. Deng, "Optimal and self-tuning weighted measurement fusion Kalman filters and their asymptotic global optimality," *Int. J. Adapt. Control Signal Processing*, vol. 24, pp. 982–1004, 2010.

[386] P. Hagander and B. Wittenmark, "A self-tuning filter for fixed-lag smoothing," *IEEE Trans. Information Theory*, vol. 23(3), pp. 377–384, 1977.

[387] P. T. K. Fung and M. J. Grimble, "Dynamic ship positioning using a self-tuning Kalman filter," *IEEE Trans. Automatic Control*, vol. 28(3), pp. 339–350, 1983.

[388] Z. L. Deng and C. B. Li, "Self-tuning information fusion Kalman predictor weighted by scalars," *Proc. the Sixth World Congress on Control and Automation*, Dalian, vol. 2, pp. 1487–1491, China, 21–23 June 2006.

[389] Z. L. Deng, Y. Gao, C. B. Li and G. Hao, "Self-tuning decoupled information fusion Wiener state component filters and their convergence," *Automatica*, vol. 44, pp. 685–695, 2008.

[390] G. Hao, G. and Z. L. Deng, "Self-tuning measurement fusion Kalman filter', *Proceeding of the Sixth World Congress on Control and Automation*, Dalian, China, vol. 2, pp. 1571–1575, June 21–23, 2006.

[391] G. Hao, G., W. J. Jia and Z. L. Deng, "Self-tuning multi–sensor measurement fusion Kalman filter," *Proc. the 25th Chinese Control Conference*, Harbin, China, pp. 395–403, August 7–11, 2006.

[392] Y. Gao, W. J. Jia, X. J. Sun and Z. L. Deng, "Self-tuning multi–sensor weighted measurement fusion Kalman filter," *IEEE Trans. Aerospace and Electronic Systems*, vol. 45(1), pp. 168–179, 2009.

[393] C. J. Ran, G. L. Tao, J. F. Liu and Z. L. Deng, "Self-tuning decoupled fusion Kalman predictor and its convergence analysis," *IEEE Sensors Journal*, vol. 9(12), pp. 2024–2032, 2009.

[394] I. Akyildiz, W. Su, Y. Sankarasubramaniam, and E. Cayirci, "Wireless Sensor Networks: a survey," *Computer Networks*, vol. 38, pp. 393–422, 2002.

[395] S. Felter, "An overview of decentralized Kalman filters," *Proc. IEEE Southern Tier Technical Conference*, Birmingham, NY, USA, pp. 79–87, 1990.

[396] S. Roy, R. Hashemi, and A. Laub, "Square root parallel Kalman filtering using reduced order local filters," *IEEE Trans. Aerospace and Electronic Systems*, vol. 27, no. 2, pp. 276–289, 1991.

[397] S. Roy and R. Iltis, "Decentralized linear estimation in correlated measurement noise," *IEEE Transactions on Aerospace and Electronic Systems*, vol. 27, no. 6, pp. 939–941, 1991.

[398] S. Shu, "Multi–sensor optimal information fusion Kalman filters with applications," *Aerospace Science & Technology*, vol. 8, no. 1, pp. 57–62, 2004.

[399] H. Durant-Whyte, B. Rao, and H. Hu, "Towards a fully decentralized architecture for multi–sensor data fusion," *IEEE Int. Conf. on Robotics and Automation*, Cincinnati, Ohio, USA, pp. 1331–1336, 1990.

[400] H. Hashmipour, S. Roy, and A. Laub, "Decentralized structures for parallel Kalman filtering," *IEEE Trans. Automatic Control*, vol. 33, no. 1, pp. 88–93, 1988.

[401] M. Hassan, G. Salut, M. Sigh, and A. Titli, "A decentralized algorithm for the global Kalman filter," *IEEE Trans. Automatic Control*, vol. 23, no. 2, pp. 262–267, 1978.

[402] R. Quirino and C. Bottura, "An approach for distributed Kalman filtering," *Revista Controle & Automatica da Sociedade Brasileira de Automatica*, vol. 21, pp. 19–28, 2001.

[403] P. Alriksson and A. Rantzer, "Distributed Kalman filter using weighted averaging," *Proc. of the 17th Int. Symp. on Mathematical Theory of Networks and Systems*, Kyoto, Japan, 2006.

[404] R. Olfati-Saber, "Distributed Kalman filtering for sensor networks," *Proc. the 46th IEEE Conf. on Decision and Control*, New Orleans, LA, USA, 2007.

[405] S. Julier and J. Uhlmann, "A non-divergent estimation algorithm in the presence of unknown correlations," *Proc. American Control Conference*, Albuquerque, New Mexico, 1997.

[406] R. Olfati-Saber, "Distributed Kalman filter using embedded consensus filters," *Proc. 44th IEEE Conf. on Decision and Control 2005 and 2005 European Control Conference (CDC-ECC–05)*, Seville, Spain, pp. 8179–8184, 2005.

[407] A. Mutambara and D.-W. H.F., "Fully decentralized estimation and control for a modular wheeled mobile robot," *Int. J. of Robotic Research*, vol. 19, no. 6, pp. 582–596, 2000.

[408] U. Khan and J. Moura, "Distributed Kalman filters in sensor networks: Bipartite Fusion Graphs," *Proc. IEEE 14th Workshop on Statistical Signal Processing*, Madison, Wisconsin, USA, pp. 700–704, 2007.

[409] R. Smith and F. Hadaegh, "Closed-Loop Dynamics of Cooperative Vehicle Formations With Parallel Estimators and Communication," *IEEE Trans. Automatic Control*, vol. 52, no. 8, pp. 1404–1414, 2007.

[410] M. P. J. Fromherz, L. S. Crawford, C. Guettier, and Y. Shang, "Distributed adaptive constrained optimization for smart matter systems," *AAAI Spring Symposium on Intelligent Distributed and Embedded Systems*, March 2002.

[411] C. Godsil and G. Royle, *Algebraic Graph Theory*, vol. 207 of Graduate Texts in Mathematics, Springer, 2001.

[412] S. Grime and H. Durrant-Whyte, "Data fusion in decentralized sensor networks," *Control Engineering Practice*, vol. 2, 1994.

[413] R. Olfati-Saber, E. Franco, E. Frazzoli, and J. S. Shamma, "Belief consensus and distributed hypothesis testing in sensor networks," *Workshop on Network Embedded Sensing and Control*, Notre Dame University, October 2005.

[414] R. Olfati-Saber and J. S. Shamma, "Consensus filters for sensor networks and distributed sensor fusion," *Proc. the 44th Conference on Decision and Control*, 2005.

[415] L. Xiao and S. Boyd, "Fast linear iterations for distributed averaging," *Systems & Control Letters*, vol. 52, pp. 65–78, 2004.

[416] Y. M. Zhu, Z. S. You, J. Zhou, K.-S. Zhang, and X. R. Li, "The optimality for the distributed Kalman filtering fusion with feedback," *Automatica*, vol. 37, pp. 1489–1493, 2001.

[417] H. K. Wang, K. Yao, G. Pottie, and D. Estrin, "Entropy-based sensor selection heuristic for target localization," *Proc. 3rd Int. Symp. Inf. Process. Sensor Netw. (IPSN)*, pp. 36–45, Paris, 2004.

[418] Liu, J., J. Reich, and F. Zhao, "Collaborative in-network processing for target tracking," *EURASIP Journal on Advances in Signal Processing*, pp. 78–391, 2003.

[419] Kaplan, L. M., "Global node selection for localization in a distributed sensor network," *IEEE Trans. Aerospace and Electronic Systems*, vol. 42 (1), pp. 113–135, 2006.

[420] T. Vercauteren and X. Wang, "Decentralized sigma-point information filters for target tracking in collaborative sensor networks," *IEEE Trans. Signal Processing*, vol. 53 (8), pp. 2997–3009, 2006.

[421] T. Rappaport, *Wireless Communications: Principles and Practice, 2nd ed*, Prentice-Hall, NJ, 2002.

[422] *IEEE Standard 802.11 - Wireless LAN Medium Access Control (MAC) and Physical Layer (PHY) Specifications*, November 1999.

[423] B. P. Crow, I. Widjaja, J. G. Kim, P. T. Sakai, "IEEE 802.11 wireless local area networks," *IEEE Commun. Magazine*, pp. 116–126, 1997.

[424] J. Weinmiller, M. Schlager, A. Festag, A. Wolisz, "Performance study of access control in wireless LANs IEEE 802.11 DFWMAC and ETSI RES 10 HIPERLAN," *Mobile Networks and Applications*, vol. 2, pp. 55–67, 1997.

[425] G. Bianchi, "Performance analysis of the IEEE 802.11 distributed coordination function," *IEEE Journal of Selected Areas in Telecommunications, Wireless Series*, Vol. 18, no. 3, pp. 535–547, 2000.

[426] Y.C. Tay and K.C. Chua, "Capacity analysis for the IEEE 802.11 MAC protocol," *ACM/Baltzer Wireless Networks*, vol. 7, No. 2, pp. 159–171, 2001.

[427] G. Bianchi, L. Fratta and M. Oliveri, "Performance evaluation and enhancement of the CSMA/CA MAC protocol for 802.11 wireless LANs," *Proc. PIMRC*, Taipei, Taiwan, pp. 392–396, 1996.

[428] F. Cali, M. Conti and E. Gregori, "Dynamic tuning of the IEEE 802.11 protocol to achieve a theoretical throughput limit," *Trans. Networking*, vol. 8, no. 6, pp. 785–799, 2000.

[429] K. C. Huang and K. C. Chen, "Interference analysis of nonpersistent CSMA with hidden terminals in multicell wireless data networks," *Proc. of IEEE PIMRC*, Toronto, Canada, pp. 907–911, 2005.

[430] H. S. Chhaya and S. Gupta, "Performance modeling of asynchronous data transfer methods of IEEE 802.11 MAC protocol," *Wireless Networks*, vol. 3, pp. 217–234, 1997.

[431] M. S. Grewal and A. P. Andrews, *Kalman Filtering: Theory and Practice Using Matlab*, John Wiley & Sons, Ltd, 2001.

[432] F. Gustafsson, *Adaptive Filtering and Change Detection*, John Wiley & Sons, Ltd, 2000.

[433] O. Doguc and J. E. Marquez, "An efficient fault diagnosis method for complex system reliability," *Proc. 7th Annual Conference on Systems Engineering Research (CSER 2009)*, 2009.

[434] D. Brambilla, L. M. Capisani, A. Ferrara and P. Pisu, "Fault detection for robot manipulators via second-order sliding modes," *IEEE Trans. Industrial Electron.*, vol. 55(11), pp. 3954–3963, 2008.

[435] J. Cusido, L. Romeral, J. A. Ortega, J. A. Rosero and A. Espinosa, "Fault detection in induction machines using power spectral density in wavelet decomposition," *IEEE Trans. Industrial Electron.*, vol. 55(2), pp. 633–643, 2008.

[436] S. A. Arogeti, D. Wang and C. B. Low, "Mode identification of hybrid systems in the presence of fault," *IEEE Trans. Industrial Electron.*, vol. 57(4), pp. 1452–1467, 2010.

[437] P. Lezana, J. Pou, T. A. Meynard, J. Rodriguez, S., Ceballos and F. Richardeau, "Survey on fault operation on multi-level inverters," *IEEE Trans. Industrial Electron.*, vol. 57(7), pp. 2207–2218, 2010.

[438] I. Morgan, H. Liu, B. Tormos and A. Sala, "Detection and diagnosis of incipient faults in heavy-duty diesel engines," *IEEE Trans. Industrial Electron.*, vol. 57(10), pp. 3522–3532, 2010.

[439] O. Poncelas, J. A. Rosero, J. Cusido, J. A. Ortega and L. Romeral, L., "Motor fault detection using a rogowski sensor without an integrator," *IEEE Trans. Industrial Electron.*, vol. 56(10), pp. 4062–4070, 2009.

[440] K. Rothenhagen and F. W. Fuchs, "Current sensor fault detection, isolation, and reconfiguration for doubly fed induction generator," *IEEE Trans. Industrial Electron.*, vol. 56(10), pp. 4239–4245, 2009.

[441] I. A. Hameed, "Using the extended Kalman filter to improve the efficiency of greenhouse climate control," *Int. J. Innovative Computing, Information and Control*, vol. 6(6), pp. 2671–2680, 2010.

[442] Palangi, H., and Refan, M. H., "Error reduction of a low cost GPS receiver for kinematic applications based on a new Kalman filtering algorithm," *Int. J. Innovative Computing, Information and Control*, vol. 6(8), pp. 3775–3786, 2010.

[443] Sanchez, M. P., Guasp, M. R., Folch, J. R., Daviu, J. A. A., Cruz, J. P., Panadero, R. P., "Diagnosis of induction motor faults in time-varying conditions using the polynomial-phase transform of the current," *IEEE Trans. Industrial Electron.*, vol. 58(4), pp. 1428–1439, April 2011.

[444] Wolbank, T. M., Nussbaumer, P., Chen, H., Macheiner, P. E., "Monitoring of rotor-bar defects in inverter-fed induction machines at zero load and speed," *IEEE Trans. Industrial Electron.*, vol. 58(5), pp. 1468–1478, May 2011.

[445] Bianchini, C., Immovilli, F., Cocconcelli, M., Rubini, R., Bellini, A., "Fault detection of linear bearings in brushless AC linear motors by vibration analysis," *IEEE Trans. Industrial Electron.*, vol. 58(5), pp. 1684–1694 , May 2011.

[446] H. Henao, S. R. Fatemi, G. A. Capolino and S. Sieg-Zieba, "Wire rope fault detection in a hoisting winch system by motor torque and current signature analysis," *IEEE Trans. Industrial Electron.*, vol. 58(5), pp. 1707–1717 , May 2011.

[447] K. Salahshoor, M. Mosallei and M. R. Bayat, "Centralized and decentralized process and sensor fault monitoring using data fusion based on adaptive extended Kalman filter algorithm," *J. Measurement and Control*, vol. 41, pp. 1059–1076, 2008.

[448] R. Isermann, "Model-based fault-detection and diagnosis-status and applications," *Annual Reviews in Control*, vol. 29, pp. 71–85, 2005.

[449] J. J. Gertler, *Fault detection and diagnosis in engineering systems*, CRC Press, New York, 1998.

[450] P. M. Frank, S. X. Ding and B. K. Seliger, "Current developments in the theory of FDI," *Proc. IFAC Symposium on Fault Detection, Supervision and Safety of Technical Processes*, Budapest, Hungary, vol. 1, pp. 16–27, 2000.

[451] S. Simani, C. Fantuzzi and R. Patton, "Model-based fault diagnosis in dynamic systems using identification techniques," *Advances in Industrial Control*, Springer, 2003.

[452] G. Q. P. Zhang, "Neural networks for classification: a survey," *IEEE Trans. Systems, Man and Cybernetics: Part C-Applications and Reviews*, vol. 30(4), pp. 451–462, 2000.

[453] Y. Wang, C. W. Chan and K. C. Cheung, "Intelligent fault diagnosis based on neuro-fuzzy networks for nonlinear dynamic systems," *Proc. IFAC Conference on New Technologies for Computer Control*, Hong Kong, pp. 101–104, 2001.

[454] L. J. Miguel and L. F. Blazquez, "Fuzzy logic-based decision-making for fault diagnosis in a DC motor," *Engineering Applications of Artificial Intelligence*, vol. 18(4), pp. 423–450, 2005.

[455] "Special section on motor fault detection and diagnosis," *IEEE Trans. Industrial Electron.*, vol. 47(5), pp. 982–1107, 2000.

[456] N. Mehranbod, "A probabilistic approach for sensor fault detection and identification," *Ph. D. Dissertation*, Faculty of Drexel University, November 2005

[457] H. Kirch and K., Kroschel, "Applying Bayesian networks to fault diagnosis," *Proc. IEEE Conference on Control Applications*, 895, 1994.

[458] C. R. Guzman and M. Kramer, "GALGO: A genetic algorithm decision support tool for complex uncertain systems modeled with Bayesian belief networks," *Proc. 9th Conf. on Uncertainty in Artif. Intell. (UAI-93)*, San Francisco, pp. 368–375, 1993.

[459] N. Santoso, C. Darken and G. Povh, "Nuclear plant fault diagnosis using probabilistic reasoning," *Power Engineering Society Summer Meeting, IEEE*, vol. 2, pp. 714–719, 1999.

[460] A. Nicholson, "Fall diagnosis using dynamic belief networks," *Proc. the 4th Pacific Rim International Conference on Artificial Intelligence (PRICAI-96)*, Springer, London, pp. 206–217, 1996.

[461] C. Bin Zhang, C. Sconyers, R. Byington, M. E. Patrick, G. Orchard and G., Vachtsevanos, "A probabilistic fault detection approach: Application to bearing fault detection," *IEEE Trans. Ind. Electron.*, vol. 58(5), pp. 2011–2018, May 2011.

[462] E. A. Wan, R. V. Merwe and A. T. Nelson, "Dual estimation and the unscented transformation," *Advances in Neural Information Processing Systems*, MIT Press, Cambridge, pp. 666–672, 2000.

[463] S. J. Julier, J. K. Uhlmann, and H. Durrant-Whyte, "A new approach for filtering nonlinear systems," *Proc. American Control Conference*, pp. 1628–1632, 1995.

[464] S. J. Julier and J. K. Uhlmann, "A new extension of the Kalman filter to nonlinear systems," *Proc. the 11th Int. Symposium on AeroSpace/Defense Sensing, Simulation and Controls*, vol. 3068, pp. 182–193, 1997.

[465] Y. Shi and H. Fang, "Kalman filter based identification for systems with randomly missing measurements in a network environment," *Int. J. Control*, vol. 83(3), pp. 538–551, 2010.

[466] S. Kluge, K. Reif and H. Brokate, "Stochastic stability of the extended Kalman filtering with intermittent observations," *IEEE Trans. Automatic Control*, vol. 55(2), pp. 514–518, 2010.

[467] R. Van der Merwe, "Sigma-point Kalman filters for probability inference in dynamic state-space models," *PhD Thesis*, Oregon Health and Science University, 2004.

[468] Q. Gan and C. J. Harris, "Comparison of two measurement fusion methods for Kalman filter-based multi–sensor data fusion," *IEEE Trans. Aerospace Electr. Systems*, vol. 37(1), pp. 273–280, 2001.

[469] C. J. Harris, X. Hong and Q. Gan, *Adaptive Modeling, Estimation and Fusion from Data: A Neurofuzzy Approach*, Springer, London, 2002.

[470] L. Dai and K. Astrom, "Dynamic matrix control of a quadruple tank process," *Proc. the 14th IFAC*, pp. 295–300, 1999.

[471] W. Tan, H. Marquez and T. Chen, "Multivariable robust controller design for a boiler system," *IEEE Trans. Control Systems Technology* vol. 10(5), pp. 735–742, 2002.

[472] H. J. Marquez and M. Riaz, "Robust state observer design with application to an industrial boiler system," *Control Engineering Practice*, vol. 13, pp. 713–728, 2005.

[473] R. H. Perry and D. W. Green, *Perry's Chemical Engineers Handbook, Seventh Ed.*, McGraw-Hill, 1997.

[474] A. Mirabadi, N. Mort, and F. Schmid, "Application of sensor fusion to railway systems," *Proc. IEEE/SICE/RSJ Int. Conf. multi–sensor Fusion Integration Intell. Syst.*, pp. 185–192, 1996.

[475] C. P. Tan, F. Crusca and M. Aldeen, "Extended results on robust state estimation and fault detection," *Automatica*, vol. 44, no. 8, pp. 2027–2033, August 2008.

[476] P. Zhang and S. X. Ding, "An integrated trade-off design of observer based fault detection systems," *Automatica*, vol. 44, no. 7, pp. 1886–1894, July 2008.

[477] B. Akin, U. Orguner, S. Choi and A. Toliyat, "A simple real-time fault signature monitoring tool for motor drive embedded fault diagnosis systems," *IEEE Trans. Industrial Electronics*, 2010.

[478] Y. Wang, W. Wang and D. Wang, "LMI approach to design fault detection filter for discrete-time switched systems with state delays," *International Journal of Innovative Computing, Information and Control*, vol. 6, no. 1, pp. 387–398, January 2010.

[479] I. M. Jaimoukha, Z. Li and V. Papakos, "A matrix factorization solution to the H_-/H_∞ fault detection problem," *Automatica*, vol. 42, no. 11, pp. 1907–1912, November 2006.

[480] A. P. Deshpande, S. C. Patwardhan and S. S. Narasimhan, "Intelligent state estimation for fault tolerant nonlinear predictive control," *J. Process Control*, vol. 19, no. 2, pp. 187–204, February 2009.

[481] C. W. Chan, H. Song and H. Y. Zhang, "Application of fully decoupled parity equation in fault detection and identification of DC motors," *IEEE Trans. Industrial Electronics*, vol. 53, no. 4, pp. 1277–1284, August 2006.

[482] G. Tortora, B. Kouvaritakis and D. W. Clarke, "Simultaneous optimization of tracking performance and accommodation of sensor faults," *Int. J. Control*, vol. 75, no. 3, pp. 163–176, 2002.

[483] Y. Zhang and S. J. Qin, "Adaptive actuator fault compensation for linear systems with matching and unmatching uncertainties," *J. Process Control*, vol. 19, no. 6, pp. 985–990, June 2009.

[484] T. J. Kim, W. C. Lee and D. S. Hyun, "Detection method for open-circuit fault in neutral-point-clamped inverter systems," *IEEE Trans. Industrial Electronics*, vol. 56, no. 7, pp. 2754-2763, July 2009.

[485] A. S. Willsky and H. L. Jones. A generalized likelihood ratio approach to the detection and estimation of jumps in linear systems, *IEEE Trans. Automatic Control*, vol. 21, no. 1, pp. 108-112, Feb. 1976.

[486] F. Gustafsson, *Adaptive Filtering and Change Detection*. West Sussex, England: John Wiley & Sons, 2000.

[487] Y. Zhang, Y. Chen, J. Sheng and T. Hesketh, "Fault detection and diagnosis of networked control system," *International Journal of Systems Science*, vol. 39, no. 10, pp. 1017–1024, October 2008.

[488] Y. Wang, S. X. Ding, H. Ye, L. Wei, P. Zhang and G. Wang, "Fault detection of networked control systems with packet based periodic communication," *Int. J. Adaptive Control and Signal Processing*, vol. 23, no. 8, pp. 682–698, August 2009.

[489] S. Ghantasala and N. H. El-Farra, "Robust diagnosis and fault-tolerant control of distributed processes over communication networks," *Int. J. Adaptive Control and Signal Processing*, vol. 23, no. 8, pp. 699–721, August 2009.

[490] X. He, Z. Wang and D. H. Zhou, "Network-based robust fault detection with incomplete measurements," *Int. J. Adaptive Control and Signal Processing*, vol. 23, no. 8, pp. 737–756, August 2009.

[491] Z. Gu, D. Wang and D. Yue, "Fault detection for continuous-time networked control systems with non-ideal QoS," *Int. J. Innovative Computing, Information and Control*, vol. 6, no. 8, pp. 3631–3640, August 2010.

[492] Y. Xia, J. Chen, G. P. Liu and D. Rees, "Stability analysis of networked predictive control systems with random network delay " *in Proc. the 2007 IEEE Int. Conference on Networking, Sensing and Control*, London, UK: pp. 815–820, 2007.

[493] G. P. Liu, Y. Xia, D. Rees and W. Hu, "Design and stability criteria of networked predictive control systems with random network delay in the feedback channel," *IEEE Trans. Systems, Man, and Cybernetics-Part C*, vol. 37, no. 2, pp. 173–184, March 2007.

[494] G. P. Liu, Y. Xia, J. Chen, D. Rees and W. Hu, "Networked predictive control of systems with random network delays in both forward and feedback channels," *IEEE Trans. Industrial Electronics*, vol. 54, no.3, pp. 1282–1297, June 2007.

[495] X. Liu, Y. Liu, M. S. Mahmoud and Z. Deng, "Modeling and stabilization of MIMO networked control systems with network constraints," *Int. J. Innovative Computing, Information and Control*, vol. 6, no. 10, pp. 4409–4420, October 2010.

[496] Y. B. Zhao, G. P. Liu and D. Rees, "Improved predictive control approach to networked control systems," *IET Control Theory and Applications*, vol. 2, no. 8, pp. 675–681, August 2008.

[497] Y. Xia, M. Fu, B. Liu and G. P. Liu, "Design and performance analysis of networked control systems with random delay," *J. Systems Engineering and Electronics*, vol. 20, no. 4, pp. 807–822, 2009.

[498] Y. Zhang and S. Li, "Networked model predictive control based on neighbourhood optimization for serially connected large-scale processes," *Journal of Process Control*, vol. 17, no. 1, pp. 37–50, January 2007.

[499] J. Huang, Y. Wang, S. Yang and Q. Xua, "Robust stability conditions for remote SISO DMC controller in networked control systems," *Journal of Process Control*, vol. 19, no. 5, pp. 743–750, May 2009.

[500] M. Vaccarini, S. Longhi and M. R. Katebi, "Unconstrained networked decentralized model predictive control," *Journal of Process Control*, vol. 19, no. 2, pp. 328–339, Sep. 2009.

[501] P. Mendez-Monroy and H. Benitez-Perez, "Supervisory fuzzy control for networked control systems," *ICIC Express Letters*, vol. 3, no. 2, pp. 233–238, June 2009.

[502] Y. Xia, Z. Zhu and M. S. Mahmoud, "H_2 control for networked control systems with Markovian data losses and delays," *ICIC Express Letters*, vol. 3, no. 3(A), pp. 271–276, September 2009.

[503] B. Sinopoli, L. Schenato, M. Franceschetti, K. Poolla, M. Jordan and S. S. Sastry, "Kalman filtering with intermittent observations," *IEEE Trans. Automatic Control*, vol. 49, no. 9, pp. 1453–1464, September 2004.

[504] H. Fang, H. Ye, and M. Zhong, "Fault diagnosis of networked control systems," *Annual Reviews in Control*, vol. 31, no. 1, pp. 55–68, 2007.

[505] H. Ye, and S. X. Ding, "Fault detection of networked control systems with network-induced delay," *Proc. the 8th International Conference on Control, Automation, Robotics and Vision (ICARCV04)*, pp. 654–659, 2004.

[506] W. Zhang, M. S. Branicky and S. M. Phillips, "Stability of networked control systems," *IEEE Control Systems Magazine*, vol. 21, no. 1, pp. 84–99, February 2001.

[507] H. Ye, P. Zhang, S. X. Ding and G. Wang, "A time-frequency domain fault detection approach based on parity relation and wavelet transform," *In Proceedings of the 39th IEEE Conference on Control and Decision (IEEE CDC 00)*, pp. 4156–4161, 2000.

[508] D. Huang and S. K. Nguang, "Robust fault estimator design for uncertain networked control systems with random time delays: An ILMI approach," *Information Sciences*, vol. 180, no. 3, pp. 465–480, February 2010.

[509] N. Xiao, L. Xie and M. Fu, "Kalman filtering over unreliable communication networks with bounded Markovian packet dropouts," *International Journal of Robust and Nonlinear Control*, vol. 19, no. 16, pp. 1770–1786, November 2009.

[510] A. Abur, "Distributed state estimation for megagrids," *15th PSCC*, Liege, 22–26 August 2005.

[511] M. Alighanbari and J. P. How, "An unbiased Kalman consensus algorithm," *Proc. American Control Conference*, Minneapolis, Minnesota, USA, June 14–16, 2006.

[512] R. Vadigepalli R., and F. J. Doyle III, "Structural analysis of large-scale systems for distributed state estimation and control applications," *Control Engineering Practice*, vol. 21, no. 11, pp. 895–905, 2003.

[513] R. Vadigepalli R., and F. J. Doyle III, "A distributed state estimation and control algorithm for plantwide processes," *IEEE Trans. Control Systems Technology*, vol. 11, no. 1, pp. 119–127, January 2003.

[514] M. S. Mahmoud, *Decentralized Control and Filtering in Dynamical Interconnected Systems*, CRC Press, 2010.

[515] R. O. Saber, "Distributed Kalman filter with embedded consensus filters," *44th IEEE Conference on Decision and Control 2005 and 2005 European Control Conference (CDC-ECC 05)*, pp. 8179–8184, December 2005.

[516] B. D. O. Anderson and J. B. Moore, *Optimal Filtering*, Prentice-Hall, Englewood Cliffs, NJ., 1979.

[517] Shalom Y., and Fortmann T. E., "Tracking and Data Association," *Academic Press*, 1988.

[518] Shalom Y. B., and Li X. R. *Multitarget-multi–sensor Tracking: Principles and Techniques*, YBS Publishing, Story, CT, 1995.

[519] J. Cortes, "Distributed Algorithms for Reaching Consensus on Arbitrary Functions," *Automatica*, October 2006.

[520] D. Fox, J., Hightower, L. Liao, D. Schulz and G. Borriello, "Bayesian filtering for Location Estimation," *IEEE Pervasive Computing*, vol. 2 no. 3, pp. 24–33, July-September, 2003.

[521] V. Gupta, D. Spanos, B. Hassibi and R. M. Murray, "On LQG Control Across a Stocjastic Packet-Dropping Link," *Proc. of the 2005 Automatic Control Conference*, pp. 360–365, June 2005.

[522] R. O. Saber, "Distributed tracking for mobile sensor networks with information-driven mobility," *Proc. of the 2006 American Control Conference*, July 2007.

[523] B. S. Y. Rao, H. F. D. Whyte and J. A. Sheen, "A Fully Decentralized multi–sensor System for Tracking and Surveillance," *Int. Journal of Robotics Research*, vol. 12, no. 1, pp. 20–44, February 1993.

[524] D. B. Reid, "An Algorithm for Tracking Multiple Targets," *IEEE Trans. on Automatic Control*, vol. 24, no. 6, pp. 843–854, December 1979.

[525] B. Sinopoli, Schenato L., Franceschetti M., Poola K., Jordan M. I., and Sastry S. S., "Kalman Filtering with Intermittent Observations," *IEEE Trans. Automatic Control*, vol. 49, no.9, pp. 1453–1464, September 2004.

[526] J. L. Speyer , "Computation and Transmission Requirements for a Decentralized Linear-Quadratic-Gaussian Control Problem," *IEEE Trans. on Automatic Control*, vol. 24, no.2, pp. 266–269, February 1979.

[527] P. Alriksson and A. Rantzer, "Experimental Evaluation of a Distributed Kalman Filter Algorithm," *Proc. 46th IEEE Conference on Decision and Control*, New Orleans, LA, December 2007.

[528] R. Carli, A. Chiuso, L. Schenato, and S. Zampieri, "Distributed Kalman Filtering Based on Consensus Strategies," *IEEE Journal on Selected Areas in communications*, vol. 26, no. 4, May 2008.

[529] D. W. Casbeer, "Decentralized Estimation using Information Consensus Filters with a Multi-Static UAV Radar Tracking System," *Ph. D. Thesis*, Department of Electrical and Computer Engineering, Brigham Young University, April 2009.

[530] F. S. Cattivelli, C. G. Lopes, and A. H. Sayed, "Diffusion Strategies for Distributed Kalman Filtering: Formulations and Performance Analysis," *Proc. Workshop on Cognitive Inf. Process.*, Santorini, Greece, June 2008.

[531] T. Jiang, Matei I., Baras J. S., "A trust based distributed Kalman filtering approach for mode estimation in power systems" , *Proc. of the First Workshop on Secure Control Systems*, 2010.

[532] M. M. Jr, A. A. Emadzadeh and S. Vakil, "Distributed Kalman filtering with embedded consensus filters," *University of California Project*, Los Angeles, Winter 2007.

[533] R. Olfati-Saber, "Distributed Kalman filter with embedded consensus filters," *Dartmouth College Thayer School of Engineering*, 2008.

[534] A. Speranzon, C. Fischione, K. H. Johansson, and Vincentelli A.S., "A Distributed Minimum Variance Estimator for Sensor Networks," *IEEE Journal on Selected Areas in Communications*, vol. 26, no. 4, May 2008.

[535] X. Dai and S. Khorram, "Data fusion using artificial neural networks: a case study on multitemporal change analysis," *Comput. Environ. Urban Syst.*, vol. 23, pp. 19–31, 1999.

[536] C. H. Adelson, and J. R. Bergen, "Pyramid methods in image processing," *RCA Eng.*, vol. 29, pp. 33–41, 1984.

[537] J. Xiang and X. Su, "A pyramid transform of image denoising algorithm based on morphology," *Acta Photon. Sin.*, vol. 38, pp. 89–103, 2009.

[538] J. Dong, D. Zhuang, Y. Huang and J. Fu, "Advances in multi–sensor data fusion: algorithms and applications," *Sensors*, vol. 9, pp. 7771–7784, 2009.

[539] B. Khaleghi, A. Khamis, F. O. Karray, and S. N. Razavi, "Multi–sensor data fusion: a review of the state-of-the-art," *Information Fusion*, vol. 14, no. 1, pp. 28–44, 2013.

[540] H. Durrant-Whyte and T. Henderson, *Multisensor Data Fusion*, Springer, New York, NY, USA, 2008.

[541] J. Llinas, C. Bowman, G. Rogova, A. Steinberg, E. Waltz, and F. White, "Revisiting the JDL data fusion model II," *Proc. the 7th International Conference on Information Fusion (FUSION 04)*, pp. 1218–1230, July 2004.

[542] Z. Pawlak, *Rough Sets: Theoretical Aspects of Reasoning about Data*, Kluwer Academic, Norwell, Mass, USA, 1992.

[543] R. Mahler, *Statistical Multisource-Multitarget Information Fusion*, Artech-House, Boston, Mass, USA, 2007.

[544] B. Khaleghi, A. Khamis, and F. Karray, "Random finite set theoretic based soft/hard data fusion with application for target tracking," *Proc. the IEEE Int. Conference on Multi–sensor Fusion and Integration for Intelligent Systems (MFI10)*, pp. 50–55, September 2010.

[545] B. S. Rao and H. F. Whyte, "Fully decentralized algorithm for multi–sensor Kalman filtering," *IEEE Proceedings*, 1991.

[546] L. Idkhajine, E. Monmasson and A. Maalouf, "Fully FPGA-based sensorless control for synchronous AC drive using an extended Kalman filter," *IEEE Trans. Industrial Electronics*, vol. 59(10), pp. 3908–3915, October 2012.

[547] W. Yen and T. L. Hua, "A high-performance sensorless position control system of a synchronous reluctance motor using dual current-slope estimating technique," *IEEE Trans. Industrial Electronics*, vol. 59(9), pp. 3411–3426, September 2012.

[548] G. Rigatos, "A derivative-free Kalman filtering approach to state estimation-based control of nonlinear systems," *IEEE Trans. on Industrial Electronics*, vol. 59(10), pp. 3987–3997, October 2012.

[549] R. Luo and C. Lai, "Enriched indoor map construction based on multi–sensor fusion approach for intelligent service robot," *IEEE Trans. Industrial Electronics*, vol. 59(8), pp. 3135–3145, August 2012.

[550] I. Sadinezhad and V. Agelidis, "Frequency adaptive least-squares-Kalman technique for real-time voltage envelope and flicker estimation," *IEEE Trans. Industrial Electronics*, vol. 59(8), pp. 3330–3341, August 2012.

[551] P. Shi, X. Luan and F. Liu, "H_∞ filtering for discrete-time systems with stochastic incomplete measurement and mixed delays," *IEEE Trans. Industrial Electronics*, vol. 59(6), pp. 2732–2739, June 2012.

[552] Y. Shalom, "On the track-to-track correlation problem," *IEEE Trans on Automatic Control*, 1981, vol. 26(2), pp. 571–572.

[553] Y. Shalom and L. Campo, "The effect of the common process noise on the two-sensor fused-track covariance," *IEEE Trans. Aerospace and Electronic Systems*, AES-22, pp. 803–805, 1986.

[554] K. C. Chang, R. K. Saha and Y. Shalom, "On optimal track-to-track fusion," *IEEE Trans. Aerospace and Electronic Systems*, vol. 33(4): 1271–1276, 1997.

[555] C. Y. Chong, K. C., Chang and S. Mori, "Distributed tracking in distributed sensor networks," *Proc. the American Control Conference*, Seattle, 1986.

[556] S. Julier and J. K. Uhlmann, "General decentralized data fusion with Covariance Intersection (CI)," *Handbook of multi–sensor data fusion*, edited by D. Hall and J. Llinas, Chapter 12, pp. 12-1–12-25. CRC Press, 2001.

[557] X. R. Li, Y. M. Zhu and C. Z. Han, "Unified optimal linear estimation fusion Part I: unified model and fusion rules," *Proc. the 3rd Int. Information Fusion Conference*, Paris, France: ISIF, pp. MoC2-10–Moc2-17, 2000.

[558] K. C. Chang, T. Zhi and R. K. Saha, "Performance evaluation of track fusion with information matrix filter," *IEEE Trans. Aerospace and Electronic Systems*, vol. 38(2), pp. 455–466, 2002.

[559] K. C. Chang, "Evaluating hierarchical fusion with information matrix filter," *Proc. the 3rd International Information Fusion Conference*, Paris, France: ISIF, 2000.

[560] C. Y. Chong and S. Mori, "Convex combination and covariance intersection algorithm in distributed fusion," *Proc. the 4th Int. Information Fusion Conference*, Montreal, Canada: ISIF. 2001.

[561] Y. B. Shalom and L. Campo, "The effect of common process noise on the two-sensor fused-track covariance," *IEEE Trans. Aerospace and Electronic Systems*, vol. 22, pp. 803–805, 1986.

[562] S. L. Sun, "multi–sensor optimal information fusion kalman filter for discrete multichannel ARMA Signals," *Proc. the IEEE Int. Symposium on Intelligent Control*, pp. 377–382, 2003.

[563] S. L. Sun and Z. L. Deng, "multi–sensor optimal information fusion Kalman filter," *Automatica*, vol. 40, pp. 1017–1023, 2004.

[564] B. F. Scala and A. Farina, "Choosing a track association method," *Information Fusion*, vol. 3, pp. 119–133, 2001.

[565] C. Y. Chong, K. C. Chang and S. Mori, "Distributed tracking in distributed sensor networks," *Proc. the American Control Conference*, Seattle, pp. 1863–1868, 1986.

[566] J. K. Uhlmann, S. Julier and H. F. Whyte, "A culminating theory in the theory and practice of data fusion, filtering and decentralized estimation," *Technical Report, Covariance Intersection Working Group*, 1997.

[567] S. J. Julier and J. K. Uhlman, "A non-divergent estimation algorithm in the presence of unknown correlations," *Proc. the American Control Conference*, pp. 2369–2373, 1997.

[568] M. Farina, Trecate, G. F., Scattolini, R., "Distributed moving horizon estimation for nonlinear constrained systems," *Int. J. Robust and Nonlinear Control*, January 2012, vol. 22(2), pp. 123–143, January 2012.

[569] G. Wen, Z. Duan, W. Yu and G. Chen, "Consensus in multi-agent systems with communication constraints," *Int. J. Robust and Nonlinear Control*, vol. 22(2), pp. 170–182, January 2012.

[570] J. Liang, Z. Wang and X. Liu, "Distributed state estimation for uncertain Markov-type sensor networks with mode-dependent distributed delays," *Int. J. Robust and Nonlinear Control*, vol. 22(3), pp. 331–346, February 2012.

[571] A. P. C. Goncalves, A. R. Fioravanti and J. C. Geromel, "Filtering of discrete-time Markov jump linear systems with uncertain transition probabilities," *Int. J. Robust and Nonlinear Control*, vol. 21(6), pp. 613–624, April 2011.

[572] J. A. Roecker and C. D. McGillem, "Comparison of two-sensor tracking methods based on state vector fusion and measurement fusion," *IEEE Trans. Aerospace and Electronic Systems*, vol. 24, pp. 447–449, 1988.

[573] K. C. Chang, Z. Tian and R. K. Saha, "Performance evaluation of track fusion with information matrix filter," *IEEE Trans. Aerospace and Electronic Systems*, vol. 38, pp. 455–466, 2002.

[574] K. C. Chang, R. K. Saha and Y. Shalom, "On optimal track-to-track fusion," *IEEE Trans. Aerospace and Electronic Systems*, vol. 33, pp. 1271–1276, 1997.

[575] Y. Shalom, "On the track-to-track cross-covariance problem," *IEEE Trans. Automatic Control*, vol. 26, pp. 571–572, 1981.

[576] Y. Shalom and L. Campo, "The effect of common process noise on the two-sensor fused-track covariance," *IEEE Trans. Aerospace and Electronic Systems*, vol. 22, pp. 803–805, 1986.

[577] Q. X. Dong and K. C. Chang, "Information matrix fusion with feedback versus number of sensors," *Proc. Fusion*, Stockholm, Sweden, vol. I, pp. 686–692, July 2004.

[578] Q. Gan and C. J. Harris, "Comparison of two measurement fusion methods for Kalman-filter-based multi–sensor data fusion," *IEEE Trans. Aerospace and Electronic Systems*, vol. 37(1), pp. 273–279, January 2001.

[579] Z. Liu, M. Wang and J. Huang, "An evaluation of several fusion algorithms multi–sensor tracking system," *Journal of Information and Computational Science*, vol. 7(10), pp. 2101–2109, 2010.

[580] A. H. Sayed, *Fundamentals of Adaptive Filtering*, Wiley, NJ, 2003.

[581] R. T. Sukhavasi and B. Hassibi, "The Kalman-like particle filter: Optimal estimation with quantized innovations/measurements," *Proc. Conference on Decision and Control*, pp. 4446–4451, 2009.

[582] B. Khaleghi, A. Khamis, F. O. Karray, S. N. Razavi, "multi–sensor data fusion: A review of the state-of-the-art," *Information Fusion*, vol. 14, pp. 28–44, 2013.

[583] D. L. Hall and J. Llinas, "An introduction to multi–sensor fusion," *Proc. the IEEE*, vol. 85(1), pp. 6–23, 1997.

[584] D. Smith and S. Singh, "Approaches to multi–sensor data fusion in target tracking: a survey," *IEEE Trans. Knowledge and Data Engineering*, vol. 18(12), pp. 1696–1710, 2006.

[585] F. E. White, "Data fusion lexicon," *Joint Directors of Laboratories, Technical Panel for C3, Data Fusion Sub-Panel*, Naval Ocean Systems Center, San Diego, 1991.

[586] L. A. Klein, *Sensor and Data Fusion Concepts and Applications, second ed.*, Society of Photo-optical Instrumentation Engineers (SPIE), Bellingham, WA, 1999.

[587] H. Bostrm, S.F. Andler, M. Brohede, R. Johansson, A. Karlsson, J. van Laere, L. Niklasson, M. Nilsson, A. Persson and T. Ziemke, "On the definition of information fusion as a field of research," *Informatics Research Centre, University of Skvde, Tech. Rep. HS-IKI-TR-07-006*, 2007.

[588] E. L. Walts, "Data fusion for C3I: a tutorial," *Command, Control, Communications Intelligence (C3I) Handbook*, EW Communications Inc., Palo Alto, CA, pp. 217–226, 1986.

[589] A. N. Steinberg, C. L. Bowman and F. E. White, "Revisions to the JDL data fusion model," *Proc. the SPIE Conference on Sensor Fusion: Architectures, Algorithms, and Applications III*, pp. 430–441, 1999.

[590] J. Llinas, C. Bowman, G. Rogova, A. Steinberg, E. Waltz and F. E. White, "Revisiting the JDL data fusion model II," *Proc. of the Int. Conference on Information Fusion*, pp. 1218–1230, 2004.

[591] B. V. Dasarathy, *Decision Fusion*, IEEE Computer Society Press, Los Alamitos, CA, 1994.

[592] I. R. Goodman, R. P. S. Mahler and H.T. Nguyen, *Mathematics of Data Fusion*, Kluwer Academic Publishers, Norwell, MA, 1997.

[593] M. M. Kokar, J. A. Tomasik and J. Weyman, "Formalizing classes of information fusion systems," *Information Fusion*, vol. 5(3), pp. 189-202, 2004.

[594] M. Kumar, D. P. Garg and R. A. Zachery, "A generalized approach for inconsistency detection in data fusion from multiple sensors," *Proc. the American Control Conference*, pp. 2078-2083, 2006.

[595] P. Smets, "Analyzing the combination of conflicting belief functions," *Information Fusion*, vol. 8(4), pp. 387–412, 2007.

[596] R. P. S. Mahler, "Statistics 101 for multi–sensor, multitarget data fusion," *IEEE Aerospace and Electronic Systems Magazine*, vol. 19(1), pp. 53-64, 2004.

[597] R. Joshi and A. C. Sanderson, *multi–sensor Fusion: A Minimal Representation Framework*, World Scientific, 1999.

[598] X. L. Dong, L. Berti-Equille and D. Srivastava, "Truth discovery and copying detection in a dynamic world," *Journal Proc. of the VLDB Endowment*, vol. 2 (1), pp. 562-573, 2009.

[599] Y. Zhu, E. Song, J. Zhou and Z. You, "Optimal dimensionality reduction of sensor data in multi–sensor estimation fusion," *IEEE Trans. Signal Processing*, vol. 53(5), pp. 1631-1639, 2005.

[600] B. L. Milenova and M. M. Campos, "Mining high-dimensional data for information fusion: a database-centric approach," *Proc. the Int. Conference on Information Fusion*, pp. 638645, 2005.

[601] Z. Chen, "Bayesian filtering: From Kalman filters to particle filters and beyond," *adaptive Syst Lab McMaster Univ Hamilton ON Canada*, pp. 9–13, 2003.

[602] S. Oh, and S. Sastry, "Approximate estimation of distributed networked control system," *Proc. American Control Conference*, 9–13 July, pp. 997–1002, 2007.

[603] S. S. Li, "Optimal multi–sensor kalman smoothing fusion for discrete multi-channel ARMA signals," *J. Control Theory and Applications*, vol. 2, pp. 168–172, 2005.

[604] S.Sun, "multi–sensor optimal information fusion Kalman filter for discrete multichannel ARMA signals," *Proc. IEEE Int. Symposium on Intelligent Control*, Chicago, Houston, pp. 377–382, October 2003.

[605] S. Sun and Z. Deng, "multi–sensor optimal information fusion Kalman filter," *Automatica*, vol. 40(6), pp. 1017–1023, 2004.

[606] Y. Luo, Y. Zhu, D. Luo, J. Zhou, E. Song and D. Wang, "Globally optimal multi–sensor distributed random parameter matrices Kalman filtering fusion with applications," *Sensors*, vol. 8, pp. 8086–8103, 2008.

[607] C. Y. Chong, K. C., Chang, and S. Mori, "Distributed tracking in distributed sensor networks," *Proc. the American Control Conference*, Seattle, 1986.

[608] H. R. Hashmipour, S. Roy and A. J. Laub, "Decentralized structures for parallel Kalman filtering," *IEEE Trans. Autom. Control*, vol. 33, pp. 88–93, 1988.

[609] R. O. Saber, and R. M. Murray, "Consensus protocols for networks of dynamic agents," *Proc. the American Control Conference*, vol. 2, pp. 951–956, June 2003.

[610] M. Mesbahi, "On state-dependent dynamic graphs and their controllability properties," *IEEE Trans. on Automatic Control*, vol. 50(3), pp. 387–392, 2005.

[611] R. O. Saber, "Ultrafast consensus in small-world networks," *Proc. American Control Conference*, June 2005.

[612] Y. Hatano, and M. Mesbahi, "Agreement over random networks," *IEEE Conf. on Decision and Control*, 2004.

[613] L. Moreau, "Stability of multiagent systems with time-dependent communication links," *IEEE Trans. on Automatic Control*, vol. 50(2), pp. 169–182, 2005.

[614] W. Ren, and R. W. Beard, "Consensus seeking in multi-agent systems under dynamically changing interaction topologies," *IEEE Trans. on Automatic Control*, vol. 50(5), pp. 655–661, 2005.

[615] L. Xiao and S. Boyd, "Fast linear iterations for distributed averaging," *Systems and Control Letters*, vol. 52, pp. 65–78, 2004.

[616] L. Xiao, S. Boyd and S. Lall, "A scheme for asynchronous distributed sensor fusion based on average consensus," *Fourth Int. Symposium on Information Processing in Sensor Networks*, April 2005.

[617] O. R. Saber, "Flocking for multi-agent dynamic systems: algorithms and theory," *IEEE Transactions on Automatic Control*, vol. 51(3), pp. 401-420.

[618] Y. M. Zhu, Z. S. You, J. Zhou, K. -S. Zhang and X. R. Li, "The optimality for the distributed Kalman filtering fusion with feedback," *Automatica*, vol. 37, pp. 1489–1493, 2001.

[619] Y. Sheng, X. Hu and P. Ramanathan, "Distributed particle filter with GMM approximation for multiple targets localization and tracking in wireless sensor networks," *Fourth Int. Symposium on Information Processing in Sensor Networks*, April 2005.

[620] H. D. Whyte, "Data fusion in sensor networks," *Fourth International Symposium on Information Processing in Sensor Networks*, April 2005.

[621] S. Grime and H. D. Whyte, "Data fusion in decentralized sensor networks," *Control Engineering Practice*, vol. 2, 1994.

[622] D. P. Spanos, R. O., Saber, and R. M. Murray, "Distributed sensor fusion using dynamic consensus," *Proceedings of IFAC World Congress*, 2005.

[623] O. Costa, and M. Fragoso, M., "Stability results for discrete-time linear systems with Markovian jumping parameters," *Journal of Mathematical Analysis and Applications*, vol. 179, pp. 154–178, 1993.

[624] E. Mosca, "Optimal, predictive, adaptive control," *New Jersey: Prentice-Hall*, 1995.

[625] M. S. Mahmoud, and M. F. Emzir, "State estimation with asynchronous multi-rate multi-smart sensors', *Information Sciences*, volume 196(1), August 2012, pp. 15–27.

[626] S. Boyd, L.E. Ghaoui, E. Feron, V. Balakrishnan, *Linear Matrix Inequalities in System and Control Theory*, SIAM Studies in Applied Mathematics, Philadelphia, 1994.

[627] F.S. Cattivelli, A.H. Sayed, "Diffusion strategies for distributed Kalman filtering and smoothing," *IEEE Trans. Automatic Control*, vol. 55 (9), pp. 1520-1533, 2010.

[628] Z. Deng, P. Shi, H. Yang, Y. Xia, "Robust H_∞ filtering for nonlinear systems with interval time-varying delays, Int. J. Innovative Comput., Inform. Control 6 (12) (2010) 5527–5538.

[629] Z. Deng, P. Zhang, W. Qi, J. Liu, Y. Gao, "Sequential covariance intersection fusion Kalman filter, Inform. Sci. 189 (4) (2012) 293-309.

[630] C. Du, L. Xie, "H_∞ Control and Filtering of Two-Dimensional Systems," Springer-Verlag, Berlin Heidelberg, 2002.

[631] E. Gershon, U. Shaked, I. Yaesh, "H_∞ Control and Estimation of State-Multiplicative Linear Systems," Springer-Verlag, London, 2005.

[632] E. Gershon, U. Shaked, I. Yaesh, "H_∞ control and filtering of discrete-time stochastic systems with multiplicative noise," Automatica 37 (3) (2011) 409-417.

[633] B. Jiang, P. Shi, Z. Mao, "Sliding mode observer-based fault estimation for nonlinear networked control systems," Circ., Syst., Signal Process. 30 (1) (2011) 1-16.

[634] B. Jiang, Z. Mao, P. Shi, "H 1 filter design for a class of networked control systems via TS fuzzy model approach," IEEE Trans. Fuzzy Syst. 18 (1) (2010) 201–208.

[635] T. Kaczorek, *Two-Dimensional Linear Systems*, Springer-Verlag, Berlin Heidelberg, 1985.

[636] S. Kar, J. M. F. Moura, "Sensor networks with random links: topology design for distributed consensus," *IEEE Trans. Signal Process.*, vol. 56 (7), pp. 3315–3326, 2008.

[637] S. Kar, J. M. F. Moura, "Distributed consensus algorithms in sensor networks with imperfect communication: link failures and channel noise," *IEEE Trans. Signal Process.*, vol. 57 (1), pp. 355–369, 2009.

[638] P. Lin, Y. Jia, L. Li, "Distributed robust H_∞ consensus control in directed networks of agents with time-delay," *Syst. Control Lett.*, vol. 57 (8), pp. 643-653, 2008.

[639] Y. Liu, Y. Jia, "H_∞ consensus control of multi-agent systems with switching topology: a dynamic output feedback protocol," *Int. J. Control*, vol. 83 (3), pp. 527–537, 2010.

[640] Y. Liu, C. Li, W. Tang, and Z. Zhang, "Distributed estimation over complex networks," *Inform. Sci.*, 2012. $http$ $://dx.doi.org/10.1016/j.ins.2012.02.008$.

[641] J. Lfberg, "YALMIP: A toolbox for modeling and optimization in MATLAB," *Proc. the CACSD Conference*, Taipei, Taiwan, 2004.

[642] M. Mahmoud, P. Shi, J. Yi, J. Pan, "Robust observers for neutral jumping systems with uncertain information," *Inform. Sci.*, vol. 176 (16), pp. 2355-2385, 2006.

[643] R. Olfati-Saber, "Distributed Kalman filtering for sensor networks," *Proc. 46th IEEE Conf. Decision and Control*, New Orleans, USA, December 2007.

[644] R. Olfati-Saber, "Kalman-consensus filter: optimality, stability and performance," *Proc. 48th IEEE Conf. Decision and Control*, Shanghai, China, December 2009.

[645] B. Shen, Z. Wang, Y. Hung, "Distributed H_1-consensus filtering in sensor networks with multiple missing measurements: the finite-horizon case," *Automatica*, vol. 46 (10), pp. 1682–1688.

[646] V. Ugrinovskii, "Distributed robust filtering with H_∞ consensus of estimations," *Proc. the American Control Conference*, Baltimore," MD, June 2010.

[647] Z. Wang, Daniel W.C. Ho, X. Liu, "Variance-constrained filtering for uncertain stochastic systems with missing measurements, *IEEE Trans. Automatic Control*, vol. 48 (7), pp. 1254-1258, 2003.

[648] Z. Wang, F. Yang, Daniel W.C. Ho, X. Liu, "Robust H_1 filtering for stochastic time-delay systems with missing measurements," *IEEE Trans. Signal Process.*, vol. 54 (7), pp. 2579-2587, 2006.

[649] F. Yang, Z. Wang, Daniel W. C. Ho, Mahbub Gani, "Robust H 1 control with missing measurements and time delays," *IEEE Trans. Autom. Control*, vol. 52(9), pp. 1666-1672, 2007.

[650] H. Yang, Y. Xia, P. Shi, "Stabilization of networked control systems with nonuniform random sampling periods," *Int. J. Robust Nonlinear Control*, vol. 21(5), pp. 501-526, 2011.

[651] R. Yang, L. Xie, C. Zhang, "H_2 and mixed H_2/H_∞ control of two-dimensional systems in Roesser model," *Automatica*, vol. 42(9), pp. 1507-1514, 2006.

[652] R. Yang, L. Xie, C. Zhang, "Generalized two-dimensional KalmanYakubovichPopov lemma for discrete Roesser model," *IEEE Trans. Circ. Syst. I, Reg.*, vol. 55(10), pp. 3223-3233, 2008.

[653] H. Yu, Y. Zhuang, W. Wang, "Distributed H_∞ filtering with consensus in sensor networks: a two-dimensional system-based approach," *Int. J. Syst. Sci.*, vol. 42(9), pp. 543-1557, 2011.

[654] L. Zhou, G. Lu, "Quantized feedback stabilization for networked control systems with nonlinear perturbation," *Int. J. Innovative Comput. Inform. Control*, vol. 6(6), pp. 2485-2496, 2010.

[655] Y. Shang, W. Ruml, Y. Zhang and Fromherz, M., "Localization from connectivity in sensor networks," *IEEE Trans. Parallel and Distributed System*, vol. 15, no. 11, November 2004.

[656] N. Patwari, J. N. Ash and S. Kyperountas, "Locating the nodes," *IEEE Signal Processing Magazine*, July 2005.

[657] D. L. Mills, "Precision synchronization of computer network clock," *ACM Computer Communication Review*, 1994.

[658] G. Di Stefano, F. Graziosi and F. Santucci, "Distributed positioning algorithm for ad-hoc networks," *Int. Workshop on Ultra Wideband Systems (IWUWBS)*, Oulu, 2003.

[659] A. Savvides, W. L. Garber, R. L. Moses and M. B. Srivastava, "An analysis of error inducing parameters in multihop sensor node localization," *IEEE Transaction on Mobile Computing*, vol. 4, no. 6, November/December 2005.

[660] L. Xiao and S. Boyd and S. Lall, "A scheme for robust distributed sensor fusion based on average consensus," *Proc. IEEE IPSN*, 2005.

[661] D. P. Bertsekas, A. Nedic, and A. E. Ozdaglar, *Convex Analysis and Optimization*, Athena Scientific, 2003.

[662] W. Yu, G. Chen, and J. L, "On pinning synchronization of complex dynamical networks," *Automatica*, vol. 45, no. 2, pp. 429–435, 2009.

[663] T. Chen, X. Liu, and W. Lu, "Pinning complex networks by a single controller," *IEEE Trans. Circuits Syst. I, Reg. Papers*, vol. 54, no. 6, pp. 13171326, June 2007.

[664] Z. Schuss, *Theory and Applications of Stochastic Differential Equations*, Wiley, New York, 1980.

[665] A. L. Barabsi and R. Albert, "Emergence of scaling in random networks," *Science*, vol. 286, no. 5439, pp. 509–512, Oct. 1999.

[666] L. O. Chua, "The genesis of Chua's circuit," *Arch. Elektr. Ubertrag.*, vol. 46, no. 3, pp. 250–257, 1992.

[667] R. Carli, A. Chiuse, L. Schenato and S. Zampieri, "Distributed Kalman filtering based on consensus strategies," *IEEE J. Selected Areas in Communications*, vol. 26(4), pp. 622–633, 2008.

[668] H. M. Chen, K. S. Zhang and X. R. Li, "Optimal data compression for multi–sensor target tracking with communication constraints," *Proc. the 43th IEEE Conference on Decision Control*, Atlantis, Bahamas, pp. 8179–8184, 2004.

[669] A. G. Dimakis, S. Kar, J. M. F. Moura, M. G., Rabbat and A. Scaglione, "Gossip algorithms for distributed signal processing," *Proc. The IEEE*, vol. 98(11), pp. 1847–1864, 2010.

[670] A. Dogandzic and B. H. Zhang, "'Distributed estimation and detection for sensor networks using hidden Markov random field models," *IEEE Trans. Signal Processing*, vol. 54(8), pp. 3200–3215, 2006.

[671] A. Giridhar and P. R. Kumar, "Towards a theory of in-network computation in wireless sensor networks," *IEEE Communications Magazine*, 44(4), pp. 98–107, 2006.

[672] S. Kar and Jose M. F. Moura, "Gossip and distributed Kalman filtering: weak consensus under weak detectability," *IEEE Trans. Signal Processing*, vol. 59(4), pp. 1766–1784, 2011.

[673] S. Kar, B. Sinopoli and Jose M. F. Moura, "Kalman filtering with intermittent observations: weak convergence to a stationary distribution," *IEEE Trans. Automatic Control*, vol. 57(2), pp. 405–420, 2012.

[674] J. L. Li and G. AlRegib, "Distributed estimation in energy-constrained wireless sensor networks," *IEEE Trans. Signal Processing*, vol. 57(10), pp. 3746–3758, 2009.

[675] Y. Liang, T. Chen and Q. Pan, "Multi-rate optimal state estimation," *Int. J. Control*, vol. 82(11), pp. 2059–2076, 2009.

[676] Y. Liang, Chen, T. W., and Q. Pan, "Multi-rate stochastic H_∞ filtering for networked multi–sensor fusion," *Automatica*, vol. 46(2), pp. 437–444, 2010.

[677] E. J. Msechu, S. I. Roumeliotis, A. Ribeiro and G. B. Giannakis, "Decentralized quantized Kalman filtering with scalable communication cost," *IEEE Trans. Signal Processing*, vol. 56(8), pp. 3727–3741, 2008.

[678] A. Ribeiro, I. D. Schizas, S. I. Roumeliotis and G. B. Giannakis, "Kalman filtering in wireless sensor networks," *IEEE Control Systems Magazine*, vol. 30(2), pp. 66–86, 2010.

[679] I. D. Schizas, G. B. Giannakis and Z. Q. Luo, Z. Q., "Distributed estimation using reduced- dimensionality sensor observations," *IEEE Trans. Signal Processing*, vol. 55(8), pp. 4284–4299, 2007.

[680] I. D. Schizas, A. Ribeiro and G. B. Giannakis, "Consensus in ad hoc WSNs with noisy links: part I: distributed estimation of deterministic signals," *IEEE Trans. Signal Processing*, vol. 56(1), pp. 350–364, 2007.

[681] S. L. Sun and Z. L. Deng, "multi–sensor optimal information fusion Kalman filter," *Automatica*, vol. 40(6), pp. 1017–1023, 2004.

[682] S. L. Sun, L. H. Xie, W. D. Xiao and Y. C. Soh, "Optimal linear estimation for systems with multiple packet dropouts," *Automatica*, vol. 44(7), pp. 1333–1342, 2008.

[683] Z. D. Wang, F. W. Yang, D. W. C. Ho and X. H. Liu, "Robust finitehorizon filtering for stochastic systems with missing measurements," *IEEE Signal Processing Letters*, vol. 12(6), pp. 437–440, 2005.

[684] L. Xiao, S. Boyd and S. J. Kim, "'Distributed average consensus with least mean-square deviation," *Proc. the 17th International Symposium on Mathematical Theory of Networks and Systems, MTNS*, Kyoto, Japan, pp. 2768–2776, 2006.

[685] J. J. Xiao, S. G. Cui, Z. Q. Luo and A. J. Goldsmith, "Power scheduling of universal decentralized estimation in sensor networks," *IEEE Trans. Signal Processing*, vol. 54(2), pp. 413–422, 2006.

[686] N. Xiao, L. H. Xie and M. Y. Fu, "Kalman filtering over unreliable communication networks with bounded Markovian packet dropouts," *Int. J. Robust and Nonlinear Control*, vol. 19(16), pp. 1770–1786, 2009.

[687] W. A. Zhang, L. Yu and H. B. Song, "H_∞ filtering of networked discrete-time systems with random packet losses," *Information Sciences*, vol. 179(22), pp. 3944–3955, 2009.

[688] H. Zhu, I. D. Schizas and G. B. Giannakis, "Power-efficient dimensionality reduction for distributed channel-aware Kalman tracking using WSNs," *IEEE Trans. Signal Processing*, vol. 57(8), pp. 3193–3207, 2009.

[689] O. R. Saber, "Distributed Kalman filtering for sensor networks," *Proc. 46th IEEE Conf. Decision and Control*, New Orleans, LA, December 2007.

[690] Lin, P., Y. Jia, and L. Li, "Distributed robust H_∞ consensus control in directed networks of agents with time-delay," *Systems & Control Letters*, vol. 57(8), pp. 643–653, 2008.

[691] T. Li, and J. Zhang, "Mean square average consensus under measurement noises and fixed topologies: necessary and sufficient conditions," *Automatica*, vol. 45, pp. 1929–1936, 2009.

[692] T. Li, and J. Zhang, "Consensus conditions of multi-agent systems with time–varying topologies and stochastic communication noises," *IEEE Trans Automatic Control*, vol. 55, pp. 2043–2057, 2010.

[693] P. Lin, and Y. Jia, "Consensus of second-order discerete-time multi-agent systems with nonuniform time-delays and dynamically changing topologies," *Automatica*, 45, 2154–2158, 2009.

[694] P. Lin, and Y. Jia, "Consensus of a class of second-order multi-agent systems with time-delay and jointly-connected topologies," *IEEE Trans. Automatic Control*, 55, 778–784, 2009.

[695] Y. Sun, and L. Wang, "Consensus of multi-agent systems in directed networks with nonuniform time–varying delays," *IEEE Trans. Automatic Control*, 54, 1607–1613, 2009.

[696] A. Tahbaz-Salehi, and A. Jadbabaie, "A necessary and sufficient condition for consensus over random networks," *IEEE Trans. Automatic Control*, 53, 791–795, 2008.

[697] Y. Zhang, and Y. Tian, "Consentability and protocol design of multi-agent systems with stochastic switching topology," *Automatica*, 45, 1195-1201, 2009.

[698] G. Miao, S. Xu, and Y. Zhou, "Necessary and sufficient conditions for mean square consensus under Markov switching topologies," *Int. J. Systems Science*, DOI:10.1080/00207721.2011.598961, 2011.

[699] Sinopoli, B., Schenato, L., Franceschetti, M., Poolla, K., Jordan, M. I., Sastry, S. S., Kalman filtering with intermittent observations. IEEE Transactions on Automatic Control, 49, 1453–1464, 2004.

[700] Hounkpevi, F. O., Yaz, E. E., Robust minimum variance linear state estimators for multiple sensors with different failure rates. Automatica, 43(7), 1274–1280, 2007.

[701] Aliyu M. and Boukas E., Mixed H_2/H_∞ nonlinear filtering, International Journal of Robust and Nonlinear Control, 19, 394–417, 2009.

[702] Y. Liu, and B. Xu, "Filter designing with finite packet losses and its application for stochastic systems," IET Control Theory and Applications, 5, 775–784, 2010.

[703] P. Shi, P., M. S. Mahmoud, S. Nguang and A. Ismail, "Robust filtering for jumping systems with mode-dependent delays," *Signal Processing*, vol. 86, pp. 140–152, 2006.

[704] B. Zhou, W. Zheng, M. Fu and G. Duan, "H_∞ filtering for linear continuous-time systems subject to sensor nonlinearities," *IET Control Theory and Applications*, vol. 5, pp. 1925–1937, 2011.

[705] G. Wei, Z. Wang Z. and Y. Hung, "Robust filtering with stochastic nonlinearities and multiple missing measurements," *Automatica*, vol. 45, pp. 836–841, 2009.

[706] H. Dong, Z. Wang, D. W. C. Ho, and H. Gao, "Variance-constrained H_∞ filtering for a class of nonlinear time-varying systems with multiple missing measurements: The finite-horizon case," *IEEE Trans. Signal Processing*, vol. 58, pp. 2534–2543, 2010.

[707] Y. Liu and B. Xu, "Filter designing with finite packet losses and its application for stochastic systems," *IET Control Theory and Applications*, vol. 5, pp. 775–784, 2011.

[708] J. Xiong and J. Lam, "Fixed-order robust H_∞ filter design for Markovian jump systems with uncertain switching probabilities," *IEEE Trans. Signal Processing*, vol. 54, pp. 1421–1430, 2006.

[709] H. Zhang, G. Feng, G. Duan and X. Lu, "H_∞ filtering for multiple time-delay measurements," *IEEE Trans. on Signal Processing*, vol. 54, pp. 1681–1688, 2006.

[710] F. Yang, Z. Wang, G. Feng and X. Liu, "Robust filtering with randomly varying sensor delay: the finite-horizon case," *IEEE Trans. Circuits and Systems I*, vol. 56, pp. 664–672, 2009.

[711] B. Shen, Z. Wang and Y. Hung, "Distributed H_∞-consensus filtering in sensor networks with multiple missing measurements: The finite-horizon case," *Automatica*, vol. 46, pp. 1682–1688, 2010.

[712] Q. Han, "Absolute stability of time-delay systems with sector-bounded nonlinearity." *Automatica*, vol. 41, pp. 2171–2176, 2005.

[713] Y. Liu, Z. Wang and X. Liu, "Synchronization and state estimation for discrete-time complex networks with distributed delays," *IEEE Trans. Systems, Man, and Cybernetics Part B: Cybernetics*, vol. 38, pp. 1314–1325, 2008.

[714] H. Khalil, *Nonlinear Systems*, Prentice-Hall, Upper Saddle River, NJ, 2002.

[715] K. Chung, *A Course in Probability Theory*, Third edition, San Diego, Academic Press, 2001.

[716] M. S. Mahmoud, *Robust Control and Filtering for Time-Delay Systems*, Marcel-Dekker, New-York, 2000.

[717] M. S. Mahmoud, *Switched Time-Delay Systems*, Springer-Verlag, New York, 2010.

[718] M. S. Mahmoud, *Decentralized Control and Filtering in Interconnected Dynamical Systems*, CRC Press, New York, 2010.

[719] K. Gu, V. L. Kharitonov, Jie Chen, *Stability of Time-Delay Systems*, Birkhauser, Boston, 2003.

[720] S. Boyd, L. El Ghaoui, E. Feron and V. Balakrishnan, *Linear Matrix Inequalities in System and Control Theory*, SIAM Studies in Applied Mathematics, Philadelphia, 1994.

[721] M. S. Mahmoud, *Resilient Control of Uncertain Dynamical Systems*, Springer, Heidelberg, 2004.

[722] W. Ren, W. and R. W. Beard, "Consensus seeking in multiagent systems under dynamically changing interaction topologies," *IEEE Trans. Automatic Control*, vol. 50, no. 5, pp. 655–661, 2005.

[723] P. Lin, Y. M. Jia and L. Li, "Distributed robust $\mathcal{H}_\infty$ consensus control in directed networks of agents with time-delay," *Systems and Control Letters*, vol. 57, no. 8, pp. 643-653, 2008.

[724] D. Angeli and P.-A. Bliman, "Convergence speed of unsteady distributed consensus: decay estimate along the settling spanning-trees," *SIAM J. Control Optimization*, vol. 48(1), pp. 1–32, 2009.

[725] J. Zhou and Q. Wang, "Convergence speed in distributed consensus over dynamically switching random networks," *Automatica*, vol. 45(6), pp. 1455–1461, 2009.

[726] R. S. Smith and F. Y. Hadaegh, "Control of deep-space formation-flying spacecraft; relative sensing and switched information," *J. Guidance and Control Dynamics*, vol. 28(1), pp. 106–114, 2005.

[727] D. S. Bernstein, *Matrix Mathematics,* Princeton University Press, Princeton, NJ, 2005.

Index

Printed and bound by CPI Group (UK) Ltd, Croydon, CR0 4YY

21/10/2024

01777095-0019